Y0-AUZ-362

CONTEMPORARY PHYSICS

CONTEMPORARY PHYSICS

AN INTRODUCTION FOR THE NONSCIENTIST

JEROME PINE
Professor of Physics
California Institute of Technology

McGRAW-HILL BOOK COMPANY

New York St. Louis San Francisco Düsseldorf Johannesburg
Kuala Lumpur London Mexico Montreal New Delhi
Panama Rio de Janeiro Singapore Sydney Toronto

CONTEMPORARY PHYSICS: AN INTRODUCTION FOR THE NONSCIENTIST

1 2 3 4 5 6 7 8 9 0 K P K P 7 9 8 7 6 5 4 3 2

This book was set in Palatino by Black Dot, Inc., and printed and bound by Kingsport Press, Inc.
The designer was Edward A. Butler;
the drawings were done by Hank Iken.
The editors were Jack L. Farnsworth and J. W. Maisel.
Peter Guilmette supervised production.

Library of Congress Cataloging in Publication Data

Pine, Jerome, 1928–
Contemporary physics.

1. Physics. I. Title.
QC21.2.P55 530 70-39178
ISBN 0-07-050071-1

CONTENTS

PREFACE

The history of physics is rich in personalities, conflicts, and fascinating philosophical questions. To the nonscientist, for whom this book is written, these aspects of the subject are often particularly interesting. However, the subject matter of physics—our basic understanding of the physical universe—is no less interesting than its cultural background.

A relatively small number of physical laws provide powerful insights into an enormous variety of the phenomena around us. Furthermore, in searching for the most basic understanding of the physical world, the physicist has been led to explore the atom and the cosmos. In these realms, one so tiny and the other so huge that both are beyond the reach of our unaided senses, nature behaves in ways far different than in the everyday world, ways which can be understood only by the greatest stretches of our imagination.

This book tries to explain some of the basic discoveries of physics and to give the reader a feeling for what physicists do and the way they do it. While the more "humanistic" aspects of the subject will occasionally be touched upon here, I hope the student will learn more in this area from his instructor and from the wealth of good books that are available. I also hope that the student will perform some experiments himself and thereby learn about the doing of physics in a way which no book can simulate.

Since the language of physics is mathematics, it is almost impossible to learn physics without using mathematics. This book relies upon arithmetic and a small but essential amount of elementary algebra. As much as possible, graphs are used to illustrate mathematical relations.

Throughout the book, the physics of atoms, nuclei, and their constituents is emphasized as much as possible. I think this is the most interesting and fundamental kind of physics, and I hope the reader will agree. Problem solving is also emphasized, since it is so much a part of real physics and since problems enable a student to discover interesting things for himself.

The book is organized so that the first eight chapters constitute a core of basic physics, built up so each chapter depends on the preceding ones. The final six chapters depend on the first eight, but each is intended to stand independent of the others. They are intended to be sampled according to the interests of students and instructors. In general, my goal has been to provide a text which embodies a useful amount of beginning material on a variety of modern topics, but which leaves room for the user to be individualistic in both the depth and breadth of his studies.

It is impossible to individually thank all those to whom I am grateful for ideas or for helpful criticism and comments. However, I would like to acknowledge particularly my debts to Edwin F. Taylor and John A. Wheeler with respect to my treatment of relativity and to Arthur Rosenfeld for that of elementary particle symmetries. In addition, Bob March, David Beckwith, and especially Hugh D. Young provided many extremely helpful suggestions for improving the original manuscript.

JEROME PINE

CONTEMPORARY PHYSICS

INTRODUCTION

1-1 FOREWORD

The goal of this book is to provide an understanding of what physics is about and how the physicist approaches his subject. In this introductory chapter, there are some comments on the nature of physics. However, a key premise of this book is that you will best learn what physics is like by doing it yourself. In the context of this text, doing physics means problem solving, but you will, it is hoped, also have an opportunity to perform some experiments.

Problem solving, reduced to its basic essential, means making *quantitative* predictions about the behavior of physical objects in particular circumstances. The abstract concepts of physics become real when they are used to solve problems, and the validity of a physical theory ultimately depends upon its usefulness in problem solving. There is a richness and a creativity in problem solving which you will discover as you participate in the process. In this book, you will find problems given throughout the text, some to test your ability at straightforward application of basic concepts, and others which are more challenging.

In addition to emphasizing problem solving, this book emphasizes the subject matter of twentieth-century physics. In this introductory chapter, the concept of the atomic nature of matter is reviewed, and some properties of atoms, such as their sizes and weights, are discussed. In the seven succeeding chapters, we will mainly be concerned with the physics which is needed for a full understanding of the structure and properties of atoms, since to a large extent modern physics is concerned with atomic and subatomic physics. In the final chapters of the book, a variety of special topics are discussed, in each case with the basic material of the first eight chapters as a foundation.

Throughout this book, the needed mathematics will be held to a minimum. Trigonometry will not be used, and complicated algebra will be avoided. However, problems requiring numerical answers will occur quite frequently. Therefore, the use of an inexpensive slide rule for ease in multiplying and dividing is strongly recommended. In this chapter, some mathematical review will be incorporated in the discussion, covering the main items of mathematical language which will be used in the text.

1-2 THE NATURE OF PHYSICS

Physics is the study of matter. And matter includes all the "stuff" of the universe: earth, stars, air, water, trees, cars, people, everything. Even heat, light, and sound, seemingly nonmaterial, are included. The study of matter of course includes not only what its static properties are but also how and why it moves, changes, and in general interacts with other matter around it. Since this is really too much, what physics really involves is the study of the *simplest things* about matter. The relation of physics to chemistry, astronomy, biology, geology, etc., is that to a large extent each of the other sciences copes more with the complication and variety of some fraction of the universe.

The physicist idealizes and simplifies in an attempt to reach the bare essentials, or at least some of them. In doing this he makes *models*, which are simplified pictures of the real world. As an example, in the nineteenth century, physicists who were trying to understand the properties of gases developed the following picture: A gas is mostly empty space except for widely separated atoms or molecules, too small to be seen, which behave like little hard spheres. The spheres are in rapid and chaotic motion, colliding with each other and with the walls of any container in which they are held. The physicists who made this model of a gas understood the laws which govern the motion and collisions

Figure 1-2-1 **A tiny volume of gas, greatly magnified, according to one possible model. The black spheres represent atoms in motion. For each atom, an arrow points in its direction of motion, and the length of the arrow is proportional to its speed.**

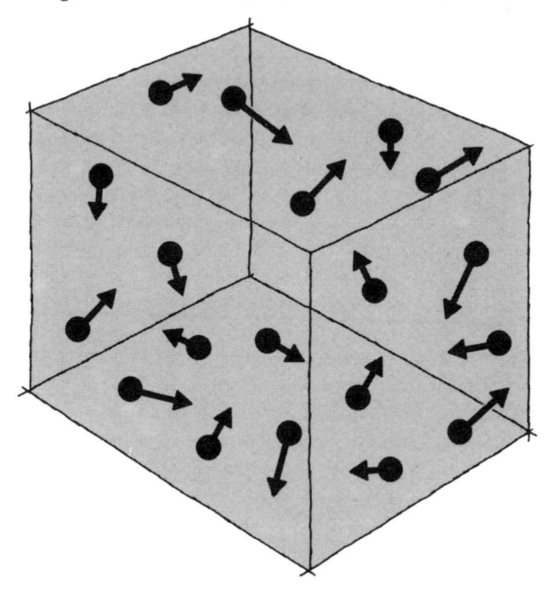

of hard spheres. The model, which is illustrated schematically in Fig. 1-2-1, was an attempt to gain an understanding of the behavior of gases in terms of known physics.

If a given weight of gas is sealed inside a container which has a volume V, one interesting property of many gases is that there is a simple relation between the volume V and the pressure of the gas, P. In terms of the model in Fig. 1-2-1, the pressure, which is the outward push of the gas against the walls of the container, results from the rebounding of atoms off the walls. (Let someone bounce a basketball off you if you need convincing that a push is exerted by a rebounding object.)

To quantitatively study the relation between the pressure P and the volume V for a fixed amount of gas, an apparatus such as that shown in Fig. 1-2-2 can be used. The sealed piston can move up or down, as shown by the arrows, changing the volume V without allowing any gas to escape. The pressure P is read on a gage. According to the model, if V is made smaller, each little sphere will rebound more often, having a shorter distance to travel between collisions. Thus, the number of rebounds per second off

Figure 1-2-2 An apparatus for varying the volume of some gas and measuring its pressure.

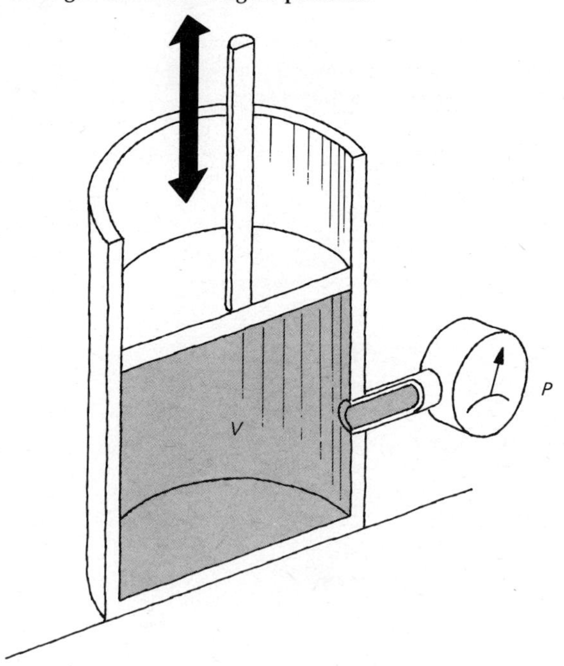

any wall will go up, and therefore P will increase. The model predicts qualitatively that pressure increases when volume decreases, and vice versa. Moreover, by assuming that the molecules or atoms have an average speed which *does not depend on the volume,* a quantitative relation was easily derived:

$$PV = k$$

where k is a constant for a given amount of gas at a given temperature (that is, k is some number which does not depend on P and V).

The equation $PV = k$ is a mathematical expression of a physical law, called Boyle's law, which was discovered experimentally in 1660. It was, however, not understood until the invention of the model we are now discussing, at a much later date. Like most physical laws it expresses a quantitative relationship between physical quantities, the pressure and volume of a gas in this case. It can be expressed in words as well as symbols: "The product of the pressure and the volume of a fixed quantity of gas is equal to a constant." Usually, however, the mathematical equation is more convenient than words for expressing and using such a quantitative relation.

Sometimes a graphical representation of a relationship such as $PV = k$ is the most illuminating. In Fig. 1-2-3 the solid line is a graph on axes labeled Pressure and Volume of the relationship $PV = k$, when k has the particular value 1.2. Any point on the curve represents a pair of values of P and V such that the product PV equals 1.2. The shape of the curve, a hyperbola, is characteristic of the form of the mathematical relationship. Other equations relating P and V would be represented by different curves on such a graph. The hyperbola is as definite a statement of Boyle's law as the algebraic equation.

Also shown on the graph of Fig. 1-2-3 are measured points which are pairs of values of P and V obtained in an experiment with an apparatus like that shown in Fig. 1-2-2. The plotting of such points on a graph facilitates comparing the experimental result with the theoretical curve. The use of graphs for presenting and interpreting data is extremely important to the physicist.

PROBLEM 1

From the graph of Fig. 1-2-3, estimate the volume if the pressure is 5 lb/in^2. Compare with the value you expect according to the equation $PV = 1.2$.

Suppose only the two experimental points corresponding to

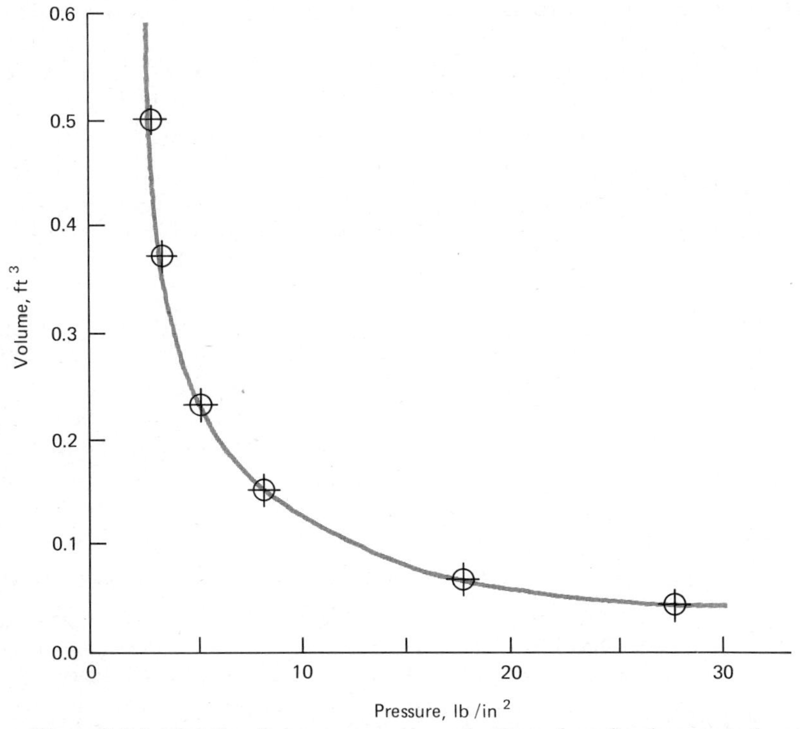

Figure 1-2-3 Relation between pressure and volume for a fixed amount of a gas. The circles are experimental points, and the curve is calculated theoretically from the model discussed in the text.

pressures of about 8 and 17 lb/in^2 in Fig. 1-2-3 were known. A simple curve which would fit the points would be a straight line through them. What would be the equation for such a straight line?

The data points shown in Fig. 1-2-3 were all obtained at a particular temperature, while points with the same gas in the same apparatus obtained at some other fixed temperature are found to lie on a hyperbola characterized by a different value of the constant k. This temperature effect can be included within the framework of the model we have been discussing if the average speed of the gas atoms is assumed to depend in a particular way upon the temperature. In fact, a refined *hard-sphere* model of an *ideal gas* correctly predicts much more about the behavior of real gases than it is appropriate to go into here. The theory based on the model we have been discussing is called the kinetic (meaning "motional") theory of gases, and it illustrates one of the most successful uses of a simple model to correctly predict a wide variety of experimentally observed phenomena.

PROBLEM 2

In an experiment some gas was held in a container with a fixed volume, and the pressure was measured at various gas temperatures. The results are tabulated below:

TEMPERATURE, °F	PRESSURE, lb/in^2
−90	10.2
80	15.0
212	18.6
450	25.2

Plot these experimental data on a graph of pressure vs. temperature. Can you guess a temperature at which the pressure might become zero?

Can you write an equation relating pressure to temperature for pressure in pounds per square inch and temperature in degrees fahrenheit? How does your equation look if temperature is measured in degrees above the temperature at which the pressure might become zero? The temperature measured in this way is called *absolute temperature.*

If a model is a success in its quantitative predictions, it is then perhaps natural that the qualitative, or conceptual, basis of the model may be taken seriously. Thus, the success of the kinetic theory of gases made the existence of atoms and molecules more believable, at a time when their reality was by no means established. We tend to get from successful models our most concrete conceptualizations of the invisible atomic and subatomic world. We may come to believe almost intuitively in hard-sphere atoms, or spongy atoms, or even perhaps some day in angry or friendly atoms, depending upon the atomic attributes involved in building models which work.

Notice that a model is judged on its ability to predict the results of experiments. It is never proven completely right, since all possible experiments are never completed. Thus, on any day a new experimental result may contradict a hitherto successful theory. Sometimes the theory must merely evolve or grow a bit, and sometimes it receives a mortal blow. In the latter case, if the theory has been important in the basic structure of physics, the greatest possible creative efforts will be mobilized, as men try to meet the challenge with new theories. In this way, physics grows as a changing fabric of theory and experiment. To some, the growth of physics may seem to be a battle in which feeble constructs of man are pitted against the infinite variety of the real universe.

1-3 ATOMIC WEIGHTS

The hypothesis that all matter is made of tiny invisible building blocks, called atoms, is credited to the Greek philosopher Democritus (in the sixth century B.C.), but over 2,000 years elapsed before science progressed to the point where the hypothesis could be tested experimentally. The eighteenth- and nineteenth-century chemists began to make atoms believable. First they noted that all the varied substances of the world could be classified as either compounds or elements. The compounds could be broken down into elements and synthesized from elements, while the elements defied all efforts to decompose them into other component substances. From a relatively few elements an enormous variety of compounds could be created. Some examples of elements are hydrogen, carbon, oxygen, iron, and gold; and the total number now known to occur in nature is 92.

By the end of the nineteenth century, mainly as a result of work by chemists, there was good, though indirect, evidence for the atomic nature of matter. There was a large body of experimental facts which could best be explained by an atomic view, summarized as follows:

1. A quantity of a pure element consists of a great number of identical atoms characteristic of that element.
2. A quantity of a pure compound consists of a great number of identical molecules characteristic of the compound. Each molecule is a particular combination of two or more kinds of atoms, or elements.

More direct evidence, accumulated in the twentieth century, supports this view of the atomic constitution of matter, with one slight addition. Careful studies have shown that slight differences can occur among the nuclei of atoms of a given element. For our present brief introduction this relatively fine point will be ignored.

In order to discover the weights of the various kinds of atoms, we can proceed in two steps. First, by chemical techniques the *relative* weights of the atoms can be found. As an example consider that most familiar compound, water, whose chemical formula H_2O states that each of its molecules consists of two hydrogen atoms and one oxygen atom. In the laboratory, water can be formed from hydrogen gas and oxygen gas, and the weights of each which combine are found to be in the ratio of 1 to 8. According to our molecular view, the weights of hydrogen and oxygen in each identical water molecule are then also in the ratio 1 to 8. Since there are two hydrogen atoms in the molecule and one oxygen atom, the weight of an oxygen atom is thus deduced to be 16 times that of a hydrogen atom. Table 1-3-1 lists the relative atomic weights of the lightest 20 elements, inferred from known chemical formulas and from the weights of combining elements, as in the example just discussed.

Table 1-3-1 The atomic weight of the 20 lightest elements in units of the weight of the hydrogen atom

ATOMIC NUMBER	NAME	SYMBOL	ATOMIC WEIGHT
1	Hydrogen	H	1.00
2	Helium	He	3.98
3	Lithium	Li	6.89
4	Beryllium	Be	8.95
5	Boron	B	10.7
6	Carbon	C	11.9
7	Nitrogen	N	13.9
8	Oxygen	O	15.9
9	Fluorine	F	18.9
10	Neon	Ne	20.0
11	Sodium	Na	22.8
12	Magnesium	Mg	24.1
13	Aluminum	Al	26.8
14	Silicon	Si	27.9
15	Phosphorus	P	30.8
16	Sulphur	S	31.8
17	Chlorine	Cl	35.3
18	Argon	A	39.6
19	Potassium	K	38.8
20	Calcium	Ca	39.8

Were the actual weight of any particular kind of atom known, then, through the relative weights, the absolute weights of all the others would also be known. In Chap. 4 we will study a direct method for determining the weight, or more specifically, the *mass*, of a single atom. In this way it is found for example that the mass of a hydrogen atom is 0.00000000000000000000000166 gram. The distinction between *weight* and *mass* is that mass is a fixed attribute of an object, which results in a force on the object, called weight, when the object is acted upon by gravity. In a given gravitational environment, such as at the earth's surface, weights are proportional to masses.

PROBLEM 3

It is found that 6 grams of carbon combine with 8 grams of oxygen to form a gas. Use the atomic weights in Table 1-3-1 to write a possible chemical formula for the gas molecule.

According to the atomic hypothesis, what mass of gas would you expect to be formed by the given masses of carbon and oxygen?

The weight of a one-gram mass is about one-fifth the weight of a dime, at the earth's surface. The weight of a hydrogen atom is that of a fantastically small mass, written out as a decimal fraction of a gram above. An alternative way to express the tiny mass of one hydrogen atom is to give the number of atoms in one gram of hydrogen, about 602,000,000,000,000,000,000,000 (602 thousand million million million atoms per gram). Since a gram of hydrogen is contained in about one-third of an ounce of water, you can see that truly enormous numbers of atoms are involved in even very small quantities of matter.

If we generally handled very small and very large numbers in the manner above, utilizing long strings of zeros, it would be very inconvenient. Instead, the zeros are normally summarized in a very compact way. For the mass of a hydrogen atom, symbolized by m_{H}, we write:

$$m_{\mathrm{H}} = 1.66 \times 10^{-24} \text{ g}$$

where g is the abbreviation for gram. Alternatively, the number of hydrogen atoms per gram, symbolized N_0, is written:

$$N_0 = 6.02 \times 10^{23} \text{ atoms per gram}$$

This is called *exponential notation.* The exponents establish a power of 10 which replaces all the zeros in the notation we used previously. A few simple review statements should enable you to do exponential arithmetic without too much trouble:

$$10^a \times 10^b = 10^{(a+b)}$$

$$10^{-b} = \frac{1}{10^b}$$

$$\frac{10^a}{10^b} = 10^{(a-b)}$$

Since atoms are so tiny, we will often need to handle numbers like m_H and N_0 in studying modern physics, so that the exponential notation is essential. If you feel unsure of the elementary aspects of exponentials summarized above, you should utilize a math text for a brief review.

The quantity N_0 is called *Avogadro's number* after the Italian physicist who "invented" it in 1811. Although the number itself wasn't known in those days, its significance was recognized as a measure of the scale of the atomic realm. The imagination fails when we try to grasp the magnitude of 6.02×10^{23}. The number of people in the world is huge, a few times 10^9. The number of seconds in a century is also a few times 10^9. Suppose we took just one drop of water and somehow divided it up among all the people of the world and they all started counting the hydrogen atoms at the rate of 10 per second. It would take about 6,000 years to count the atoms, if everyone worked 24 hours a day!

PROBLEM 4

Estimate very approximately how many people it would take to count individually the blades of grass on a lawn measuring 15×30 ft in 1 hr.

1-4 ATOMIC SIZES

Can you imagine how small an atom is? Iron is a typical element whose atomic weight of 55.5 is neither exceptionally high nor exceptionally low. We can do a little model making and estimate the size of an iron atom. A block of iron does not squash easily and certainly cannot be compressed like a gas into a small fraction of its original volume. It might be that in solid iron the atoms are packed about as tightly as they can be. Figure 1-4-1 shows some spheres packed not quite as closely as possible, but in a nice simple way. For a rough estimate, suppose the atoms in a block of iron look like this.

A cube of iron 1 in on a side is easy to think about. Its mass is about 130 g. How many atoms is this? If we divide 130 g by the

Fig. 1-4-1 Part of a block of iron, greatly magnified, according to a simple model. The spheres represent iron atoms.

mass of one atom, we get $130/(9.14 \times 10^{-23}) = 1.42 \times 10^{24}$ atoms. (We found the mass of one atom from the product of m_H and the atomic weight of 55.5.) In the model each atom occupies a little cube of space, and thus it takes 1.42×10^{24} such cubes to make one cubic inch. Each atomic cube is then $1/(1.42 \times 10^{24}) = 7.05 \times 10^{-25}$ cubic inches in volume.

If the side of a cube has length d, then the volume is d^3. For the iron atom, we have $d^3 = 7.05 \times 10^{-25}$ in^3, and therefore $d = 8.90 \times 10^{-9}$ in. Just as we measured the mass in grams, we will prefer to measure the diameter in centimeters (abbreviated cm). One centimeter is about 0.4 inches, or there are about 2.5 centimeters in 1 inch. Thus, the diameter of the iron atom is about $(2.5) \times (8.90) \times 10^{-9} = 2.3 \times 10^{-8}$ cm.

Our model was very crude. Hard spheres can be nested together more compactly. More important, atoms are not really hard spheres, even though they occasionally behave as if they are. The size depends somewhat on the atom's environment. An atom packed in a solid occupies a somewhat smaller volume than when it is essentially all alone, as in a gas. Atoms seem to be a bit spongy, so an atom's size isn't a perfectly well-defined number, but it doesn't change by more than a factor of 2 in different circumstances.

The same kind of calculation we did for iron can be carried out for other elements that form solids, and we find that the diameters don't differ very much. Assuming solids are close-packed atoms, the diameters are all of order 10^{-8} cm. The phrase "of order" is very useful for giving the result of a rough estimate, so useful that there is a symbol for it: $\sim$. Its meaning is, by its nature, not very precise, but roughly it means *equal to within a factor of 10*. The diameter of all atoms is $\sim 10^{-8}$ cm, including those which don't happen to form solids at ordinary temperatures. When considering the size of atoms or the diameter of the universe such an *order-of-magnitude* estimate is very informative, though it might not be acceptable on an income tax form.

PROBLEM 5

If a tiny quantity of a compound called stearic acid is placed on a water surface it is observed to float as a thin film whose area is proportional to the weight of stearic acid used. The hypothesis is proposed that the stearic acid is forming a layer one molecule thick (called a monolayer). Explain why this is consistent with the facts. Is it proven? Assuming the stearic acid forms a monolayer, can you figure out a way you might determine the thickness of the monolayer and thus get an idea of the size of one molecule? Such a layer is too thin to measure directly, so you will have to be clever.

It is hard to appreciate actually how small atoms are. For example, it takes about 10^{14} atoms to cover the top of a pinhead. Suppose that we somehow magnified a picture of a pinhead so that one atom was about 0.1 cm in diameter and we could see it. Then the picture of the entire pinhead would be 6 miles across! Even the most powerful light microscope comes far from being able to see an individual atom. The best electron microscopes come closer, and Fig. 1-4-2 shows an electron micrograph of some molecules which are about 10^{-6} cm in diameter.

Just a few years ago a very special instrument, the field-ion microscope, was perfected to the point where individual atoms, or at least their locations, could be seen. Figure 1-4-3 shows a view looking toward the tip of an almost incredibly sharp tungsten needle, magnified by a factor of about 10 million. The point is really a round dome formed out of atoms stacked in flat layers, and the edge and corner atoms of each layer are especially prominent. The figure shows a similar array of cork spheres to help you visualize what the tungsten point looks like on an atomic scale.

Fig. 1-4-2 An electron micrograph of part of a protein crystal, showing the individual molecules to be roughly spherical and neatly stacked. Each molecule is about 10^{-6} cm in diameter. *(Courtesy of Prof. R. W. G. Wyckoff.)*

PROBLEM 6

Assuming all atoms have diameters of order 10^{-8} cm, estimate how many atoms are in each molecule of Fig. 1-4-2.

1-5 UNITS

Most of the world uses the metric system of units, in which the basic distance is the meter, about 39 inches, and the basic mass is the kilogram, whose weight at the earth's surface is about 2.2 pounds. From the meter and the gram, other units more convenient in various circumstances are derived, by taking them to be factors of 10, 100, 1,000, etc., smaller or larger. In the everyday world, the centimeter, 1/100 of a meter, has a role similar to the inch.

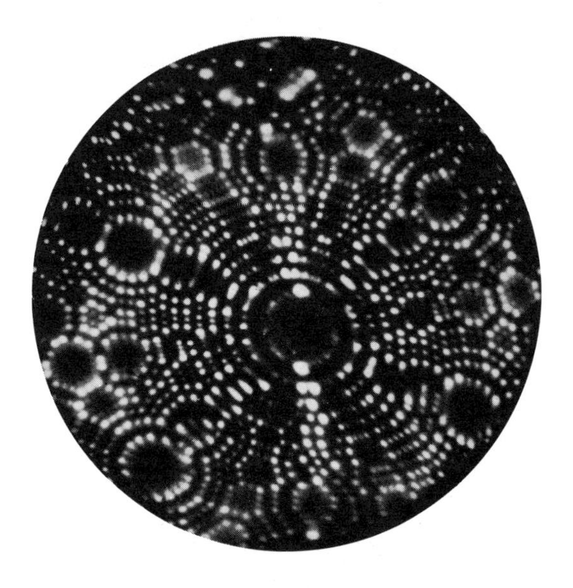

Fig. 1-4-3 At top, a field-ion microscope picture looking toward the tip of a tungsten needle. Below, an ordinary photo of cork spheres arranged in neatly packed layers. Each layer closer to the camera has a smaller area so that the entire assembly of cork spheres is dome-shaped. Note that the edge and corner "atoms" of the cork-sphere model form patterns similar to the bright spots of the field-ion micrograph. *(Courtesy of Prof. Erwin Mueller.)*

Scientists use the metric system for convenience. Factors of 10 are easy to multiply and divide by, while in the English system we need a factor of 16 to go from ounces to pounds, a factor of 12 to go from inches to feet, a factor of 3 for feet to yards, a factor of 5,280 for feet to miles, and so on. Tradition also has helped establish the metric system, since the early growth of the physical sciences was dominated by Europeans, who have used metric units as everyday standards since the French Revolution.

Table 1-5-1 summarizes the metric units of length and mass which will be used in this book and gives conversion factors to inches and pounds. Our unit of time will be the second. As we go along, there will be other units to consider for quantities such as force, electric charge, etc.

PROBLEM 7

While traveling with a friend in his European car, I see that the speedometer reads 100 kilometers per hour. The speed limit is 50 miles per hour. Is he speeding?

Table 1-5-1 Units

LENGTH	
Kilometer (km)	$= 10^3$ m
Meter (m)	
Centimeter (cm)	$= 10^{-2}$ m
Millimeter (mm)	$= 10^{-3}$ m $= 10^{-1}$ cm
Micron (μ)	$= 10^{-6}$ m
Millimicron (mμ)	$= 10^{-9}$ m
Angstrom (Å)	$= 10^{-10}$ m $= 10^{-8}$ cm
MASS	
Kilogram (kg)	
Gram (g)	
Milligram (mg)	$= 10^{-3}$ g
Microgram (μg)	$= 10^{-6}$ g
CONVERSION FACTORS	
1 in	= 2.54 cm
1 cm	= 0.394 in
1 kg	= 2.20 lb
1 lb	= 0.454 kg

PROBLEM 8
How many inches are in a mile? How many millimeters are in a kilometer? What decimal fraction of a yard is an inch? What decimal fraction of a meter is a centimeter?

1-6 WHAT LIES AHEAD

Modern physics is largely the study of the structure and interactions of atoms. Far from being hard spheres, atoms are actually complex objects consisting of electrons, protons, and neutrons held together by electrical and nuclear forces. To study the atom and its component parts, we shall need to learn *dynamics*, the physical laws that govern the motion of matter. *Newtonian dynamics*, the oldest and simplest laws of motion, will be important and useful, but as we look more deeply into the atom and its nucleus, quantum mechanics and the theory of relativity will be needed.

Dynamics permits us to find the motion of matter, if we know the forces acting on it. A weight hanging by a string (a pendulum) will swing to and fro if we pull it off to one side and let go. The forces exerted by gravity and by the string combine to cause the motion. From dynamics we can calculate the motion if we know the forces. Fortunately, as far as we now know, there are only four basic kinds of forces in the universe, the electrical, gravitational, "strong" nuclear, and "weak" nuclear forces. For the pendulum, we have both gravitational and electrical forces. The string acts ultimately through electrical forces because its atoms are held together by such forces.

Of the four forces, we can say that gravity and electricity are well "understood." In a given situation we can calculate the gravitational forces and the electrical forces from the masses and charges that are present, and make quantitative predictions that agree with the observed facts. For example, planetary motions or the motion of rockets and satellites can be very precisely calculated from the known effects of gravity. However, if we ask *why* there is gravity, or electricity, the answer is "That's the way the world is." Ultimately, physics *describes* nature but does not *explain* it.

With regard to the nuclear forces, our understanding is imperfect, but we can approximately calculate the effects of these forces in many cases. As a result many of the phenomena of nuclear forces are at least roughly understood. We also know that the effects of the nuclear forces are almost entirely limited to the tiny subatomic realm of the nucleus, and that these forces are *not* important in understanding much of the physics of atoms.

Although in its barest essentials physics may seem to be a narrow study of the four types of forces, its ultimate goal, to try to understand the infinite variety of the universe, is hardly narrow. The physicist observes nature, discovers (or invents?) her generalizing principles (*physical laws*), and then applies these basic principles. He may then be able to interpret in a simple and even beautiful way some hitherto mysterious phenomenon, such as the motion of the planets. Or he may discover a new possibility in the behavior of nature, like the laser, the transistor, or the transformation of mass into energy. The succeeding chapters will try to introduce you to the varied aspects of physics through some of its observational data, through discussions of the most important physical laws, and through the application of these laws to the understanding of a few selected physical phenomena.

SUMMARY

The goal of the physicist is to *quantitatively* predict the behavior of nature. To do this, he usually makes simplified *models.* When they are successful in making predictions, the models are often assumed to be true representations of nature.

Matter is made up of *atoms* of the *elements*, which can combine to form *molecules* of *compounds.* The formulas of compounds can be deduced from chemical experiments in which the amounts of different elements which combine to form various compounds are measured. From this information *relative masses* of the atoms of the different elements can be found. The mass of the lightest atom, hydrogen, is known from physical evidence, which will be discussed in later chapters, to be

$$m_H = 1.67 \times 10^{-24}\ g$$

While the size of an atom is not a precisely defined quantity, all atoms are of order 10^{-8} *cm* in diameter.

Metric units are used in physics, and the standards of length and mass are the *meter* and *kilogram.* The meter is a length of roughly one yard (actually 39.37 inches), and the kilogram is a mass whose weight at the earth's surface is 2.20 pounds. The unit of time is the *second.* Prefixes are used to develop other units from the basic ones, such as *centimeter* for 10^{-2} meter and *gram* for 10^{-3} kilogram.

All the interactions of matter which have so far been studied can be ascribed to the action of four types of *forces.* These are the *gravitational, electrical, strong nuclear,* and *weak nuclear* forces. Much of physics is based upon obtaining a description of how these forces act. Given this description, *dynamical theories* then describe the effects of the forces on the motion of matter. *Physical laws* are basic principles which can be used to understand the behavior of nature. Examples of physical laws are the description of the gravitational force and the dynamical theory of Isaac Newton.

ELECTRIC CHARGE

2-1 POSITIVE AND NEGATIVE CHARGE

Sometimes when you run a comb through your hair, there is a crackling sound, and the comb becomes *electrified.* It will pick up small bits of thread or paper. Also, after walking across a carpeted room, you may have drawn a spark from your fingertip and felt a shock when touching a doorknob or another person. These phenomena and a multitude of others (of which a lightning flash is the most dramatic) are generally called manifestations of *static electricity.* The ancient Greeks observed similar effects, especially when rubbing amber, and the word electricity is derived from the Greek for amber.

Some revealing experiments can be done with very simple apparatus. Figure 2-1-1*a* shows two light metal spheres hanging from stands by silk threads. (The spheres can conveniently be styrofoam painted with aluminum paint, which provides a thin metal coating.) Suppose we rub a glass rod with a piece of silk cloth, touch it to each sphere, as in Fig. 2-1-1*b*, and then move the spheres closer together. They repel each other, as in Fig. 2-1-1*c*. After a process that smacks of magic, something has happened to the spheres. They have become *charged.*

Figure 2-1-1 Repulsion between two similarly charged spheres.

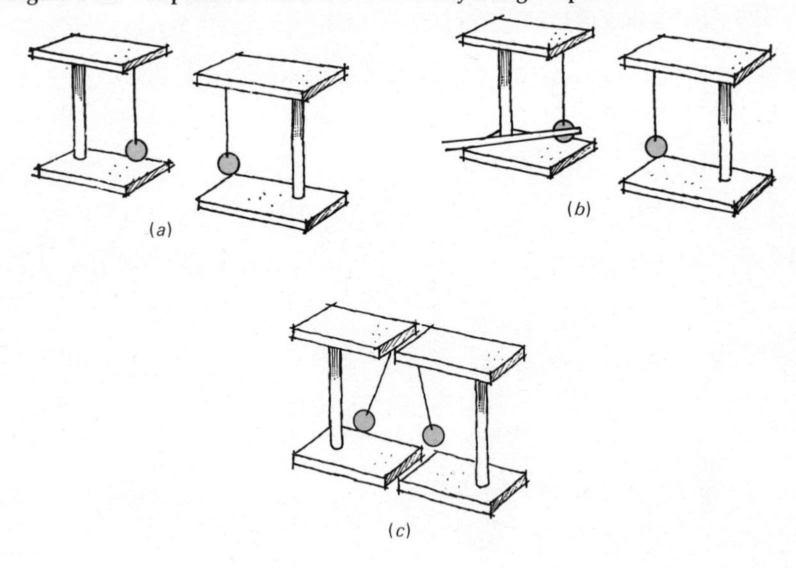

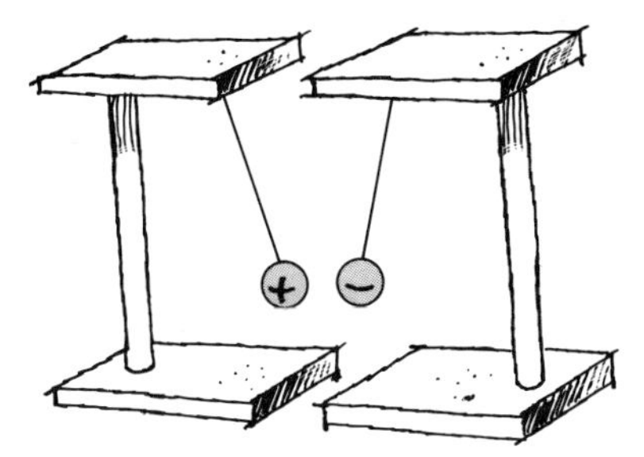

Figure 2-1-2 Attraction of oppositely charged spheres.

Suppose instead of rubbing glass with silk we rub a plastic rod with fur and repeat the simple experiment with the two spheres. The result is similar; they again repel each other. However, if we charge one sphere with a glass rod that has been rubbed with silk and the other with a plastic rod that has been rubbed with fur, they attract each other, as in Fig. 2-1-2. When we try experiments with all the combinations of materials we can find, there appear to be two, and only two, kinds of charge. The types of charge are called *plus* and *minus*, with plus defined arbitrarily to be the kind that came from the glass rod and minus the kind that came from the plastic one. Two objects which are positively charged repel each other, as do two negatively charged objects. Objects which are oppositely charged, plus and minus, attract each other.

The modern explanation of electrical phenomena utilizes the knowledge that each individual atom of any kind of matter contains positively charged protons and negatively charged electrons. The charge of one proton is precisely equal in magnitude, and opposite in sign, to the charge of an electron. The actual structure of atoms—how the electrons and protons are arranged inside them—will be a subject for later chapters. However, each atom contains equal numbers of protons and electrons, and thus equal amounts of positive and negative charge. Looked at from outside, the atom appears uncharged. When we speak of charged objects we really mean objects with a net charge, i.e., an excess of either plus or minus charge.

The positively charged protons are very firmly locked into the atom, but some of the negatively charged electrons can be removed relatively easily. When objects are rubbed together, electrons can be dislodged at the touching surfaces and can end up being transferred from one object to the other. The ease with which electrons can be transferred depends upon the materials.

When a glass rod is rubbed with silk, more electrons are transferred from the rod to the silk than from the silk to the rod. As a result, the rod ends up positively charged, having an excess of protons because it has lost some electrons. For a plastic rod rubbed with fur, the fur suffers a net loss of electrons, and the plastic has a net gain. Thus, the plastic becomes negatively charged.

One predicted consequence of the explanation given above is that when a glass rod is charged by rubbing with silk, the charge on the silk is exactly equal in magnitude and opposite in sign to the charge on the rod. This can be experimentally confirmed by observing that the net charge of silk plus rod is indeed zero.

To return to the charging of the spheres, when a positively charged rod is touched to an uncharged sphere, some of the electrons from the sphere are attracted to the rod. The sphere loses some electrons and becomes positively charged while the rod in general retains a net positive charge. The electrons which flow from the sphere are not sufficient to completely neutralize the charge of the rod. Thus, the sphere flies away from the rod soon after it has been touched because of the repulsion of its positive charge by the charge on the rod.

The plastic rod can be used to put negative charge on a sphere because the negative charges on the surface of the plastic repel each other and would like to escape. At the point of contact between rod and sphere the electrons on the rod can flow to the sphere, thus charging it negatively. Once again the sphere and rod repel each other after a momentary contact and the sphere moves away.

2-2 ELECTRICAL CONDUCTION

In solid matter the atoms are not free to move, and the positive charge, locked in the atom, is thus immobile. However, some of the atomic electrons can move around inside a solid, with an ease which depends on the particular material. In metals, electrons can move very freely but cannot readily escape from the surfaces, and we call metals good electrical *conductors.*

Figure 2-2-1*a* shows two spheres like those in Fig. 2-1-2. Suppose they have equal and opposite charges. Figure 2-2-1*b* shows the spheres after they have been connected by a fine wire, and

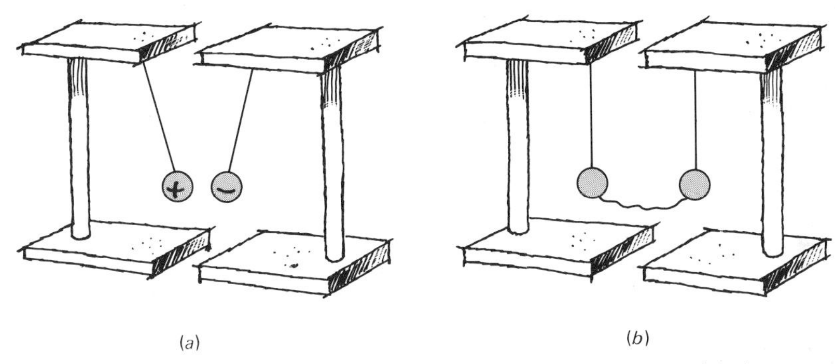

Figure 2-2-1 (*a*) **Two spheres with equal and opposite charges.** (*b*) **The same sphere after being connected by a wire.**

they now appear uncharged. The excess electrons on the negatively charged sphere have flowed through the wire to the positively charged sphere with the result that the net charge on either sphere is now zero. Electrons flow through a wire between oppositely charged conductors partly because the excess electrons on the negatively charged conductor repel each other and thus push electrons into one end of the wire. In addition, the positively charged conductor attracts electrons out of the other end of the wire.

We have assumed from the start that the spheres in our experimental apparatus have conducting surfaces. This helps in charging them, since touching one at a single point can cause a surplus or deficiency of electrons to be spread over the whole surface. Note also that if a conducting sphere has a surplus of negative charge, the extra electrons repel each other, and in trying to drive each other apart they end up on the surface. If there is a shortage of electrons on a conductor, electrons flow from the surface so as to neutralize any net positive charge in the interior of the conductor. Thus, a net positive charge is left on the surface. Any net charge on a conductor always lies on its surface.

In the same sense that the whole surface of a conductor can be charged by touching it at one point, all the net charge can be removed by touching at one point. This is why the conducting wire in Fig. 2-2-1*b* is effective in transferring all the excess electrons from the negatively charged sphere to the other one.

While metals are solid materials in which electrons are very free to move, there are other materials in which the electrons are nearly immobile. These materials, like glass, paper, silk, and plastic, are called *insulators.* The spheres in our apparatus are suspended with silk threads, which do not conduct charge off the spheres to an appreciable extent.

PROBLEM 1

Suppose you have two identical conducting spheres on insulating stands that can be positioned as you wish. If initially one has charge $+Q$ and the other is uncharged, describe how to get a charge $+Q/2$ on each. Given many such spheres, how could you get one with charge $+Q/16$? (Hint: Look for a method based on symmetry.)

An electric battery has terminals labeled + and −. An interesting experiment is to connect a battery with wires to two of our light conducting spheres, as shown in Fig. 2-2-2*a*. The spheres

Figure 2-2-2 (*a*) **Two conducting spheres connected to a battery.** (*b*) **The same spheres with a wire between them. The wire is hot enough to ignite a piece of paper.**

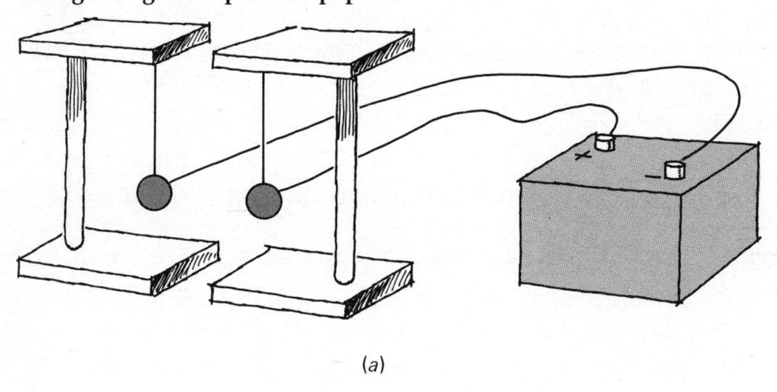

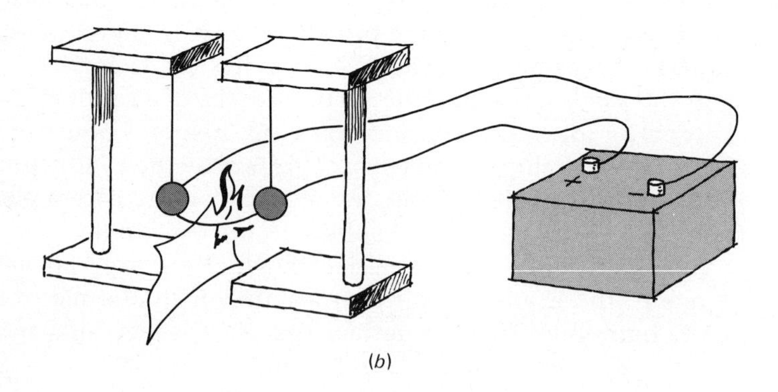

do not appear to attract each other, but if we again connect a wire between the two spheres, as in Fig. 2-2-2*b*, the wire gets hot. It may even get hot enough to glow. The heat comes from the flow of electrons through the wire, and the + and − signs on the battery do indeed tell us the signs of charge that are put on the spheres when they are connected to the battery terminals. However, for any ordinary battery, the charges on the spheres are too small to produce a noticeable attraction.

The important attribute of a battery is that chemical reactions inside it keep the spheres charged even when they are connected by the conductor, as in Fig. 2-2-2*b*. Therefore, while the battery is connected, a steady flow of electrons is maintained through the wire. We say that the wire carries a steady electrical current. In contrast, after the spheres are charged with glass and plastic rods that have been rubbed with silk and fur, a large current flows through a wire at the instant it is connected between the spheres, but after a very short time the current ceases to flow.

The battery keeps its own terminals charged + and −, so that if we want to produce a steady current, the spheres are not necessary and a wire can be connected directly from one battery terminal to the other. Electrons flow through the wire, through the battery, and back around through the wire, steadily. The battery is like a pump for maintaining the current.

The modern experimental physicist uses batteries or fancier electronic equipment to charge conductors and to produce electric currents. This discussion has been merely the most superficial introduction to electric circuits, to make it easier to discuss some important experiments later on. Chapter 11 provides a detailed discussion of electric circuits and of elementary electronics.

PROBLEM 2

Three conducting spheres are attached to insulating stands, and they are shown in Fig. P-2-2 at four successive times t_1, t_2, t_3, and t_4. At time t_1 sphere A has a net charge $+Q$, and spheres B and C are uncharged. At time t_4 sphere A has been taken far away, and spheres B and C are found to have equal and opposite charges. Explain.

2-3 FORCE

A major purpose of this chapter is to describe quantitatively the interaction between charges at rest; this interaction is called the

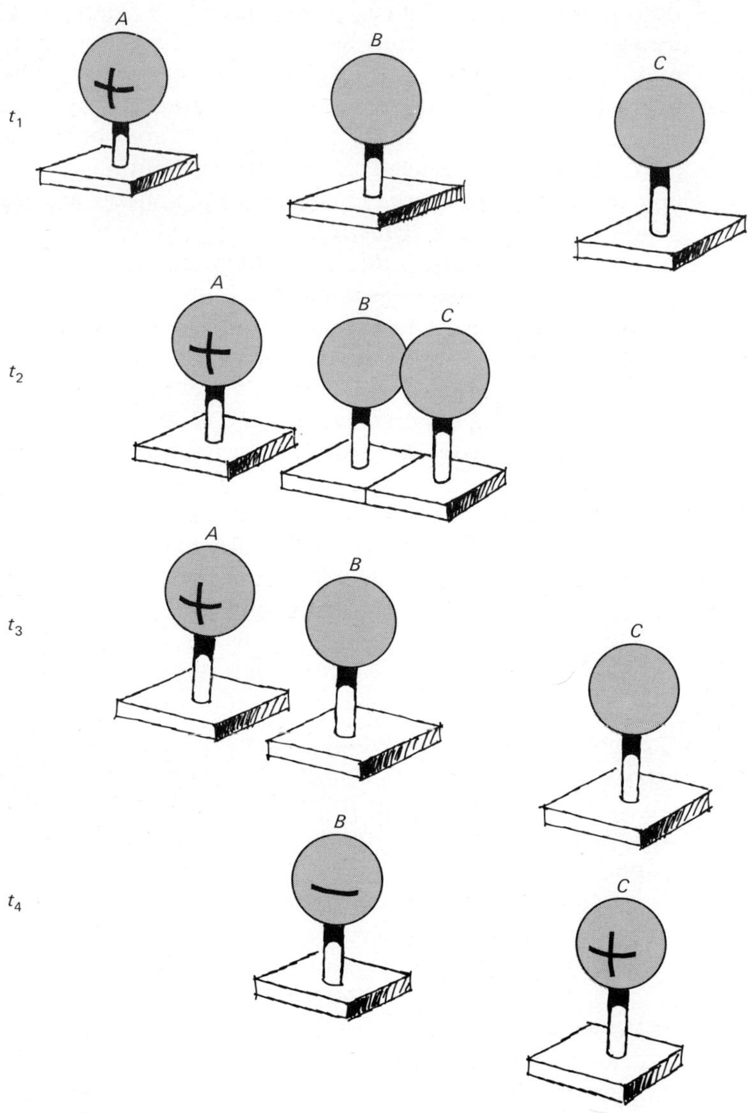

Figure P-2-2

electrostatic force. For charges in motion there is an additional magnetic force which will be discussed somewhat later. But first, what do we mean by a *force*?

In everyday language a force is an influence which tends to produce some effect. We speak of the force of evil, the force of a person's will, a force for peace. Also, we force open doors, force people to act, and so on. The colloquial use of force is varied, and it is not simple to give a single precise definition. In physics we use the word force in only one sense: It is a push (or pull) on an object.

A single object may be acted upon by more than one force, as two people may simultaneously pull on your arms. If two or more forces act on a body, they produce an effect which we describe in terms of a single *resultant* force or *net* force acting on the body. However, two or more forces may sometimes exactly cancel, combining to produce zero net force. Suppose you are standing still and two friends pull equally hard, one on each of your arms. If one pulls directly to your right and the other directly to your left, your body will remain stationary. The two equally strong forces in opposite directions cancel with regard to their effects on your body as a whole.

If your two friends pull as hard as before but both try to make you move in the same direction, then your body will indeed move. Thus, in determining the effect of forces, it is important to know their *magnitudes* (sizes) and also their *directions.*

The specific test for whether there is a net force on a body is implied by *Newton's first law of motion,* which may be stated as follows: *If there is no net force on a body, it remains at rest or moves with constant speed in a fixed direction.* If there *is* a net force on a body, then the body moves with changing speed, or in a changing direction, or both. Newton's second law describes quantitatively such changes in speed or direction, and will be discussed in the next chapter.

A quantity, such as force, which must be specified by giving its magnitude and direction is called a *vector quantity* or often simply a *vector.* A vector can be conveniently represented graphically by an arrow which points in the direction of the vector and which has a length proportional to the magnitude of the vector. (Note the velocity vectors in Fig. 1-2-1.) Not all physical quantities are vectors. Temperature, for example, has no direction and thus can be represented simply by a numerical value, like 68°F. Such a quantity is called a *scalar* and has a magnitude but no direction.

As you have seen, letters are often used as symbols of physical quantities in mathematical equations and in written text. We may speak of the temperature T or the force $\vec{F}$, and a letter with an arrow over it is used to stand for a vector quantity. A letter without an arrow stands for a scalar quantity or sometimes for the mag-

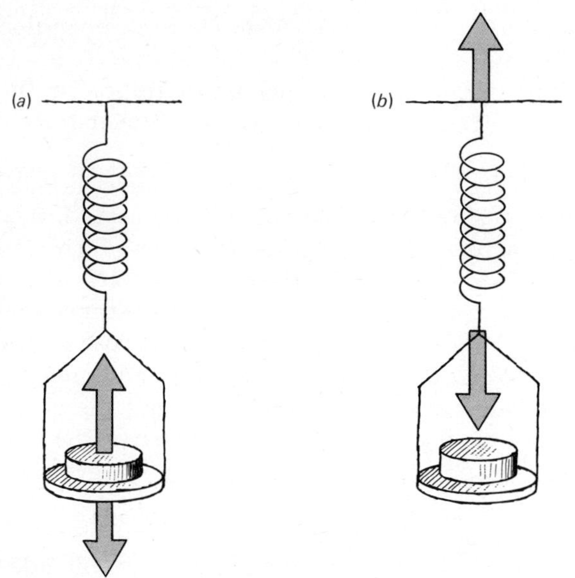

Figure 2-3-1 A weight hanging on a scale. (*a*) Forces on the weight; (*b*) forces on the spring.

nitude of a vector, as F might stand for the magnitude of $\vec{F}$. Sometimes the magnitude of a vector is written $|\vec{F}|$, where the vertical lines mean "the magnitude of."

How can we measure a force? One way is with a spring, and to begin a study of the spring scale, consider Fig. 2-3-1*a*, which shows a mass on a scale and two force vectors. The downward vector is the force exerted on the mass by gravity, its weight. When the mass is first placed on the scale, it moves as the spring stretches. However, ultimately it remains stationary, and therefore the net force on it must be zero. Thus the scale must pull up on the mass with a force equal in magnitude to the force of gravity, and opposite in direction, as shown by the upward vector of Fig. 2-3-1*a*.

In Fig. 2-3-1*b* we see the forces on the stationary spring. Again they are equal and opposite, and the *net* force is zero. However, the forces on the spring obviously tend to stretch it, and Fig. 2-3-2 shows the scale when four different weights are on it. The distances s_1, s_2, and s_4 show how much the spring is stretched by 1-, 2-, and 4-kilogram masses. The different loads on the spring correspond to different forces with known ratios, since the force of gravity on 2 kilograms is twice that on 1 kilogram, and on 4 kilograms is four times that on 1 kilogram.

The results shown in Fig. 2-3-2 are plotted on a graph in Fig. 2-3-3. A table is also shown on the graph which gives the numerical values of s_1, s_2, and s_4 in centimeters. The circled points on the graph correspond to the data in the table, and the light vertical and horizontal lines are to remind you of how the points were plotted from values of load and stretch. A straight line is drawn on the graph and passes through all four circled points. We can say that a straight line fits the data very well.

A reasonable hypothesis would be that the size of an unknown force could be determined by measuring how much it stretched this spring, using the straight line in Fig. 2-3-3 as a *calibration curve* for the spring. For example, if the stretch were 5.1 cm the force would be equal to the gravitational force on 3 kg. The dashed lines on the graph are to remind you of how to find the load that corresponds to a 5.1-cm stretch by using the calibration curve. It is

Figure 2-3-2 A spring scale loaded with different weights. A reference line shows the height of the pan when the scale is unloaded, and the distances s_1, s_2, and s_4 show the amount the spring stretches when 1, 2, or 4 kg are placed on the scale.

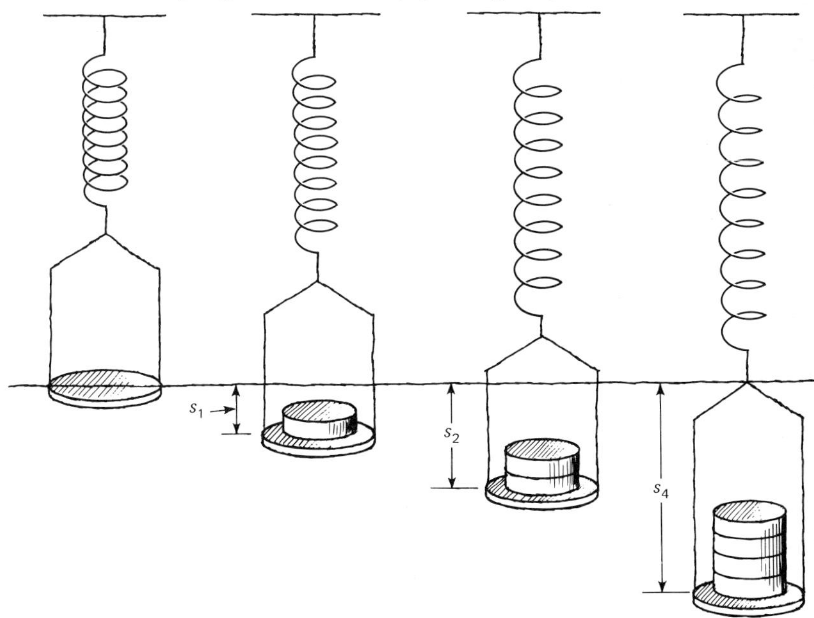

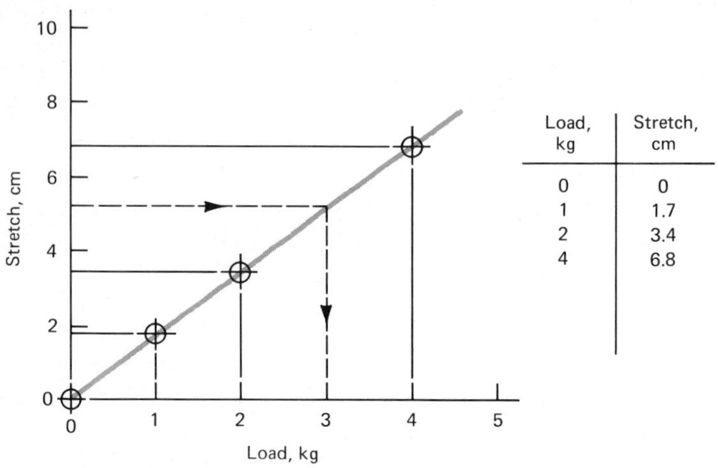

Figure 2-3-3 Stretch plotted against load.

in fact true that the stretch of a spring is related to the size of the equal and opposite forces on its ends by a straight line calibration curve, as long as the spring is not stretched too far. Thus, a spring is a convenient device for measuring forces.

The unit of force which we will use in this book is the *newton,* abbreviated N. A force of one pound is equal to 4.45 newtons, and the gravitational force on a kilogram mass at the surface of the earth equals 9.80 newtons. A mass of one kilogram is defined as a mass equal to that of a certain platinum-iridium block owned by the French Bureau of Standards. Masses can be compared by comparing the gravitational force on them at any given location. Note that the gravitational force on a 1-kg mass is 9.80 N in our laboratory but that this force varies according to the distance of the mass from the earth. Somewhere far away from the earth the gravitational force on a 1-kg mass could be only 1 N.

We could have expressed the relation between the stretch of our spring and the load as follows: The stretch in centimeters is equal to 1.7 times the load in kilograms. This is precisely the relation specified by the straight-line calibration curve. To express the stretch properly in terms of the force on the spring, we would have converted to newtons and concluded from the data of Fig. 2-3-3 that the stretch in centimeters is equal to 0.174 times the force in newtons. This is the proper calibration of the spring as a force-measuring device, and it does not depend on gravity.

PROBLEM 3

A boy who weighs 50 lb sits in a swing supported by two vertical chains, and the seat of the swing weighs 10 lb. What is the force in newtons exerted on the seat by each chain?

PROBLEM 4

A rope, two pulleys, and a weight are shown in Fig. P-2-4. One end of the rope and one pulley are attached to a ceiling. When the force exerted downward on the free end of the rope is 350 N, the weight remains stationary. What is the net force on the weight in this case? Find the gravitational force on the weight and its mass in kilograms. (Hint: For a rope passing over ideal pulleys, the tension in the rope stays constant.)

Figure P-2-4

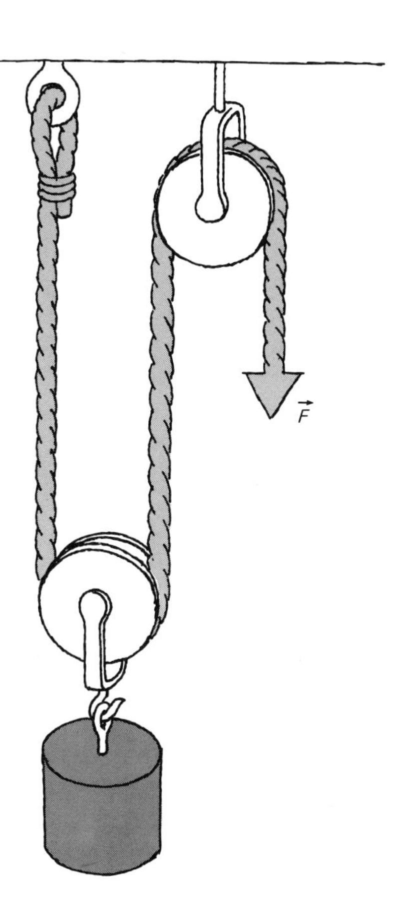

2-4 COULOMB'S LAW

In the mid-eighteenth century Charles Augustin Coulomb, a French physicist, measured how the force which one charged sphere exerts on another depends on the distance between them and on their charges. To measure the force he used a sophisticated version of the spring scale we discussed above, better adapted to measuring small forces. Although he used spheres, we always speak of his result as the force law between *point charges.* A point charge is an idealization which you can imagine as being a charged sphere so small that its size does not affect the physics in question. Coulomb designed his experiment so that the finite sizes of his charged spheres introduced negligible effects.

First, Coulomb found that for two charges of magnitude Q_1 and Q_2, a fixed distance r apart, the force depended on the product of the two amounts of charge according to the relation:

$$F \propto Q_1Q_2$$

where the sign $\propto$ means *is proportional to.* [We adopt the common practice of omitting multiplication signs. Symbols next to each other are to be multiplied, while numbers to be multiplied are placed in "touching" parentheses ( )( ).] F describes the force on either charge. It is in the direction away from the other charge if Q_1 and Q_2 are both positive or both negative, so that they repel. It is toward the other charge if Q_1 and Q_2 have different signs and attract.

Next, Coulomb measured how the force on two given charges depended on the distance between them. The graph of Fig. 2-4-1 shows the relation he found. Doubling the separation reduced the force by a factor of 4; tripling it gave a factor of 9; and so on. The final algebraic formula which summarizes all his results is called Coulomb's law:

$$F \propto \frac{Q_1Q_2}{r^2} \qquad \text{2-4-1}$$

or

$$F = k\frac{Q_1Q_2}{r^2} \qquad \text{2-4-2}$$

Any time we have a relation with a proportionality sign we can convert it into an equality by inserting a constant, like k above, since that is just the meaning of proportionality. The constant k

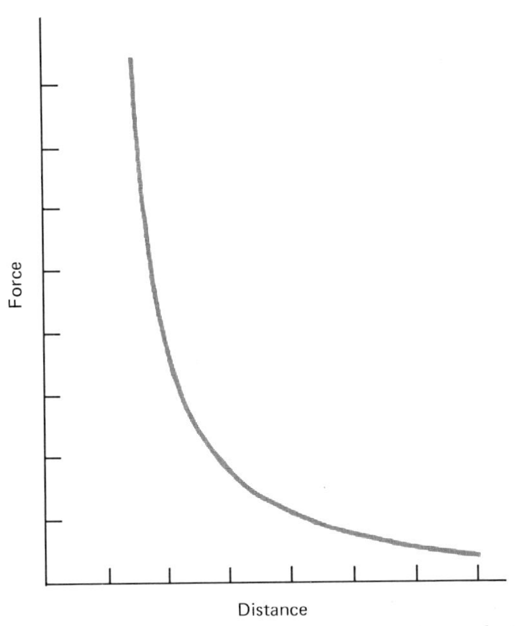

Figure 2-4-1 The way the force on one charge due to a second charge depends upon the distance between them. No numerical values of force or distance are given, but the shape of the curve shows that the force is inversely proportional to the square of the distance.

in Coulomb's law is a number which stays the same no matter what values Q_1, Q_2, and r have.

The value of k depends on the units in which the quantities in the equation are measured. You can see how this goes in a very simple case. If you drive a car north at a velocity of 50 miles per hour for 2 hours, you'll go 100 miles. If your velocity doesn't change either in magnitude or direction (it is a vector), you can write a general formula:

Distance in miles = (velocity in miles per hour)(hours driven)

But you could also write:

Distance in feet = (5,280)(velocity in miles per hour)(hours driven)

In the first equation the constant is concealed; it equals unity. In the second equation, 5,280 is the value of k that comes from measuring distance in feet and velocity in miles per hour. In either case we could write, of course,

Distance $\propto$ (velocity)(time)

but if we want to get real numbers we have to choose a system of units, write an equality using the constant k, and find out what k is.

The system of units we will use throughout this book is called the *mks system*, which means that the meter, kilogram, and second are the units of length, mass, and time. In this system the unit of force is the newton, equal to about 0.22 pounds. All equations in this book will be written with appropriate constants for all quantities to be in mks units, unless there is a specific statement to the contrary.

If there were no established unit of electric charge, then Coulomb's law, Eq. 2-4-2, could be used to define the unit of charge so that the constant k in this equation were 1. By this we mean that one unit of charge could be defined as that amount which produces a force of one newton between equal unit charges one meter apart. However, in the mks system there is another electrical law, involving currents, which is used to choose a convenient unit of charge. The unit of charge defined in this way is called the *coulomb,* abbreviated C, and is *not* the right-sized unit to make Coulomb's law simple. (About one coulomb of charge per second flows through a 100-watt household light bulb.) For mks units, with charge in coulombs, Coulomb's law is

$$F = (9.0)(10^9)\frac{Q_1 Q_2}{r^2} \qquad \text{2-4-3}$$

PROBLEM 5

Each of two conducting spheres has a net charge $-Q$, and the radius of each sphere is 1 cm. When the centers of the spheres are 3 cm apart, the force between them is found to be smaller than predicted by Coulomb's law for point charges 3 cm apart. Explain. What if each sphere had a charge $+Q$? What if the spheres had charges of opposite signs?

PROBLEM 6

Three point charges are in a line as shown in Fig. P-2-6, with $Q_1 = 10^{-6}$ C, $Q_2 = 2 \times 10^{-6}$ C, and $Q_3 = 2 \times 10^{-6}$ C. What is the force on Q_2?

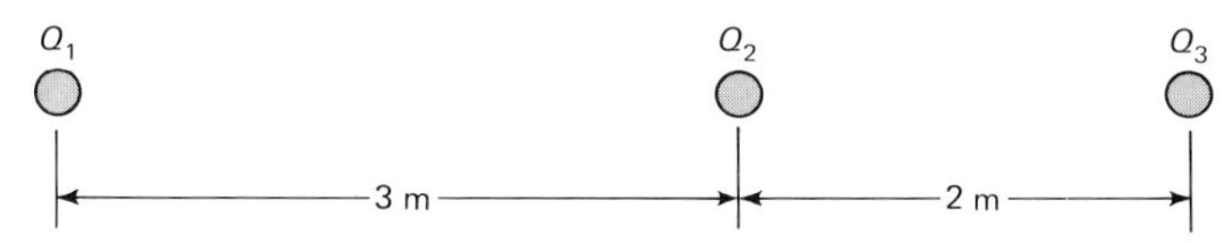

Figure P-2-6

Can you find a place along the line between Q_1 and Q_3 where the force on Q_2 would be zero? At this point the force on any charge would be zero. Explain.

2-5 SUPERPOSITION AND VECTOR ADDITION

Suppose we consider three point charges Q_1, Q_2, and Q_3 lying in a horizontal plane. Figure 2-5-1 shows a top view of them. To have a concrete example, the amounts of charge, their signs, and the distances between them are also given on the figure. Now consider the problem of determining the force on Q_3. Two vectors $\vec{F}_{13}$ and $\vec{F}_{23}$ are shown, the first being the force on Q_3 that would come from Q_1 only. (The subscript signifies this and is read "one, three," not "thirteen.") The force on Q_3 which would come from Q_2 alone is labeled $\vec{F}_{23}$.

The most important fact we want to present in this section is

Figure 2-5-1 A view looking down on three point charges Q_1, Q_2, and Q_3 lying in a horizontal plane. The vectors $\vec{F}_{13}$ and $\vec{F}_{23}$ show the forces exerted on Q_3 by the other two charges.

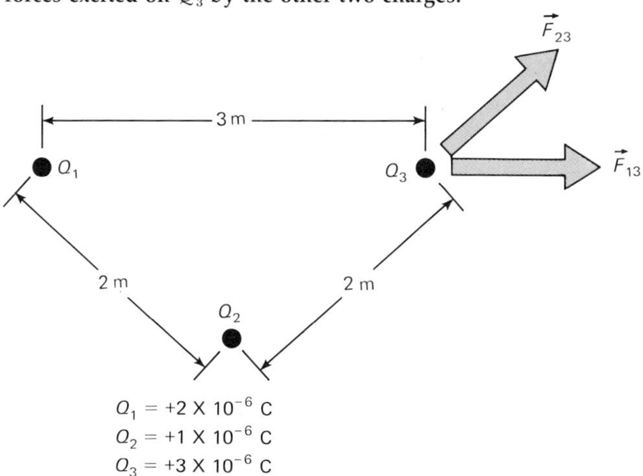

that the presence of Q_2 doesn't change the force on Q_3 due to Q_1, or vice versa. The net electric force on Q_3 is found by combining the force $\vec{F}_{13}$, calculated as if Q_2 didn't exist, and the force $\vec{F}_{23}$, calculated as if Q_1 didn't exist. This is called the *principle of superposition.* Although we have considered the superposition of only two forces for simplicity, the principle applies to any number. It applies to all electrical forces in a vacuum, and under most conditions in matter. The cases where it does not apply are so special we will never meet any in this book. The superposition principle also applies as far as we know to gravitational forces, and it permeates all physics. Were this not true, physical phenomena and the theories which describe them would be a great deal more complex, and the physical world would be very different from the world we know.

Let us now discuss how to find the net force (or total force) on Q_3 from $\vec{F}_{13}$ and $\vec{F}_{23}$. Figure 2-5-2 shows one way: Lay the two vectors down head to tail with their proper lengths and directions, and draw the new vector that extends from the tail of the first to the head of the second. This vector, which we will call simply $\vec{F}$, is the force on Q_3 from Q_1 and Q_2, the *resultant* of the individual forces $\vec{F}_{13}$ and $\vec{F}_{23}$.

Naturally, if Q_3 doesn't move, $\vec{F}$ must be exactly opposed by a force from whatever holds Q_3 at rest, so that finally the net force on Q_3 is zero. However, we are only interested now in electric

Figure 2-5-2 The vector $\vec{F}$ is the resultant, or the vector sum, of $\vec{F}_{13}$ and $\vec{F}_{23}$. The figure shows how $\vec{F}_{23}$ has been moved so its tail is at the head of $\vec{F}_{13}$ and its direction is unchanged. You can see that the same resultant would be given if $\vec{F}_{13}$ had been moved so its tail were at the head of the original vector $\vec{F}_{23}$ and its direction kept unchanged.

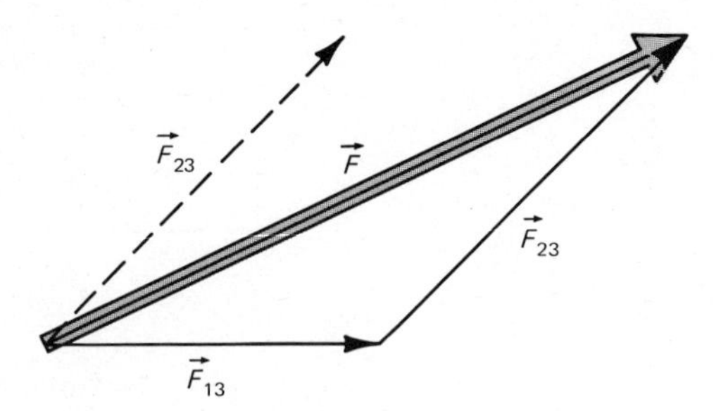

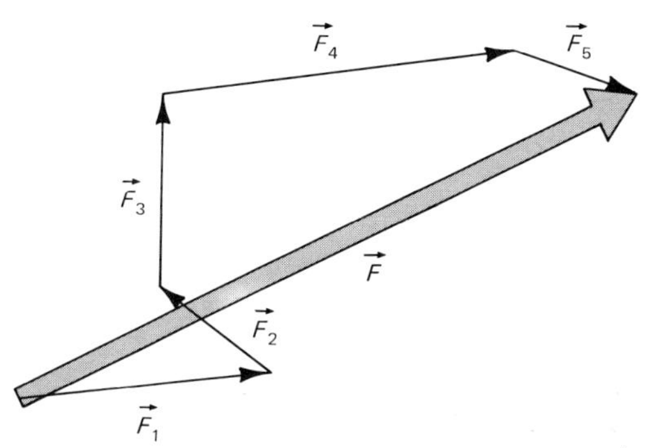

Figure 2-5-3 The vector sum of five vectors $\vec{F}_1$ through $\vec{F}_5$ to give the resultant $\vec{F}$.

forces on charges at rest, so we will ignore the forces needed to keep the charge from moving.

To show how things go if we have to combine more than two forces, Fig. 2-5-3 illustrates a more complicated case. As you might guess, we just lay down all the force vectors head to tail and then draw in the resultant. (The rule works even for vectors which don't all lie in a plane, and you can think of some of the vectors in the figure as pointing toward or away from you.) This process is called *vector addition*, and the resultant $\vec{F}$ is the *vector sum* of the individual force vectors. This is a general technique used for all vector quantities, for example, velocities and displacements as well as forces.

One way of adding vectors is graphically, as in Fig. 2-5-2, by making an accurate scale drawing of the directions and relative magnitudes of the vectors. You can check that the magnitudes and directions of $\vec{F}_{13}$ and $\vec{F}_{23}$ in Fig. 2-5-2 are consistent with the sizes and locations of the charges in Fig. 2-5-1. Another way is to use a lot of trigonometry to avoid actually making a scale drawing, but we will not get involved in that technique here.

Sometimes instead of combining two or more vectors to form a resultant, it is useful to start with one vector and break it into two or more *components*. For example, Fig. 2-5-4 illustrates three ways in which a particular vector $\vec{F}$ can be expressed as a sum of two components, taking the components to be in various directions. If the directions of the components are given, their magnitudes can easily be found graphically. This technique is often particularly useful in physical problems where the component of a vector in a particular direction is important. For ex-

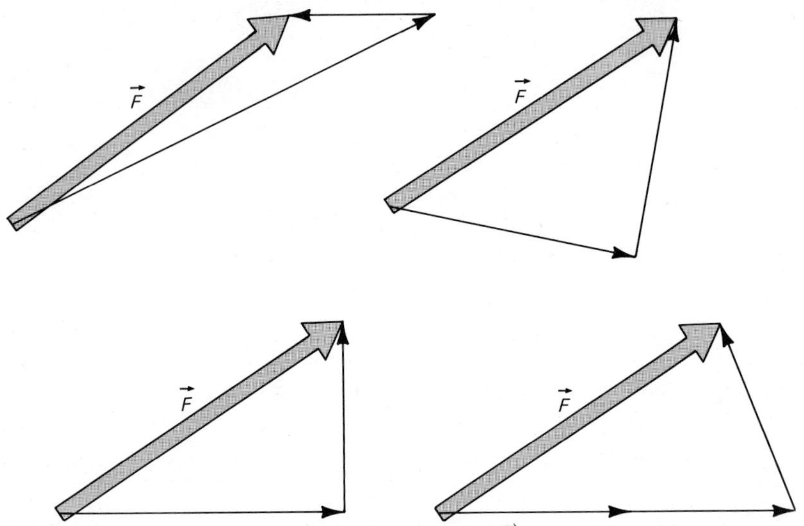

Figure 2-5-4 A few ways in which a given vector $\vec{F}$ can be expressed as a sum of two vectors.

ample, if $\vec{F}$ were an aerodynamic force on an airplane wing, horizontal and vertical components, like those in the third sketch of Fig. 2-5-4, would be used to separately determine drag and lift.

PROBLEM 7

What is the magnitude of the force $\vec{F}_{13}$ in Fig. 2-5-1? Assuming that Fig. 2-5-2 is an accurate scale drawing, use the answer you just found to find the magnitude of the net force $\vec{F}$ on Q_3 in Fig. 2-5-1.

PROBLEM 8

Two forces act on a body, 300 N downward and 400 N sideways. Find the magnitude of the resultant force graphically, and then find it from the pythagorean formula for a right triangle.

PROBLEM 9

A 100,000-lb jet airplane is acted upon by a horizontal thrust force from its engines of 10,000 lb, a horizontal drag force of 5,000 lb, and a vertical lift force of 105,000 lb. Find the magnitude and direction of the net force on the plane.

PROBLEM 10

A 10-kg mass is supported by two ropes as shown in Fig. P-2-10. What is the vertical component of the force exerted on the mass

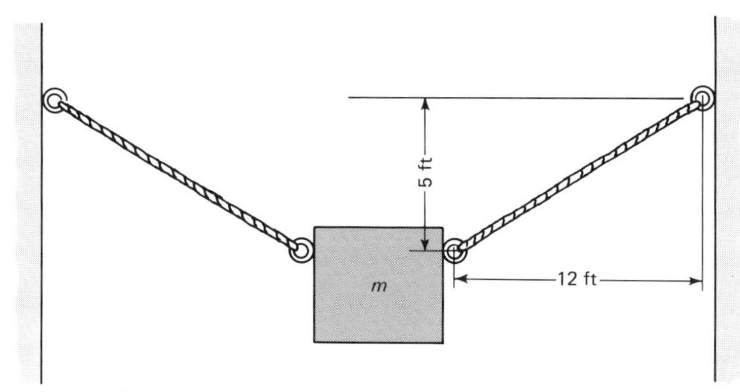

Figure P-2-10

by each rope? What is the magnitude of the total force exerted on the mass by each rope? (Hint: A rope can only exert a force along the direction it is stretched.)

2-6 THE ELECTRIC FIELD

Suppose in the example illustrated in Fig. 2-5-1 that Q_3 were +1 C. Just as above, we could find $\vec{F}$, the vector sum of $\vec{F}_{13}$ and $\vec{F}_{23}$, which would now be the force on 1 C located where Q_3 is. Would the direction of $\vec{F}$ be different? No, because $\vec{F}_{13}$ and $\vec{F}_{23}$ would both be changed by the same factor (note from Coulomb's law that these forces are both proportional to Q_3). Thus, our old vector diagrams will still be good with the scale changed. Since Q_3 was originally $(3)(10^{-6})$ C and is now assumed to be 1 C, all the vectors now stand for forces whose magnitudes are a factor $(3.33)(10^5)$ larger than they were before. The directions remain the same. Now, the force on 1 C of charge located at some point in space is called the *electric field* $\vec{E}$, at that point. It has a magnitude and a direction—it is a vector—and from it we can find the force on any quantity of charge Q located at that point,

$$\vec{F} = Q\vec{E} \qquad 2\text{-}6\text{-}1$$

The unit of $\vec{E}$, by definition, is newtons per coulomb.

The force on a charge $+Q$ has the direction of $\vec{E}$ and is Q times as big in magnitude. But what if Q is a negative charge? Then the force is in the opposite direction to $\vec{E}$ and again its magnitude is Q times as large as the magnitude of E. We will often have occasion to multiply vectors by some number (some *scalar*), and the general rule is as follows: If $\vec{A}$ is some vector, then $k\vec{A}$ is k times as long and in the same direction as $\vec{A}$ for positive k and in the opposite direction for negative k.

Why bother with $\vec{E}$? It breaks down the problem of finding the force on a charge like Q_3 into two parts. First, we find the electric field where Q_3 is located, and then we multiply by Q_3. That may sometimes help our thinking, but it isn't the crucial reason. Another convenience is that in studying the action of an electric force on some interesting system, say, for example, a hydrogen atom, we can begin by saying "there is an upward field $\vec{E}$ at the atom . . . ," knowing how such a field might be created and not wanting to go into a detailed description of where some charges are placed so as to obtain $\vec{E}$.

However, the true importance of $\vec{E}$ is that it is "real." Suppose we go back to the simplest of all possible cases, the force of one charge, Q_1, on another, Q_2, some distance away. If Q_1 is very suddenly moved to another location nearby, should the force on Q_2 instantly adjust itself to the right value, from Coulomb's law, for the new position of Q_1? No, for no observable physical effect can propagate from one place to another faster than the speed of light. This is the keystone of the *special theory of relativity.* Therefore, Q_2 doesn't "know" that Q_1 has jumped until some time afterward: the time it takes light to travel from Q_1 to Q_2. The original electric field of Q_1 persists at the location of Q_2 after Q_1 has moved, and *it* is what matters in determining the force at Q_2.

Actually, if a charge oscillates rapidly up and down, its electric field propagates out into space in a very different manner than if it stays motionless or moves slowly. Radio transmission works this way. Moving charges in the antenna at the transmitter generate traveling field-wiggles (radio waves) that in turn act on charges in the home radio antenna to produce a small but accurate replica of what the transmitter is broadcasting. Radio wave propagation can be analyzed, with great difficulty, by working directly with forces between charges and correcting for the finite velocity of light, but the alternative description in terms of the creation of an electric field by a charge and its subsequent action on some other charge is much more straightforward.

There are many instances in electromagnetism where believing in a "real" field seems most natural. Even if it may be difficult to absolutely prove that the field idea is the only really correct one, it leads to such effective models of the more sophisticated aspects of physics that it has become indispensable.

Now we want to describe some specific electrostatic fields set up by stationary charges. Most basic is the electric field of a point charge Q_1:

$$|\vec{E}| = (9.0)(10^9)\frac{Q_1}{r^2} \text{ newtons/coulomb} \qquad 2\text{-}6\text{-}2$$

in the direction radially outward from Q_1 if Q_1 is positive, and radially inward if it is negative. Since $\vec{E}$ is a vector, we must give its magnitude and direction. The force on a charge Q_2 a distance r away from Q_1 is then $\vec{F} = Q_2\vec{E}$, which reproduces Coulomb's law with the above formula for $\vec{E}$. In terms of fields, our original three-charge problem becomes the problem of finding $\vec{E}$ at the location of Q_3 from the fields of Q_1 and Q_2. We thus find the vector sum of two point-charge fields: $\vec{E} = \vec{E}_{13} + \vec{E}_{23}$. The principle of superposition applies to electric fields just as it does to electric forces.

PROBLEM 11
Two point charges are arranged as shown in Fig. P-2-11. Find the magnitude and direction of the electric field at point A. From the field, find the force on a charge of -1.5×10^{-9} C at A.

Figure 2-6-1 shows a kind of picture of the field of a point charge. The *field lines* point in the direction of $\vec{E}$. There is a convention that the spacing between field lines indicates the strength of the field (the magnitude of $\vec{E}$), $\vec{E}$ being large when they are close together and small when they are far apart. In general, we will not try to draw field lines so this is precisely true, but the

Figure P-2-11

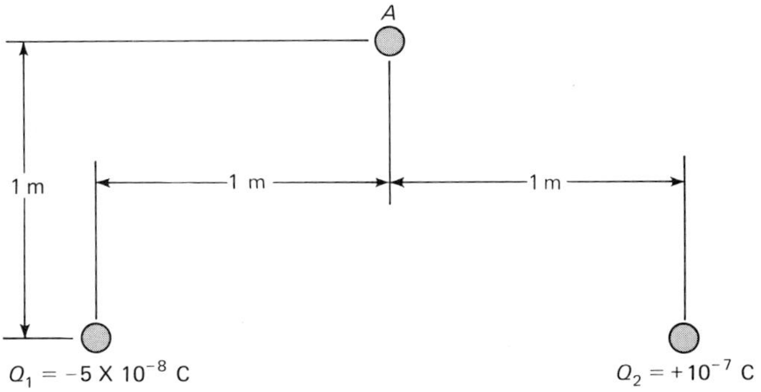

A

1 m

1 m

1 m

$Q_1 = -5 \times 10^{-8}$ C

$Q_2 = +10^{-7}$ C

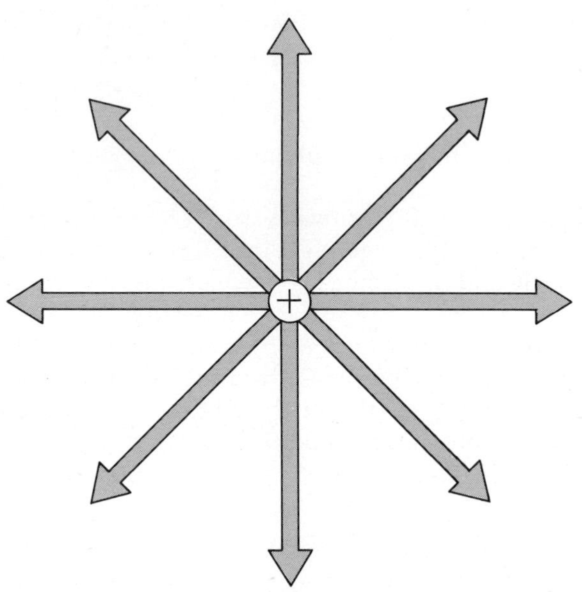

Figure 2-6-1 Electric field lines of a positive point charge.

field lines can give a qualitative feel for how the electric force on a given charge would depend on its location. In Fig. 2-6-2, a number of fields are shown which result from the presence of more than one charge. You can check that the field lines for the two-charge examples are consistent with what you would get by roughly sketching in the vectors $\vec{E}_1$ and $\vec{E}_2$ from each of the charges and finding their vector sum $\vec{E}$. One such example is shown in Fig. 2-6-2*a*.

PROBLEM 12

Two point charges are 1 cm apart, and the charges have opposite signs. The magnitude of the + charge is twice that of the − charge. Try to qualitatively sketch the electric field lines in a plane that contains the two charges, out to distances of at least 10 cm from the charges. What must the field be like very close to either charge? Can you guess what the field must be like very far away from the charges? The answers to these questions may help in making the sketch.

The two fields we will have occasion to use most are that of a point charge and the uniform field, which is shown in Fig. 2-6-2*c* and *d*. In Fig. 2-6-2*c* imagine a flat, uniformly charged sheet seen

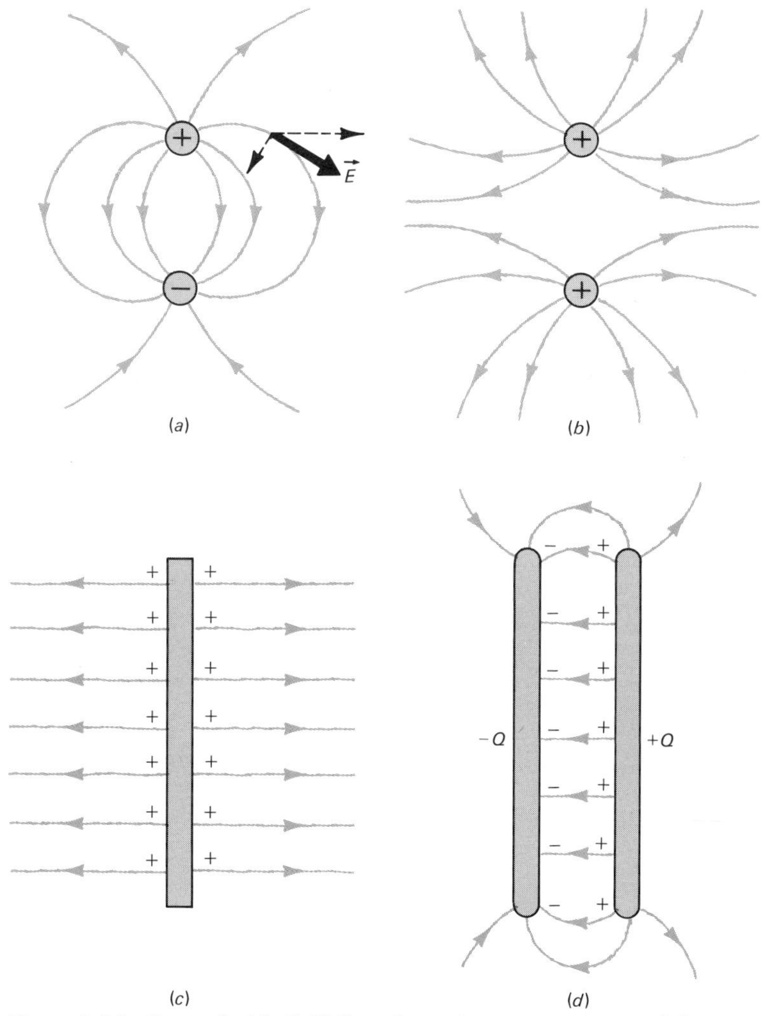

Figure 2-6-2 Some electric field lines for various arrangements of charges. (*a*) Equal and opposite point charges. (*b*) Identical positive point charges. (*c*) A uniformly charged sheet. (*d*) Two parallel charged plates, with equal and opposite charges.

from the edge. Also imagine that it is a very large sheet and that the edge view shown refers to the field lines in the region near its center. We will not rigorously calculate the field of such a charge distribution, but we will try to make the result plausible. [One way to calculate the field would be to let the very large sheet contain an astronomical number of small point charges uniformly laid out on its surface, and to add up (vectorially) all their fields. That's the kind of job the integral calculus was invented for—actually to add up the effects of an infinite number of infinitesimal things.]

Suppose the sheet is so large that the charges near its edges create negligible fields near the center. Then the shape of the sheet, i.e., where its edges really are, doesn't matter. Therefore, the field lines ought to be perpendicular to the sheet since there is no reason for them to slant toward any edge. Also, if we stay some fixed distance away from the sheet and ask how the field changes as we move parallel to it, there is no reason for it to change as long as it doesn't "know" where the edges are. Finally, if we move away from the sheet along any field line, the field strength remains constant as long as we don't go too far away. (The field lines are parallel and their spacing stays constant.) Thus, over some region of space we can in this way manufacture a uniform field in a certain direction and with some fixed magnitude.

In practice, a more convenient way to produce a uniform field is shown in Figure 2-6-2*d*. This represents two parallel conducting plates, each of area A, with equal and opposite charges. If their dimensions are much bigger than their spacing, then the field between them is uniform to within about one gap width of the edges. In order to show that this arrangement produces a strong uniform field between the plates, use the principle of superposition. Apply the field picture shown in Figure 2-6-2*c* to the two oppositely charged plates of Fig. 2-6-2*d* (reversing the field direction for the negatively charged plate). When the fields for the two sheets of charge are added vectorially, they cancel except in the region between the plates, where they are both in the same direction. The quantitative answer for the uniform field region of Fig. 2-6-2*d* is

$$|\vec{E}| = (4\pi)(10^9)\frac{Q}{A} \text{ newtons/coulomb}$$

PROBLEM 13

An uncharged flat metal plate is inserted between and parallel to a pair of charged plates arranged as in Fig. 2-6-2*d*. After a short time the electric field inside this plate is zero. Explain this, utilizing the fact that a metal contains electrons which are free to move.

2-7 THE MILLIKAN OIL-DROP EXPERIMENT

About 1910 Robert A. Millikan devised an elegant experiment which we now are in a position to understand. He succeeded in

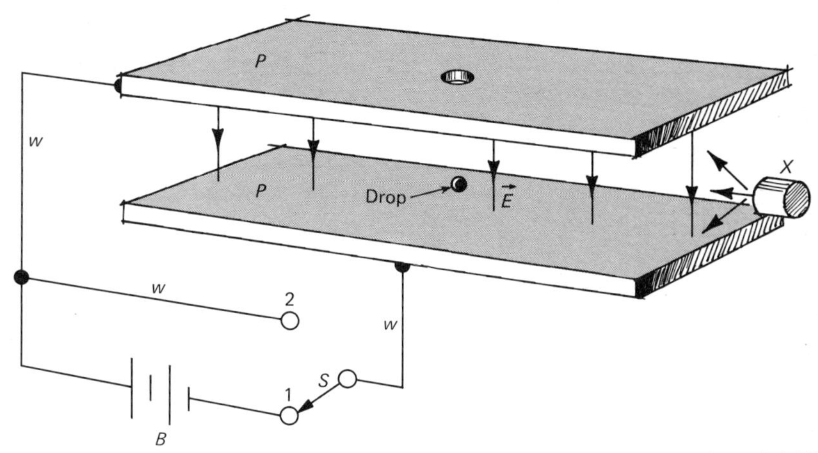

Figure 2-7-1 Schematic view of a Millikan oil-drop experiment. A uniform field E is established between the two horizontal plates P with the battery B and the switch S shown symbolically as on drawings of electrical circuits. Oil drops can be introduced through the small hole in the top plate and viewed with a telescope. The drops are illuminated by a strong spotlight not shown in the drawing. The x-ray source X could be turned on to change the charge on a given drop.

measuring, to an accuracy of better than 1 percent, the tiny charge of a single ionized atom. An ionized atom or *ion* is an atom which has gained or lost one or more electrons so that it has a net charge. The apparatus he used is shown schematically in Fig. 2-7-1.

With the battery B he could establish a uniform $\vec{E}$ field in the region between the plates P when the switch S was in position 1. The field could be made zero by putting the switch in position 2. The lines marked w symbolize copper wires for conducting charges $+Q$ and $-Q$ to the two plates with the battery connected. From the known battery voltage and the size and spacing of the plates it is possible to calculate Q and the field $\vec{E}$. The plate separation was about 1 cm.

Drops of oil from the spray of an atomizer could fall into the space between the plates through the small hole in the top plate. Only drops so small that they fell very slowly were used, and these were observed with a telescope as sharp points of light, like stars, when illuminated by a strong spotlight. The small oil drops often were charged in the process of formation at the atomizer, and the apparatus was designed to measure the amount of charge on a single droplet.

The basic principle of the experiment was to deduce the charge on a droplet by comparing its downward velocity when acted upon only by gravity with its downward velocity when acted upon by the combined forces of gravity and a known electric field. The very small droplets used in the experiment moved at constant velocity because of the viscous air drag acting on them. The upward drag force on a sufficiently small drop falling at velocity v is given by the expression

$$F_{\text{drag}} = Cav \qquad \text{2-7-1}$$

where C is a known constant which depends upon the properties of air, a is the drop radius, and v is its speed.

When falling under the influence of gravity, a droplet quickly reached a constant velocity v_g, such that the drag force was equal and opposite to the gravitational force. (Recall Newton's first law: When the net force is zero the velocity is constant.) The gravitational force is given by the expression

$$F_{\text{grav}} = mg \qquad \text{2-7-2}$$

where m is the mass in kilograms of the drop and g is a constant equal to 9.80 N/kg, determined by the strength of the earth's gravitational attraction. An equation for zero net force under the action of gravity and viscous drag can now be written

$$mg = F_{\text{drag}} = Cav_g \qquad \text{2-7-3}$$

When falling under the influence of both gravity and the electric field, the downward force acting on a charged drop is $mg + qE$. In this case the equation for zero net force on the drop when the upward force F_{drag} is considered is

$$mg + qE = F_{\text{drag}} = Cav_e \qquad \text{2-7-4}$$

where the symbol v_e represents the constant drop velocity in this case. The velocities v_g and v_e were measured in Millikan's apparatus by observing the drop with the telescope and timing its fall through a known distance. By means of an auxiliary source of electric field, not shown in Fig. 2-7-1, the drop could be raised to the top of the region between the plates of the apparatus whenever desired, and thus measurements of v_g and v_e could be made on a given drop during two or more successive downward traversals of the region between the plates.

Equations 2-7-3 and 2-7-4 can easily be combined to obtain a relation for the charge on a drop in terms of the measured velocities v_g and v_e, the known constant C, the known field E, and the drop radius a. Substituting for mg in Eq. 2-7-4 its value from Eq. 2-7-3,

$$Cav_g + Eq = Cav_e \qquad 2\text{-}7\text{-}5$$

This equation can then be solved for q:

$$q = \left(\frac{Ca}{E}\right)(v_e - v_g) \qquad 2\text{-}7\text{-}6$$

The drop radius, which must be substituted in Eq. 2-7-6 in order to find q, could not be found by direct measurement. The drops were of order 10^{-4} cm in diameter, far too small to accurately measure directly by any means available to Millikan. However, Eq. 2-7-3 could be used to find the radius from v_g. The mass m is given by the product of the volume of the drop, $4/3\ \pi a^3$, times the known density of the oil from which the drop was formed. By expressing m in terms of the radius a in this way, Eq. 2-7-3 can be solved for the radius in terms of v_g. The result is

$$a = C'\sqrt{v_g} \qquad 2\text{-}7\text{-}7$$

where C' is a known constant found from C, g, and the oil density.

By using Eq. 2-7-6 to find the charges on a large number of drops, the drops with the smallest net charge were found to have magnitudes of q which seemed to correspond to one or a few times a fixed "smallest unit" of charge. This might be interpreted as the charge of a *single electron or proton.* To refine the experiment so as to best determine a fundamental unit charge, a whole series of measurements were made on each single drop while the charge on the drop was made to change between measurements by irradiating the gas around it with x-rays. As a consequence, Eq. 2-7-6 could be applied for successive values of q with a given value of a characteristic of one single drop. For a change in charge $(q_2 - q_1)$ we can find from Eq. 2-7-6 an expression for the change in velocity

$(v_e)_2 - (v_e)_1$,

$$(v_e)_2 - (v_e)_1 = (q_2 - q_1)\frac{E}{Ca} \qquad 2\text{-}7\text{-}7$$

Table 2-7-1 shows some data from one of Millikan's early experiments for a single drop whose charge was varied. Figure 2-7-2 is a graph which shows the observed change in v_e for each successive run, obtained from the last column of the table. The graph shows strikingly that the charge changed only in small multiples

Table 2-7-1 Millikan's data for one oil drop

RUN NO.	t_e sec	v_e cm/sec	CHANGE IN v_e cm/sec
1	12.5	0.0417	—
2	12.4	0.0420	+0.0003
3	21.8	0.0239	−0.0181
4	34.8	0.0149	−0.0090
5	84.5	0.0062	−0.0087
6	85.5	0.0061	−0.0001
7	34.6	0.0151	+0.0090
8	34.8	0.0150	−0.0001
9	16.0	0.0326	+0.0176
10	34.8	0.0150	−0.0176
11	34.6	0.0151	+0.0001
12	21.9	0.0238	+0.0187

of an amount which changed v_e by close to 0.009 cm/sec. Millikan wrote, in commenting on these data and other measurements which he had made:

> Relationships of this sort have been found to hold absolutely without exception, no matter in what gas the drops have been suspended, or what sort of droplets were used upon which to catch the ions. In many cases a given drop has been held under observation for five or six hours at a time and has been seen to catch not eight or ten ions, as in the experiment above, but hundreds of them. . . . Here, then, is direct, unimpeachable proof that the electron is not a statistical mean but rather the electrical charges found on ions all have exactly the same value or else small multiples of that value.

In addition to establishing the existence of a fundamental unit of charge, Millikan could find its magnitude. His early data were subject to considerable errors but gave a result within a few percent of the best modern value of the charge of the electron. The value of this fundamental unit of charge, called e, is

$$e = 1.60 \times 10^{-19}\ \text{C}$$

Having guessed that all atoms might contain electrons, physicists who were trying to understand the structure of atoms needed to know the charge, mass, and other properties of these tiny particles. Millikan's experiment was therefore very important,

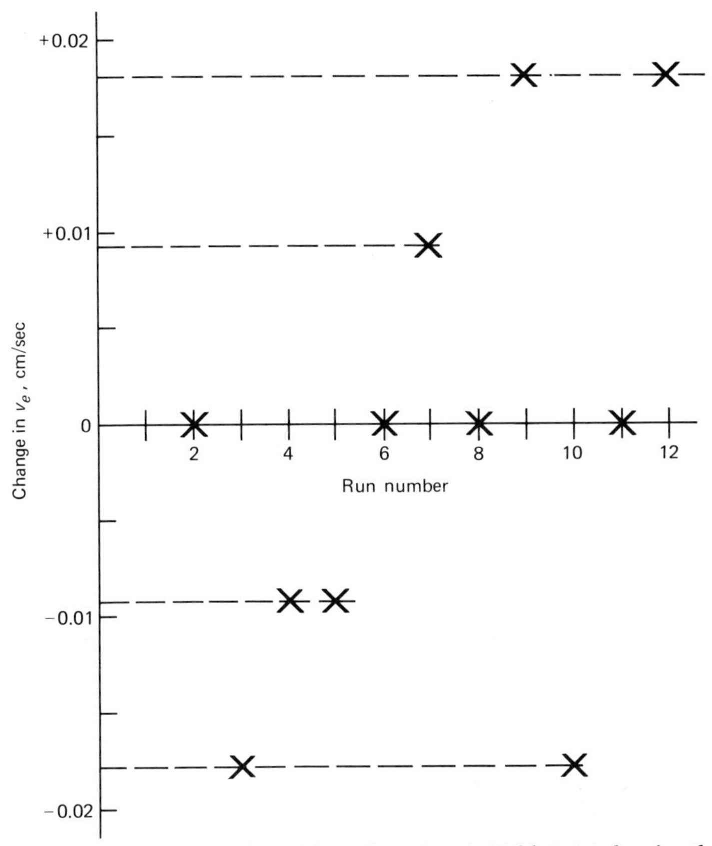

Figure 2-7-2 A graph plotted from data given in Table 2-7-1 showing the various measured changes in v_e, the drop velocity with the electric field on. The crosses are experimental values. Note from the dashed lines that the changes occurred in small multiples of 0.0090 cm/sec.

along with a few others which we will also describe, in the development of atomic physics. It is also an experiment which is well worth looking back on to appreciate the ingenuity with which he made an extremely difficult measurement with such simple equipment.

PROBLEM 14

A charged table-tennis ball is supported in midair, against the pull of gravity, by an electric field. The mass of the ball is 5 g. The charge on it is $+2 \times 10^{-6}$ C. What is the magnitude and direction of the electric field? At what distance from a point charge of $+3 \times 10^{-6}$ C would the field have this magnitude?

PROBLEM 15

Suppose each iron atom had a small net charge equal to $10^{-9}\ e$, where e is the charge of the electron. What would the electrostatic force be between two iron spheres with mass 1 kg each placed 1 m apart?

If the force between two such spheres is known to be less than 10^{-3} N, what limit can you place on the net charge per iron atom?

SUMMARY

There are *positive* and *negative* electric *charges.* In matter, *protons* are the basic source of plus charge and *electrons* are the basic source of minus charge. The *net charge* of a body is the excess of plus over minus charge. The electric force on a body depends on the net charge, being zero for zero net charge. Charges can move in *conductors*, giving rise to *electric currents.*

Like charges repel each other, and unlike charges attract each other, with an electric *force.* Force is a *vector* quantity, with *magnitude and direction.* The net force on a body can be found by *vector addition.* For zero net force, *Newton's first law* is true: If there is no net force on a body it remains at rest or moves with constant speed in a given direction. Conversely, if there is a net force, then the speed, or direction of motion, or both, change.

The magnitude of the electric force between two charges Q_1 and Q_2 is described by *Coulomb's law*:

$$F = (9.0)(10^9)\frac{Q_1Q_2}{r^2} \text{ newtons}$$

where all quantities must be in mks units, including charge in coulombs. The direction of the coulomb force is found from the signs of the charges: repulsive for like charges and attractive for unlike ones.

The *electric field* $\vec{E}$ at a particular place in space is defined as the force which would be observed on a one coulomb positive *test charge.* The force $\vec{F}$ on any charge q is then given by

$$\vec{F} = q\vec{E}$$

The field of a point charge Q is radial and decreases with distance from Q according to the relation

$$E = (9.0)(10^9)\frac{Q}{r^2} \text{ newtons/coulomb}$$

A uniform electric field can be obtained in a region between two parallel plates with opposite charges.

In the *Millikan oil-drop experiment*, the charge of the electron was inferred from the electric force on minute droplets containing one or a few net electron charges. The modern value for the magnitude of the electron's charge, e, is

$$e = 1.60 \times 10^{-19} \text{ C}$$

MOTION

3-1 INTRODUCTION

In the previous chapter the concept of force was introduced. In this chapter *classical dynamics* will be discussed, which describes how a force on an object influences its motion. To do this we will first describe *displacements*, *velocity*, and *acceleration*, the quantities which are used to describe motion. Then, after presenting Newton's second law, which is the fundamental equation of classical dynamics, we will apply it to a few important examples.

In spite of its consistency with all experiments during a period of hundreds of years, classical dynamics was shown at the beginning of the twentieth century to be only an approximation, valid for physical systems much larger than atomic dimensions and for objects moving with velocities much less than the speed of light. However, classical dynamics is still basic to most of physics. The newer dynamical theories, quantum mechanics and relativity, are built upon foundations from classical dynamics. They extend, rather than replace, newtonian mechanics.

In this chapter, and throughout this book, we will primarily be interested in the motion of *particles*, objects which can be treated as "points with mass," for which spinning or tumbling can be neglected. Sometimes we will deal with cars, blocks, etc., for convenience, but we will consider only those simple motions which are like those of a point particle.

3-2 VELOCITY AND DISPLACEMENT

Imagine a car which starts up, travels a short distance in a straight line, and then stops. One way to describe its motion would be to specify how its speed varied with time. The speedometer and a clock could be used to measure the speed at different instants, and the data could be presented in the form of a graph, such as the one shown in Fig. 3-2-1*a*.

First, let us examine Fig. 3-2-1*a* to see if it makes physical sense. At $t = 0$, when the car is assumed to start moving, the velocity v is equal to zero. During the next five seconds v changes from 0 to 10 m/sec, as the car speeds up. From $t = 5$ sec to $t = 12.5$ sec the velocity remains constant, and from $t = 12.5$ sec to $t = 15$ sec, when the brakes are applied, the velocity decreases to zero. The graph describes a plausible physical situation, provided 10 m/sec is not a ridiculous speed, which you can check.

A second way to describe the motion of the car is to describe

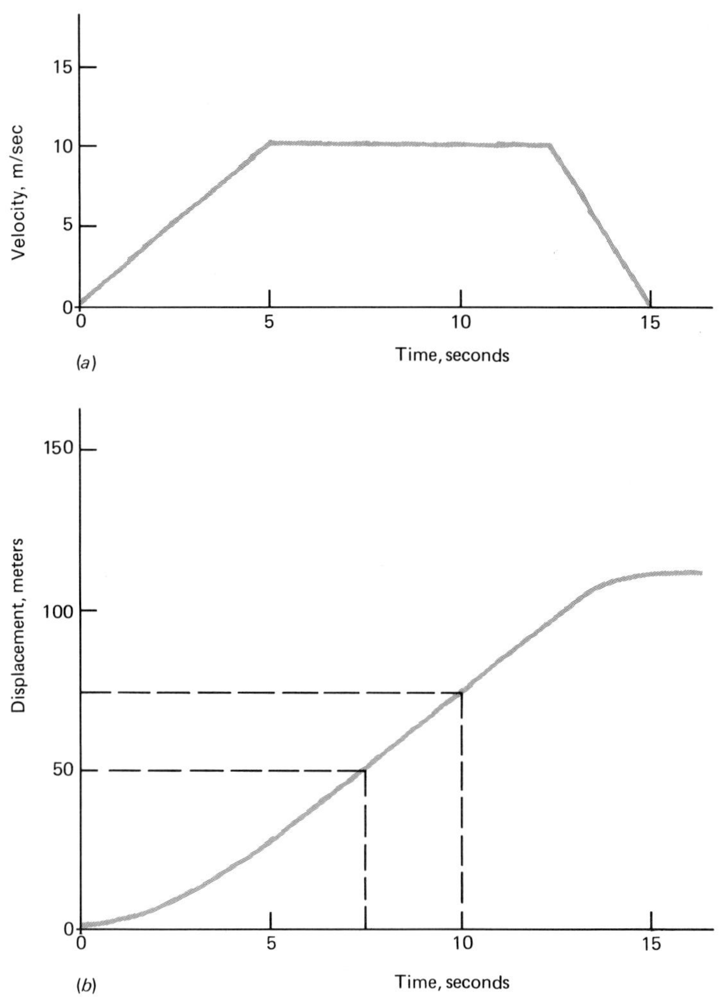

Figure 3-2-1 Variation of velocity and distance with time for an example of motion of a car in a straight line.

how its position changes with time. First we imagine an x axis along the direction of motion of the car. Then the motion of the car can be described by how the x coordinate of some given part of the car, like the rear bumper, changes with time. The graph of Fig. 3-2-1b shows the motion of the car in this way, and the value of the x coordinate is called the displacement, or alternatively, the position. The displacement is taken to be zero at $t = 0$, when the car is at the starting point. From $t = 0$ to $t = 15$ sec the car is

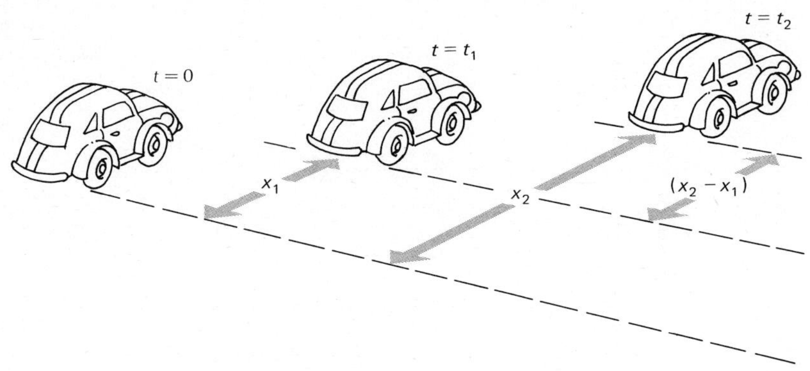

Figure 3-2-2 Positions of a car at the times $t = 0$, $t = t_1$, and $t = t_2$.

moving, so that we see the displacement changing continually with time. At $t = 15$ sec, the displacement x is equal to 112.5 meters. After $t = 15$ sec, x remains at this value since the car is standing still.

In order to describe the motion, is it necessary to have both graphs in Fig. 3-2-1, v vs. t (vs. means "plotted against") and x vs. t? The answer is no, and we will now see that either graph can be derived from the other. The velocity, *by definition*, tells how much the displacement is changing *per second*. If v remains fixed at 10 meters per second, then x changes by 10 meters in 1 second, 15 meters in 1.5 seconds, 20 meters in 2 seconds, etc. This can be summarized in a formula relating the change in position to the change in t for constant velocity. Assuming motion along an x axis,

$$x_2 - x_1 = v(t_2 - t_1) \qquad \text{3-2-1}$$

In this formula x_1 is the value of x at time t_1, and x_2 is the value of x at t_2. The quantity $(x_2 - x_1)$ is the change in position, or the distance traveled, during the time interval $(t_2 - t_1)$. Figure 3-2-2 shows the car at $t = 0$, $t = t_1$, and $t = t_2$.

Equation 3-2-1 applies to the car example during the time when the velocity is constant, between $t = 5$ sec and $t = 12.5$ sec, and can be used to find part of the curve in Fig. 3-2-1*a* from the curve in Fig. 3-2-1*b*. If we solve Eq. 3-2-1 for v,

$$v = \frac{x_2 - x_1}{t_2 - t_1} \qquad \text{3-2-2}$$

From Fig. 3-2-1*b*, x changed from 50 to 75 meters during the time interval from 7.5 to 10 sec. If the velocity is constant, Eq. 3-2-2 applies, and substituting in that equation,

Figure 3-2-3 Enlarged views of two parts of Fig. 3-2-1*b*.

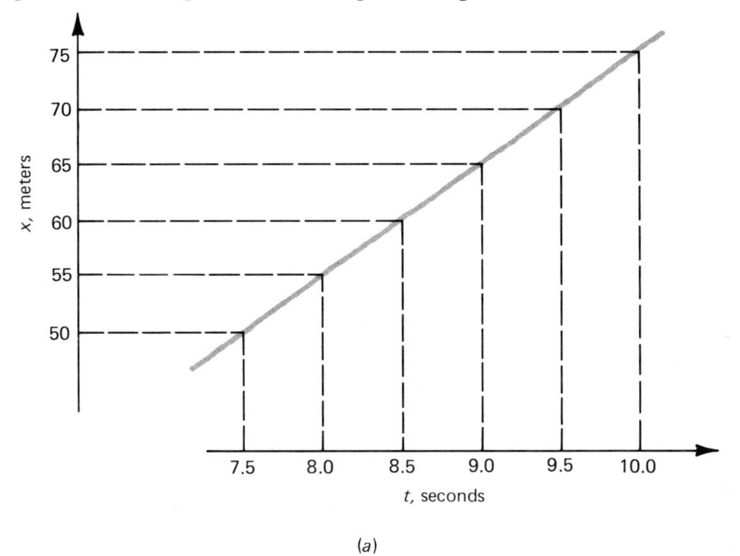

(*a*)

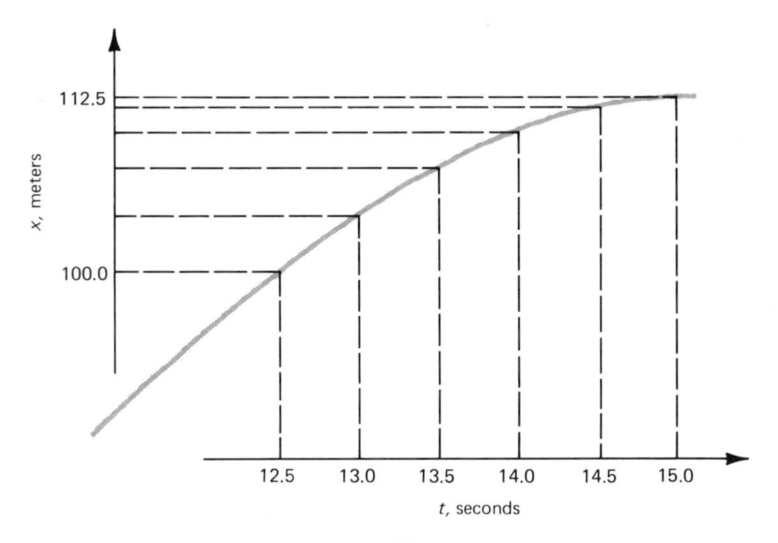

(*b*)

$$v = \frac{75 - 50}{10 - 7.5} = 10 \text{ m/sec} \qquad 3\text{-}2\text{-}3$$

This is the velocity shown in Fig. 3-2-1*a* during this time interval.

PROBLEM 1

A car travels north at 45 miles per hour for 2 hours and at 65 miles per hour for 1 hour. How far does it go?

Actually, substituting a distance interval and a time interval in Eq. 3-2-2, as we have done, gives the average velocity during that interval, which is the actual velocity throughout the interval if the velocity is constant. To check whether the motion is really at constant velocity, we can find the average velocity during each of several successive time intervals. Figure 3-2-3*a* shows an enlarged view of the part of Fig. 3-2-1*b* from which we have just calculated the average velocity. For each half-second time interval, x changes by the same amount, 5 meters. Applying Eq. 3-2-2 to any of these intervals, the value for v is found to be 10 m/sec, so the velocity is indeed constant.

PROBLEM 2

In the example described by Fig. 3-2-1, what is the average velocity of the car during the first 5 seconds? During the first 10 seconds?

Figure 3-2-3*b* shows an enlarged view of the x vs. t graph, Fig. 3-2-1*b*, for times between 12.5 and 15 sec. For each successive half-second time interval the change in x is smaller. In fact, you can see that application of the formula given in Eq. 3-2-2 gives a steadily decreasing velocity. By doing the calculation for several short time intervals between 12.5 and 15 sec, it is possible to find that the velocity is decreasing in just the way shown in Fig. 3-2-1*a*.

Looking back at the way the curve of x vs. t has been used to find points on the curve of v vs. t, perhaps you have noticed that a straight line part of the x vs. t curve led to a constant value of v while a curved part of the x vs. t curve led to a changing value of v. This correspondence is true in general. During the time interval from $t = 0$ to $t = 5$, when the graph of x vs. t is curved, we thus expect to find a changing velocity. By choosing a number of short time intervals in this region it is possible to reproduce the first part of the v vs. t curve, thereby completing the task of deriving Fig. 3-2-1*a* from Fig. 3-2-1*b*.

When v is changing, we speak of the *instantaneous velocity*, the value of v at some instant of time. What this means in terms of an experimental measurement is that for a very short time interval, which we shall call Δt ("delta t"; Δ is the Greek letter delta), we determine the small change in position Δx. Applying Eq. 3-2-2 to these quantities

$$v = \frac{\Delta x}{\Delta t} \tag{3-2-4}$$

The instantaneous velocity is the velocity calculated in this way, when Δt is made so short that the velocity is essentially constant during the time interval. Equation 3-2-4, with the understanding that Δt is such a very short time interval, is the precise definition of v. Because v is the ratio of the change in x to the change in t, it is called the time rate of change of position or simply the *rate of change of position.*

The quantity $\Delta x/\Delta t$ is called the *slope* of the curve of x vs. t. Referring back to Fig. 3-2-1*b*, we might intuitively (and correctly!) describe the slope of the curve as increasing between $t = 0$ and $t = 5$, then constant from $t = 5$ to $t = 12.5$, and then decreasing between $t = 12.5$ and $t = 15$. After $t = 15$ the curve is level, or the slope is zero. The foregoing description of how the slope behaves may be compared with the velocity curve, Fig. 3-2-1*a*. During each time interval the velocity in Fig. 3-2-1*a* indeed behaves exactly like the slope in Fig. 3-2-1*b*. Merely by carefully observing the shape of a curve of x vs. t we can infer the shape of the corresponding v vs. t curve. Of course, for quantitative work we have to use Eq. 3-2-4 with numerical values of Δx and Δt.

We now want to show how to derive a curve of x vs. t from a curve of v vs. t, the converse of the procedure discussed above. First, it will be convenient to rewrite the fundamental relation, Eq. 3-2-4, in the form

$$\Delta x = v(\Delta t) \tag{3-2-5}$$

Now we will apply this equation to Fig. 3-2-4, which shows the first part of the original v vs. t curve, Fig. 3-2-1. Short time intervals of 0.5 sec are marked by vertical dashed lines on Fig. 3-2-4. Consider one such interval, from 5 to 5.5 sec, which corresponds to the shaded area of the figure. The quantities v and Δt which would go into Eq. 3-2-5 for this time interval are shown on the figure. Notice that the area of the shaded rectangle, its height times width, is equal to the product $v(\Delta t)$, which, according to Eq. 3-2-5, is equal to the change in x, (Δx).

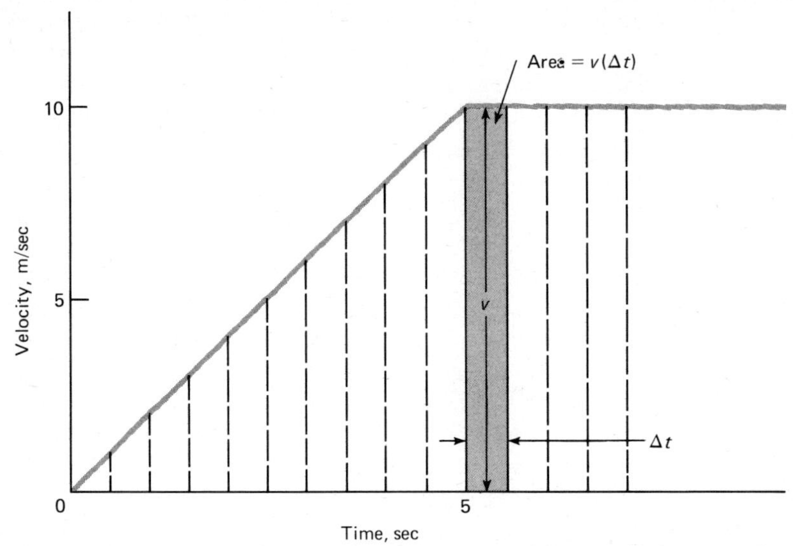

Figure 3-2-4 An enlarged view of the first part of the graph shown in Fig. 3-2-1*a*, with successive half-second time intervals shown.

In the discussion above, we have discovered a general relation: The change in x between any two times is equal to the area of the v vs. t curve between those times. If we take as the early time $t = 0$, then the area of the curve up to any later time will be the value of x at that time. For example, the area between $t = 0$ and $t = 5$ is the area of a triangle, given by $\frac{1}{2}$(base)(height), equal to $\frac{1}{2}$(5 sec)(10 m/sec) = 25 m. We therefore predict that at $t = 5$ the displacement x is equal to 25 m. This checks, according to Fig. 3-2-1*b*.

This completes an introduction to the relationship between position and velocity. We have seen that one is the time rate of change of the other, and as a result, a knowledge of the way either varies in time can be used to find the behavior of the other. Many other important physical quantities are related to each other like velocity and position. In discussing the mathematical connections between such a pair of quantities in terms of either slopes or areas of curves, we have touched the foundations of differential and integral calculus, which are used to quantitatively relate such pairs of quantities. The differential calculus may be characterized as the mathematics of slopes, and the integral calculus as the mathematics of areas.

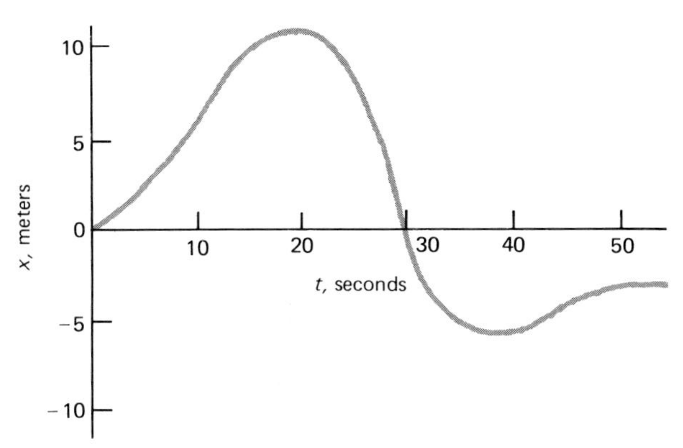

Figure P-3-3

PROBLEM 3

A curve of x vs. t is shown for a particle which moves along the x axis. When is it farthest from the starting point? When does it return to the starting point? During what times is its velocity in the $+x$ direction? During what times is it moving in the $-x$ direction? When is the speed greatest? When is the speed zero?

PROBLEM 4

A curve of v vs. t is shown for a particle which moves along the x axis. The particle is at $x = 0$ when $t = 0$. When has it moved the farthest in the $+x$ direction? Estimate approximately the value of x at this time. After it stops moving, is the particle at a position which corresponds to x greater than zero (positive) or x less than zero (negative)? Justify your answer.

Figure P-3-4

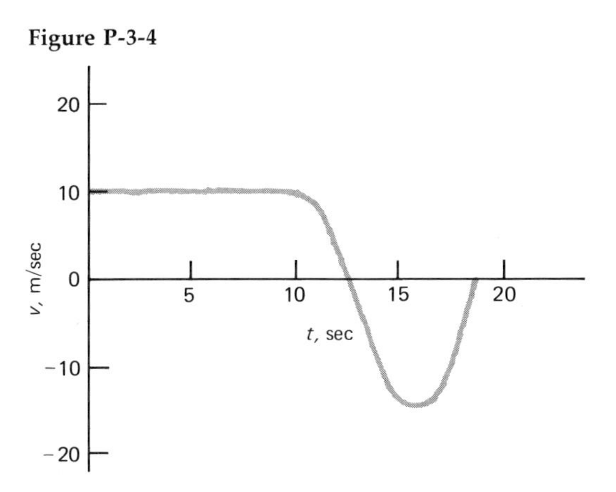

3-3 ACCELERATION

The *acceleration* is defined as the *rate of change of velocity*. Like velocity and displacement, acceleration is a vector quantity, with a direction. Up to now, because of our one-dimensional example, we have not emphasized the vector nature of velocity and displacement. In one-dimensional motion along an x axis, the only possible vector directions are along and opposite to the direction of increasing x, in what are called the $+x$ or $-x$ directions. Therefore, a scalar quantity with either a plus or minus sign suffices to describe any vector in one-dimensional motion. A negative value of either x or v in the discussion preceding would have indicated a displacement or a velocity to the left rather than the right, in the $-x$ direction. For simplicity, such values were not introduced in the example. In this section, we will continue to discuss the same one-dimensional example, but we will, however, find acceleration vectors in both the $+x$ and $-x$ directions.

The defining equation for v is $v = \Delta x/\Delta t$. Similarly, the acceleration is defined by

$$a = \frac{\Delta v}{\Delta t} \qquad \text{3-3-1}$$

As before, it is understood that Δt is a very short time interval—in this case an interval when the acceleration changes by such a small amount that it can be assumed constant.

If there is a long time, say from t_1 to t_2, during which the acceleration is in fact constant, then Eq. 3-3-1 can be used for the entire time interval and written in the form which corresponds to constant, or uniform, acceleration:

$$a = \frac{v_2 - v_1}{t_2 - t_1} \qquad \text{3-3-2}$$

Equations 3-3-2 and 3-3-1 for acceleration are exactly analogous to Eqs. 3-2-2 and 3-2-4, respectively, for velocity. The unit of acceleration is the unit of velocity per second, or meters per second per second, often abbreviated m/sec^2.

Figure 3-3-1 shows three curves for the car example we have been discussing, the two curves of x vs. t and v vs. t previously drawn in Fig. 3-2-1 and a curve of a vs. t. During the time from 0 to 5 sec we can pick an interval Δt as shown in Fig. 3-3-1b, find the change in velocity Δv, and apply the formula in Eq. 3-3-2 to find the acceleration. However, because the curve of v vs. t is a straight line, the result for any time interval between 0 and 5 sec will be $a = 2$ m/sec^2, as shown on the graph of a vs. t.

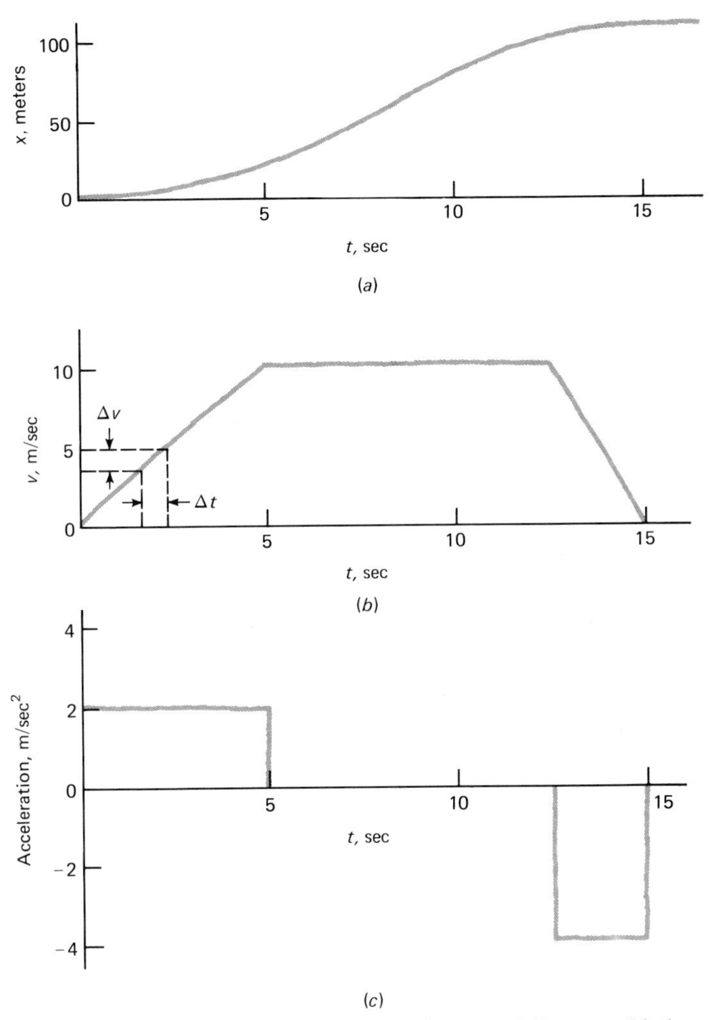

Figure 3-3-1 Variation of acceleration, velocity, and distance with time, for the example of motion of a car in a straight line.

We have previously learned that the slope of the curve of x vs. t tells us the behavior of the rate of change of x, or the velocity. Similarly, the slope of the curve of v vs. t tells us the behavior of its rate of change, the acceleration. Notice that the curve of v vs. t consists of just three straight line segments, having constant slopes. Therefore, there are three time intervals during each of which the car is moving with uniform (constant) acceleration. From $t = 5$ sec to $t = 12.5$ sec the velocity is not changing, the slope equals zero, and the acceleration is thus equal to zero. We have noted the positive acceleration between $t = 0$ and 5 when the velocity is increasing. From $t = 12.5$ to 15, with v decreasing, the acceleration is constant and negative. (We interpret the downward slope of this part of the v vs. t curve as a negative slope.)

Previously, we were able to derive the time variation of x from the time variation of v, which is its rate of change, by using the area under the curve of v vs. t. We can, in complete analogy, relate the time variation of v to the area under the curve of a vs. t. Since v is zero at $t = 15$, the total area under the a curve between $t = 0$ and $t = 15$ must also be zero. The area for the negative value of a between $t = 12.5$ and $t = 15$ is taken negative and indeed exactly cancels the other positive area from $t = 0$ to $t = 5$.

So far we have treated acceleration more or less as an abstract idea, but everyday experience gives us a good feeling for accelerations. The accelerations of a car may produce the sensation of being pushed forward when the car is slowing down, or sideward when it is rounding a curve. When riding in a fast elevator, the accelerations at the start and stop of a downward trip are usually felt strongly, as are the vertical accelerations in an airplane traveling through rough weather. If you think about situations in which your velocity is changing fairly rapidly, you will notice that each involves some sensation that arises from the acceleration.

PROBLEM 5

A curve of v vs. t is shown for a particle moving along the x axis. Sketch the curve of acceleration vs. time, showing the times when the acceleration changes, and also showing the magnitude of the acceleration correctly.

Figure P-3-5

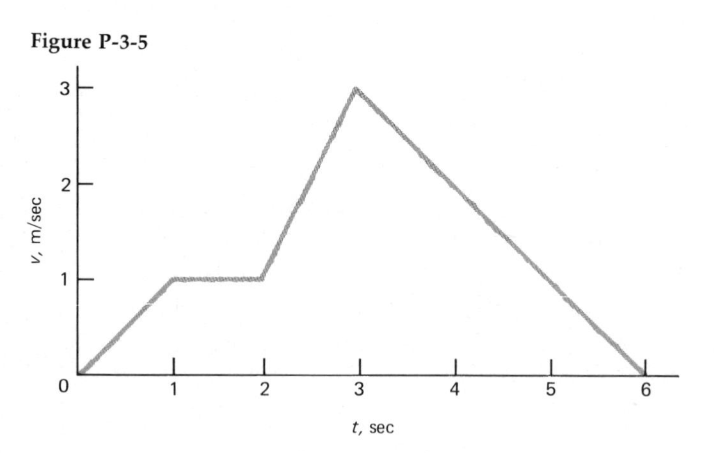

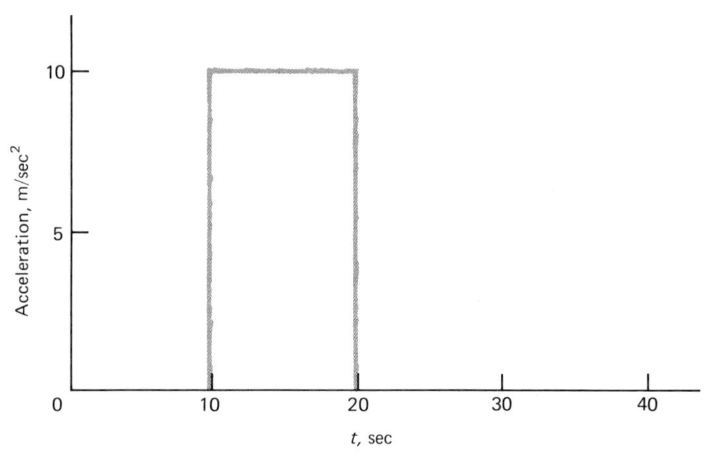

Figure P-3-6

PROBLEM 6

A graph of acceleration vs. time is shown for a particle moving along the x axis. When does the particle begin to move? How fast is it moving at $t = 11$ sec? How fast is it moving at $t = 30$ sec? (Let $v = 0$ at $t = 0$.)

3-4 NEWTON'S SECOND LAW

You may recall, from the previous chapter, the statement of *Newton's first law:* If there is no net force on a body, it remains at rest or it moves with constant speed in a fixed direction. This law correctly describes the motion in those simple cases where the net force is zero. If the net force on a body is not zero, then its motion is described by *Newton's second law*:

$$\vec{F} = m\vec{a} \qquad \text{3-4-1}$$

This simple equation summarizes the second law concisely, and with it we can calculate the motion of any object on which there is a net force $\vec{F}$. It is a vector equation, specifying that $\vec{a}$ is in the direction of $\vec{F}$, with magnitude F/m.

The quantity m is called the *inertial mass* or, most commonly, the *mass*. The inertial mass of a body is in fact defined from the second law, a mass of 1 kilogram being that which is accelerated at 1 m/sec^2 by a force of 1 newton. The inertial masses of different objects can be compared by comparing their accelerations under the influence of equal forces.

The experimental observations made by Newton and others,

which led Newton to formulate the second law, were studies of motion under the influence of known forces. Perhaps the simplest example is the motion of a body under the influence of gravity. The gravitational force on a body, its weight, is essentially the same everywhere within a room, or even within a large building. (A spring scale could be used, for example, to check this.) To study the motion of a body under the influence of this force, we need only to drop it and record how its position changes with time.

The motion of a body falling freely under the influence of gravity is found to exhibit constant acceleration, provided air resistance is negligible. The constant acceleration of a freely falling body acted upon by a constant force partly establishes the form of Newton's second law, but it does not check that the acceleration is proportional to the force. To verify the proportionality between force and acceleration, it is necessary to change the magnitude of the net force exerted on a given body and to observe the resultant change in its acceleration. One simple apparatus for this type of experiment consists of a body sliding perfectly down an inclined plane. An ice cube on smooth metal would be quite good, but somewhat inconvenient. A little cart with wheels works well if the wheels and the surface on which they roll are smooth, and if the wheel bearings turn very freely.

A cart such as this is shown schematically in Fig. 3-4-1, held at rest on an inclined plane by a string. The force $\vec{F}_s$ exerted by the string is shown, along with a number of other force vectors. The vector $\vec{W}$, for weight, represents the downward gravitational force,

Figure 3-4-1 A cart on an inclined plane.

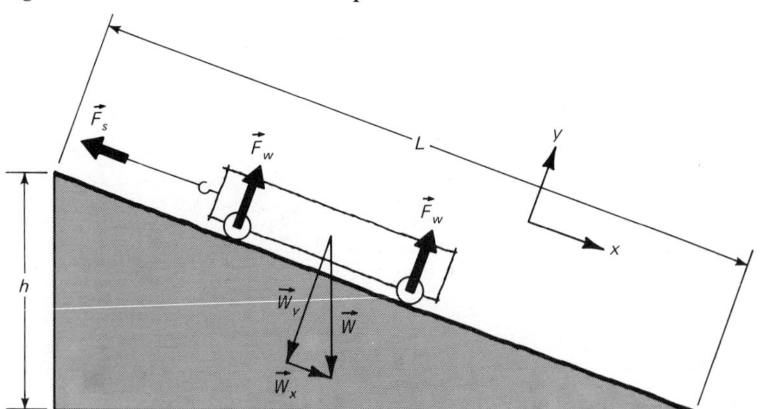

and it has been resolved into two components, $\vec{W}_x$ and $\vec{W}_y$. The component $\vec{W}_x$ is along the direction of motion, parallel to the inclined plane on which the cart rolls, while the component $\vec{W}_y$ is perpendicular to the surface of the inclined plane.

The essential feature of the inclined-plane experiment is that when the string is released and the cart rolls down the plane, the cart is accelerated only by the component W_x of the force W. The component W_y acts in a direction perpendicular to the surface of the plane, a direction in which motion is impossible. W_y must therefore be exactly balanced by forces F_W acting on the wheels of the cart, so that there is indeed no net force and no acceleration in the direction perpendicular to the surface of the plane.

When the string is released, as we have seen, the net force on the cart is W_x. For various inclinations of the plane this force can be varied from zero to the full weight W. Thus, it is possible to measure the acceleration of the cart for different forces and to indeed verify the proportionality between a and F asserted by Newton's second law. Both Newton and Galileo, his predecessor in the development of classical dynamics, were familiar with the results of such inclined-plane experiments. Like almost all theoretical laws, the second law was formulated with a view toward explaining existing experimental facts. The major breakthrough was in correlating acceleration with force, rather than velocity, for example, a step which required sophisticated thought and the stimulus of reliable experimental data.

PROBLEM 7

A cart of weight W is on an inclined plane which consists of a plank 10 ft long. One end of the plank is raised 4 ft higher than the other. Draw a side view of the plank and of the vector W, and graphically resolve W into components along the plank and perpendicular to it. Find the ratio of the component along the plank to the total weight W.

Inclined-plane experiments show that the one-dimensional version of the second law is correct. Other experiments verify the fact that for the general motion of a body under the influence of a net force, the more general vector equation $\vec{F} = m\vec{a}$ is correct. For "large" bodies moving at speeds much less than the speed of light there are only very minute deviations from Newton's second law. With it, the motion of baseballs, cars, rockets, tops, water waves, etc., can be very accurately calculated, if the forces are known.

3-5 INERTIAL AND GRAVITATIONAL MASS

At the beginning of the preceding section, the proportionality constant m in Newton's second law was described as the *inertial mass.* In the first chapter, it was stated that masses of different bodies can be compared by measuring the gravitational forces acting on them at some fixed location, as in some laboratory at the earth's surface. If W represents weight, the following two equations correspond to these two statements about mass.

$$F = m_i A \tag{3-5-1}$$

$$W = m_g g \tag{3-5-2}$$

We are now distinguishing between two possible kinds of mass, m_i, signifying inertial mass, and m_g, signifying gravitational mass. The mass m_i can be found for any body by measuring its acceleration under the influence of a known net force and using Eq. 3-5-1 in the form $m_i = F/a$. The mass m_g is defined quite differently in Eq. 3-5-2. The constant g represents the *gravitational field* at the earth's surface, which, acting on the mass m_g, produces the force W. This equation is analogous to the equation for the electric force F_e on a charge q in an electric field E; $F_e = qE$.

The gravitational mass is a kind of "gravitational charge" acted upon by gravitational fields. Moreover, the analogy with electric charge is quite complete, since gravitational fields are set up by gravitational masses, just as electric charges set up electric fields.

Suppose we consider the very simple experiment in which an object is allowed to fall freely under the influence of gravity. The force in Eq. 3-5-1 is then W, the weight, so that the second law equation is

$$W = m_i a \tag{3-5-4}$$

Now we can substitute the value of W from Eq. 3-5-2 into Eq. 3-5-4 to obtain the equation

$$m_g g = m_i a \tag{3-5-5}$$

If we rearrange this equation, we can find an expression for the ratio of the gravitational mass to the inertial mass:

$$\frac{m_g}{m_i} = \frac{a}{g} \tag{3-5-6}$$

The remarkable experimental fact is that the acceleration in free fall *is the same* for all bodies at a given location. At any given place the gravitational field g is a constant, so if the acceleration is the same for all bodies, the ratio of the gravitational mass to the inertial mass is the same for all bodies.

If this is an accident, it is a remarkable one. Why should the quantity of "stuff" in a body which is acted upon by a gravitational field be exactly proportional to the quantity of stuff which gives the body *inertia*, i.e., makes it necessary to exert a force on it to change its velocity? Very careful experiments have verified that this proportionality is exact to an accuracy of one part in 10^{10}. Most physicists feel sure that there must therefore be a deep connection between gravity and inertia, which is not yet understood.

The unit chosen for both gravitational and inertial mass is the *kilogram*, defined as the mass of a particular metal block owned by the French Bureau of Standards. Choosing the same unit for both kinds of mass makes the proportionality constant which relates them equal to 1 and makes the constant g, often called the *acceleration of gravity*, equal to the acceleration of a body in free fall at the earth's surface, 9.80 m/sec^2. The gravitational force, or weight of objects at the earth's surface, is 9.80 N/kg.

PROBLEM 8

A man buys a scale and stands on it while riding down in an elevator. When the elevator starts to descend he is amazed to find that the scale reads only half his normal weight. Explain.

3-6 UNIFORMLY ACCELERATED MOTION

Motion of a body under the influence of a constant force leads to uniform acceleration. Rearranging the equation of the second law,

$$\vec{a} = \frac{\vec{F}}{m} \tag{3-6-1}$$

Uniformly accelerated motion is sufficiently important that it is worthwhile to derive the mathematical relations between acceleration, velocity, and distance for one-dimensional uniformly accelerated motion.

We can assume without loss of generality that the motion to be considered begins at $x = 0$ at time $t = 0$. At $t = 0$ the velocity need not be zero, however, and will be taken to be v_0. Applying Eq. 3-3-2, which follows from the definition of acceleration, between times $t = 0$ and $t = t$, when the velocity is v_0 and v, respectively,

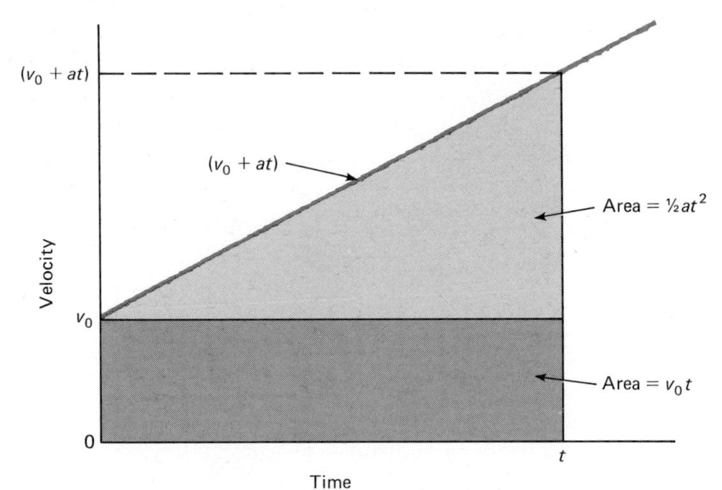

Figure 3-6-1 A graph of v vs. t, where $v = v_0 + at$.

$$a = \frac{v - v_0}{t} \qquad 3\text{-}6\text{-}2$$

Rearranging this equation,

$$v = v_0 + at \qquad 3\text{-}6\text{-}3$$

This is the desired equation for the variation of v with time. Graphically, this equation is illustrated by the straight-line graph shown in Fig. 3-6-1.

The shading in Fig. 3-6-1 will be helpful in finding the behavior of the displacement x with time. Recall from the latter part of Sec. 3-2 that we have already seen that the change in x between two times is equal to the area enclosed under the curve of v vs. t between those times. The change in x between $t = 0$ (when $x = 0$) and $t = t$ (when $x = x$) is therefore the sum of the areas of the shaded rectangle and the shaded triangle of Fig. 3-6-1. Thus the value of x at time t is given by

$$x = v_0 t + \frac{1}{2}at^2 \qquad 3\text{-}6\text{-}4$$

where the first term is the area of the rectangle and the second is that of the triangle.

Equations 3-6-3 and 3-6-4 fully describe uniformly accelerated motion. Sometimes, for convenience it is useful to have an expression for v in terms of the displacement x, which takes a simple form if v_0 is equal to 0. From Eq. 3-6-4, with $v_0 = 0$, $t = \sqrt{2x/a}$. Substituting this value of t into Eq. 3-6-2,

$$v = \sqrt{2ax} \qquad \text{when } v_0 = 0 \qquad \text{3-6-5}$$

PROBLEM 9
An electron is released with zero velocity in a uniform electric field with $E = 5 \times 10^5$ N/C. Its mass is 9.1×10^{-31} kg. Assuming the second law is correct, how long does it take the electron to reach a speed of 3×10^8 m/sec? How far does it travel in this time?

PROBLEM 10
A ball that is thrown vertically is released 2 m above the ground with an upward velocity of 20 m/sec. Assuming that the ball is thrown at $t = 0$, when does its velocity equal zero? How high does it go? How long does it stay in the air if it is caught 2 m above the ground? How fast is it moving when it is caught?

PROBLEM 11
A car with mass 1,000 kg travels down a uniformly sloping road which drops 100 m/km. Suppose the car's brakes fail a half-kilometer from the bottom of the hill, when it has a speed of 10 m/sec. Neglecting air resistance and friction, what would the speed of the car be at the bottom of the hill?

PROBLEM 12
Suppose in the above problem that the frictional force remains negligible but that the force due to air resistance is given by

$F_{\text{air}} = (1.50)v^2$ newtons, for v in m/sec

Find the maximum velocity the car can reach on a grade which drops 100 m/km.

3-7 MOTION IN TWO DIMENSIONS

A simple example of motion in two dimensions is illustrated in Fig. 3-7-1. A particle of mass m is moving horizontally at time $t = 0$ with velocity v_0. The only force acting on it is the gravitational force mg (where $g = 9.80$ N/kg). Were no gravitational force acting, we would expect the particle to continue to move at velocity $\vec{v}_0$ in a straight line along the x axis. The gravitational force will accelerate it downward, so we expect that the particle will follow some curved trajectory, as shown by the dashed line.

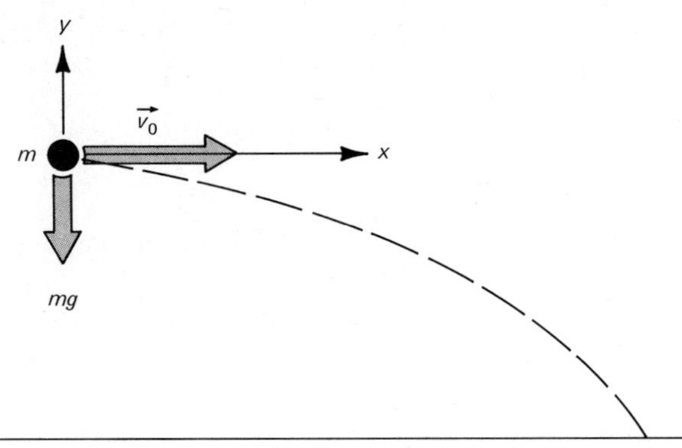

Figure 3-7-1 A particle moving horizontally at $t = 0$. The dashed line illustrates its motion at later times.

The figure shows a *coordinate system*, meaning a set of x and y axes, with its origin, the point $x = 0$, $y = 0$, at the position of the particle at $t = 0$. The motion in this example, which could correspond, for example, to the motion of a bullet or a baseball, can be calculated entirely from the results for uniformly accelerated motion derived in the preceding section. First, we write the vector equation of the second law, $\vec{F} = m\vec{a}$, as two equations for the x and y components of the vectors:

$$F_x = ma_x \tag{3-7-1}$$

$$F_y = ma_y \tag{3-7-2}$$

For a vector equation to hold, it is obeyed by the components of the vectors along any direction. When we resolve vectors into components, we almost always choose directions at right angles to each other, like the x and y axes of this problem. The convenience of such *orthogonal*, or right-angled, axes is that each component of the vector is independent of the other. (In three dimensions we would take x, y, and z axes, all mutually perpendicular.)

The components of force are known: $F_x = 0$ and $F_y = -mg$, where the minus sign corresponds to a force in the $-y$ direction. Substituting these values into Eqs. 3-7-1 and 3-7-2, we obtain

$$0 = ma_x \tag{3-7-3}$$

$$-mg = ma_y \tag{3-7-4}$$

These equations can immediately be solved for the components of the acceleration $\vec{a}$:

$$a_x = 0 \qquad 3\text{-}7\text{-}5$$

$$a_y = -g \qquad 3\text{-}7\text{-}6$$

The motion of the particle is a combination of two uniformly accelerated motions, in the x and y directions, and the x motion is very simple because $a_x = 0$. The formulas given in Eqs. 3-6-3 and 3-6-4 rewritten in terms of x components are

$$v_x = v_{0_x} + a_x t \qquad 3\text{-}7\text{-}7$$

$$x = v_{0_x} t + \frac{1}{2} a_x t^2 \qquad 3\text{-}7\text{-}8$$

Substituting $a_x = 0$ and $v_{0_x} = v_0$, the x component of the motion is described by the two equations

$$v_x = v_0 \qquad 3\text{-}7\text{-}9$$

$$x = v_0 t \qquad 3\text{-}7\text{-}10$$

The x component of the motion is seen to proceed as if there were no gravitational force acting.

Writing Eqs. 3-6-3 and 3-6-4 for the y components,

$$v_y = v_{0_y} + a_y t \qquad 3\text{-}7\text{-}11$$

$$y = v_{0_y} t + \frac{1}{2} a_y t^2 \qquad 3\text{-}7\text{-}12$$

Substituting $a_y = -g$ and $v_{0_y} = 0$,

$$v_y = -gt \qquad 3\text{-}7\text{-}13$$

$$y = -\frac{1}{2} g t^2 \qquad 3\text{-}7\text{-}14$$

The motion of the particle has now been completely described. If we want to calculate its position at any time, Eqs. 3-7-10 and 3-7-14 give the x and y coodinates. From the position at several times, the path of the particle can be drawn. The components of velocity at any time can also be found, from Eqs. 3-7-9 and 3-7-13.

The two-dimensional motion has been found to break up into two simultaneous one-dimensional motions. Each motion is independent of the other, which makes them simple to calculate. Whenever the force on a body depends only on its position, and does not depend on its velocity, this simple separation into independent component motions can be effected. These conditions are fulfilled in many physical situations.

PROBLEM 13
A gun shoots a bullet at a velocity of 200 m/sec and is tipped upward at an angle of 45° to the ground. Neglect air resistance. What are the vertical and horizontal components of velocity when the bullet leaves the gun? (Find them graphically if necessary.) How long after it leaves the gun does the bullet have zero vertical velocity? What is its horizontal velocity at this time? How far does it travel horizontally if it hits a target at the same height as the gun barrel?

3-8 CIRCULAR MOTION

If a particle of mass m goes around a circle of radius R at constant speed, is it accelerating? Figure 3-8-1 shows the particle and its velocity vector at two times. While the magnitude of $\vec{v}$ is constant, *its direction is not*, so the particle *is* being accelerated. Consider a short time interval Δt, during which $\vec{v}$ changes from $\vec{v}_1$ to $\vec{v}_2$, as the particle goes a short distance along the circle. Figure 3-8-2 illustrates this case. For small Δt the vectors $\vec{v}_1$ and $\vec{v}_2$ differ by the small amount of $\Delta\vec{v}$, shown in the figure. The acceleration is, by definition,

$$\vec{a} = \frac{\Delta\vec{v}}{\Delta t} \tag{3-8-1}$$

As Δt is made shorter and shorter, $\Delta\vec{v}$ is more and more nearly perpendicular to $\vec{v}_1$. Thus, the acceleration anywhere along the path of the particle is perpendicular to $\vec{v}$. Therefore, it is directed *radially inward* toward the center of the circle.

In order to determine the magnitude of $\vec{a}$, we first note that in

Figure 3-8-1 Position and velocity at two different times for a particle moving in a circle.

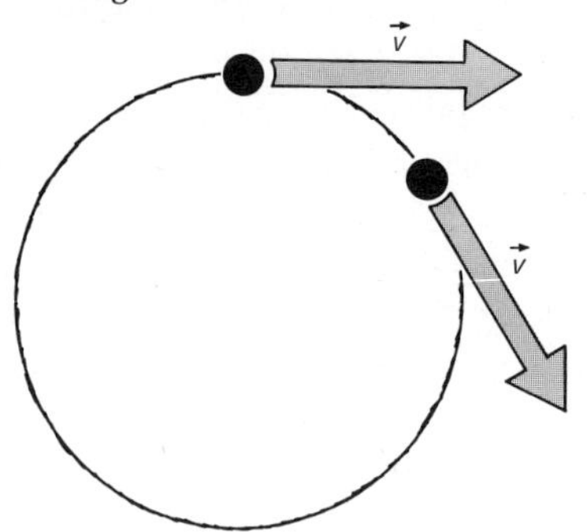

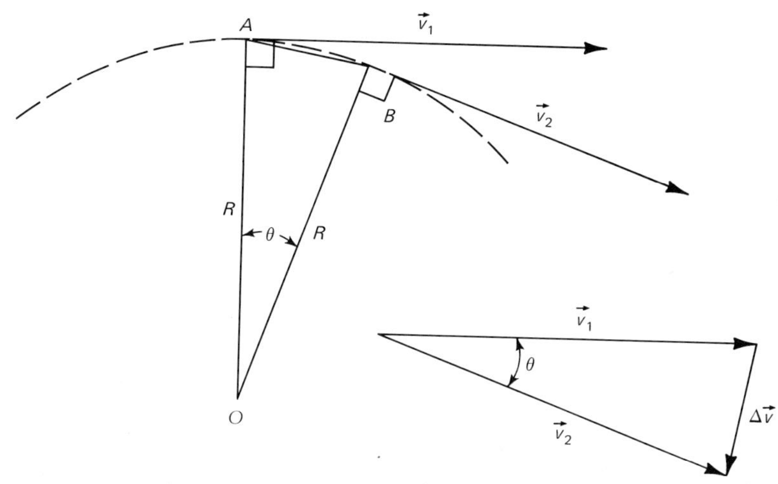

Figure 3-8-2 Changes in position and velocity during a short time interval Δt, for a particle moving in a circle of radius R.

Fig. 3-8-2 the triangle formed by the velocities $\vec{v}_1$, $\vec{v}_2$, and $\Delta\vec{v}$ is similar to triangle *OAB*. Denoting the length *AB* by *s*, and using the fact that the triangles are similar,

$$\frac{s}{R} = \frac{\Delta v}{v} \tag{3-8-2}$$

(The ratios of corresponding sides of similar triangles are equal.)

In a time Δt the particle moves a distance $v(\Delta t)$ along the circle, and for a short time interval, as in Fig. 3-8-2, the straight line distance s can be taken equal to the distance traveled along the circle. Substituting $v(\Delta t)$ for s in Eq. 3-8-2,

$$\frac{v(\Delta t)}{R} = \frac{\Delta v}{v} \tag{3-8-3}$$

Equation 3-8-3 can be solved for the acceleration as follows:

$$\Delta v = \frac{v^2}{R}\Delta t$$
$$\frac{\Delta v}{\Delta t} = a = \frac{v^2}{R} \tag{3-8-4}$$

We have now found the magnitude and direction (inward toward the center) of the acceleration on a particle moving in a circle of radius R with constant speed v. If the particle has a mass m, we can find the inward-directed, or *centripetal*, force required to produce this acceleration by using Newton's second law:

$$F = ma = \frac{mv^2}{R} \qquad \text{3-8-5}$$

As an example of a calculation involving circular motion, we can now consider an artificial satellite. Suppose the satellite moves above the earth at a constant radius R from the earth's center. We require that the gravitational force on the satellite, which is directed toward the center of the earth, provide the required centripetal force for its motion.

For a satellite relatively near the earth's surface—100 miles up, for example—the gravitational force is about the same as at the earth's surface, that is, mg, with $g = 9.8$ N/kg. Setting this force equal to the centripetal force given by Eq. 3-8-5,

$$mg = \frac{mv^2}{R} \qquad \text{3-8-6}$$

This equation can only be satisfied if v has a particular value, such that

$$\frac{v^2}{R} = g \qquad \text{or} \qquad v^2 = gR \qquad \text{or} \qquad v = \sqrt{gR} \qquad \text{3-8-7}$$

For a satellite near the earth's surface, R is about 6.5×10^6 m and g is about 9.8 m/sec^2. Equation 3-8-7 then gives the result $v = 8.0 \times 10^3$ km/sec, or about 5 mi/sec. Notice that v does not depend upon m, the mass of the satellite. All that is necessary to put any object in orbit is to lift it above the earth's atmosphere, give it a horizontal velocity of about 5 mi/sec, and release it. This is the job done by the launch rocket, and we will see in a later chapter why it isn't easy.

PROBLEM 14

An airplane dives and then pulls up along an arc of a circle until it is climbing. At the moment when the airplane is moving horizontally, the pilot estimates he is pushing down on his seat with twice the normal force of gravity. If the plane is flying at 600 miles per hour, and the pilot's estimate is right, what is the radius of the circle along which he is flying?

PROBLEM 15

A communications satellite is in a circular orbit above the earth's equator. It always remains above the same point on the earth's surface, so its period of rotation must be 24 hours. Find its distance from the center of the earth.

PROBLEM 16
Suppose an atom consists of a single electron moving in a circular orbit about a stationary proton, and that the electron and proton have equal and opposite charges. Assume that the electric force is all that holds the electron in its orbit. If the radius of the orbit is 10^{-8} cm, how many revolutions per second does the electron make? (The electron mass is 9.11×10^{-31} kg.)

PROBLEM 17
An astronaut in orbit around the earth finds that objects can float freely in his space capsule, apparently weightless. Try to explain this.

3-9 THE GRAVITATIONAL FORCE

One of the most interesting applications of the dynamics of circular motion was also one of the earliest. Newton realized that the motion of the moon in a nearly circular orbit around the earth required that a centripetal force act on it. If you believe the story, it was one fine day when an apple dropped on his head that he was moved to wonder whether the same gravitational attraction that made objects fall at the surface of the earth might also reach out as far as the moon and provide the force needed to hold it in its orbit.

In Newton's day astronomical data on the motion of the moon were crude, but he utilized knowledge of the moon's distance and its period of revolution to estimate from an equation like Eq. 3-8-7 the required value of the constant g at the moon's orbit. The result was suggestive. After many years of effort, during which he had to invent and use the integral calculus to convince himself of the validity of his idea, he announced the hypothesis that any two point masses exerted an attractive force on each other according to the relationship

$$F = G\frac{m_1 m_2}{r^2} \qquad \text{3-9-1}$$

Through his use of the calculus, Newton was able to show that the point mass equation (3-9-1) could be used to find the force exerted by the earth on a mass m at its surface. It was valid to substitute for m_1 the mass of the earth m_e, for r the radius of the earth r_e, and for m_2 the mass m of a body at the earth's surface. The resultant formula is

$$F = \left(G\frac{m_e}{r_e^2}\right)m \qquad 3\text{-}9\text{-}2$$

The quantity in parentheses could be identified with the gravitational acceleration at the earth's surface, a distance r_e from its center. For the gravitational acceleration at the moon, it was only necessary to replace r_e with the radius of the moon's orbit. The reduction in g by the square of the ratio of the radii fit the astronomical data on the moon quite well, but not perfectly. However, the trouble ultimately was traced to the astronomers and not to Newton's theory.

Since Newton's time, his hypothesis of a gravitational attraction between any pair of masses, sometimes called *universal gravitation*, has been extremely well verified by precise astronomical observations and by terrestrial measurements. Astronomically, the motions of all the planets have been successfully derived from the gravitational force law. On earth, experiments using apparatus similar to Coulomb's have verified Eq. 3-9-1 and led to determination of the constant G. The force between laboratory-sized masses is so small that the experiments to determine G were much more delicate than those used to study the electrostatic force between charges. In mks units, G has the value $6.67 \times 10^{-11}\ \mathrm{N \cdot m^2 \cdot kg^{-2}}$.

Note that the gravitational force varies inversely as the square of the distance between the masses, just as the electrostatic force between point charges was later found to vary. For a while it was believed that all physics would be found to involve forces which varied like $1/r^2$. Magnetic forces do vary in this way, but unfortunately the other known forces, the nuclear forces, have not lived up to this expectation.

PROBLEM 18

Suppose the gravitational force between the earth and a particle at the earth's surface is the same as if all the earth's mass were concentrated at a point at the center of the earth. Show, using Eq. 3-5-2, that the constant g is equal to Gm_e/r_e^2, where m_e is the mass of the earth and r_e is its radius. The constant G has been measured to be 6.67×10^{-11} in mks units. The radius of the earth is 4,000 miles. Find the mass of the earth.

SUMMARY

The description of motion, called *kinematics*, involves the three vector quantities: *displacement*, *velocity*, and *acceleration*. Velocity is the *time rate of change* of displacement, and acceleration is the time rate of change of velocity. For one-dimensional motion with constant velocity.

$$v = \frac{x_2 - x_1}{t_2 - t_1}$$

For constant acceleration,

$$a = \frac{v_2 - v_1}{t_2 - t_1}$$

For motion with varying velocity or acceleration, the above formulas give average values. *Instantaneous values* are found from changes during a very short time interval, Δt:

$$v = \frac{\Delta x}{\Delta t} \qquad a = \frac{\Delta v}{\Delta t}$$

Newtonian (or classical) dynamics is based on *Newton's second law*:

$$\vec{F} = m\vec{a}$$

which is a vector equation. Each *component* of the vectors of a vector equation satisfies its own equation, so that, for example,

$$F_x = ma_x \qquad F_y = ma_y$$

For *uniformly accelerated motion* in one dimension,

$$v = v_0 + at \qquad x = v_0 t + \frac{1}{2} at^2$$

For *circular motion at constant speed* at a radius R, there is a centripetal acceleration with constant magnitude, given by

$$a = \frac{v^2}{R}$$

The *gravitational force* between two point masses is attractive and depends upon $1/r^2$, like the electrical force:

$$F = G\frac{m_1 m_2}{r^2}$$

where $G = 6.67 \times 10^{-11}\ \text{N} \cdot \text{m}^2 \cdot \text{kg}^{-2}$.

EXPERIMENTS WITH PARTICLE BEAMS

4-1 INTRODUCTION

In 1897, J. J. Thomson discovered the electron, while studying streams of particles known then as cathode rays. As a result of Thomson's work these rays were identified as electrons, and some of their most basic properties were studied. In Thomson's experiment, the ratio of the charge to the mass of individual cathode-ray particles was inferred from the way they were acted upon by electric and magnetic forces. However, rather than trying to produce and detect the particles one at a time, Thomson found that a steady stream of them, in the form of a narrow beam, could be used to good advantage.

To help appreciate the philosophy of Thomson's experiment, we can consider a stream of machine-gun bullets fired horizontally from a fixed gun. To the unaided eye, the fast-moving bullets are as invisible as electrons. Yet, by seeing them hit a target we can easily infer their presence. Furthermore, if we measure the amount the bullets have fallen below a straight line in going from gun to target, we can see the effect of the gravitational force on the individual bullets. From the observed gravitational deflection, plus our knowledge of newtonian mechanics and of the gravitational force, we could in fact rather easily infer the muzzle velocity with which the bullets left the gun (assuming they do not change their speed appreciably from gun to target).

The Thomson experiment utilized the action of electric and magnetic forces on a beam of electrons, not to find the speed of the particles, as in the example above, but to determine the ratio e/m of the charge to the mass of each particle. Just as in the machine-gun example, the action of a force on invisible particles was inferred by seeing where a beam of the particles struck a target.

Since atomic and subatomic particles are so small, and hence difficult to study singly, the techniques introduced by Thomson have remained important from his day to the present. In this chapter we will first study Thomson's experiment in some detail, for its own interest as well as for its significance as a prototype for many other important particle-beam experiments. We will then study another very important, and somewhat different, early particle-beam experiment, in which the atomic nucleus was discovered.

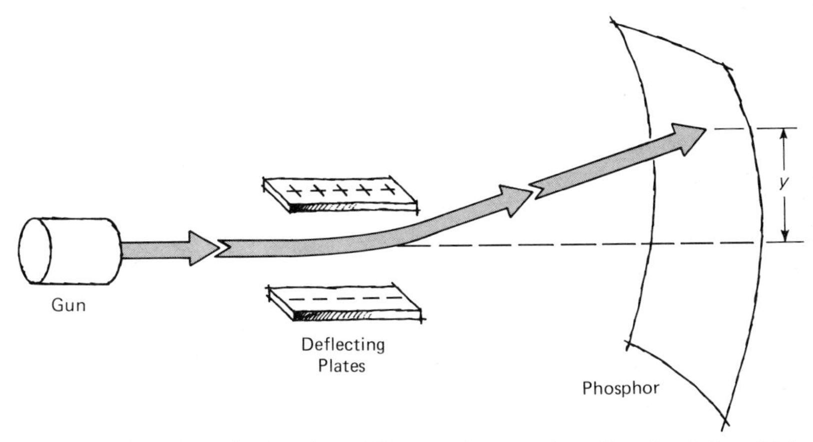

Figure 4-2-1 Schematic drawing of Thomson's apparatus. The glass tube which encloses the whole apparatus is not shown, nor are the electric wires which connect the gun and deflecting plates to batteries.

4-2 THE CATHODE-RAY TUBE

Thomson's apparatus is shown schematically in Fig. 4-2-1. The gun produced a beam of cathode rays, now known to be electrons, all of which had the same speed. There were two deflecting plates which could be charged, as shown in the figure, so as to produce a vertical electric field in the region between them. Such a field would then cause the electron beam to be bent from a straight line path, striking the phosphor shown on the right side of the figure a distance y away from where it would hit in the absence of the electric field. A phosphor is a chemical, such as zinc sulfide, which emits light when struck by fast electrons.

The whole apparatus was contained in a glass envelope from which almost all the air was pumped out. As a result, electrons could travel from gun to phosphor with a very small chance of hitting a gas molecule. The glass envelope containing a gun, deflecting electrodes, and a phosphor screen constituted Thomson's *cathode-ray tube.* The phosphor was in the form of a thin layer on one end of the tube, and the spot of light emitted where the electron beam hit was visible through the end glass. Wires were led through vacuum-tight seals in the tube so that the deflecting plates and the gun electrodes could be connected to external batteries.

The gun of Thomson's cathode-ray tube is shown schematically

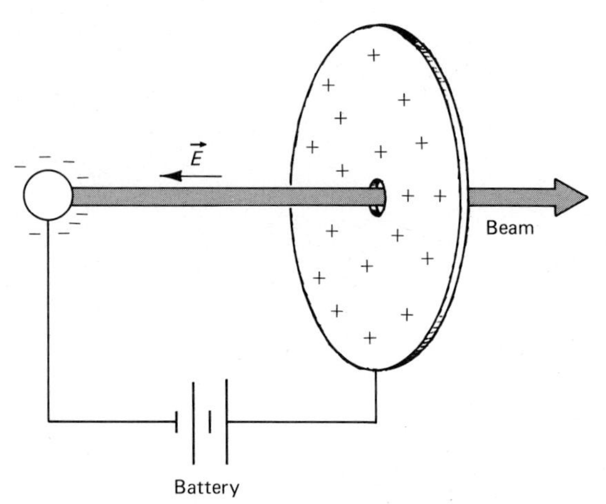

Figure 4-2-2 Schematic view of Thomson's electron gun.

in Fig. 4-2-2. The two electrodes were charged + and − by a high-voltage battery, and a very strong electric field was set up between them. As a result of the action of this strong field on the residual gas in the tube, a gas discharge took place, with the result that positive ions were formed and attracted to the negative electrode. When these hit the negative electrode at high speed, electrons were ejected which were accelerated toward the positive electrode.

Those electrons that passed through the hole in the center of the positive electrode constituted the beam of the cathode-ray tube. They were called cathode rays because they came from the negatively charged electrode of a gas-discharge tube, which was traditionally known as the cathode of the tube. Each electron gained most of its speed by being accelerated by the electric field as it traveled between the cathode and the positive electrode. Since each electron was similarly accelerated, they all had very nearly identical speeds.

PROBLEM 1

In a particular electron gun, the electric field has a constant value everywhere along the path of the electrons. Assume the electrons are emitted from one electrode with essentially zero velocity and accelerated toward the other. Show that after an electron has traveled a distance d from one electrode to the other its velocity is

$$v = \sqrt{2E_0 d \frac{e}{m}}$$

where E_0 is the strength of the electric field and e/m is the ratio of the electron's charge to its mass.

Thomson's tube embodies the main features of today's cathode-ray tubes. However, the source of electrons used in modern guns is usually a hot filament. The oscilloscope and the TV set are very common modern applications of the cathode-ray tube. Each capitalizes on the fact that the electrons in the beam can be deflected very rapidly to any place on the phosphor screen by suitably applied electric or magnetic fields. For the oscilloscope, electric fields are used on two pairs of deflection plates, each pair similar to Thomson's. One pair is arranged to produce vertical deflections and the other to produce horizontal deflections. For the TV picture tube, two sets of coils which carry electric currents are used to set up magnetic fields inside the tube. One set of coils produces a magnetic field which causes vertical deflections of the beam and the other produces horizontal deflections. The main virtue of magnetic deflection for TV picture tubes is that it is particularly effective for bending the beam through large angles, thereby making possible a relatively short picture tube with a large screen.

When a television set is operating properly, the electron beam in the picture tube is being very rapidly deflected back and forth and more slowly up and down, so as to sweep over the face of the picture tube in 525 horizontal lines (which you can see if you look closely). In about one-thirtieth of a second the beam "paints" these lines and then starts over again, with the result that the screen appears to glow steadily, or with only a little flicker. Sometimes just after a set is turned off you can see a single bright spot at the center of the screen, made by the undeflected electron beam.

4-3 ELECTRIC DEFLECTION

In this section the path of a charged particle which is electrically deflected will be calculated with the aid of Newton's second law, in order to quantitatively understand Thomson's experiment. Suppose, as in Fig. 4-3-1, there is a particle with mass m, charge $-e$, and velocity v_0, traveling horizontally until it is deflected by an electric field. It then moves along the path ABC, where C is the point at which it strikes a screen which may be coated with a phosphor. The distances d and D specify important dimensions of the apparatus. The particle could be one of the electrons in Thomson's beam.

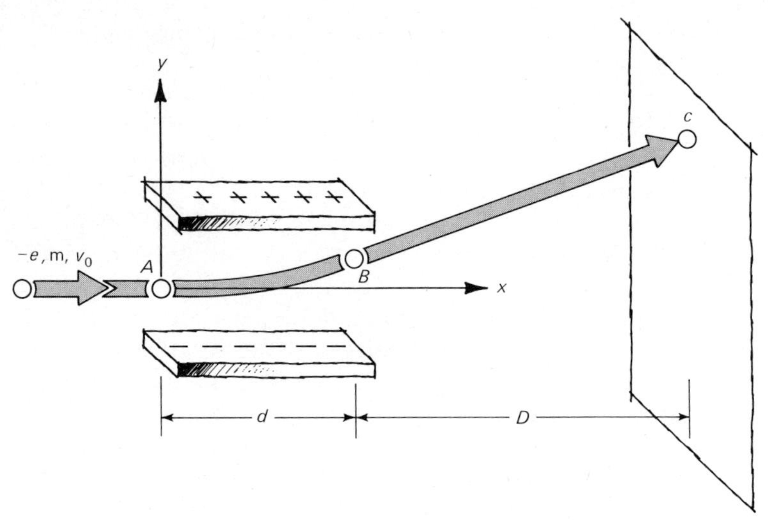

Figure 4-3-1 The path of a charge $-e$ in Thomson's apparatus.

In order to simplify the calculation, we will assume that there is a vertical electric field between the deflecting plates with a constant value E and that the field is zero everywhere else. If the length of the deflecting plates d is much greater than their separation, this assumption is a good one, for the field is essentially constant except near the edges. Also, the gravitational deflection of the particle will be neglected, which is justified if it is traveling very fast.

The motion of the charged particle can now be described very simply. The net force on it is zero except when it is between the plates, so it moves in a straight line until it reaches the point A and it again moves in a straight line from B to C. The path from A to B is an example of uniformly accelerated motion in two dimensions, which has been discussed in Sec. 3-7. Consider x and y axes with origin at A, as drawn in Fig. 4-3-1. At point A the vector velocity can be given in terms of its components:

$$v_x = v_0$$
$$v_y = 0 \qquad 4\text{-}3\text{-}1$$

In the region along the path from A to B the force on the particle is due to the field E and can also be described in terms of its components,

$$F_x = 0$$

$$F_y = eE \tag{4-3-2}$$

Newton's second law, $\vec{F} = m\vec{a}$, enables us to write the components of the acceleration in terms of the components of $\vec{F}$,

$$a_x = \frac{F_x}{m} = 0$$
$$a_y = \frac{F_y}{m} = \frac{eE}{m} \tag{4-3-3}$$

We see that there is no acceleration along the x direction. Thus, at the point B, v_x is the same as at the point A:

$$v_x = v_0 \tag{4-3-4}$$

The effect of the electric field is to produce uniform acceleration in the y direction. Since the y component of velocity at point A is initially zero, the y component of velocity at point B is simply

$$v_y = a_y t = \frac{eE}{m} t \tag{4-3-5}$$

where t is the time it takes the particle to go from A to B.

The x component of the displacement from A to B is the length of the deflecting plates d, and the x component of the particle velocity is constant and equal to v_0. Thus,

$$t = \frac{d}{v_0} \tag{4-3-6}$$

If the value of t from this equation is substituted into Eq. 4-3-5, we have a final result for the y component of velocity at B in terms of e, m, E, d, and v_0,

$$v_y = \frac{e}{m} \frac{Ed}{v_0} \tag{4-3-7}$$

Since both components of the vector velocity at B are now known, from Eqs. 4-3-4 and 4-3-7, we know the direction of motion of the particle at B. Figure 4-3-2 shows the velocity at B and the straight line path along which the particle moves from B to C. Points D and O are on the path of the particle if the electric field between the plates is zero.

We can see in Fig. 4-3-2 that the triangle $BO'C$ is similar to the triangle formed by $\vec{v}$, $\vec{v}_x$, and $\vec{v}_y$. Therefore, the ratio s/D is given by

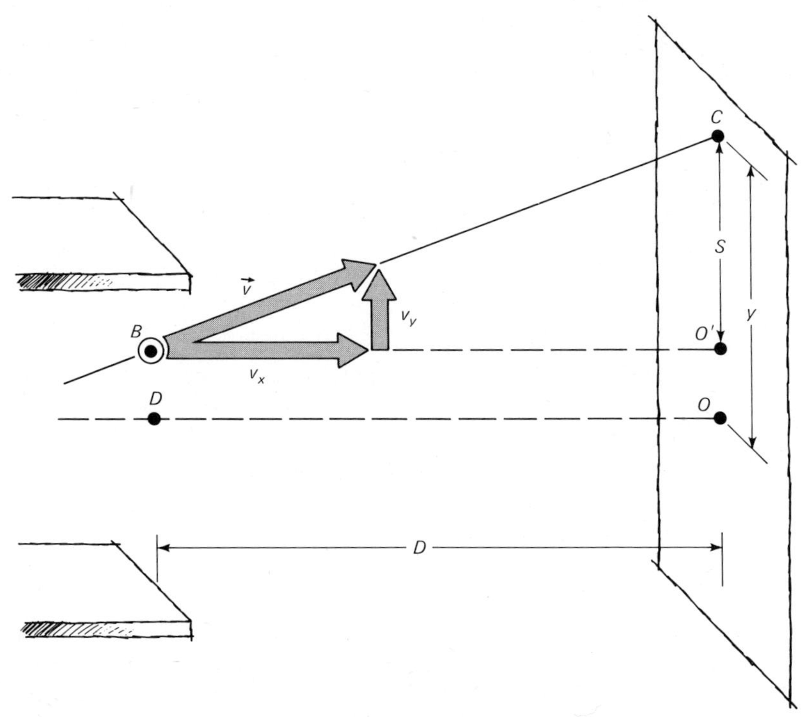

Figure 4-3-2 Enlarged view of Thomson's apparatus showing the motion of a particle after it passes between the deflecting plates.

$$\frac{s}{D} = \frac{v_y}{v_x} = \frac{e}{m}\frac{Ed}{v_0^2} \qquad \text{4-3-8}$$

or, rearranging,

$$s = \frac{e}{m}\frac{EdD}{v_0^2} \qquad \text{4-3-9}$$

Thus the distance s is proportional to e/m, and the proportionality constant, EdD/v_0^2, depends on v_0 and the presumably known quantities E, D, and d.

In a cathode-ray tube where D is much greater than the length of the plates, which is commonly true, the observed deflection of the beam, labeled y in Fig. 4-3-2, is only a little larger than s. We can write approximately, for $D >> d$,

$$y \doteq s = \frac{e}{m}\frac{EdD}{v_0^2} \qquad \text{4-3-10}$$

where the symbol $\doteq$ means "almost exactly equal to."

In order to get an exact formula for y, it is only necessary to add to the value of s given by Eq. 4-3-9 the vertical displacement at point B. The exact result is

$$y = \frac{e}{m}\frac{Ed(D + d/2)}{v_0^2} \qquad \text{4-3-11}$$

It is now easy to understand how J. J. Thomson used an electric deflection apparatus to measure e/m for an electron beam. Referring again to Fig. 4-3-2, point O corresponds to the point where an electron beam would strike a screen coated with phosphor if there were no electric field. Point C is the point where the beam would hit if there were a field E between the deflection plates. The deflection y, corresponding to the two spot positions, was easily measured. From Eq. 4-3-11, e/m is given, in terms of the measured deflection and presumably known quantities, by

$$\frac{e}{m} = \frac{yv_0^2}{Ed(D + d/2)} \qquad \text{4-3-12}$$

In fact, there is no simple way to find the speed v_0 from the known electric fields in the gun if the ratio e/m is not known. In order to find e/m from Eq. 4-3-12, it was necessary for Thomson to infer v_0 from data in which magnetic and electric deflection were combined in his cathode-ray tube. Later in this chapter, magnetic deflection, and the determination of v_0, will be discussed. The other quantities on the right-hand side of Eq. 4-3-12 are directly determined by the geometry of the apparatus and the voltage of the battery connected to the deflection plates.

PROBLEM 2

Show that the distance BD in Fig. 4-3-2 is given by

$$\overline{BD} = \frac{E}{2}\left(\frac{e}{m}\right)\left(\frac{d}{v_0}\right)^2$$

PROBLEM 3

In a particular apparatus like that for Thomson's experiment the distances D and d are 30 and 5 cm, respectively. When the field between the plates is 2,000 N/C, the deflection at the phosphor is 10 cm. Find the electron velocity assuming e/m is known. If an electron starts out traveling horizontally with this velocity, how far does it fall as a result of gravity in traveling a distance D?

4-4 THE MASS OF THE ELECTRON AND OF IONS

By using the method described above, Thomson's first result was $e/m = 2.6 \times 10^{11}$ C/kg. The best modern value for this ratio is $e/m = 1.760 \times 10^{11}$ C/kg, which is about 30 percent lower than Thomson's value. Nonetheless, his experiment was remarkably good for its day.

After Thomson's original measurement, his technique and methods were used to determine q/m, the charge to mass ratio, for negative particles extracted in various ways from a great variety of materials. The ratio q/m in all these cases was the same as for cathode rays within the possible range of error of the experiments, and this evidence strongly supported the idea that negatively charged particles called electrons were constituents of all atoms. Thomson "discovered" the electron because he measured its value of e/m and was thereby able to identify it unambiguously as a single universal constituent of matter.

While the charge of the electron was not well known in Thomson's time, it had been estimated from some ingenious experiments which were the forerunners of Millikan's work. Given Thomson's value of e/m and the value of e, it is of course possible to find the mass of the electron. In terms of the presently accepted values of e/m and e, the mass is given by

$$m = 9.11 \times 10^{-31} \text{ kg} \qquad \text{4-4-1}$$

Even in Thomson's day, his data could be used to learn that the electron mass was much smaller than any atomic mass, thus bolstering the idea that the electron was a building block of atoms.

Soon after Thomson's measurement of e/m, his methods were applied to the measurement of the ratio q/m for positive ions. For the positive ions of hydrogen, q/m is found to be 1,836 times smaller than for electrons. Suppose hydrogen ions are atoms from which one electron has been removed, a plausible assumption that has proven to be true. Then the charge of these ions is $+e$, and the mass of a hydrogen atom must be 1,836 times the electron mass. The first accurate determinations of many atomic masses were made in this way, by assuming that ions with the smallest observed charge carried the charge $+e$. The best current value for the mass of a hydrogen atom is

$$m_H = 1.673 \times 10^{-27} \text{ kg} \qquad \text{4-4-2}$$

From this value and the relative atomic weights measured by

chemists, the actual masses of other atoms can also be found. For example, from the atomic weights in Table 1-3-1 you could calculate the masses of all the elements listed. However, a very interesting discovery was made when techniques were developed for accurately measuring q/m for ions. The ions of some elements were found to have two or more different values of q/m. For these elements, there would be two discrete deflected spots when the field was turned on in an apparatus like that discussed in the previous sections.

Chlorine, element number 17, provides a striking example of this phenomenon. The ions are found to consist of two kinds, one with a mass 34.7 times the mass of a hydrogen atom and the other with a mass 36.7 times the mass of hydrogen. No ordinary chemical methods can break up chlorine into more basic chemical constituents, so it is indeed a chemical element and not a compound of two or more elements. Furthermore, ordinary chemistry is useless for separating the two kinds of chlorine whose ions have different masses. Finally, the charges on the two kinds of ions are not different, and the masses determined by assuming that each ion has a charge $+e$ are indeed correct.

We are forced to the conclusion in the case of chlorine, as for many other elements, that the pure element is made up of atoms which are chemically identical but can have two or more discrete mass values. An atom of an element with a particular mass value is called an atom of an *isotope* of that element, and the preceding paragraph describes the two common isotopes of chlorine. The existence of isotopes can be understood with our current knowledge of the atomic nucleus, but when they were first discovered isotopes presented a surprising mystery which took many years to unravel. Later on we will come back to them.

4-5 MAGNETIC DEFLECTION

Up to now we have assumed that a beam of electrons or ions of known velocity has been available for electric deflection experiments. In fact, determining the velocity of a beam of fast atomic particles is not simple. One technique, which utilizes magnetic forces, was used by J. J. Thomson to measure the velocity of the electrons in the same apparatus with which he measured e/m. The principle is so effective that it is often used in modern laboratories to obtain particle beams with known velocities.

The force between two charges which are in motion consists

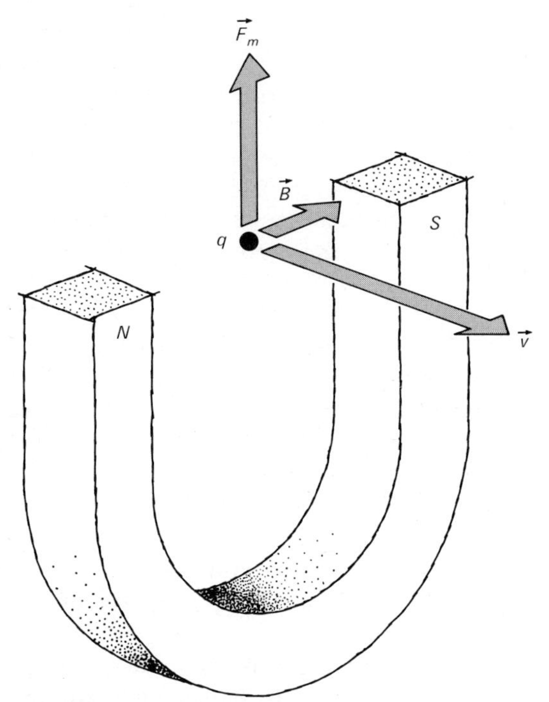

Figure 4-5-1 A charge q moving between the poles of a horseshoe magnet. The force $\vec{F}_m$ is perpendicular to both the velocity of the particle and the direction of the magnetic field $\vec{B}$.

of two parts, an electric force and a magnetic force. We describe the combined effect of both as the *electromagnetic force.* This section will introduce the *magnetic force*, which is just as basic a part of physics as the electric force. Both forces can be calculated from fields, and we are familiar with the equation for the electric force on a charge q in a field $\vec{E}$, denoted here by $\vec{F}_e$:

$$\vec{F}_e = q\vec{E} \tag{4-5-1}$$

Remember that the field $\vec{E}$ is set up by all the charges other than the one on which we are calculating the force. We know how to calculate $\vec{E}$ from a single point charge, and by vector addition we can find $\vec{E}$ for any number of point charges.

If a charge is moving, we may observe that there is a force on it even if $\vec{E} = 0$. For example, if the charge is moving between the poles of a horseshoe magnet, as in Fig. 4-5-1, it experiences a force as shown. This is an example of the action of a magnetic force, which we will call $\vec{F}_m$. This force $\vec{F}_m$ can also be found from a magnetic field, denoted by $\vec{B}$, with the magnitude of the magnetic force given by

$$F_m = qvB_\perp \tag{4-5-2}$$

Notice that the magnitude of $\vec{F}_m$ is proportional to v, so there is no magnetic force on a charge at rest. Also, the magnitude of $\vec{F}_m$ is proportional to the component of the magnetic field $\vec{B}$ *perpendicular* to the direction of motion of the particle. This is signified by the quantity $B_\perp$ in Eq. 4-5-2, and it may seem quite strange until you get used to the idea. The answer to the question "Why is the magnetic force the way it is?"—like the answer to the same question about electric forces—is "That's the way the world is." The magnetic force is so different in behavior from the electric force that it took physicists quite a lot longer just to describe it.

The direction of $\vec{F}_m$ is best understood from Fig. 4-5-2. For a positive charge the force points in a direction *perpendicular to both* $\vec{v}$ *and* $\vec{B}$, as in the figure. You can think of the figure as showing $\vec{v}$ and $\vec{B}$ in a horizontal plane and $\vec{F}_m$ pointing upward. Were the charge negative, the force $\vec{F}_m$ would point in the opposite direction, downward.

In this section, and throughout this book, we will only need to consider examples where $\vec{B}$ is perpendicular to $\vec{v}$. This results in

Figure 4-5-2 Magnetic force $\vec{F}_m$ on a charge q moving with velicity $\vec{v}$ in a magnetic field $\vec{B}$. It is assumed that q is positive. You can visualize $\vec{v}$ and $\vec{B}$ lying in a horizontal plane, with $\vec{F}_m$ vertical.

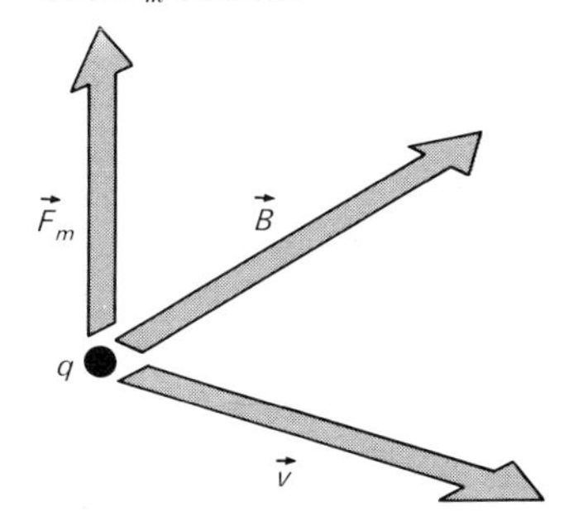

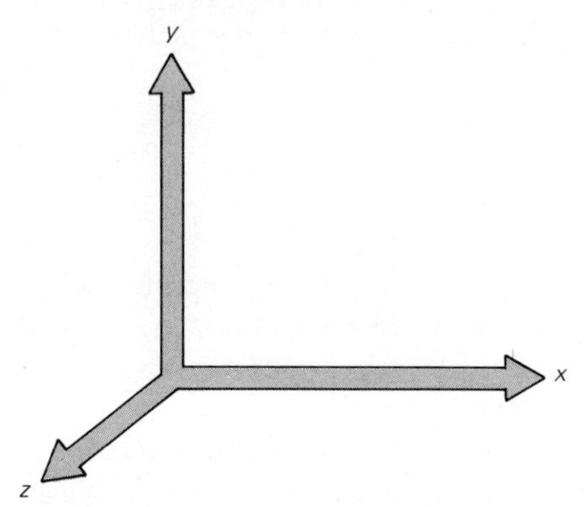

Figure 4-5-3 Three mutually perpendicular axes.

a maximum magnetic force, since $B_{\perp}$ is then equal to B. The direction of $\vec{F}_m$ is usually found by a *right-hand rule*, which is also simple to use when $\vec{B}$ and $\vec{v}$ are perpendicular. To learn the rule, first refer to Fig. 4-5-3, which shows three axes, all perpendicular to each other, as if they were drawn on a corner of a cube.

To use the right-hand rule, arrange your right hand so that the thumb points along the direction of the x axis in Fig. 4-5-3, your first finger (i.e., the one next to your thumb) points along the y axis, and your second finger points along the z axis. It can be done! Once your fingers are in this position, you can move your hand around so that any given finger points in a variety of directions but the three fingers remain the same with respect to each other. The right-hand rule is

Point the thumb along $\vec{v}$.

Point the first finger along $\vec{B}$.

The force $\vec{F}_m$ is in the direction of the second finger.

For negative charges, find $\vec{F}_m$ as above and then reverse its direction.

PROBLEM 4

The mks unit of magnetic field is "webers per square meter," abbreviated w/m^2. There is a uniform B field in the x direction with magnitude 0.1 w/m^2. Find the magnitude and direction of the force on an electron moving with a velocity of 3.0×10^7 m/sec in the y direction. In the x direction.

Magnetic fields are produced by moving charges, i.e., electrical currents. In the case of permanent magnets, the magnetic field is the result of a great many tiny atomic currents. We won't describe in detail how the magnetic field at some point in space can be calculated from the currents which give rise to it. However, in Fig. 4-5-4 there are sketches of two ways in which a fairly uniform magnetic field can be established in a region between two coils of wire. Batteries are shown connected to the coils so as to produce steady currents, and the magnetic field can be varied by varying these currents.

The lines of $\vec{B}$ in the figure indicate the magnetic field direction (tangent to the field lines) at the various points in space. This is the same as for lines of $\vec{E}$. Also as for $\vec{E}$ lines, the closer the spacing between lines of $\vec{B}$, the stronger the field. Remember, however, that the direction of the magnetic force is *not* along the direction of $\vec{B}$, but in a direction at right angles to both $\vec{B}$ and the velocity of a moving charge. Thus, if we think of the field lines in the gap of the iron magnet of Fig. 4-5-4*b* as vertical, the force on a moving charge due to this field must be horizontal. Its direction cannot be fully specified until we know the direction of the particle's velocity.

Figure 4-5-4*b* illustrates the way in which an iron "core" can be used to help establish a field which is very uniform and well localized in a particular region of space. The very uniform field in the gap of the iron is the magnetic analog of the electric field between two parallel plates. In an example, when a uniform magnetic field is assumed to exist, you can suppose it is made in this way.

Figure 4-5-5 shows a charge $+q$ moving with velocity $\vec{v}$ along the x axis. Suppose it is in a region of uniform electric field in the y direction and uniform magnetic field in the z direction, as indicated by the $\vec{E}$ and $\vec{B}$ vectors in the figure. Then the magnetic and electric forces will point as shown, opposing each other. If their magnitudes are equal, the *net* force on the moving charge will be zero. The equation for zero net force is simply $qE = qvB$, or

$$v = \frac{E}{B} \qquad \text{4-5-3}$$

Thus, for a particle with a given velocity there is a particular ratio E/B for which the net force is zero, provided the directions of $\vec{E}$ and $\vec{B}$ are chosen properly. J. J. Thomson measured the velocity of the electrons in his beam by using this equation. He

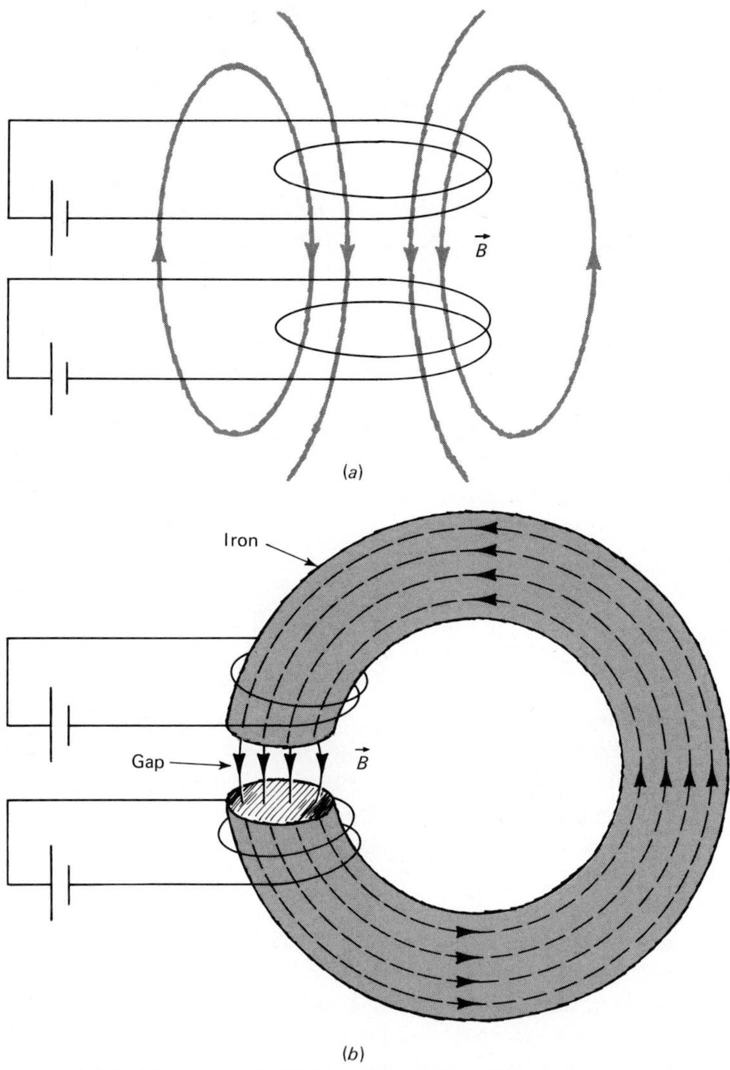

Figure 4-5-4 The magnetic field which can be set up by currents flowing in two coils of wire. The field with and without an iron "core" is shown. In the region of space between the coils the field is approximately uniform.

arranged for a uniform magnetic field in the region between the deflecting plates where the uniform $\vec{E}$ was established, with $\vec{B}$ small everywhere else. The field $\vec{B}$ was perpendicular to $\vec{E}$, in a

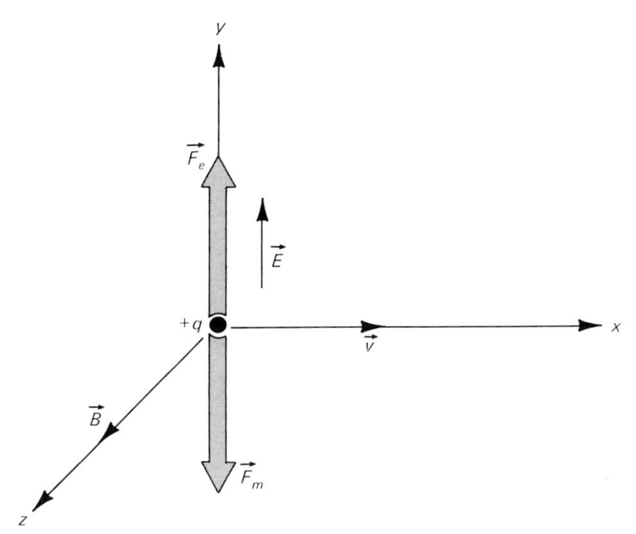

Figure 4-5-5 Forces on a positive charge moving in a region where there are $\vec{E}$ and $\vec{B}$ fields in the directions indicated.

direction such that the electric and magnetic forces would oppose each other. He then varied the magnitudes of $\vec{E}$ and $\vec{B}$ until the electron beam was undeflected.

If the beam were not deflected, he reasoned that the electric and magnetic forces must be equal in magnitude, and the velocity of the electrons could be found from the known values of E and B by using Eq. 4-5-3. If Thomson had turned off the $\vec{E}$ field, then of course there would have been magnetic deflection. In modern particle beams, as in the TV picture tube, magnetic fields are often used alone to produce deflections, as well as in combination with electric fields to measure velocities.

PROBLEM 5

A particle with charge $+q$ is moving with a velocity of 1.5×10^7 m/sec in the $+x$ direction. It moves in a uniform E field of 10^6 N/C in the $+z$ direction. Find the magnitude and direction of a uniform B field which will produce a force on the particle equal and opposite to the electric force. Show the directions of $\vec{v}$, $\vec{E}$, and $\vec{B}$ with a sketch. How does your answer change if the particle has charge $+2q$? If it has a charge $-q$?

PROBLEM 6
A charged particle is moving horizontally in a region where there is a uniform magnetic field in the vertical direction. Explain why the magnetic force causes the particle to move in a circle at constant speed. Show that the radius of the circle R is given by

$$R = \frac{mv}{qB}$$

PROBLEM 7
In the problem above, what is the time required for the particle to make one revolution around the circle? Note that it does not depend upon v.

4-6 THE RUTHERFORD ATOM

The study of atomic structure has been a most important branch of physics from J. J. Thomson's day to the present, and probably the most important single advance in our knowledge of the atom was made by Ernest Rutherford in 1911. Much of what was known about the atom in his time has been presented in the previous parts of this chapter and in the chapters preceding. Atoms were known to be small, with diameters $\sim 10^{-8}$ cm, according to estimates like the one we made in Chap. 1 for the iron atom. They were known to contain both positive and negative charges. The negative charge appeared to be in the form of very light electrons, the mass of one electron being almost 2,000 times less than the mass of hydrogen, the lightest atom. Thus, the positive stuff in atoms seemed likely to be responsible for most of their mass.

There had been one other important discovery about atoms that has not yet been mentioned. Antoine Henri Becquerel, a French physicist, found out accidentally, in 1896, that some substances emit invisible radiation capable of penetrating appreciable thicknesses of matter. The phenomenon was called radioactivity and could be traced to the individual atoms of radioactive substances. It was soon learned that radioactive atoms actually emit three kinds of invisible rays, which were called alpha, beta, and gamma radiation after the first three letters of the Greek alphabet. (In their search for letters to be used as symbols, physicists have taken over all the letters of the Greek alphabet which are not confusable with Latin letters.)

By 1911, the alpha radiation was known to consist of helium ions with a charge of $+2e$, which are now commonly called alpha particles. These ions are emitted from some kinds of radioactive atoms with such high speeds that they can pass through thin layers of metal. Some radioactive elements emit alpha particles, all of which have identical speeds. For example, the atoms of an element called polonium are alpha emitters of this kind. The question of what happens to an atom that has emitted an alpha particle, and why it does so, as well as many other interesting aspects of radioactivity, will not be discussed now. We will be concerned here with one particular kind of experiment which was done by Geiger and Marsden and which utilized a beam of alpha particles.

Geiger (who later invented the Geiger counter) and Marsden were working in Rutherford's laboratory in Cambridge, England, studying how alpha particles passed through matter. Their experimental apparatus is schematically shown in Fig. 4-6-1. An alpha-particle beam was formed by using a polonium source and a collimator, which was a sheet of metal thick enough to stop the alpha particles unless they passed through a small hole at its center. The beam which emerged from the hole in the collimator then struck a very thin gold foil. The experiment was a measurement of the likelihood that alpha particles were "scattered" or deflected through various angles in passing through the foil. The detector was a small screen coated with zinc sulfide, a phosphor which emits light when struck by fast particles. This screen could be moved around so as to detect particles scattered at any angle θ, such as in Fig. 4-6-1.

Figure 4-6-1 An apparatus for alpha-particle scattering experiments. The source S and collimator C produce a beam which is incident on a thin foil F. Particles scattered at an angle θ strike the detector D.

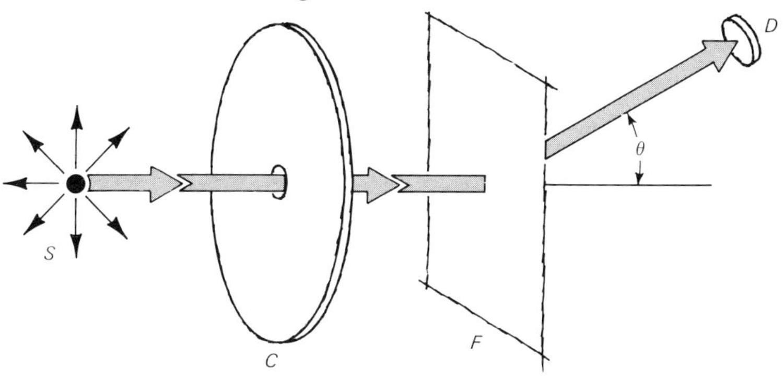

Most of the alpha particles passed through the foil without being deflected significantly. However, very rarely an alpha particle was scattered at some reasonably large angle. For the beam used by Geiger and Marsden, the number of alpha particles striking their small screen per minute was small when the screen was set for any large angle. They were able to investigate the likelihood of scattering at various angles by actually counting the weak light flashes from their screen as it was struck by individual alpha particles.

In trying to understand the experimental data on the likelihood of large-angle scatterings, Rutherford was struck by the fact that, although such scatterings were rare, they occurred much too often to be explained by any atomic models invented up to then. One such model, proposed by J. J. Thomson, is shown in Fig. 4-6-2. The electrically neutral atom is envisioned as a glob of positively charged matter, with a diameter equal to that of the atom, in which electrons are embedded. Rutherford could show that the results obtained by Geiger and Marsden could not be explained by the action of purely electrical forces between such atoms and alpha particles.

Led by his intuition, Rutherford guessed that an atomic model like that shown in Fig. 4-6-3 might predict at least qualitatively the kind of results obtained by Geiger and Marsden. The main feature of Rutherford's atom is that it contains a small *nucleus* surrounded by electrons. The cloud of electrons gives the atom its

Figure 4-6-2 The Thomson, or raisin-pudding, model of the atom.

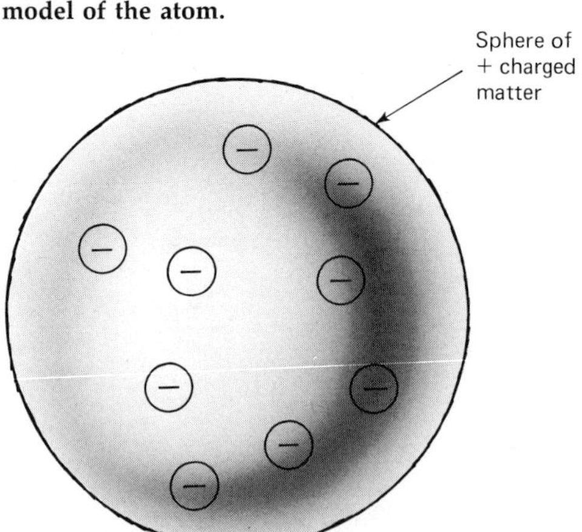

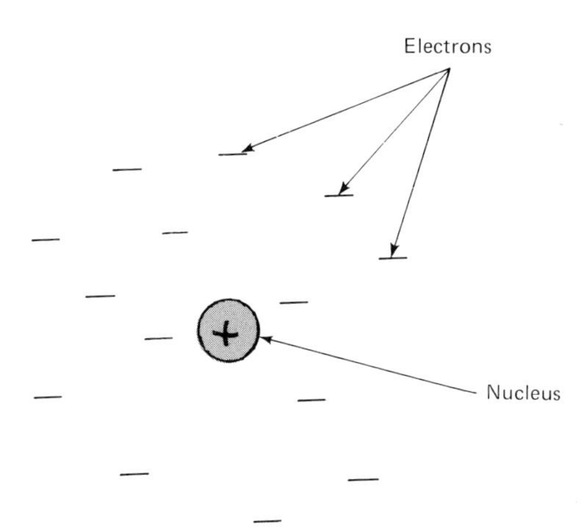

Figure 4-6-3 The Rutherford model of the atom. If the number of electrons is Z, the nuclear charge is $+Ze$.

size, but almost all the mass of the atom, and its positive charge, is in the small nucleus. If Z stands for the number of electrons, then the nuclear charge must be $+Ze$, so that the net charge of the atom will be zero.

A fast alpha particle, whose mass is more than 7,000 times the electron mass, could plow through the electrons in Rutherford's atom much as a 16-pound bowling ball would roll through toy foam-plastic pins. However, if the alpha particle happened to pass close to the heavy nucleus, the repulsion between the positive charge on the alpha particle and the positive charge of the gold nucleus could result in a large-angle scattering. The atomic weight of gold is about 50 times that of the alpha particle, so alpha particles could bounce off a gold nucleus like a tennis ball bouncing off a bowling ball.

According to Rutherford's model, alpha particles passing through a gold foil might behave as shown schematically in Fig. 4-6-4. The heavy positive nuclei are shown in the figure and the electrons are omitted on the assumption that they don't produce large-angle scatterings. The foils used in the experiments which led to Rutherford's idea, and in the further experiments done later to check whether it was correct, were actually a few thousand atomic diameters thick. But Fig. 4-6-4 properly illustrates Rutherford's picture of what happened to alpha particles traversing even such a comparatively thick foil.

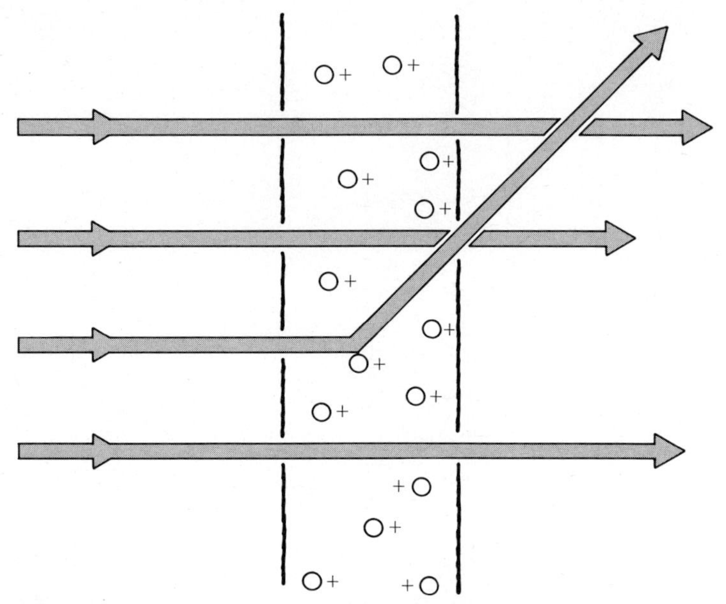

Figure 4-6-4 Schematic view of a few alpha particles traversing a thin gold foil. The points labeled + represent gold nuclei and the electrons in the foil are not shown. One particle approaches so close to a nucleus that it is scattered through a large angle.

4-7 RUTHERFORD SCATTERING

In this section we will see how to predict the result of an alpha-particle scattering experiment on the basis of Rutherford's model. First, suppose we could see a single nucleus and a single alpha particle and actually measure the path of the alpha particle as it was scattered by the nucleus. Figure 4-7-1 shows how an alpha particle approaching the nucleus along a path shown by the dashed line would be deflected by the nuclear electric field and scattered through an angle shown as θ on the figure.

The quantity b shown in Fig. 4-7-1 is called the *impact parameter* and is a measure of how "close" the collision is. For $b = 0$ the collision is head-on and the alpha particle will be scattered straight backward. For large b, the alpha particle suffers a grazing encounter with the nucleus and is only very slightly deflected. For a given b, we can calculate the alpha-particle trajectory and the resultant scattering angle θ, assuming that the only force acting between the alpha particle and the nucleus is the electric repulsion of two point charges. This is what Rutherford did.

It is actually not possible to observe the alpha-particle trajectory in full detail. However, it is possible to determine the scattering angle θ, as shown in Fig. 4-6-1. To compare theory with experiment, the first requirement is to find the relationship between impact parameter b and scattered angle θ. The problem is not in principle very different from that of calculating the deflection angle of the electron beam in J. J. Thomson's cathode-ray tube. In his case, charges on two deflection plates changed the direction of electrons. In the present case, the electric field of a point charge (the nucleus) deflects alpha particles. However, because the point-charge field varies in space, in contrast with the uniform field of the Thomson apparatus, the mathematics for alpha-particle scattering is much more complicated.

The detailed calculation of how θ depends on b for alpha-particle scattering won't be discussed here. The result is a formula which gives the tangent of half the scattering angle in terms of b and known quantities shown in Fig. 4-7-1:

$$\tan\frac{\theta}{2} = (9.0)(10^9)\frac{qQ}{m{v_0}^2}\frac{1}{b} \tag{4-7-1}$$

It won't be necessary to use this expression in detail, so that if $\tan(\theta/2)$ is not a quantity you wish to cope with, don't be concerned. Equation 4-7-1 can be represented on a graph such as that shown in Fig. 4-7-2, which shows the relation between θ and b well enough for our purposes.

PROBLEM 8

In a typical experiment in which alpha particles are scattered off gold nuclei, the velocity of the alpha particles is 1.5×10^7 m/sec.

Figure 4-7-1 Scattering of a particle with charge $+q$ by a stationary charge $+Q$. The dashed line is the path the particle would take if there were no electric repulsion. The distance b is the impact parameter, and the angle θ is the scattering angle.

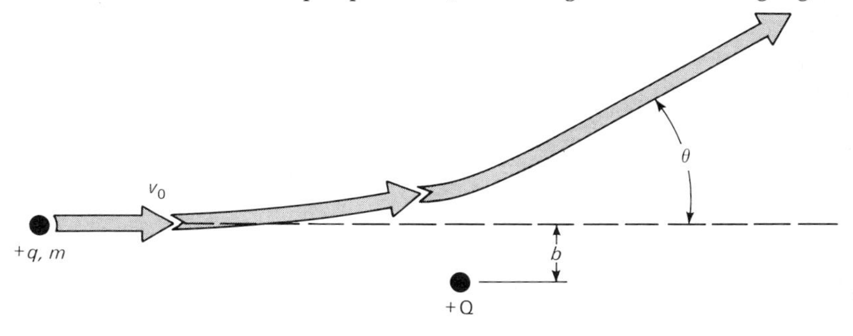

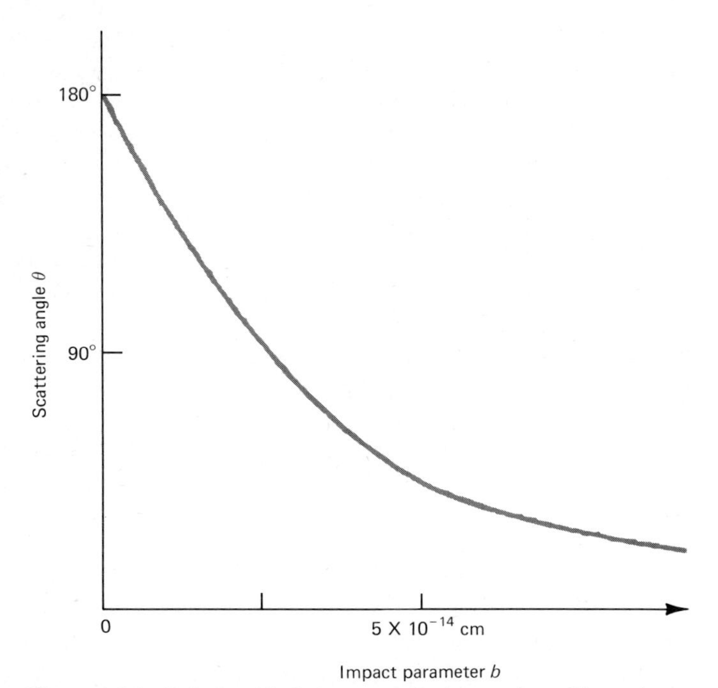

Figure 4-7-2 Relationship between scattering angle and impact parameter for polonium alpha particles scattered by gold, according to Rutherford's model.

> The charge of the gold nucleus is $+89e$, where e is the magnitude of the electron charge. If tan $(\theta/2)$ in Eq. 4-7-1 equals 0.1, what is the impact parameter?

The relation given in Eq. 4-7-1 specifically results from the model in which the alpha particle and the nucleus interact as point charges according to Coulomb's law. A different kind of force law, or a large cloud of positive charge (as in Thomson's model), would lead to a different relationship. Thus the test of Rutherford's model is to determine whether Eq. 4-7-1 correctly predicts the result of an actual alpha-particle scattering experiment.

If an experiment could be set up so that the impact parameter b were known for individual alpha scattering events, Eq. 4-7-1 could be directly compared with experiment. However, the impact parameters of interest are a small fraction of an atomic diameter, so that it does not seem technically feasible to directly determine b. More fundamentally, the basic laws of the quantum theory actually forbid an accurate knowledge of b in scattering processes like the one we are considering here.

The way to verify or to disprove Rutherford's model is to calculate the result of an experiment such as Geiger and Marsden's, shown in Fig. 4-6-1. In such an experiment, there is a beam striking the target over an area which is enormous compared to the diameter of a single atom, and the number of nuclei of the target which are exposed to the beam is also enormous. It is impossible to know the impact parameter for any particular alpha particle, but it is still possible to predict the likelihood that it will scatter through an angle greater than any particular specified angle. We will now describe how this can be done.

While Fig. 4-6-4 shows a schematic side view of an alpha-particle beam traversing a thin foil, Fig. 4-7-3 shows a view of the foil as seen by an approaching beam particle. The positively charged nuclei are shown as dots surrounded by shaded circles, and you should imagine the beam particles as raining down on the paper. All the nuclei are shown, as if the target were transparent, and we continue to ignore the electrons.

Around each dot which represents the location of a nucleus in Fig. 4-7-3 there is a shaded circle of radius b_1. Consider one beam

Figure 4-7-3 "Beam's eye view" of a thin gold foil. The points labeled + and surrounded by shaded circles correspond to gold nuclei. All the nuclei throughout the thickness of the foil are shown, as if it were transparent.

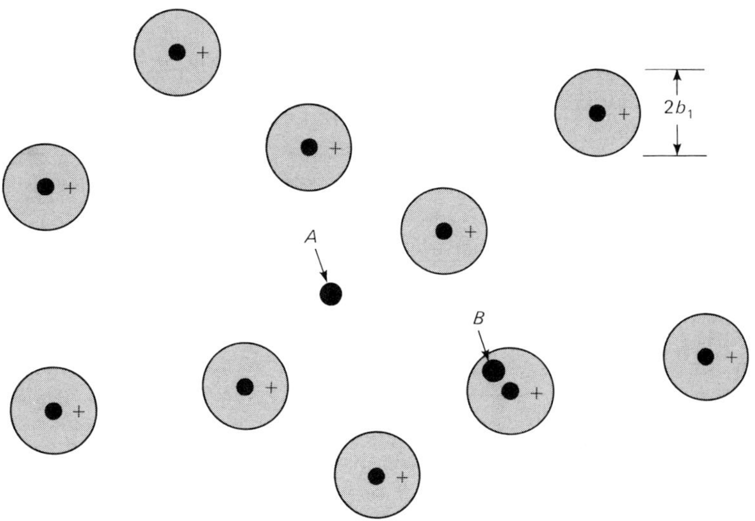

particle moving toward the target. Its path might be aimed at point *A* of the figure, in the clear area, or at point *B* in one of the shaded circles. If θ_1 is the scattering angle which corresponds to an impact parameter b_1, then a particle aimed at point *A*, or any other point in the unshaded area, will scatter through an angle less than θ_1 since its impact parameter with respect to any nucleus is greater than b_1. On the other hand, if the particle is aimed at a point inside any shaded circle, such as point *B*, its impact parameter with respect to one nucleus is less than b_1, and it will scatter through an angle greater than θ_1.

We have assumed that the scattering angle is determined entirely by the smallest single impact parameter between the path of the particle and any nucleus. This *single scattering approximation* is valid if the experiment involves only impact parameters which are much smaller than the average distance between nuclei in Fig. 4-7-3. In a large-angle scattering experiment with a thin target, like that of Geiger and Marsden, the single scattering approximation is a good one.

Suppose the shaded circles in Fig. 4-7-3 cover 10 percent of the total target area in the beam's eye view. If beam particles are aimed completely at random, there should then be a 10 percent chance that a specific particle hits inside a shaded circle. Therefore, on the average, 10 percent of the beam particles will be expected to hit inside a shaded circle. Such a hit corresponds to a scattering through an angle greater than θ_1, so that we could predict that 10 percent of the time such a scattering would occur.

The way in which a particular relation between b and θ can be used to predict the results of an actual scattering experiment is through the assumption of random aiming discussed above. For a given connection between b and θ the *probability* of scattering through an angle greater than θ_1 can be calculated. For different values of θ_1 the probabilities will differ in a way which depends upon how b and θ are related.

The equation for the probability P of a scattering through an angle greater than θ_1, according to Rutherford's model, is

$$P = K\left(\frac{qQ}{mv_0^2}\frac{1}{\tan(\theta_1/2)}\right)^2 \qquad \text{4-7-2}$$

where the constant K depends upon known properties of the target, q and Q are the charges of alpha particle and target nucleus, m is the mass of the alpha particle, and v_0 is the alpha-particle velocity. Figure 4-7-4 shows a graph of how P varies with θ_1 ac-

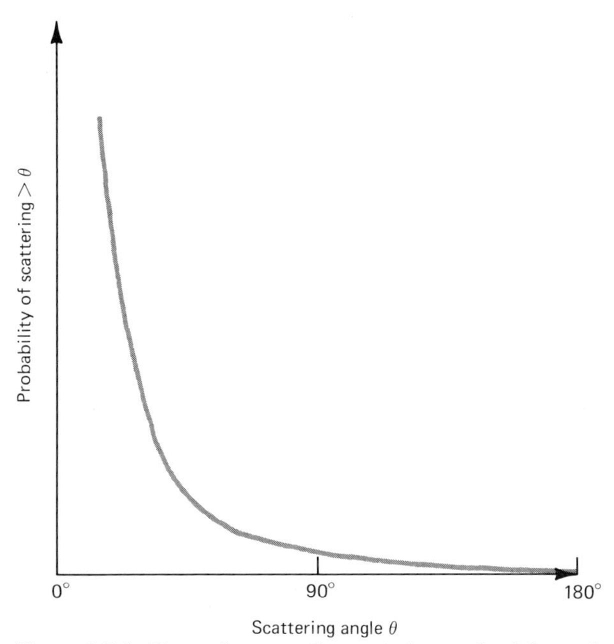

Figure 4-7-4 Dependence on the scattering angle of the probability of scattering, according to the Rutherford model.

cording to Eq. 4-7-2. After Rutherford's hypothesis, Geiger and Marsden did a careful series of experiments specifically intended to check the predictions embodied in Eq. 4-7-2. They observed the predicted variation of scattering probability with angle exactly as shown in the graph of Fig. 4-7-4, and they verified that the scattering probabilities depended on v_0^2 in the way predicted by Eq. 4-7-2. They also found that the data could be well fitted if the nuclear charge Q was assumed to be $+e$ times the atomic number of the scatterer.

The invention of Rutherford's model of the atom and its subsequent verification are beautiful examples of the interplay of experiment and theory in the progress of physics. As a particle-beam experiment, it provides an interesting contrast to J. J. Thomson's e/m experiment. In the former the properties of the beam particles were studied, while in the alpha scattering experiments a known beam was used to probe the nature of a target. This second category of experiment typifies the techniques used to study nuclear and elementary particle physics up to the present day.

In order to probe inside the tiny nucleus, giant particle accelerators are now built to obtain more powerful beams than the natural alpha particles used by Geiger and Marsden. However, measuring the probability of scattering through various angles for different kinds of beam particles and different targets remains the most important single tool in studying the properties of nuclei and of their constituents. We will return to this subject in more detail in later chapters.

SUMMARY

For an electron beam of known velocity, *electric deflection* in a cathode-ray tube can be used to determine the *charge to mass ratio* e/m. The modern value of e/m is:

$$\frac{e}{m} = 1.760 \times 10^{11}\ \mathrm{C/kg}$$

From the Millikan oil-drop experiment and other sources,

$$e = 1.602 \times 10^{-19}\ \mathrm{C}$$

Combining the values of e and e/m, the *mass of an electron* can be deduced:

$$m_e = 9.11 \times 10^{-31}\ \mathrm{kg}$$

The *magnetic field* $\vec{B}$ is used to describe the magnetic force exerted on a moving charge by other moving charges. (This force acts in addition to the electric force described by a field $\vec{E}$.) The magnetic force on a charge q moving with velocity $\vec{v}$ is given by

$$F_m = qvB_\perp$$

where $B_\perp$ is the component of the vector $\vec{B}$ perpendicular to $\vec{v}$, and where the direction of the force $\vec{F}$, perpendicular to both $\vec{v}$ and $\vec{B}$, is given by a *right-hand rule*.
Applications of magnetic force are:

1 The *velocity selector:* $v = E/B$, for no net force on a moving charge, given proper directions of the fields.

2 *Circular orbits:* $R = mv/qB$, if $\vec{v}$ is perpendicular to $\vec{B}$.

Scattering experiments, using alpha-particle beams, can be interpreted in terms of the *nuclear model* of the atom:

1 The atom has a very small *positively charged nucleus* which contains most of its mass.
2 Negatively charged *electrons surround the nucleus,* giving the atom its size and leading to a net charge of zero for the atom.
3 The nuclear charge—and thus the number of electrons—determines the type of atom, i.e., which chemical element it is.

MOMENTUM AND ENERGY

5-1 CONSERVATION OF MOMENTUM

In the past two chapters we used Newton's second law, $\vec{F} = m\vec{a}$, to study the motion of individual particles under the action of various forces, and in the last chapter we considered a two-body collision between an alpha particle and a nucleus. However, we simplified the collision to a problem involving only the motion of the alpha particle by assuming that the nucleus remained fixed during the collision. Now we will consider two-body collisions in more general terms.

Suppose two particles are initially moving directly toward each other and that there is a repulsive force between them when they get close together. Figure 5-1-1 shows the particles at three different times, t_1, t_2, and t_3. At t_2, the forces arising from the repulsive interaction between the two particles are shown, while at t_1 and t_3 the particles are assumed to be so far apart that the force between them is negligible. Figure 5-1-1 may be described as showing a head-on collision, in which particles with initial velocities v_1 and v_2 collide and then rebound with final velocities v_1' ("v-one, prime") and v_2'.

Newton hypothesized that in the interaction between two bodies, as at time t_2 in our example, the force on one body is equal and opposite to the force on the other. He stated this as his *third,*

Figure 5-1-1 Two particles making a head-on collision. At times t_1 and t_3 the forces on them are negligible. At time t_2, when they are close to each other, the forces are shown. The velocities are not shown for this time but they need not be zero.

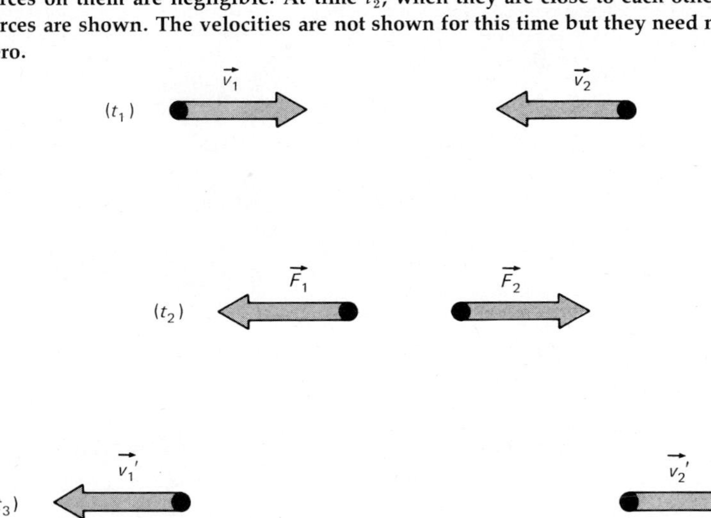

and final, *law of motion*: For every force there is an equal and opposite force. In the example shown in Fig. 5-1-1 the forces might arise from the repulsion of like charges on the two particles. Alternatively, the "particles" might be rubber balls, and the forces might arise when they touch and squash each other. In these cases, and many others, Newton's third law is experimentally verified. He assumed, as we will for the moment, that the law is true in general.

We now want to state the second law in a new way, expressing it in terms of a quantity called *momentum*. The momentum $\vec{p}$ of a particle is defined to be the product of its mass and velocity:

$$\vec{p} = m\vec{v} \tag{5-1-1}$$

Since the acceleration is by definition the rate of change of velocity, $\vec{a} = \Delta\vec{v}/\Delta t$, the second law may be written in the form

$$\vec{F} = m\vec{a} = \frac{m(\Delta\vec{v})}{\Delta t} \tag{5-1-2}$$

In the time interval Δt, the velocity changes by an amount $\Delta\vec{v}$, from $\vec{v}$ to $\vec{v} + \Delta\vec{v}$. In the time interval Δt, the momentum therefore changes from $m\vec{v}$ to $m(\vec{v} + \Delta\vec{v})$. The change in momentum, $\Delta\vec{p}$, is thus $m(\Delta\vec{v})$. The second law, as expressed in Eq. 5-1-2, can now be written in the form

$$\vec{F} = m\vec{a} = \frac{\Delta\vec{p}}{\Delta t} \tag{5-1-3}$$

Thus, according to the second law, the net force on a particle is equal to the *rate of change of its momentum*. Note that momentum is a vector, which points in the direction of $\vec{v}$, and with a magnitude equal to m times the magnitude of $\vec{v}$. Suppose we now write Eq. 5-1-3 for particles 1 and 2 at time t_2,

$$\vec{F}_1 = \frac{\Delta\vec{p}_1}{\Delta t} \qquad \vec{F}_2 = \frac{\Delta\vec{p}_2}{\Delta t} \tag{5-1-4}$$

According to the third law, $\vec{F}_1 = -\vec{F}_2$, so that Eq. 5-1-4 leads to the result

$$\frac{\Delta\vec{p}_1}{\Delta t} = -\frac{\Delta\vec{p}_2}{\Delta t} \tag{5-1-5}$$

This equation states that during any interval of time Δt, the interaction between the two particles leads to equal and opposite changes in their momentum

$$\Delta \vec{p}_1 = -\Delta \vec{p}_2 \qquad \text{5-1-6}$$

Since the change in $\vec{p}_1$ is equal and opposite to the change in $\vec{p}_2$, the sum $\vec{p}_1 + \vec{p}_2$ does not change at all, so that

$$\vec{p}_1{}' + \vec{p}_2{}' = \vec{p}_1 + \vec{p}_2 \qquad \text{5-1-7}$$

This result is known as the law of *conservation of momentum.* The total momentum before the collision is equal to the total momentum after the collision, or total momentum is "conserved."

PROBLEM 1

At $t = 0$, a particle of mass 2 g is shot horizontally at a speed of 50 m/sec. It is free to fall under the action of gravity. Find the horizontal and vertical components of its momentum at $t = 2$ sec.

PROBLEM 2

A particle moves along the x axis so its momentum varies with time as shown in Fig. P-5-2. Plot a graph of the force vs. time.

The importance of a conservation law such as this one is that it can give information about the result of an interaction even if the details are unknown. If two particles interacted by means of the "strong nuclear force," it would not be possible to write down the force between them when they were some given distance apart, since the force law for nuclear forces is not known. How-

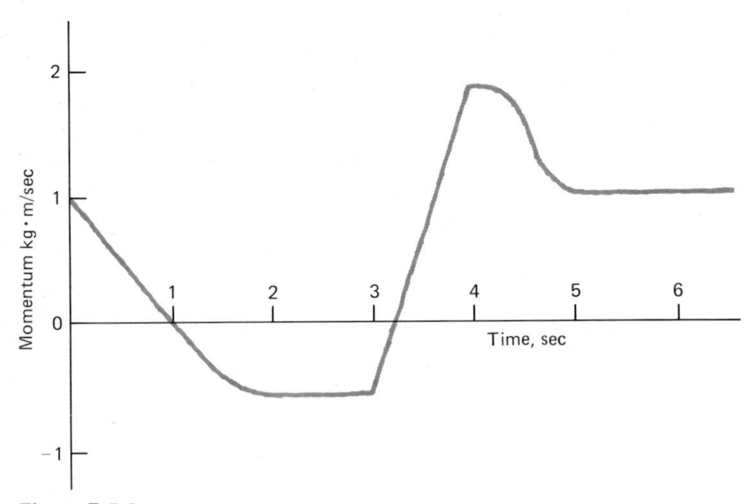

Figure P-5-2

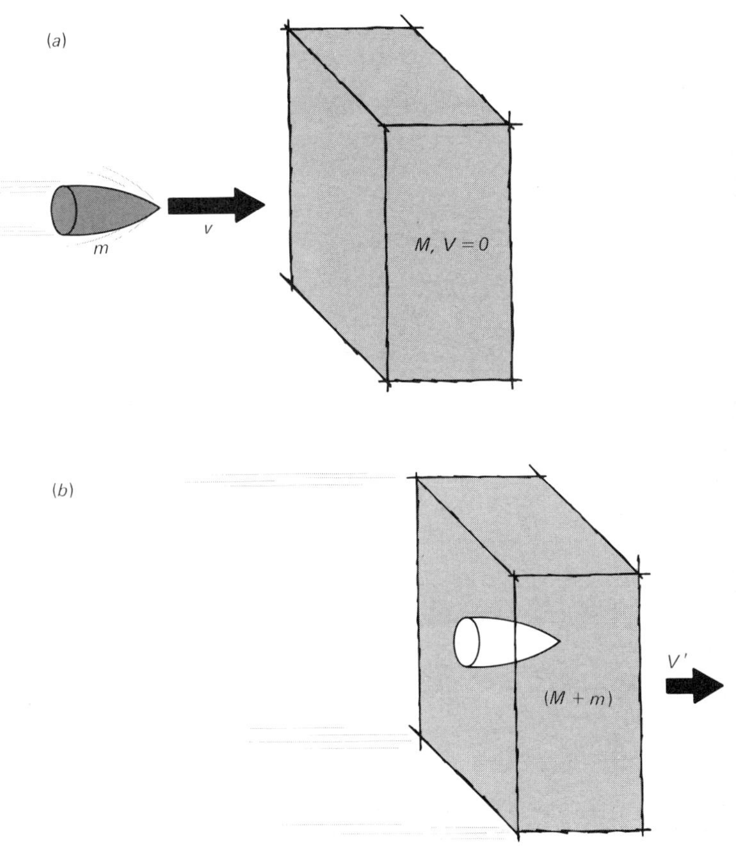

Figure 5-1-2 Collision of a bullet with a stationary block. After the bullet hits the block they form a single mass, moving with velocity V'.

ever, Eq. 5-1-7, the law of conservation of momentum, still supplies an extremely useful relation between the momenta before and after the collision.

To illustrate the use of momentum conservation in a simple example, consider the interaction of a bullet and a wooden block, shown schematically in Fig. 5-1-2. The total momentum before the bullet strikes the block is simply given by the momentum of the bullet:

$$p_i = mv \tag{5-1-8}$$

where the subscript i stands for "initial." The block is stationary, with $V = 0$, so its momentum is initially zero. After the collision, the bullet and block are joined, forming a mass $(M + m)$ which moves with velocity V'. The momentum after the collision is thus

$$p_f = (M + m)V' \qquad 5\text{-}1\text{-}9$$

where the subscript f stands for "final."

The interaction which takes place when the bullet strikes the block is very complicated. However, by using the law of momentum conservation, $p_i = p_f$, we can find the final velocity after the collision, assuming m, M, and v are known. From Eq. 5-1-8 and 5-1-9, plus momentum conservation,

$$mv = (M + m)V' \qquad 5\text{-}1\text{-}10$$

Therefore, rearranging Eq. 5-1-10,

$$V' = \frac{mv}{M + m} \qquad 5\text{-}1\text{-}11$$

The law of conservation of momentum can be generalized to include any number of interacting particles. The result will be given now, without proof. Suppose, as in Fig. 5-1-3, there are a number of particles in a region of space such as that bounded by the dashed line. The particles may exert forces on each other, but if no net force is exerted on the particles from outside the dashed line, then their total momentum is constant. The total momentum $\vec{P}$ for n particles is the vector sum of the individual momenta,

Figure 5-1-3 A number of particles in a region of space. Their total momentum is constant if no forces act from outside the dashed line.

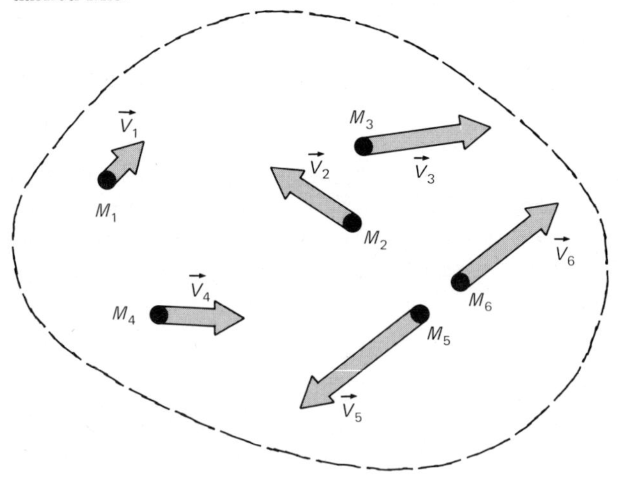

$$\vec{P} = m_1\vec{v}_1 + m_2\vec{v}_2 + \cdots + m_n\vec{v}_n \qquad 5\text{-}1\text{-}12$$

The second law can be used to relate the rate of change of $\vec{P}$ to the net force exerted from outside, $\vec{F}_{\text{ext}}$, where "ext" stands for external:

$$\vec{F}_{\text{ext}} = \frac{\Delta\vec{P}}{\Delta t} \qquad 5\text{-}1\text{-}13$$

If $\vec{F}_{\text{ext}}$ is zero, $\Delta\vec{P}/\Delta t$ is zero; hence $\vec{P}$ is constant, and the total momentum is conserved. Equation 5-1-13 can be shown to be correct even if the individual forces between particles do not obey the third law, and even if effects arising from the special theory of relativity are included. As far as we now know, the law of conservation of momentum is always valid.

PROBLEM 3

A wooden block is free to slide on a smooth table. It is struck by a bullet traveling horizontally, which stops in the block. After the bullet hits it, the block has a velocity of 10 m/sec. The mass of the bullet is 4.0 g, and the mass of the block is 0.5 kg. How fast was the bullet traveling when it hit the block?

PROBLEM 4

A radioactive atom is at rest just before it emits an alpha particle with velocity 1.6×10^7 m/sec. Take the atomic weight of the alpha particle to be 4.0 and of the original atom to be 226. After the alpha particle is emitted, assume the rest of the atom remains intact. If it is free to move, what is its velocity?

5-2 ROCKET PROPULSION

Rocket propulsion provides an interesting example of the application of two conservation laws, conservation of momentum and *conservation of mass.* The law of conservation of mass states that the total mass of a system of particles remains constant, regardless of the interactions between them, or of any forces exerted on them from outside. For example, when carbon combines with oxygen, it "burns" to form a gas, carbon dioxide. According to the law of conservation of mass, the mass of the carbon dioxide which is formed equals the combined masses of the carbon and oxygen which combined in the burning.

The law of conservation of mass was formulated mainly as a result of chemical evidence, which showed that when elements

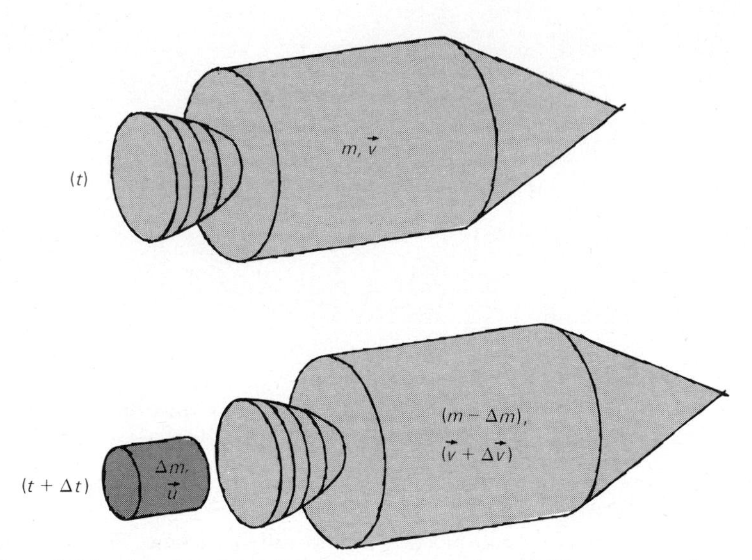

Figure 5-2-1 A rocket at two times a short interval apart. The masses and velocities are shown by labels, and the small mass with velocity u is the fuel ejected during the time interval.

combined to form compounds, or compounds were decomposed into elements, no net mass either appeared or disappeared. In fact, this conservation law is now known to be an approximation which is very good for large objects, and even for atoms in chemical reactions. In this realm, mass is conserved to an accuracy of better than one part in 10^8, or to a millionth of a percent. For nuclear physics, the conservation law must be corrected, but we will not need to deal with the corrections now.

Figure 5-2-1 shows a rocket at two closely spaced times, t and $t + \Delta t$. During the time interval Δt, a mass of fuel Δm has been burned. The fuel consists of at least two chemicals, and they react, or burn, without involving any material from outside the rocket. The chemical reaction produces hot gases which can be channeled through a properly designed nozzle so that they leave the rocket at high speed. According to the law of conservation of mass, the mass of the ejected gases is equal to Δm, the mass of fuel burned.

We will assume that when the rocket engine is operating, the gases leave with a constant speed w *with respect to the rocket.* The relation between $\vec{u}$, $\vec{v}$, and $\vec{w}$ is shown in Fig. 5-2-2, and the vector equation relating them is

$$\vec{u} = \vec{v} + \vec{w} \qquad 5\text{-}2\text{-}1$$

Notice that $\vec{u}$ may be positive or negative, depending upon whether v is greater or less than w. Figure 5-2-2 is drawn as if v were greater than w, which is not true when the rocket is just starting its flight, but is often true when the rocket is in motion.

Now, we will use the law of conservation of momentum to find the rocket thrust force. First, equate the total momentum of rocket plus fuel at time t to the total momentum at time $t + \Delta t$, with Δm the mass of fuel expelled in the interval Δt:

$$m\vec{v} = (m - \Delta m)(\vec{v} + \Delta\vec{v}) + (\Delta m)\vec{u} \qquad \text{5-2-3}$$

Multiplying out the quantities in parentheses, this equation becomes

$$m\vec{v} = m\vec{v} + m(\Delta\vec{v}) - (\Delta m)\vec{v} - (\Delta m)(\Delta\vec{v}) + (\Delta m)\vec{u} \qquad \text{5-2-4}$$

The terms $m\vec{v}$ on both sides cancel, and Eq. 5-2-4 can be rearranged in the form

$$m(\Delta\vec{v}) = (\Delta m)(\vec{v} - \vec{u} + \Delta\vec{v}) \qquad \text{5-2-5}$$

From Eq. 5-2-1, $\vec{v} - \vec{u} = -\vec{w}$, which may be used to simplify Eq. 5-2-5, leading to the equation

$$m(\Delta\vec{v}) = (\Delta m)(-\vec{w} + \Delta\vec{v}) \qquad \text{5-2-6}$$

If Δt is a very short time interval, $\Delta\vec{v}$ is negligible in comparison with $\vec{w}$. Then, $\Delta\vec{v}$ can be dropped from the right-hand side of Eq. 5-2-6, where the sum $(-\vec{w} + \Delta\vec{v})$ occurs. However, $\Delta\vec{v}$ appears alone on the left-hand side and must be kept. Equation 5-2-6 then may be written

$$m(\Delta\vec{v}) = -\vec{w}(\Delta m) \qquad \text{5-2-7}$$

According to the second law the force accelerating the rocket is equal to $m\vec{a}$, or $m(\Delta\vec{v}/\Delta t)$. Dividing both sides of Eq. 5-2-7 by Δt,

$$\vec{F} = m\left(\frac{\Delta\vec{v}}{\Delta t}\right) = -\vec{w}\left(\frac{\Delta m}{\Delta t}\right) \qquad \text{5-2-8}$$

Figure 5-2-2 Vector addition of the rocket velocity v and the exhaust velocity w, to give the velocity u of the ejected mass.

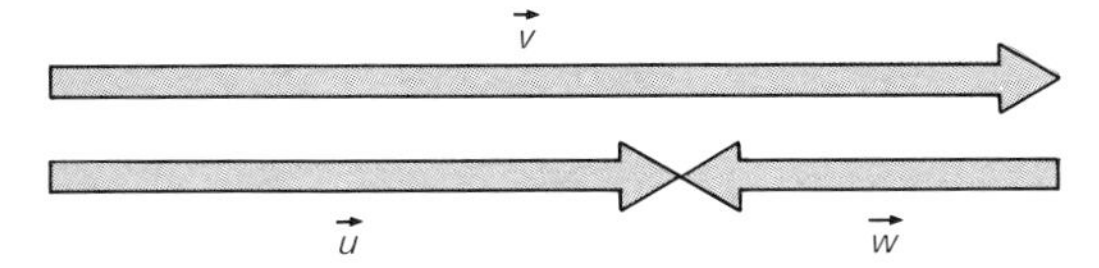

Equation 5-2-8 gives the final expression for the thrust force which accelerates a rocket. The direction of the force is opposite to the direction of $\vec{w}$, and its magnitude is proportional to the product of w and the rate at which fuel is burned ($\Delta m/\Delta t$).

PROBLEM 5

A jet engine takes in air and, by burning fuel, heats it and ejects it at high velocity with respect to the engine. If the exhaust velocity of a particular engine is 1,000 ft/sec and its thrust is 20,000 lb, what is the mass of air ejected per second? (You can neglect the mass of fuel, in contrast to a rocket engine.)

5-3 ROCKET MOTION

For real rockets it is often a good approximation to take w and $\Delta m/\Delta t$ to be constant during the entire time when the engine is on. From Eq. 5-2-8, this leads to a constant thrust force, but note that the acceleration of the rocket is *not* constant. The acceleration is given by $a = F/m$, and m is the total mass of rocket plus fuel. For an interesting rocket, capable of reaching a high enough speed to launch a satellite, almost all the mass when it starts its flight consists of fuel. Thus, as the flight progresses the mass decreases, and at the time when the fuel is all consumed the mass is a small fraction of the original mass. The acceleration continually increases as the mass decreases.

Suppose we want to know the final speed of a rocket which starts from rest with mass M_i and has mass M_f at "burnout" when all the fuel has been used. (Subscripts i and f stand for initial and final.) Assume the rocket travels in a straight line vertically, and that the gravitational force can be neglected in comparison with the thrust force. Even such a simple one-dimensional problem in nonuniformly accelerated motion is mathematically too difficult to handle here. However, we can make an approximate analysis by using a method in which the motion of the rocket is broken up into successive time intervals, and by assuming that during each interval the mass has some constant average value. A great majority of the problems physicists try to solve cannot be handled in a rigorously correct mathematical fashion. The artful use of approximations is therefore often very important for doing good physics.

Table 5-3-1 shows our approximate solution for the ratio v/w at various times, where v is the rocket velocity and w is the en-

Table 5-3-1 A stepwise calculation of rocket motion

TIME	MASS	m_{av}	$\Delta m/m_{av}$, $= \Delta v/w$	v/w
t_0	M_i			0
		$(3/4)M_i$	2/3	
t_1	$M_i/2$			2/3
		$(3/8)M_i$	2/3	
t_2	$M_i/4$			4/3
		$(3/16)M_i$	2/3	
t_3	$M_i/8$			6/3
		$(3/32)M_i$	2/3	
t_4	$M_i/16$			8/3

gine exhaust velocity. At time t_0 the rocket is ignited with $v = 0$. Then, successive times t_1, t_2, t_3, etc., have been chosen, such that at each time the mass of the rocket plus fuel is *half* the mass at the preceding time. The first two columns of Table 5-3-1 list the times and masses.

In order to compute the motion of the rocket most conveniently, we have rearranged Eq. 5-2-7 in the form

$$\frac{\Delta v}{w} = \frac{\Delta m}{m} \qquad 5\text{-}3\text{-}1$$

Equation 5-3-1 relates the magnitudes of the vectors in Eq. 5-2-7. For each time interval, $\Delta v/w$ will be given by $\Delta m/m$. An average mass can be used for m, calculated by taking half the sum of the mass at the beginning of the interval and the mass at the end of the interval. The third column of Table 5-3-1 lists the average mass m_{av}, and the fourth column lists the ratio $\Delta m/m_{av}$, equal to $\Delta v/w$.

In the last column of Table 5-3-1 the ratio v/w is given at each time, calculated by adding up the changes $\Delta v/w$ for all the intervals preceding. Note that the value of $\Delta v/w$ is 2/3 for every time interval. This is because in each interval there is the same fractional change in mass $\Delta m/m$. Picking the times so this would be true made the table very simple.

Actually, the simple form of Table 5-3-1 enables us to immediately guess at a general formula which gives v/w for any time at which the mass is $(1/2)^n$ times the initial mass,

$$\frac{v}{w} = \frac{2}{3}n \qquad 5\text{-}3\text{-}2$$

For example, at t_4 the mass is 1/16, or $(1/2)^4$, of M_i. For $n = 4$, Eq. 5-3-2 gives the result $v/w = 8/3 = 2.67$, as found in the table. The formula works for all the times in the table. It is possible with

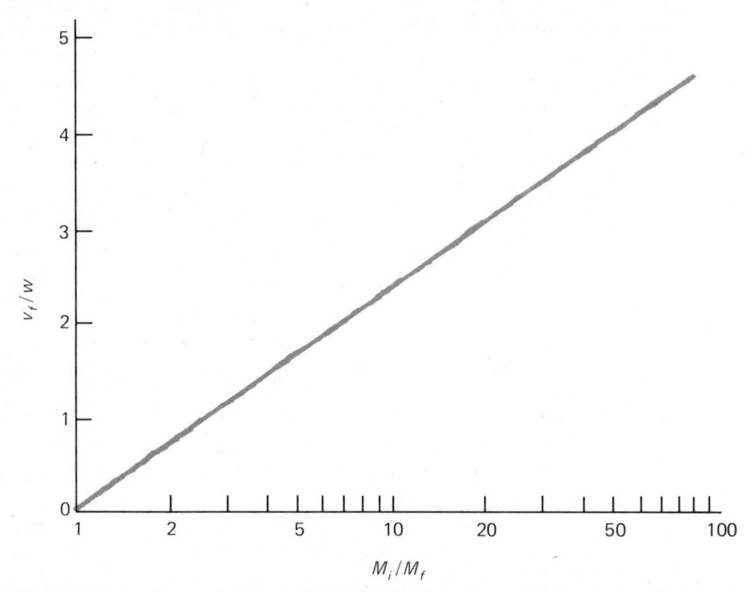

Figure 5-3-1 A graph from which the final rocket velocity, v_f, can be found for given values of M_i/M_f and w. Points calculated from Eq. 5-3-2 fall on a straight line on this semilogarithmic graph, and the line has been drawn in.

the aid of calculus to find exactly the final rocket velocity in the problem we've been considering. The result corresponds to Eq. 5-3-2 with the factor 2/3 replaced by 0.691, which is not very different!

Equation 5-3-2 can be used to find v_f, the velocity at burnout, for various values of M_i/M_f, the ratio of the initial mass to the mass at burnout. Figure 5-3-1 shows a graph of v_f/w vs. M_i/M_f in which the scale for M_i/M_f is nonuniform in the same way that the A, B, C, and D scales on a slide rule are nonuniform. The relation is a straight line on this kind of graph (called a semilogarithmic graph).

Suppose v_f is taken to be 8.0 km/sec, which corresponds to the velocity of a satellite in orbit near the earth, and suppose w is taken to be 2.0 km/sec. This value of w corresponds to the exhaust velocity achieved by the engines of the German V2 rocket, built near the end of World War II. For these values, $v_f/w = 4.0$, and, from Fig. 5-3-1, M_i/M_f is about 50. Therefore, in order for such a rocket to be launched into orbit, it would have to carry a mass of fuel at least 50 times the mass of the empty rocket. Even with the most modern engineering and materials such a requirement cannot be met.

In the estimate made above, no allowance was made for the added fuel that would be necessary to turn the rocket, so it was moving horizontally at 8 km/sec, as a satellite moves. Furthermore, the simple theory assumed that the rocket was accelerated by the thrust force only, ignoring the opposing gravitational force. Lastly, we have neglected air resistance, which may be important during the early part of the rocket's flight. Thus, the required ratio M_i/M_f would have to be even greater than was estimated.

In the past 20 years, improvements in rocketry have made it possible to launch large satellites and to achieve velocites significantly greater than 8 km/sec. While there has been some increase in w, the engine exhaust velocity, the major factor has been the use of multistage rockets. With several stages, each with a moderate mass ratio, it is possible to accelerate a small fraction of the total initial mass to the required velocity. The Apollo/Saturn V rocket combination can put about 4 percent of its initial mass into earth orbit, utilizing two stages and part of the fuel of its third stage. Using three stages, about 1.5 percent of its initial mass can be sent to the moon.

PROBLEM 6

A rocket is moving upward at 8 km/sec, with momentum $\vec{p}$. Show with a vector diagram the momentum which would be needed to turn it so it is moving horizontally at 8 km/sec. Do you think satellites are launched this way? Can you think of a more efficient scheme?

PROBLEM 7

The specifications of a two-stage rocket are as follows:

Booster:	$M_i = 10^5$ kg	$M_f = 10^4$ kg
Second stage:	$M_i = 10^4$ kg	$M_f = 10^3$ kg

The engines of both stages have exhaust velocities of 3 km/sec. With the rocket at rest, the booster engine is started and accelerates the booster plus second stage until its fuel is used up. Then the booster and second stage are disconnected, and the second stage engine is started. Assume the flight proceeds in a straight line and neglect the effects of gravity and air resistance. Find the velocity when the second stage engine is started, and the final velocity of the second stage.

5-4 KINETIC ENERGY

The *kinetic energy* of a particle of mass m moving with velocity $\vec{v}$ is defined by the following relation, where T stands for kinetic energy:

$$T = \frac{1}{2}mv^2 \qquad \text{5-4-1}$$

The kinetic energy is the first of several forms of energy which we will discuss. The word kinetic comes from the Greek word for motion, and kinetic energy is "energy of motion." It is a *scalar* quantity, and v in Eq. 5-4-1 stands for the magnitude of the velocity.

From the units of the quantities on the right-hand side of Eq. 5-4-1, the unit of T is kg-m^2-sec^{-2}. Because T is so important, and this unit is so cumbersome, it has been given a special name, the *joule* (pronounced as if it were spelled "jewel"). James Joule was an English physicist who, in the early nineteenth century, did important experiments involving energy. The joule is the unit for any form of energy.

PROBLEM 8

A particle of mass m moves in the xy plane with velocity v. Prove that the kinetic energy $\frac{1}{2}mv^2$ is equal to $\frac{1}{2}mv_x^2 + \frac{1}{2}mv_y^2$ where v_x and v_y are the x and y components of v.

Figure P-5-9

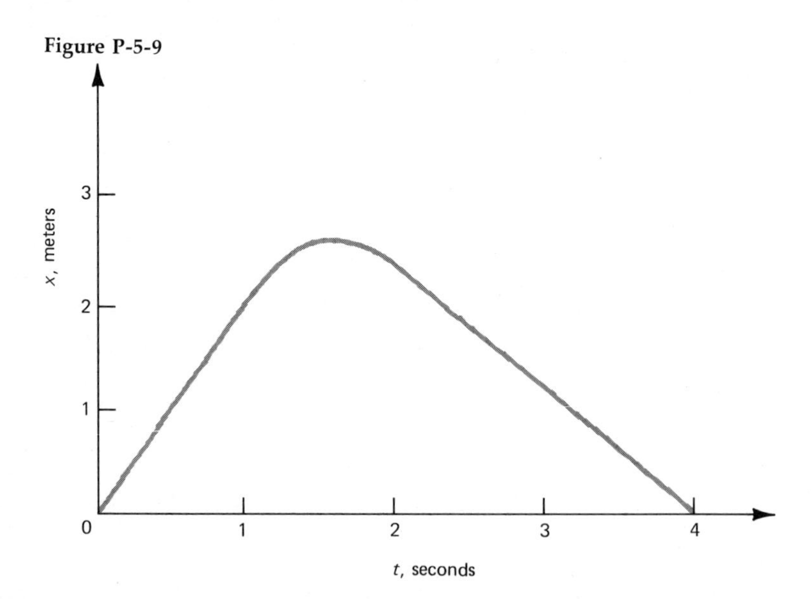

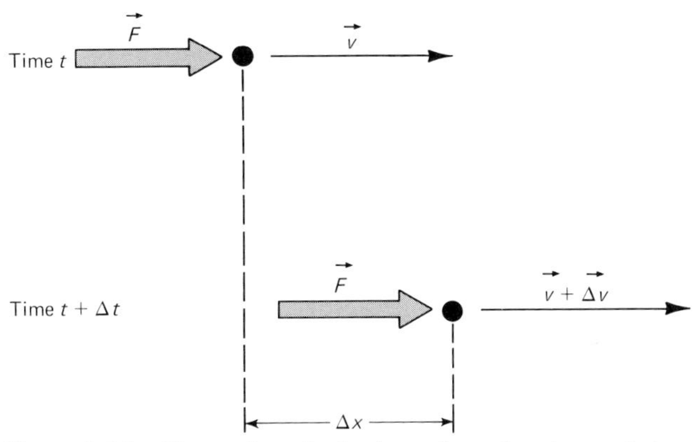

Figure 5-4-1 Change in velocity in a short time interval, for a particle acted upon by a force along its direction of motion.

PROBLEM 9

A particle of mass 2 g moves along the x axis. The curve in Fig. P-5-9 shows x vs. t. Find curves of v vs. t and T vs. t.

Kinetic energy is important because it is involved in a powerful conservation law. To begin to understand the significance of T, consider a particle moving in the $+x$ direction, acted upon by a force $\vec{F}$ along the direction of motion. Figure 5-4-1 shows the particle at time t and a short time later at $(t + \Delta t)$. During the time interval Δt, the velocity changes from v to $v + \Delta v$. While the force F may depend on time, or upon the position of the particle, we will assume Δt so small that F may be assumed constant during that interval.

We will now examine the change in T during the interval Δt and show that it is equal to the quantity $F(\Delta x)$. From the definition of T,

$$\Delta T = \frac{1}{2}m(v + \Delta v)^2 - \frac{1}{2}mv^2 \qquad \text{5-4-2}$$

Multiplying out the term $\frac{1}{2}m(v + \Delta v)^2$, this equation becomes

$$\Delta T = \frac{1}{2}mv^2 + mv(\Delta v) + \frac{1}{2}m(\Delta v)^2 - \frac{1}{2}mv^2 \qquad \text{5-4-3}$$

Notice that the terms involving $\frac{1}{2}mv^2$ cancel, so that Eq. 5-4-3 can be written

$$\Delta T = mv(\Delta v) + \frac{1}{2}m(\Delta v)^2 = m(\Delta v)\left(v + \frac{\Delta v}{2}\right) \qquad \text{5-4-4}$$

For a short time interval, Δv is much less than v and can be neglected in comparison with v in the parentheses on the right-

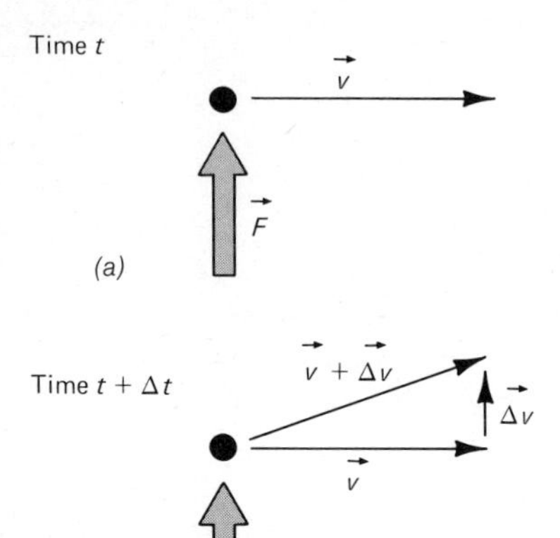

Figure 5-4-2 Change in velocity in a short time interval, for a particle acted upon by a force perpendicular to its direction of motion.

hand side of Eq. 5-4-4. The result for ΔT is then

$$\Delta T = mv(\Delta v) \qquad 5\text{-}4\text{-}5$$

From Newton's second law, $F = m(\Delta v/\Delta t)$, which can be rearranged in the form

$$F(\Delta t) = m(\Delta v) \qquad 5\text{-}4\text{-}6$$

The quantity Δt can be eliminated from this equation by using the definition of v, $v = \Delta x/\Delta t$. Therefore, $\Delta t = \Delta x/v$, and substituting this value for Δt in Eq. 5-4-6 leads to

$$F\left(\frac{\Delta x}{v}\right) = m(\Delta v) \qquad 5\text{-}4\text{-}7$$

Equation 5-4-3 can be rearranged in the form

$$F(\Delta x) = mv(\Delta v) \qquad 5\text{-}4\text{-}8$$

Comparing Eq. 5-4-8 with Eq. 5-4-5, we can see that $F(\Delta x)$ and ΔT are both equal to $mv(\Delta v)$. Therefore,

$$F(\Delta x) = \Delta T \qquad 5\text{-}4\text{-}9$$

Equation 5-4-9 relates the change in kinetic energy to the product of the force acting on the particle and the distance the particle has moved. It was, however, calculated for the special case in which $\vec{F}$ was in the same direction as $\vec{v}$. Now, suppose $\vec{F}$ is not in the direction of $\vec{v}$. We resolve $\vec{F}$ into two components, one along $\vec{v}$, as in the case we have considered, and one perpendicular to $\vec{v}$, as in Fig. 5-4-2. We will consider the effect of each component separately.

A force perpendicular to $\vec{v}$, acting for a short time interval, produces a change $\Delta\vec{v}$ at right angles to $\vec{v}$, as in Fig. 5-4-2*b*. Notice that the magnitude of $\vec{v}$ remains the same if $\Delta\vec{v}$ is small, as it will be for the short time interval we are considering. Thus, the component of $\vec{F}$ at a right angle to $\vec{v}$, called $F_\perp$ ("F perpendicular"), does not produce any change in kinetic energy. The component along $\vec{v}$, $F_{||}$ ("F parallel"), does change T, as in the discussion which led to Eq. 5-4-9. Therefore, that equation can be generalized to include all directions of $\vec{F}$ by simply replacing F by $F_{||}$:

$$F_{||}(\Delta x) = \Delta T \qquad 5\text{-}4\text{-}10$$

Finally, we can rewrite Eq. 5-4-10 for a particle moving in any direction instead of along the x axis as assumed so far. Let Δs stand for a short distance traversed by the particle, and let $F_{||}$ be the component of force *along the direction of motion*. Then,

$$F_{||}(\Delta s) = \Delta T \qquad 5\text{-}4\text{-}11$$

For example, Fig. 5-4-3 shows a curved path from point A to point B, along which a particle is assumed to move. At two places along the path a short distance Δs has been marked off. For each Δs an assumed force $\vec{F}$ has been shown and $F_{||}$ has been drawn. For either distance Δs, Eq. 5-4-11 can be used to find the change in kinetic energy, ΔT.

Figure 5-4-3 A path from point A to point B, along which a particle moves. For two short distances, $(\Delta s)_1$ and $(\Delta s)_2$, along the path, the forces on the particle, F_1 and F_2, are shown. For each force, the component $F_{||}$ along the direction of motion, is shown.

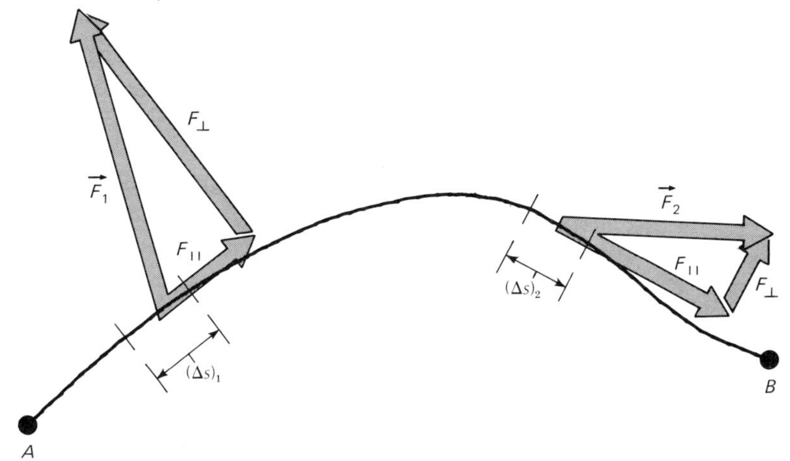

The quantity $F_{||}(\Delta s)$, which appears in Eq. 5-4-11, is called the *work* W done by the force on the particle in the distance Δs. The *total work* done on the particle when it moves from A to B, called $W_{A\to B}$, can be found by breaking up the path from A to B into a large number of successive small pieces, $(\Delta s)_1$, $(\Delta s)_2$, $(\Delta s)_3$, . . . , $(\Delta s)_n$. The total work done is equal to the sum of all the little amounts for each small part of the path,

$$\begin{aligned} W_{A\to B} &= F_{||1}(\Delta s)_1 + F_{||2}(\Delta s)_2 + \cdots + F_{||n}(\Delta s)_n \\ &= \Delta W_1 + \Delta W_2 + \cdots + \Delta W_n \end{aligned} \qquad 5\text{-}4\text{-}12$$

Along each small piece of path the change in kinetic energy ΔT is equal to the work done by the force, ΔW. Thus, the *total* change in kinetic energy of the particle as it moves from A to B is equal to the *total* work done,

$$W_{A\to B} = T_B - T_A = \frac{1}{2}mv_B^{\ 2} - \frac{1}{2}mv_A^{\ 2} \qquad 5\text{-}4\text{-}13$$

This is an important relation in its own right, and it constitutes the first step toward the law of conservation of energy.

Note that in physics the term *work* is used in a much more restricted way than in everyday language. If we are trying to start a car by pushing it, we may "work hard," and we are also doing work in the way symbolized by the letter W, provided the car is *moving*. We are *exerting a force on a body along its direction of motion.* However, in many everyday uses of the word *work*, the physical quantity W is zero. For example, holding up a heavy suitcase, or carrying it on level ground, may be "hard work" but in either case $W = 0$. Why?

PROBLEM 10

Explain why the magnetic force, discussed in the previous chapter, can do no work on a moving charge. From this result, prove that a particle moving solely under the influence of a magnetic force has a constant speed.

PROBLEM 11

A curve of F_x vs. x is shown in Fig. P-5-11, where F_x is the x component of the force on a particle. A particle moves from $x = 0$ to $x = 2$ m. How much work is done on the particle by the force? How much work would be done if the particle moved from $x = 0$ to $x = 3$ m?

As a simple specific example of how Eq. 5-4-13 can be used, we

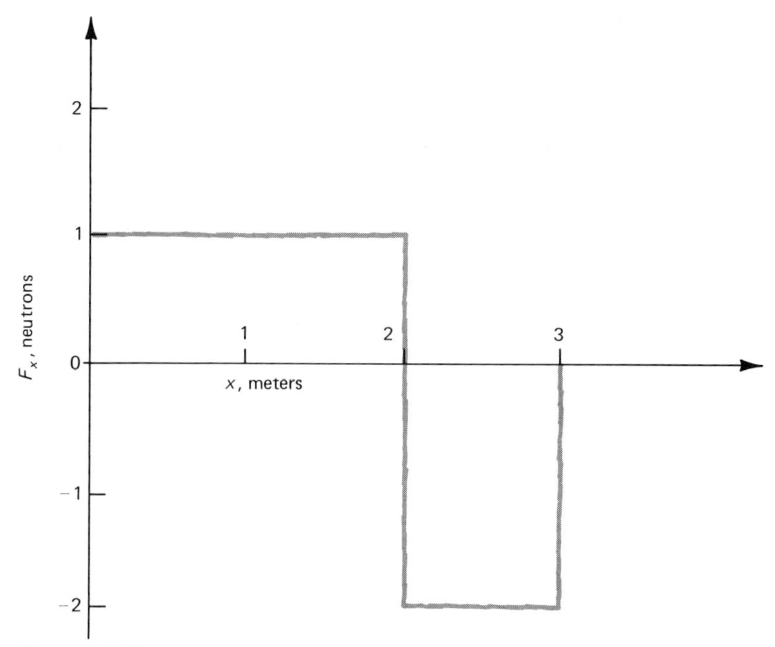

Figure P-5-11

can find the speed with which a particle dropped from a height h hits the ground. In Fig. 5-4-4, a particle of mass m is shown falling from point A to point B under the action of the gravitational force $F = mg$. For any small piece of the path (Δs), the force is the same, mg, *directed along the direction of motion*. So the work done by the gravitational force on the particle when it moves the distance Δs is $mg(\Delta s)$. The total work done by gravity when the particle moves from A to B is given by

$$W_{A \to B} = mg(\Delta s)_1 + mg(\Delta s)_2 + \cdots + mg(\Delta s)_n \tag{5-4-14}$$

where the equation is written as if the path were broken up into n pieces, $(\Delta s)_1$, $(\Delta s)_2$, $(\Delta s)_3$, . . . , $(\Delta s)_n$. For every piece, $F_{||}$ is the same so that the right-hand side of Eq. 5-4-14 is easy to add up,

$$W_{A \to B} = mg[(\Delta s)_1 + (\Delta s)_2 + \cdots + (\Delta s)_n] = mgh \tag{5-4-15}$$

Since $W_{A \to B}$ is equal to the change in kinetic energy,

$$T_B - T_A = \frac{1}{2}mv_B^2 - \frac{1}{2}mv_A^2 = mgh \tag{5-4-16}$$

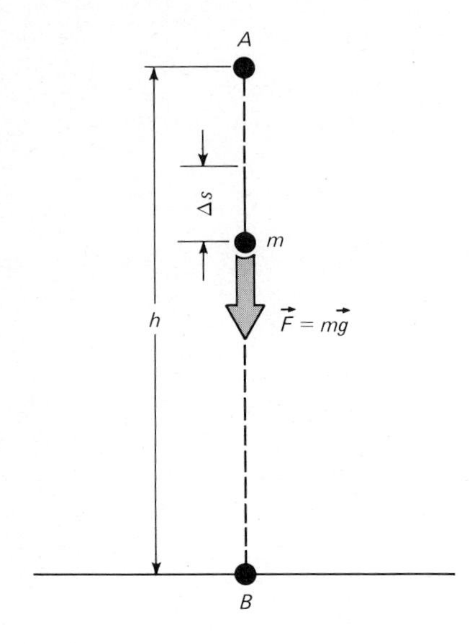

Figure 5-4-4 A particle falling a height h, from point A to point B, under gravity.

If the particle is dropped from rest, so that $v_A = 0$,

$$\frac{1}{2}mv_B^2 = mgh \qquad \text{5-4-17}$$

This equation may now be solved to give the answer for v_B,

$$v_B = \sqrt{2gh} \qquad \text{5-4-18}$$

You can check that this is the same result that can be obtained, with somewhat more effort, by using the formulas for uniformly accelerated motion given in Chap. 3.

PROBLEM 12

From the equations for uniformly accelerated motion, find the result given by Eq. 5-4-18.

PROBLEM 13

A charge $+q$ is released from rest in a uniform electric field E_0. It travels a distance d. Find the work done on the particle by the electric force, assuming the particle moves along the direction of $\vec{E}$. Find its kinetic energy and momentum after it has traveled the distance d. Suppose the particle is released with initial velocity $\vec{v}_0$. Find the kinetic energy after it has traveled the distance d, if $\vec{v}_0$ is along the direction of the field, or perpendicular to it.

5-5 POTENTIAL ENERGY

At the end of the last section we solved a free-fall problem by utilizing the concepts of work and kinetic energy. The basic relation was

$$W_{A\rightarrow B} = T_B - T_A \qquad 5\text{-}5\text{-}1$$

In Fig. 5-5-1 the free-fall example is redrawn to emphasize the essentials. There is a uniform gravitational field $\vec{g}$, which produces a force $m\vec{g}$. In the example the particle moved from A to B, and its change in kinetic energy was found by calculating the work done on the particle by this gravitational force.

Another way of describing what happens when the particle moves from A to B is to say that at point A it has potential energy V_A and at point B it has potential energy V_B. In moving from A to B some of the particle's potential energy is said to be converted into kinetic energy. The idea is to define the potential energy V of the particle at any point, such that when a particle moves from one point to another the increase in its kinetic energy is precisely equal to the decrease in its potential energy. In a way, potential energy is like money in the bank, and kinetic energy is like cash in hand.

If potential energy is defined as described above, then

$$V_A - V_B = T_B - T_A \qquad 5\text{-}5\text{-}2$$

Equation 5-5-2 is a mathematical statement of what has been said in words. The decrease in V is equated to the increase in T. Comparing Eq. 5-5-2 with Eq. 5-5-1, the decrease in V is also equal to the work done by the force on the particle,

$$V_A - V_B = W_{A\rightarrow B} \qquad 5\text{-}5\text{-}3$$

If we can write a correct formula for V at any point in space, we will be able to easily find the difference in kinetic energy for a particle moving between any two points. For the example shown in Fig. 5-5-1, there is such a formula:

$$V = mgy \quad \text{joules} \qquad 5\text{-}5\text{-}4$$

The unit of V, like the unit of T, is the joule. The coordinate y is to be measured upward, as in the coordinate system drawn in Fig. 5-5-1. Equation 5-5-4 is given without proof; we will justify it by testing whether it works.

According to Eq. 5-5-4, the potential energies at points A and B of Fig. 5-5-1 are

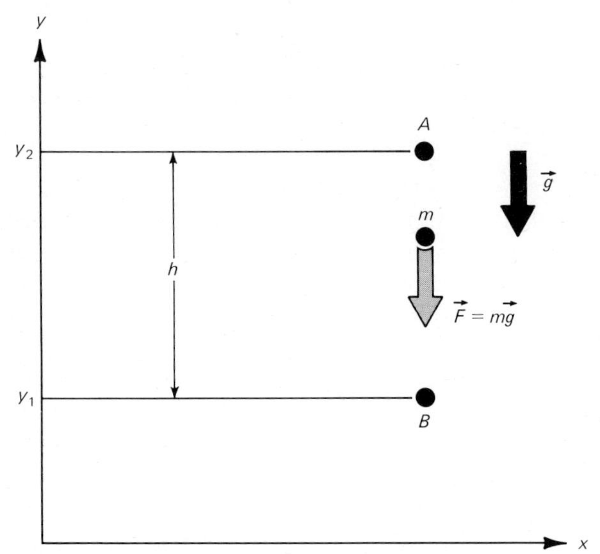

Figure 5-5-1 A particle of mass m falling freely in a uniform gravitational field. In the coordinate system shown, point A is at $y = y_2$, and point B is at $y = y_1$.

$$V_A = mgy_2$$
$$V_B = mgy_1 \qquad 5\text{-}5\text{-}5$$

For these values,

$$V_A - V_B = mg(y_2 - y_1) = mgh \qquad 5\text{-}5\text{-}6$$

which is indeed the result for $W_{A \to B}$ found in the previous section.

Since $V_A = mgy_2$, its value depends upon the location of the origin of our coordinate system. If the height of the origin is changed, the value of y_2 will change. However, notice that moving the origin must change y_2 and y_1 by the same amount, so that $y_2 - y_1$ remains equal to h. Therefore, the *difference* $(V_A - V_B)$, which matters in Eq. 5-5-6 or Eq. 5-5-3, does not depend upon the location of the origin of the coordinate system. A physical result, such as a change in T, never depends upon the actual value of a potential energy V, but only upon the *difference* in V between two points. In this sense, the analogy between V and money in the bank fails.

The formula for V given in Eq. 5-5-4 correctly gives $W_{A \to B}$ for the simple example we've been discussing. Equations 5-5-2 and

5-5-4 can thus be used for this example, but we shall now see that these equations provide a powerful method for solving a wider variety of problems. Figure 5-5-2 again shows the points A and B in a uniform gravitational field, with three different paths from A to B. Suppose a particle of mass m goes from A to B along the path ACB. According to Eq. 5-5-4 the potential energy at C is the same as at B (points B and C are both at $y = y_1$). If this formula applies to motion along the path AC, it therefore predicts that $W_{A \to C} = W_{A \to B} = mgh$. It is easy to verify that this is true, and it is left as a problem.

PROBLEM 14

Make a direct calculation of the work done by gravity on a particle of mass m moving along the line AC in Fig. 5-5-2. Prove in this way that $W_{A \to C}$ is equal to mgh.

Along the line BC the gravitational force is always perpendicular to the direction of motion. Thus, $F_{||} = 0$ everywhere along this line, and from the definition of work, $W_{C \to B} = 0$. This result is also given by Eqs. 5-5-3 and 5-5-4. Thus, for motion along the indirect path ACB the work done by the gravitational force is the same as along the direct path AB.

Is the work done along *any* path from A to B, like the path ADB,

Figure 5-5-2 Three paths from A to B in a uniform gravitational field.

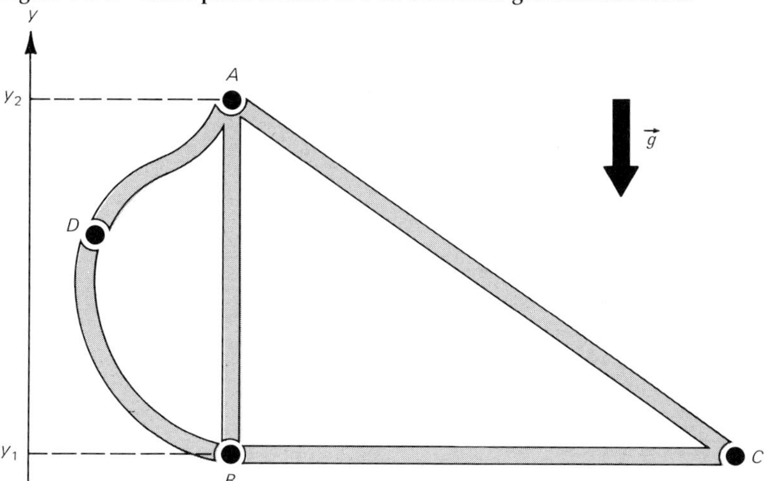

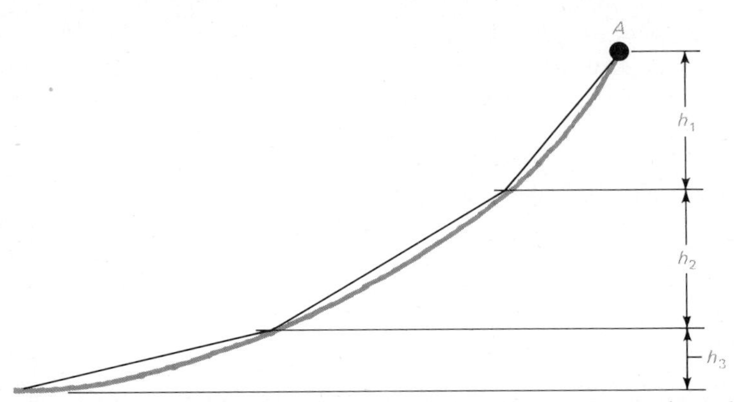

Figure 5-5-3 The beginning of the path *ADB* in Fig. 5-5-2, approximated by three straight line segments. While moving along the line segments a particle falls the distances h_1, h_2, and h_3.

always equal to mgh? To answer this question, see Fig. 5-5-3, which shows an enlarged view of the part of path *ADB* near point *A*. The curved path can be approximated as well as we like by taking a series of sufficiently short straight lines, and Fig. 5-5-3 shows three such straight line segments. From the result already obtained for the slanted path *AC*, the work done by the force when the particle moves along the top straight line is mgh_1. For the next two segments, the work done is mgh_2 and mgh_3. Thus, the work done on the particle when it moves from *A* along the part of the curve *ADB* shown in the figure is just $mg(h_1 + h_2 + h_3)$. But $(h_1 + h_2 + h_3)$ is just the total height dropped, so the vertical height the particle drops is all that matters in determining the work done by the gravitational force.

By extending the argument above to the entire curve *ADB*, it is easy to see that the total work done is mgh. Thus, for the uniform force $\vec{F} = m\vec{g}$, we can use Eq. 5-5-4 to define a potential energy at any point, such that from the potential-energy difference between any two points we can find the work done by the force on a particle which moves from one point to the other. The work done is *independent of the path* taken by the particle.

For some forces, it is not possible to find a formula for potential energy V such that $W_{A \to B}$ is independent of the path from A to B and simply equal to $V_A - V_B$. For example, friction produces a force opposite to the direction of motion of a sliding block. Whether the block moves up or down an inclined plane, or on a level surface, the frictional force will do negative work, tending to reduce its kinetic energy. We cannot find a formula for V which depends on the block's position and which always leads to this negative work.

PROBLEM 15

Two equal masses are released from rest at the same time, when each is a distance h above the ground. One falls freely in the vertical direction. The other slides without friction on an inclined plane of length $3h$. Compare their kinetic energies when they reach the ground. Find the ratio of the times they take to reach the ground.

For a particular kind of force, if we can find a formula for V which correctly gives $W_{A \to B}$ in every case, we call the force a *conservative* force. Friction is a nonconservative force, while gravity is conservative. The adjective *conservative* is used, because in such a case we can very simply state a law of conservation of energy. Just as the uniform gravitational field leads to a conservative force on a mass m, a uniform electric field leads to a conservative force on a charge q. The situation is exactly analogous.

The force between point charges is also conservative. The potential energy for two point charges, Q_1 and Q_2, a distance r apart is given by

$$V = (9.0)(10^9)\frac{Q_1 Q_2}{r} \qquad 5\text{-}5\text{-}7$$

Suppose Q_1 and Q_2 have the same sign. Then, Fig. 5-5-4 shows V plotted against r according to Eq. 5-5-7. In this graph, Q_1 is at $r = 0$, and the curve gives the potential energy if Q_2 is at any distance r away from Q_1. If Q_1 is held fixed and Q_2 is released from rest a distance r_A away from Q_1, the repulsive force will accelerate Q_2 away from Q_1. It will ultimately move to a large value of r, for which V is approximately zero. When this happens, it will have kinetic energy equal to the decrease in potential energy, or closely equal to V_A. The smaller r_A, the greater the initial repulsive force, the greater the value of V_A, and the greater the final value of T.

In this discussion of point charges, when the charges are at rest and close to each other, the Coulomb force is like a compressed spring between the two charges, trying to push them apart. In order to place the charges close together, we had to push them toward each other and "compress the spring." We did work, and it was stored in the form of potential energy. When Q_1

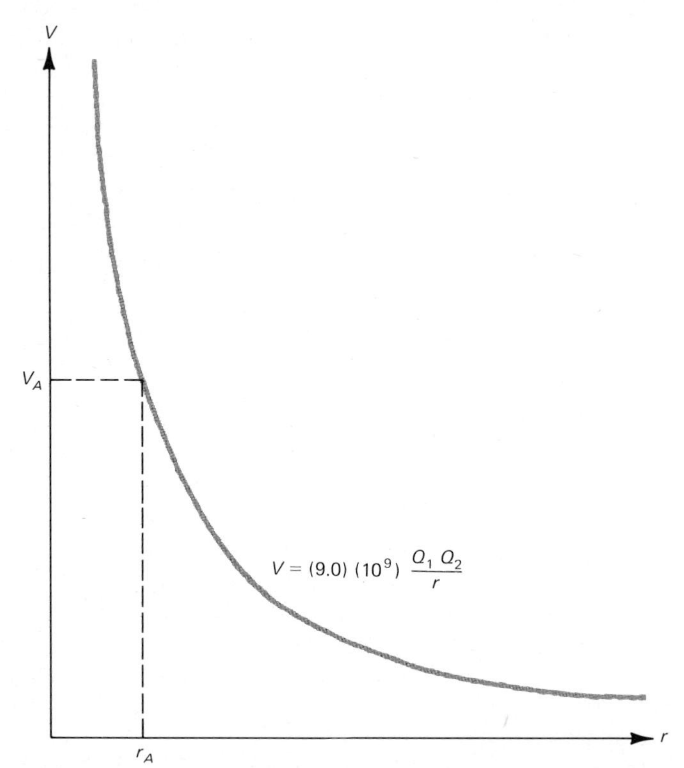

Figure 5-5-4 The potential energy V for two point charges with like signs a distance r apart.

is held fixed and Q_2 is released, this potential energy is returned in the form of kinetic energy of Q_2. Similarly, if we lift a mass up off the ground, against the gravitational force, we do work which is equal to the increase in potential energy. If we release the mass and let it fall freely to the ground, the potential energy we added in lifting it is returned in the form of kinetic energy.

PROBLEM 16

A point charge of 1.60×10^{-19} C is released from rest at a distance of 10^{-8} cm from a fixed point charge of 3.20×10^{-19} C. What is its kinetic energy when it is at a distance of 2×10^{-8} cm from the fixed charge? If the moving charge is an electron, what is its velocity at this distance?

5-6 CONSERVATION OF ENERGY

For motion of a particle under the influence of a conservative force, we have found that a change in kinetic energy can be related to a change in potential energy according to Eq. 5-5-2,

$$V_A - V_B = T_B - T_A \qquad \text{5-6-1}$$

This equation can be rearranged in the form

$$T_B + V_B = T_A + V_A \qquad \text{5-6-2}$$

Equation 5-6-2 says that the sum of the kinetic energy plus the potential energy at point B equals the sum at point A. Points A and B can be *any* two points along the path of a moving particle. Thus, no matter where or when we look at the particle, the sum of T plus V is a constant, called the *total energy* U,

$$T + V = \text{const} = U \qquad \text{5-6-3}$$

Equation 5-6-3 is the law of *conservation of energy* for motion of a particle under the influence of a conservative force. The energy which is "conserved," U, remains the same throughout the motion of the particle.

Energy conservation is a powerful concept. Suppose, as a simple first example of its usefulness, we consider the motion of a child down a slide, shown schematically in Fig. 5-6-1. The child, symbolized as a mass m in the figure, is assumed to start from rest at the top of the slide and to sit on a piece of waxed paper so that friction can be ignored. We will try to find the speed of the child at the bottom of the slide.

Only the gravitational force is assumed to be involved in our example, for which we know that the potential energy of a mass m is given by

$$V = mgy \qquad \text{5-6-4}$$

The coordinate y is to be measured upward, and in the figure the top of the slide is at $y = y_2$ and the bottom is at $y = y_1$. Equating the total energy $T + V$ at the top of the slide to the total energy at the bottom, we find, with $T = 0$ at the top,

$$mgy_2 = mgy_1 + \frac{1}{2}mv^2 \qquad \text{5-6-5}$$

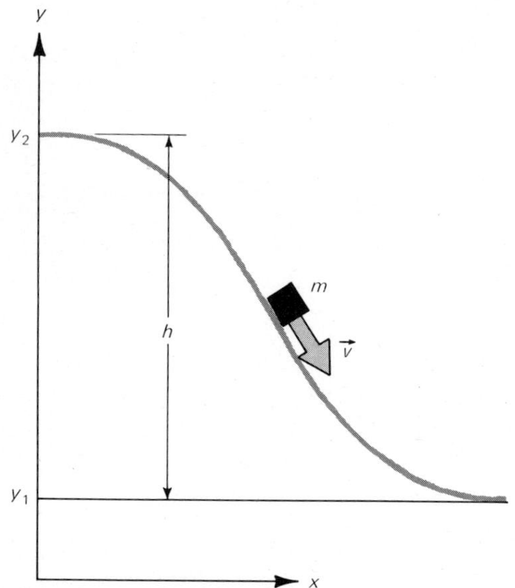

Figure 5-6-1 Schematic view of a child of mass m moving down a slide of height h.

In Eq. 5-6-5, v is the velocity at the bottom.

Equation 5-6-5 can be rearranged and solved for v, as follows:

$$\frac{1}{2}mv^2 = mgy_2 - mgy_1$$

$$\frac{1}{2}mv^2 = mg(y_2 - y_1) = mgh$$

$$v^2 = 2gh$$

$$v = \sqrt{2gh} \qquad 5\text{-}6\text{-}6$$

Notice that this result is independent of whatever complicated curve the slide may have. To completely solve the problem, finding the motion of the child all along the slide might be tedious and difficult, but energy conservation leads very easily to a final answer, and one which is valid for *any* shape slide.

Of course, the speed of a child at the bottom of an actual slide must be lower than that given by Eq. 5-6-6 because of friction. Nevertheless, this result may be a useful approximation for designing a slide. Moreover, in the motion of atomic particles frictional forces are not present, and the use of energy conservation in the form given by Eq. 5-6-2 leads to exact results. As another example very much like that of the child on a slide, consider an alpha particle released from rest a distance r from an atomic nucleus represented as a point charge Ze. We will try to determine its velocity at a great distance from the charge.

Figure 5-6-2 shows a potential energy curve for the alpha particle (whose charge is $2e$) like the one we have previously seen in Fig. 5-5-4. Its motion can be likened to "sliding down" this curve. (We could try to make a gravitational model of the alpha particle motion by releasing a ball on the side of a steep hill. Of course the force is really electric, so the "slide" of the problem is really imaginary or based on the gravitational analogy.) The conservation of energy equation can be written as follows:

$$T_1 + V_1 = T_2 + V_2$$

Figure 5-6-2 The potential energy of an alpha particle, with charge $2e$, and a nucleus, with charge Ze, separated by a distance r. An alpha particle released from rest near the nucleus seems to "slide down" this curve.

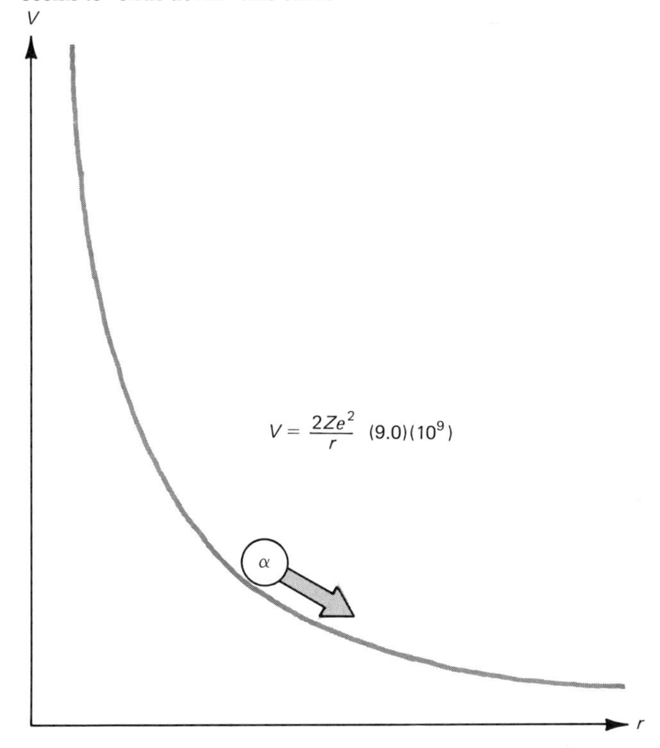

or

$$\frac{(9.0)(10^9)(Ze)(2e)}{r} = \frac{1}{2}mv^2 \qquad \text{5-6-7}$$

In Eq. 5-6-7 we have let $T_1 = 0$ because the alpha particle is assumed to be released at rest, and $V_2 = 0$ because finally the alpha particle is assumed to be a great distance from the charge. In addition, we have taken T_2 equal to the kinetic energy of the alpha particle alone, as if the charge Ze is held so that it does not move. Actually, if the mass of this charge is much greater than the alpha particle mass, we can neglect its kinetic energy even if it is free to move. The proof of this, using the laws of conservation of energy and momentum, is left as a problem.

PROBLEM 17

Use the conservation laws of energy and momentum to justify neglecting the kinetic energy of the nucleus in Eq. 5-6-7, provided the nuclear mass is much greater than the alpha particle mass.

The result given in Eq. 5-6-7 can be applied in the real world. Although we will leave a more detailed discussion of alpha emission by radioactive decay until Chap. 9, alpha particles can be assumed to start approximately from rest just outside the nucleus from which they are emitted. Even though the nucleus has a finite size, the electric field outside of it is the same as for a point charge. Thus, Eq. 5-6-7 can be applied to alpha emission, and, for example, from a knowledge of the alpha particle velocity far from the nucleus one can find the distance at which it would have started from rest. This can be used to get an approximate value for the tiny nuclear radius, and an example of such a calculation is left as a problem.

PROBLEM 18

Radium, with atomic number Z equal to 88, is found to emit alpha particles whose kinetic energy is about 8×10^{-13} joules. Obtain a rough estimate of the nuclear radius from this, assuming the alpha particles are released near the surface of the radium nucleus.

PROBLEM 19

A proton strikes an alpha particle and combines with it to form a single particle of mass five times the proton mass. The speed of the proton is initially 10^7 m/sec, and the alpha particle is initially at rest. What is the speed of the final particle? Suppose the "mass 5" particle formed in the "nuclear reaction" described above stops and decays into two pieces with masses two and three times the proton mass. If the kinetic energy of the lighter fragment is 10^{-13} joules, what is the energy of the other one?

5-7 ENERGY DIAGRAMS

Figure 5-7-1 shows a graph of V vs. x for the example of the child on the slide discussed in the previous section. The x axis points along the ground directly under the slide, with the origin directly under the top of the slide. Since V is proportional to the height, the shape of this graph is exactly like the shape of the slide.

In Fig. 5-7-2 the total energy U, the potential energy V, and the kinetic energy T are all drawn on the same graph. The curve for U is a horizontal line, since energy is conserved, and therefore U is constant. At point A, the kinetic energy is zero, so that $U = V$, as shown on the graph. Since $T + V$ must always equal U, the kinetic energy is always given by the distance upward from the curve for V to the curve for U. The kinetic energy T_B, when the child is at point B, is shown on the graph, as is the potential

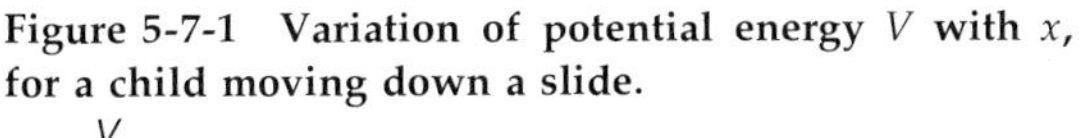

Figure 5-7-1 Variation of potential energy V with x, for a child moving down a slide.

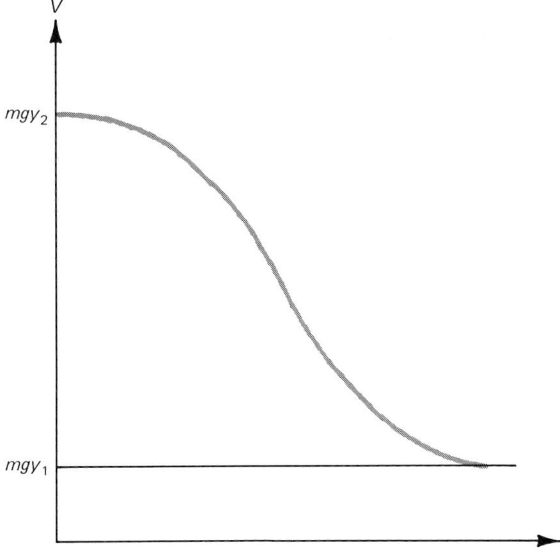

energy V_B. Notice that the two curves for T and V are drawn so that $T + V = U$ everywhere.

Diagrams like that of Fig. 5-7-2 are called *energy diagrams* and are very useful for representing the motion of an object when energy is conserved. We can now turn to a more complicated example, similar to the slide, of a cart rolling on the track of a roller coaster. Figure 5-7-3 represents part of a roller coaster which lies along a straight line on the ground, our x axis. The potential energy curve V vs. x has the shape of the roller coaster. This is because we are again considering motion under gravity, so that V at any point along the track is proportional to the height.

Suppose the cart starts from rest at point A. Let the track have a slight slope there, so it starts to roll to the right. Soon afterward the cart will be moving as shown schematically for the mass m on the figure. (For convenience the cart is shown rolling down the graph of potential energy instead of rolling on an exactly similar curve which could represent the roller coaster track.) How far will the cart go? Since it started from rest its total energy is U_1, equal to V at the starting position point A. When the cart

Figure 5-7-2 Potential energy V and total energy U plotted against x, for a child on a slide. At point A it is assumed that the child starts from rest, with kinetic energy T equal to zero. The kinetic energy and potential energy are shown at an arbitrary point B.

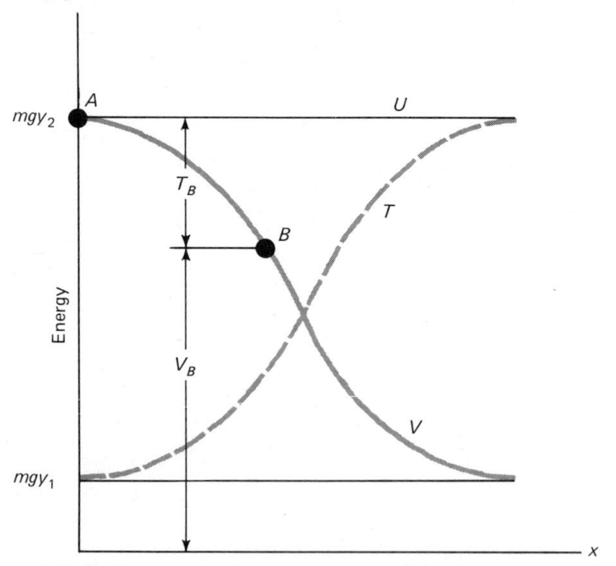

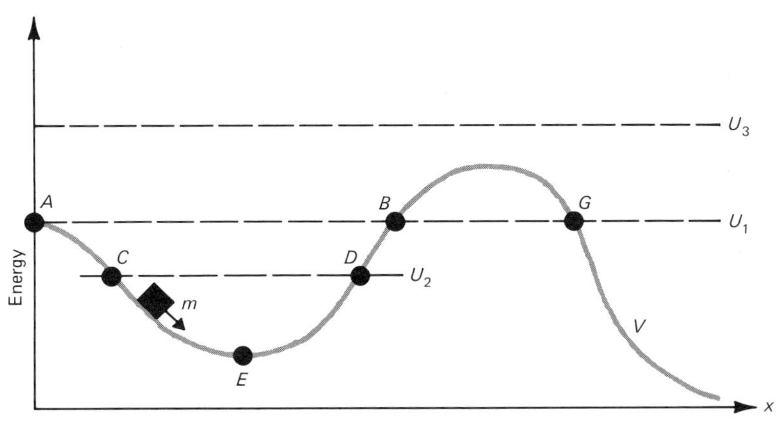

Figure 5-7-3 Energy diagram for motion of a cart on a roller coaster. The potential energy V is shown, along with three different total energies—U_1, U_2, and U_3.

has traveled to the point B, you can see from the graph that once again V and U_1 are equal. Therefore, T must be equal to zero. The cart will therefore stop there and begin rolling back toward A, since it is headed uphill. If there were no friction, the cart would roll back and forth between A and B forever. However, it will gradually show down, and after a while its total energy might be U_2. Then it will oscillate between points C and D. Ultimately it will end up at the bottom of the valley, at point E.

Whenever U and V cross, the kinetic energy must be zero. Suppose the cart entered the section of track shown in Fig. 5-7-3 from the right end instead of the left end, again with total energy U_1. How far would it get? In this case it would reach point G, where $V = U_1$. Notice that between points B and G the potential energy is greater than U_1. For energy to be conserved in this region, a cart with total energy U_1 would have to have *negative* T. However, T is equal to $\frac{1}{2}mv^2$, which is always *positive.* Therefore, the region between points B and G is *forbidden* to carts with total energy U_1, at least according to Newton's mechanics.

Of course, the part of the roller coaster between points B and G is accessible to carts with sufficient total energy. If the cart has total energy U_3 shown on Fig. 5-7-3, then it can roll over the peak of the mountain and still have some kinetic energy at that point.

PROBLEM 20

In a child's game, a marble of mass 20 g is rolled along a track. If the track is as shown in Fig. P-5-20, what is the minimum kinetic energy of a marble at point A which can travel to point B?

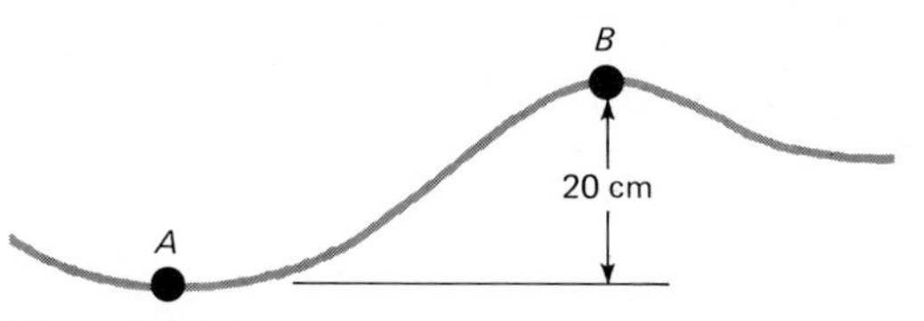

Figure P-5-20

PROBLEM 21
Suppose a marble rolls on a track with a loop in it, as shown in Fig. P-5-21. What is the minimum velocity at point A if the marble makes the loop without losing contact with the track?

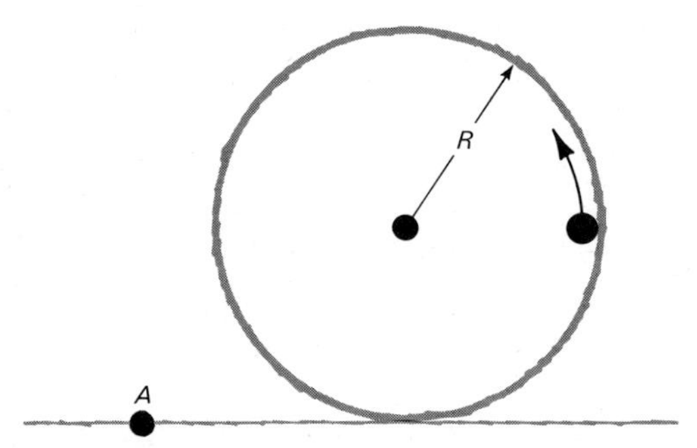

Figure P-5-21

The curve of V vs. x shown in Fig. 5-7-3 can also be used to find the force on the roller coaster along its direction of motion when it is at any point on the track. Figure 5-7-4 is an enlarged view of the portion of the curve near point B in Fig. 5-7-3. To find the force at point P in Fig. 5-7-4, first mark off a small distance Δs, as shown. For this change in position, Δs, we can find from the graph the corresponding change in potential energy, ΔV.

From the definition of work, the work ΔW done by the force on the cart when it moves from point P along the short distance Δs is given by

$$\Delta W = F_{||}(\Delta s) \qquad \text{5-7-1}$$

Also, the work done by the force is equal to the *decrease* in potential energy:

$$\Delta W = -\Delta V \qquad \text{5-7-2}$$

The minus sign appears because the potential energy actually *increases* when the particle moves from P along Δs. This is a "negative decrease."

We can now substitute ΔW from Eq. 5-7-1 into Eq. 5-7-2, to find

$$F_{||}(\Delta s) = -\Delta V \qquad \text{5-7-3}$$

Solving this equation for $F_{||}$, the result is

$$F_{||} = -\frac{\Delta V}{\Delta s} \qquad \text{5-7-4}$$

Thus, from the curve of V vs. x we can find the value of $F_{||}$, the component of force which changes the speed of the cart.

The fact that $F_{||}$ came out negative in the example of Fig. 5-7-4 means it points in the opposite direction to the assumed direction of motion, which was from P along Δs. This means the force points downhill, as we expect. By using Eq. 5-7-4 it is possible to find $F_{||}$ at any point along the roller coaster. Qualitatively, when the curve of V vs. x is steepest, the force has the largest magnitude, and its direction is always downhill.

Figure 5-7-4 Enlarged view of a region near point B on the curve of V vs. x shown in Fig. 5-7-3. The two points, P and $P + \Delta s$, differ in potential energy by an amount ΔV.

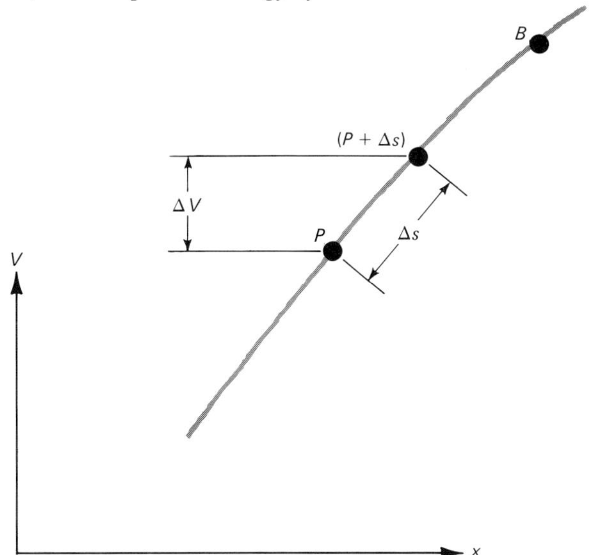

It is also possible to find a curve of V vs. x from knowledge of $\vec{F}$ vs. x. If a formula for V valid at any point in some region of space is to be found, it is necessary to know $\vec{F}$ everywhere in that region. If a simple formula giving $\vec{F}$ everywhere is known, then sometimes a fairly simple formula for V can be found. This is the case for the two equations in the preceding section, which give V for the uniform gravitational force and the Coulomb force.

The potential energy often presents information about the force in a way that is more useful than simple knowledge of the force. Energy diagrams provide one example of the use of this information. They are very nice for roller coaster problems, and indispensable in atomic and nuclear physics.

5-8 SIMPLE HARMONIC MOTION

In the last section, while considering motion of the roller coaster car we noticed an oscillating motion that resulted when it was caught in a dip. For example, referring back to Fig. 5-7-3, a cart with total energy U_2 in the region on the left part of the figure is trapped between points C and D. At each of these two points its potential energy equals its total energy; its kinetic energy must therefore be zero; and it finds itself acted upon by a force toward

Figure 5-8-1 Two simple oscillating physical systems, with the oscillatory motion along the direction of the double arrows.

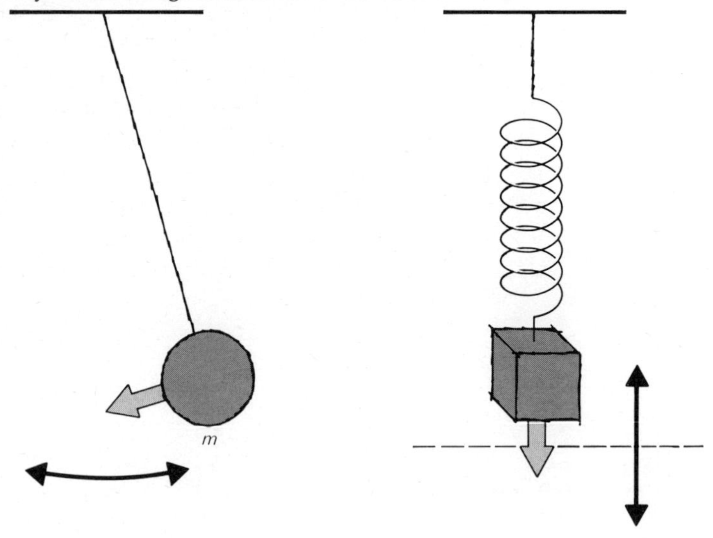

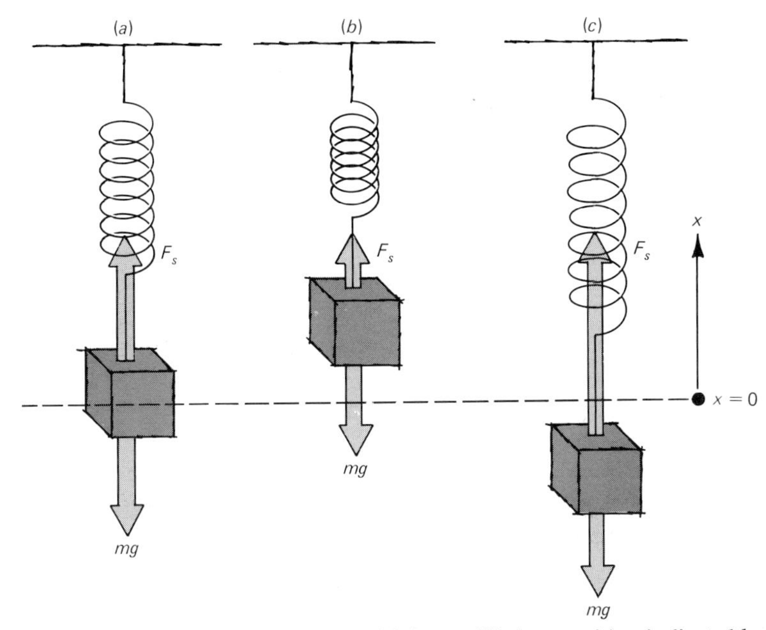

Figure 5-8-2 A mass on a spring, with its equilibrium position indicated by the dashed line. In (*a*) there is no net force, while in (*b*) and (*c*) the net force tends to return the mass to the equilibrium line. If the x coordinate is measured upward from the equilibrium position, the net force is given by $F = -kx$.

the bottom of the dip, point E. The cart oscillates back and forth forever if we neglect friction.

Oscillatory motion occurs in many physical systems. All the swinging or vibrating objects in everyday life are examples of such motion. This section will discuss a particularly important kind of oscillatory motion called *simple harmonic motion.* Many physical systems naturally oscillate in this particular way, and Fig. 5-8-1 shows two simple examples, a swinging pendulum and a mass oscillating on a spring. For other physical systems, an approximate description of the motion as if it were simple harmonic is often very useful. Finally, some oscillatory motions, like that of a vibrating guitar string, can be accurately represented by a combination of a few simple harmonic motions all going on at once. (The details of how motions are combined in such a case will be discussed in the next chapter, in connection with wave motion.)

As a particular example, consider the mass oscillating on a spring. Figure 5-8-2*a* shows the mass in its *equilibrium position,* when the spring force F_s on it is equal to mg. The net force on the mass in this case is zero, so it can remain motionless at this equilibrium position. In Fig. 5-8-2*b* the mass is shown displaced upward from its equilibrium position, and as a result the spring force is reduced because the spring is not stretched as far. The gravitational force acting downward on the mass is then greater than the upward spring force, leading to a net force downward. In Fig. 5-8-2*c* the mass is displaced downward from its equilibrium position; the spring force is now greater than mg because its stretch is increased, and there is a net upward force. We can see illustrated in this example the qualitative behavior needed to cause oscillatory motion: There is an equilibrium position, and for displacement away from it there is a *restoring force* back toward the equilibrium position.

Simple harmonic motion results when the restoring force is exactly proportional to the displacement from equilibrium. For the mass on a spring, the motion is one-dimensional, and we can conveniently characterize the motion of the mass by the change in its x coordinate, measured upward from the equilibrium position, as shown in Fig. 5-8-2. The fact that the spring force is proportional to the amount it is stretched leads to a quantitative equation which describes how the *net* force on the mass depends upon x:

$$F = -kx \qquad \text{5-8-1}$$

The quantity k is a positive constant which depends upon the spring stiffness; the stiffer the spring, the greater k is. The negative sign indicates that F is a restoring force, pointing in the opposite direction to the displacement x.

Motion under the influence of a force described by Eq. 5-8-1 can be described with an energy diagram, where the potential energy is given by the relation

$$V = \frac{1}{2}kx^2 \qquad \text{5-8-2}$$

It is left as a problem (with some hints) to verify that Eq. 5-8-2 implies the force given in Eq. 5-8-1.

PROBLEM 22

Show that Eq. 5-8-2 implies the force law of Eq. 5-8-1. Find the change in V, ΔV, when x changes from x to $x + \Delta x$. Then, relate this to work done in going from x to $x + \Delta x$, and thus to $F(\Delta x)$. You should find that $F = -kx$.

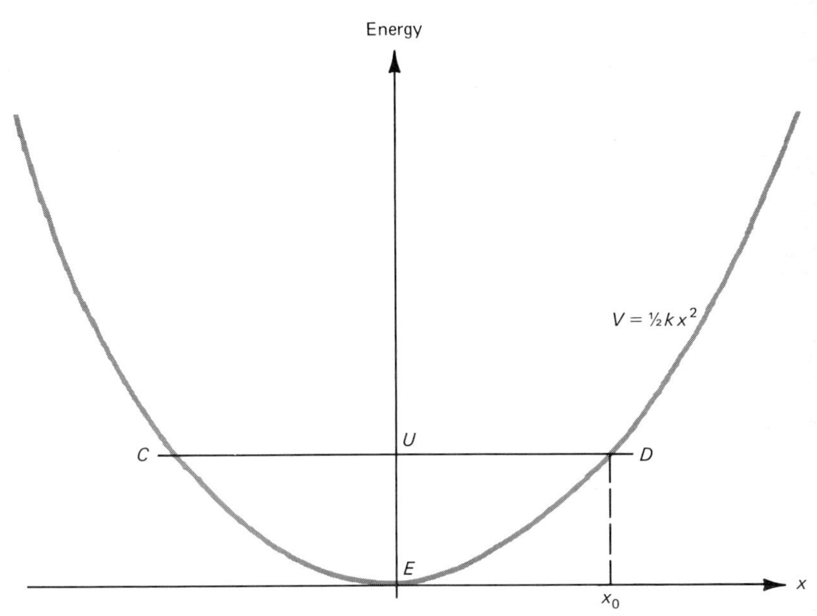

Figure 5-8-3 Energy diagram for a particle executing simple harmonic motion, with total energy U.

Figure 5-8-3 shows a potential energy curve (which is a parabola) described by Eq. 5-8-2. Suppose the mass is displaced to some position x_0 and released from rest. Its motion corresponds to motion with the total energy U indicated on the figure, where U is equal to the value of V at x_0, as shown. Points C and D represent the extremes of the oscillatory motion, where the kinetic energy is zero. Each time the particle reaches either of these turning points, it is accelerated back toward the origin, labeled E, where the net force on the mass changes sign and begins to slow it down. It is as if the mass were a marble rolling back and forth inside a bowl, or like the roller coaster cart of the previous section on a piece of track shaped like a parabola.

The variation of x with time for the motion of the mass described above is shown in Fig. 5-8-4. At $t = 0$ the mass is released at $x = x_0$, and the curve shows how the x coordinate then oscillates back and forth about zero in a particular way. A curve with this particular shape, called a *sine curve*, characterizes simple harmonic motion. It results from a parabolic potential energy curve or, equivalently, from a restoring force which is exactly proportional

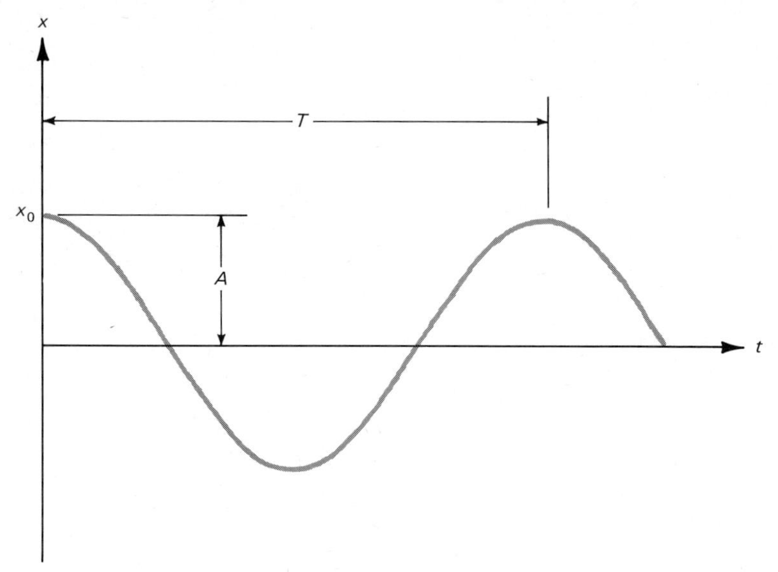

Figure 5-8-4 Variation of x with time for the particle motion corresponding to Fig. 5-8-3. The curve is called a *sine curve*, with *amplitude* A and *period* T.

to the displacement from equilibrium. We will not derive the result shown in Fig. 5-8-4. The motion is, however, the same as the x component of motion of an object moving in a circle at uniform speed, as shown in Fig. 5-8-5. Therefore, a convenient example of simple harmonic motion is a side view of an object which is resting on a rotating phonograph turntable.

The sine curve of Fig. 5-8-4 is characterized by its *amplitude* and *period*, indicated by the quantities A and T on Fig. 5-8-4. The amplitude describes how large the variation of x is, and the period tells the time to make one *cycle* of motion, to go through the basic oscillation which repeats to constitute the motion for all time. No account is taken in this ideal case of frictional forces which might cause the oscillations to die out. The ideal sine wave has a constant amplitude and is a good approximation to real motions provided friction is sufficiently small. Such a sine wave describes a swinging pendulum very effectively but not the up and down motions of a car with good shock absorbers (frictional dampers).

A complete study of the motion of the mass on a spring leads to the expression for the period T,

$$T = 2\pi\sqrt{\frac{m}{k}} \qquad 5\text{-}8\text{-}3$$

The most striking aspect of this result is that T is independent of the amplitude. In our example, if x_0 is made larger, the accelerating force when we release the mass is larger, and it ends up moving at a higher average speed during its oscillations. The speed increase exactly compensates for the greater distance traveled, leaving the period the same. The fact that the simple pendulum also exhibits simple harmonic motion (provided it does not swing so it reaches large angles with the vertical) makes its period also independent of amplitude. As a result, a pendulum clock keeps good time for a wide range of amplitudes.

As one of the many instances where simple harmonic motion is a useful approximation, consider the potential energy curve shown in Fig. 5-8-6. This curve describes the potential energy of two hydrogen atoms a distance r apart, bound together in a hydrogen molecule (as determined from a quantum-mechanical calculation). Given the curve, classical mechanics can give us an approximate value for the vibrational period of hydrogen molecules. The spacing r_0 corresponds to an equilibrium spacing, where V has a minimum value. Although the shape of the curve near $r = r_0$ is not exactly parabolic, it can be approximated by a parabola, as shown by the dashed curve on the figure. For a molecule vibrating with small amplitude about the equilibrium spacing r_0, we can estimate the period from the result for simple harmonic

Figure 5-8-5 Graphical construction of x vs. t for a particle moving in a circle at constant speed. In the plane view, circular motion is indicated, with equal time intervals between each pair of numbered points along the circle. These 12 points are projected to the "edge view" and are plotted at equal time intervals, leading to the sine curve which is shown.

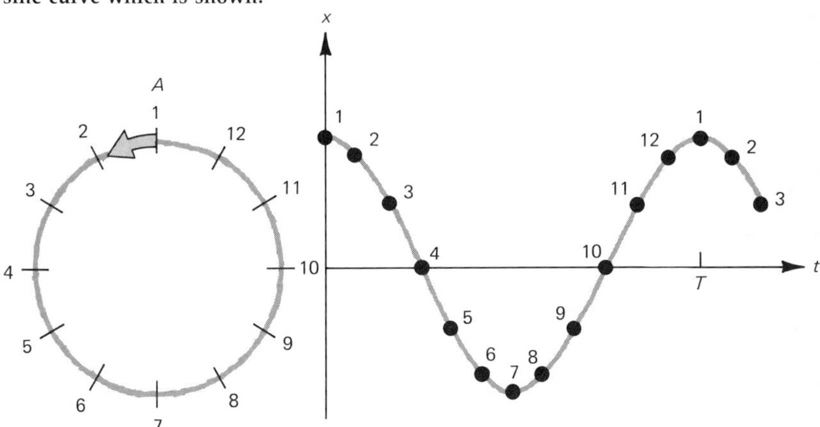

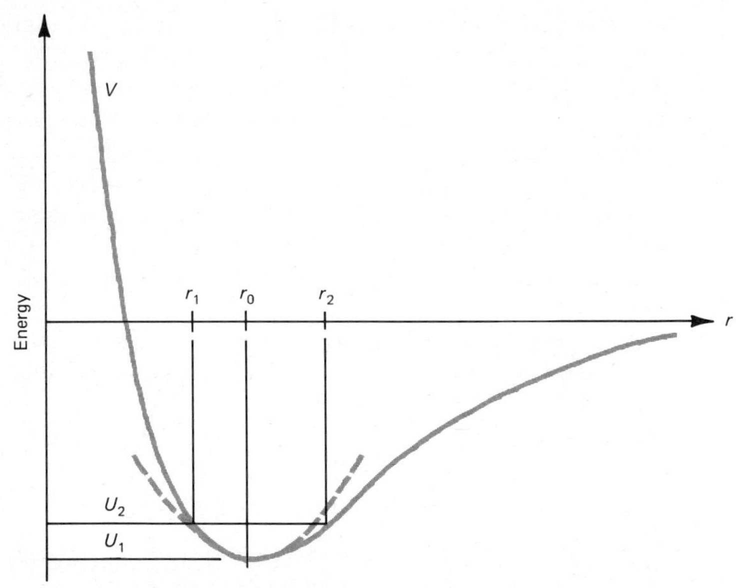

Figure 5-8-6 Energy diagram for two neutral hydrogen atoms. Total energy U_1 corresponds to the atoms at rest, separated by a distance r_0. Total energy U_2 corresponds to vibration, with the distance of separation varying from r_1 to r_2. The dashed curve is the potential energy curve for simple harmonic motion about the equilibrium position r_0.

motion, where k must be extracted from the curve and m from the atomic masses. The resultant period is about 8×10^{-15} sec, in rough agreement with the lowest observed vibration frequency of the molecule. Quantum mechanics must be used to correctly study the molecular vibrations, but even then we tend to use the approximate technique of assuming that the potential energy curve is parabolic. This leads to a quantum-mechanical form of simple harmonic motion which we will discuss in Chap. 7.

5-9 HEAT ENERGY

We have been discussing, in the preceding sections, some examples of motion where total energy U, defined as the sum of kinetic and potential energies, is conserved.

$$U = T + V = \text{constant} \qquad 5\text{-}9\text{-}1$$

Now, returning to a simple example discussed in the first section of this chapter, we will study a case where U, as defined above, is *not* conserved. The example is that of a bullet striking a stationary wood block, with the two recoiling together afterward. Figure 5-1-2 shows the collision schematically. We have already found (see Eq. 5-1-11), from momentum conservation, that the recoil velocity V' is given by the expression

$$V' = \frac{mv}{m + M} \qquad \text{5-9-2}$$

where m and v are the mass of the bullet and its initial velocity, respectively, and M is the mass of the block.

Now let us find the ratio of final total energy U' to initial total energy U for the bullet-block collision. Since there is no change in potential energy, the total energies are given by the initial and final kinetic energies:

$$U = \frac{1}{2}mv^2 \qquad U' = \frac{1}{2}(m + M)V'^2 \qquad \text{5-9-3}$$

After substituting the value of V' given in Eq. 5-9-2, the ratio U'/U can be easily found to be

$$\frac{U'}{U} = \frac{m}{m + M} \qquad \text{5-9-4}$$

The result, given in Eq. 5-9-4, is that U' is always *less than* U, and when the mass M is much larger than that of the bullet, U' is much smaller than U. Where has the missing energy gone? It has been transformed into internal kinetic energy of the wood block and bullet, into chaotic *vibrational energy* of their atoms. We call this vibrational energy *heat energy*. If we measure the temperature of the bullet and block after the collision, we find an increase. A thermometer is a device which measures the atomic kinetic energy of the material with which it is in contact. The more vibrational energy, the higher the temperature.

We often represent heat energy by the symbol Q, and energy *is* conserved in all collisions, provided we *redefine* the total energy U to include the heat energy:

$$U = T + V + Q \qquad \text{5-9-5}$$

The type of collision exemplified by the bullet-block example, where some kinetic energy is transformed into heat energy, is called an *inelastic collision*. The transformation from kinetic to heat energy takes place by friction, as the bullet buries itself in the wood in our example. In collisions which are essentially frictionless, such as between billiard balls, the heat energy does not change, and the kinetic energy is the same before and after the collision. These are called *elastic collisions*.

Rather than emphasize collisions, we want instead to pursue a bit further the physics of heat. This is a large subject, generally described by the term thermodynamics, and here we will have to limit ourselves to stating without proof one very important result: At any given temperature, the *average kinetic energy* of chaotic motion of each atom or molecule of a physical system which is free to move is given by:

$$\frac{1}{2}m(v^2)_{\text{av}} = \frac{3}{2}kT \qquad \text{5-9-5}$$

where k is a constant, called *Boltzmann's constant*, and T is the *absolute temperature.* In Chap. 1 we briefly discussed the kinetic theory of gases. Equation 5-9-5 correctly describes the mean kinetic energy of the disordered motion of gas molecules. It also describes vibrational kinetic energy of the atoms bound in a solid.

The absolute temperature is measured from a zero of temperature at which, according to classical thermodynamics, the kinetic energy would become zero. As estimated in a problem in Chap. 1, this temperature is −460°F. To use Eq. 5-9-5 with temperature in absolute degrees fahrenheit, we add 460° to the normal Fahrenheit temperature above zero. Room temperature would be about $460 + 70 = 530$°F absolute. Boltzmann's constant is

$$k = 7.7 \times 10^{-24} \text{ joules/°F absolute} \qquad \text{5-9-6}$$

Therefore, the energy corresponding to room temperature is

$$\frac{3}{2}kT = 6.1 \times 10^{-21} \text{ joules at 70°F} \qquad \text{5-9-7}$$

Although the energy given in Eq. 5-9-7 is small, it applies, for example, to each individual vibrating atom of a solid. For 10^{23} atoms, a reasonable number in a macroscopic quantity of matter, the energy is 600 joules which equals the total power radiated by a 60-watt light bulb in 10 sec. We will find in later chapters a number of instances where this thermal kinetic energy has an important influence on the behavior of atomic systems.

SUMMARY

The *momentum* $\vec{p}$ of a particle with mass m and velocity $\vec{v}$ is a *vector* quantity defined by $\vec{p} = m\vec{v}$. In terms of $\vec{p}$, Newton's second law is $\vec{F} = \Delta\vec{p}/\Delta t$. If no net force acts on a system of particles with total momentum $\vec{P}$, then $\vec{P}$ remains constant, regardless of changes in the motion of individual particles of the system. We say that *momentum is conserved.* For example, in two-body collisions with momenta $\vec{p}_1$ and $\vec{p}_2$ before the collision and $\vec{p}_1'$ and $\vec{p}_2'$ afterward, $\vec{p}_1 + \vec{p}_2 = \vec{p}_1' + \vec{p}_2'$. Rocket propulsion is an example of momentum conservation. The thrust force F_t is given by $F_t = -\vec{w}(\Delta m/\Delta t)$ where $\vec{w}$ is the exhaust velocity and $\Delta m/\Delta t$ is the rate at which mass is ejected in the exhaust.

Kinetic energy T of a particle with mass m and velocity v is a *scalar* quantity defined by $T = \frac{1}{2}mv^2$. *Work* is done on a particle by a net force $F_{||}$ along its direction of motion. Motion through a small distance Δs results in work ΔW given by $\Delta W = F_{||}(\Delta s)$. For motion between points A and B, the work done $W_{A\to B}$ is the sum of all the increments (ΔW) along the path. The concept of work is important because the *work done is equal to the increase in kinetic energy:* $W_{A\to B} = T_B - T_A$.

For a particle acted on by a *conservative force, a potential energy* V can be assigned to the particle, which depends only on its position. For motion from point A to B the decrease in V is equal to the work done by the force, and therefore to the increase in T:

$$V_A - V_B = W_{A\to B} = T_B - T_A.$$

This equation can be rearranged in the form of a statement of *conservation of energy.* The total energy $T + V$ remains constant:

$$T_A + V_A = T_B + V_B$$

Energy diagrams are based upon a curve showing how the potential energy V varies in space. A horizontal line indicates a fixed value of *total energy* U. Since $T = U - V$, the diagram also indicates how the kinetic energy varies. The slope of the potential energy curve indicates the magnitude and direction of the force.

One-dimensional *simple harmonic motion* results when a particle is acted upon by a *restoring force,* $F = -kx$. The potential energy for this force is a *parabolic well* centered at $x = 0$, $V = \frac{1}{2}kx^2$. Motion of a particle in this potential is described by a *sine curve* for the variation of x with time. The sine curve is a particular kind of *repeating* curve, characterized by its special shape, its *amplitude,* and its *period.*

Heat energy is the energy associated with the disordered motion of the atoms or molecules which are free to move in matter. At a given *absolute temperature* T, the mean kinetic energy of each moving atom is $\frac{3}{2}kT$, where k, Boltzmann's constant, is 7.7×10^{-24} joules/°F absolute.

LIGHT

6-1 INTRODUCTION

During the preceding five chapters we have emphasized the study of particle motion and limited our discussions to analysis of the motion by means of classical (newtonian) mechanics. In this chapter and the next, quantum mechanics will be introduced. Experiments involving light were extremely important in the birth of quantum theory. Furthermore, an understanding of wave motion, which is dramatically demonstrated by optical phenomena, is essential to an understanding of quantum mechanics. In this chapter, light will be studied for its own interest, as an example of wave motion, and as a subject whose puzzling phenomena provide a beginning for the development of quantum mechanics.

By Newton's time, during the seventeenth century, a great deal was known about the properties of light. First, from everyday experience, it was inferred that light travels in straight lines. For example, on a sunny day we can observe shadows of objects that can easily be explained by assuming that light rays travel in straight lines from the sun. The shadow is the area where the light rays cannot hit because they have been intercepted by the object.

If we look carefully, the edges of sun shadows are not really

Figure 6-1-1 Shadow cast by a light source with an appreciable size.

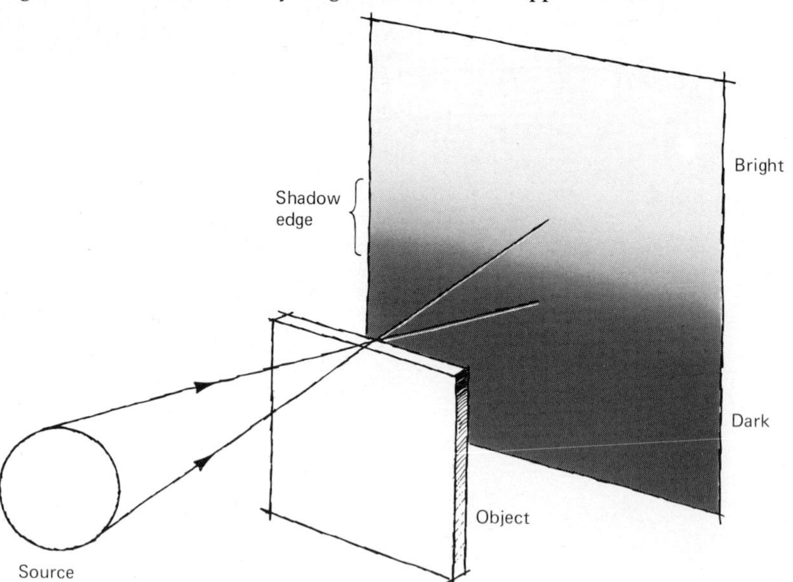

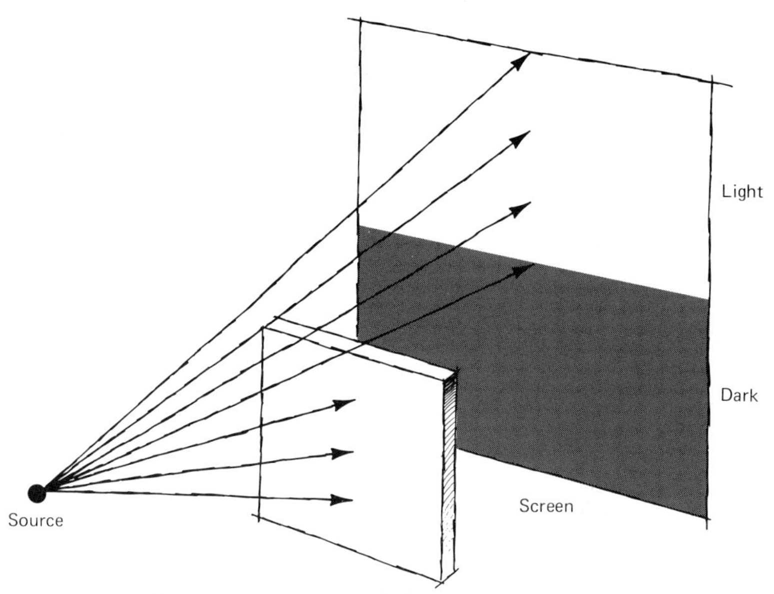

Figure 6-1-2 Shadow cast by a point light source.

perfectly sharp; but since the sun is not merely a bright point in the sky, they are not expected to be. Figure 6-1-1 shows schematically why the shadows cast by a light source with an appreciable size are fuzzy. The figure shows a spherical source of light which partially illuminates a screen. Part of the screen is in the shadow of an object with a sharp edge. Two light rays which originate from the extreme edges of the light source are shown. Between the places where these two rays strike the screen, it is illuminated by part of the source. The edge of the shadow thus has a width equal to the spacing between these two rays and the shadow appears to be fuzzy by this amount.

Figure 6-1-2 shows the sharp shadow created by rays from a point source of light. Although a perfect point source is not really possible to obtain, it can be well approximated by a tiny light bulb or by more sophisticated optical techniques. Except for observations made very carefully under special conditions, the idea that light consists of *something* traveling in straight lines is well verified by experiment. Light *rays* are often drawn to represent this fact and to aid in visualizing the operation of optical instruments. The rays hardly convey much information about the true nature of light, but they represent a kind of model of its behavior.

In Newton's time it was also known that light travels at very high speed, but not instantaneously. In 1666, a Danish astronomer, Ole Roemer, first clearly demonstrated this fact. Figure 6-1-3 shows a diagram which illustrates Roemer's observations. The sun, the earth, the planet Jupiter, and one of Jupiter's several moons are shown, as well as their orbits. The diagram is not to scale. With a modest telescope, this moon of Jupiter can be seen, and it appears bright because it is illuminated by the sun. However, once each revolution it passes through the shadow of Jupiter and is *eclipsed*, as at point A in Fig. 6-1-3.

The time for one revolution of this moon is about 43 hours, and once every revolution, when it is eclipsed, it disappears from view. Thus the moon can be used as a kind of clock, with "ticks" every 43 hours. In observing this "clock" throughout a year, Roemer was at first surprised to find that if observations were begun when the earth was at point B, the close was about 22 minutes slow 6 months later, when the earth was at point C. However, the clock caught up during the next 6 months and was all right when the earth returned to point B after a year. Roemer finally interpreted this phenomenon correctly as being the result of a

Figure 6-1-3 **The sun S, the earth E, the planet Jupiter J, and one of Jupiter's moons M. The dashed lines are orbits. The orbit of Jupiter is really much larger, in comparison with the size of the earth's orbit, than shown here. During one revolution of the earth around the sun, Jupiter moves only a small fraction of the total distance around its orbit.**

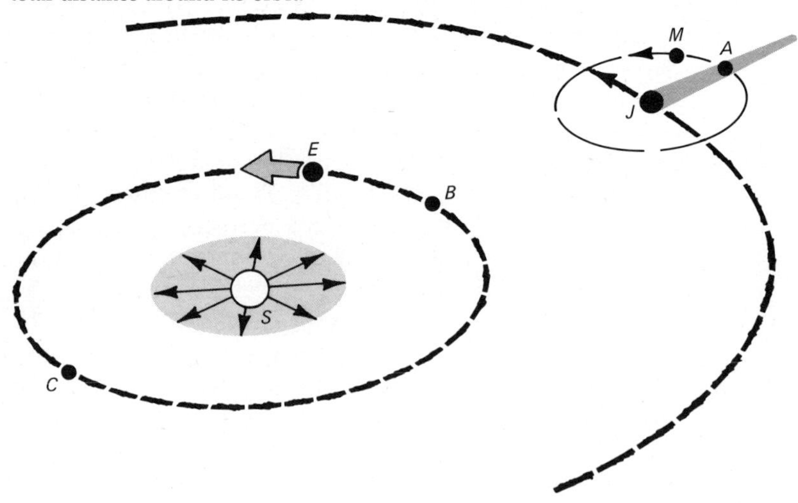

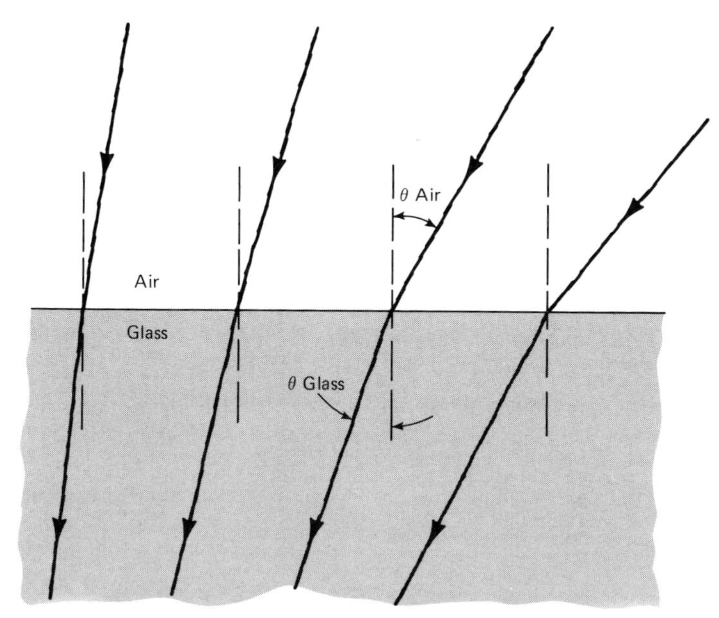

Figure 6-1-4 Refraction of light rays at a plane surface between air and glass.

finite time needed for light to travel the diameter of the earth's orbit. He inferred that the clock was 22 minutes slow at point C because light required that much time to travel the distance from B to C.

Although Roemer's measured time delay of 22 minutes was somewhat in error, and the diameter of the earth's orbit was not accurately known in his day, he could estimate that the velocity of light was roughly 2×10^8 m/sec. Nearly 200 years later it was finally possible to measure this velocity accurately on the earth, essentially by making very short light flashes and directly measuring the time it took for them to travel a distance of many kilometers. The velocity of light in vacuum is now known with great accuracy, and because it is a fundamental constant it has been given a symbol, c. The value of c to sufficient accuracy for our purposes is

$$c = 3.00 \times 10^8 \text{ m/sec} \qquad 6\text{-}1\text{-}1$$

Although light rays travel in straight lines in air or any other transparent material, the rays are bent, or *refracted*, at a surface between two different materials. Figure 6-1-4 shows a number of

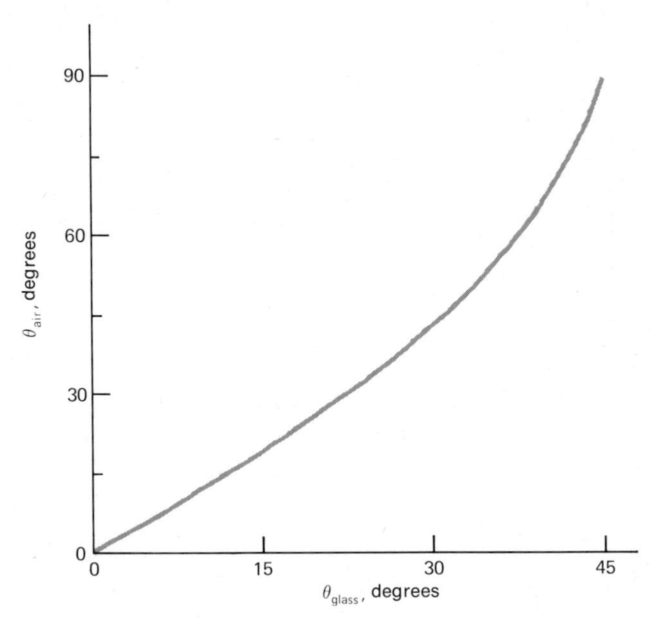

Figure 6-1-5 The law of refraction at an air-glass surface. The angles are measured with respect to a line perpendicular to the surface, as in Fig. 6-1-4.

rays passing from air to glass. If the rays go from glass to air, the paths are the same as indicated, with the directions of the arrows reversed. Figure 6-1-5 shows graphically the relation between the two angles, θ_{air} and θ_{glass}, shown in Fig. 6-1-4.

PROBLEM 1

A small "point" light source is 10 cm below the surface of a pool of water. Assume light rays from the source are refracted at the surface between water and air in the same way as at a glass-air surface. Thus, you can use the data of Fig. 6-1-5 to trace rays from the source to the air above. Trace about five or six rays. Can you understand now why an object under water, viewed from the air above, appears to be at a smaller depth than it really is?

Figure 6-1-6 shows two wedge-shaped pieces of glass, called *prisms*, each of which is shown refracting a ray of light so that when it emerges from the prism it has been bent by a total angle θ. The two prisms produce different angles of bend in accordance

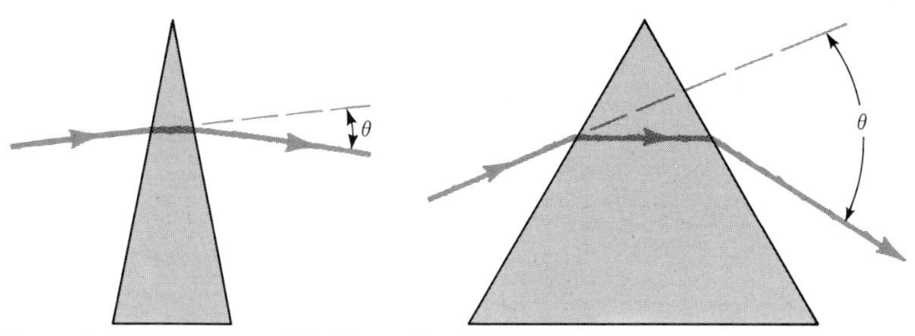

Figure 6-1-6 Bending of light rays by two different prisms.

with the law of refraction illustrated in Fig. 6-1-4. When a ray is moving very obliquely to the surface between air and glass, it is bent more than when it is moving nearly at right angles to the surface.

It is not a very large step from the prisms of Fig. 6-1-6 to a simple lens, as shown in Fig. 6-1-7. A point source of light is shown, and each ray from the source is bent differently, depending on where it strikes the lens. Because of the way the lens is curved, rays passing through near the center of the lens are deflected very little, as for the "thin" prism in Fig. 6-1-6, while rays passing near the edges of the lens are bent more, as for the

Figure 6-1-7 Refraction by a lens, so as to form a point image on a screen from a point light source.

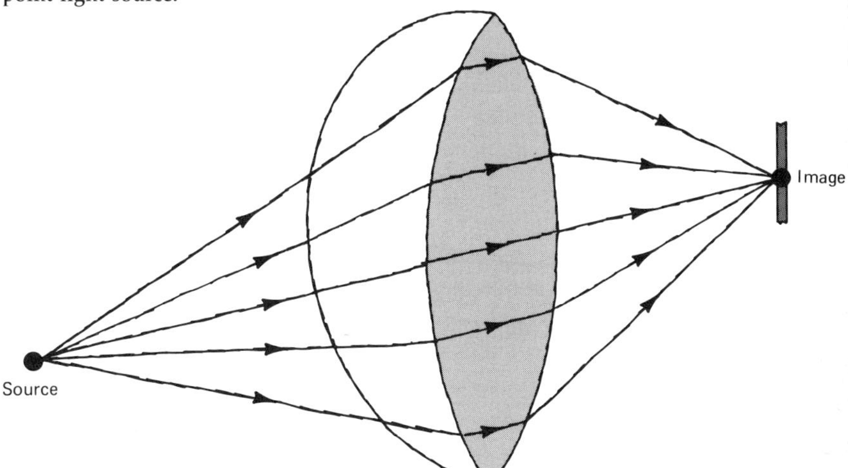

"fat" prism. If the lens is properly curved, all the rays shown will come together at a point, which is called an *image* of the light source.

A camera records an image of an object by focusing light from each point of the object onto a photographic film. Figure 6-1-8 illustrates the formation of an image of an arrow, by tracing the rays from the head and the tail through a lens. Although the arrow is not a light source in the usual sense, when it is illuminated each point of it scatters some light toward the camera lens. Just as in Fig. 6-1-7, light from each point of an object is focused by the lens to an image point. Figure 6-1-8 shows this for the head and tail, and a total image is formed by rays emerging from all along the arrow. In the human eye, a similar process takes place except that an image is formed on the retina, a layer of highly specialized cells, instead of on film. The optic nerves sense this image and almost miraculously convey to the brain the enormous richness of detail and color that we see.

In Fig. 6-1-8, if the distance from the arrow to the lens is changed, then the distance from the lens to the image changes. When we focus a lens, we adjust its location so that the image is where we want it to be. While lenses, and more complex optical instruments, are important and interesting tools, studying them will lead us away from discussing the basic nature of light. Therefore, we will postpone further discussion of lenses to Chap. 12, which is specifically devoted to optics.

Figure 6-1-8 Image of an arrow formed by a lens. The dashed lines show rays from the tail, and the solid lines show rays from the head.

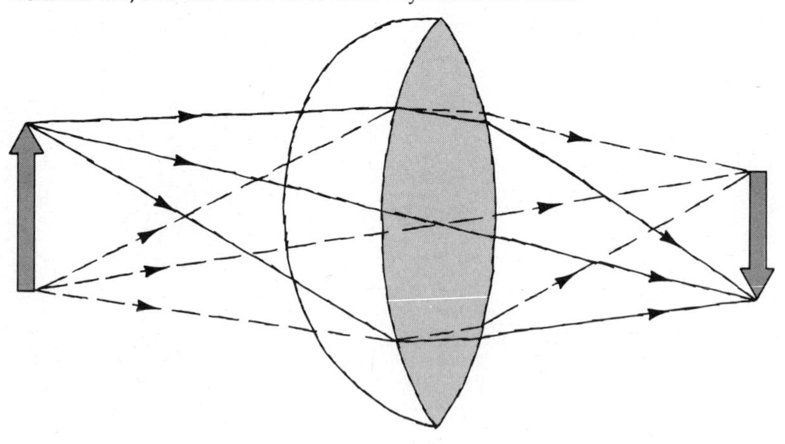

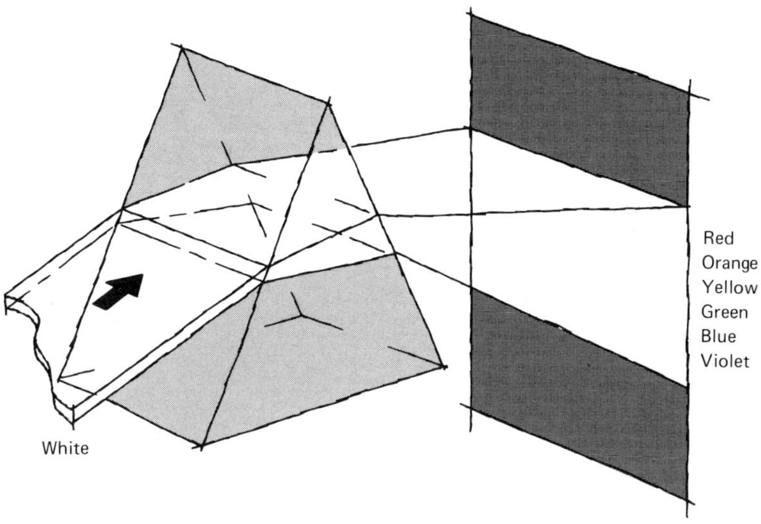

Figure 6-1-9 Refraction of white light by a prism, leading to a *spectrum* of colors.

If a narrow beam of sunlight is passed through a prism, a rather spectacular display of color results. Figure 6-1-9 shows a beam of white light, from the sun or from an incandescent light bulb, and the band of colors, or *spectrum*, which appears on a screen if the beam is passed through a prism. White light apparently contains all colors of the rainbow, and the prism bends the rays of each color a different amount.

Newton attempted to explain all the phenomena we have introduced so far in this section in terms of a *corpuscular*, or *particle*, model of light. Refraction, for example, was assumed to result from forces between glass and the light corpuscles. As a result of the behavior of a prism with white light, he was forced to assume different corpuscles for different colors. The model became a bit inelegant and complicated.

A great Dutch optician, Huygens, who was a contemporary of Newton's, took quite a different view. He believed that light was a wave motion, something like waves on water. A controversy that lasted two and a half centuries began between proponents of the two viewpoints, although for a long time it appeared as if Huygens was right. Finally, at the beginning of the twentieth century, physicists were forced by experimental evidence to accept the view now held, that both Newton and Huygens were correct. This discovery of the dual behavior of light, as both particles and waves, marked the beginning of the development of the quantum theory. The next two sections will be devoted to a discussion of wave motion, after which we will be ready to discuss this particle-wave duality in more detail.

6-2 TRAVELING WAVES

Light, sound, and radio signals are among the many examples of traveling wave phenomena. A very simple physical system with which to illustrate the behavior of traveling waves is a taut rope, as illustrated in Fig. 6-2-1. The figure shows a coordinate system, with x measured along the rope and y at right angles to the rope. The point A is assumed to remain at all times at $x=0$, but, in order to send waves down the rope, this point will be moved along the y direction. The weight and pulley are shown to emphasize that there is some fixed tension in the rope.

Suppose the point A is at rest at $x = 0$, $y = 0$, except during a time interval from t_1 to t_2, when it is wiggled by an up and down motion along the y direction. Figure 6-2-2 shows a curve of y vs. t for the point A, which is an example of such a wiggle. Perhaps your intuition tells you that shaking one end of the rope in this way will cause a "wave" to travel down the rope from point A toward the pulley. That is indeed what will happen, and Fig. 6-2-3 shows the rope at two times, t_3 and t_4, after the point A has been wiggled as in Fig. 6-2-2.

Figure 6-2-1 A rope with one end at $x=0$, $y=0$. The tension in the rope is maintained by a mass and pulley.

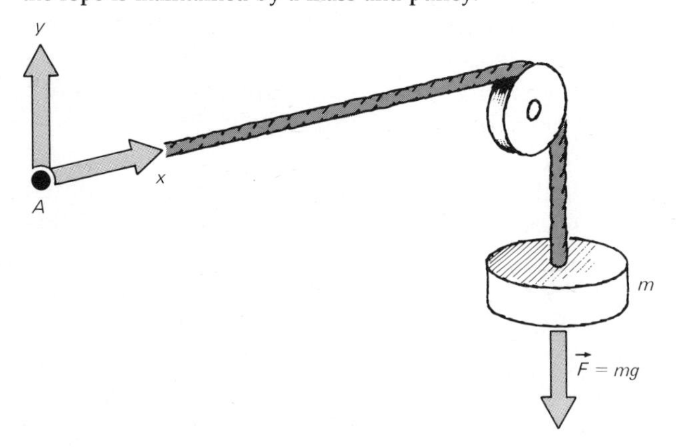

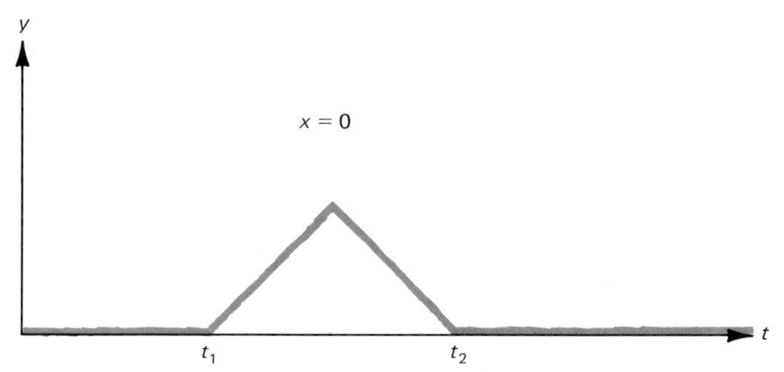

Figure 6-2-2 The end of the rope, at $x=0$, being wiggled between times t_1 and t_2 by motion in the y direction.

According to the physics of wave propagation in a rope, if a wave is generated by shaking one end, it travels down the rope at a speed v which depends only on the tension in the rope and the

Figure 6-2-3 The shape of a rope, y vs. x, at two times after one end has been wiggled as in Fig. 6-2-2.

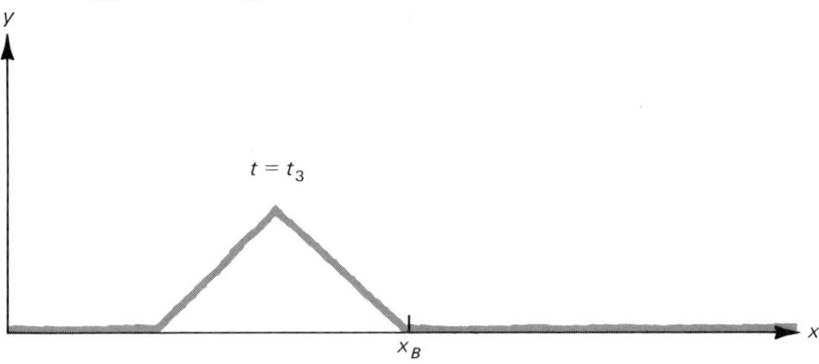

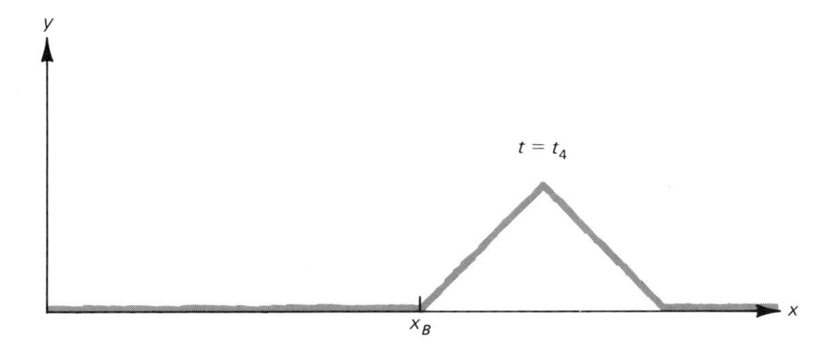

mass per unit length of the rope and does not depend on the shape of the wave. The two curves of Fig. 6-2-3 are like two frames out of a movie which would show the wave traveling at constant speed toward the pulley.

The times t_3 and t_4 have been chosen specifically to be the times between which the wave originating from point A at the end of the rope passes some point B, located part way down the rope from A, at $x = x_B$. Figure 6-2-4 shows how y varies with time at this point, and the striking fact is that the motion at point B is the same as at point A, except that it occurs later. The delay between the motion at B and the motion at A is equal to the time it takes the wave pulse to get from A to B. If the wave travels at constant speed v, then the time difference $t_3 - t_1$ is equal to x_B/v.

PROBLEM 2

Sketch a graph of the shape of the rope, like those in Fig. 6-2-3, which shows its shape at $t = t_2$, the time when the motion at $x = 0$ has just ceased.

One way in which energy could be transported from A to B would be to shoot a particle with kinetic energy T toward B from point A. When the particle arrived at B, it could hit something so as to give up the kinetic energy it carried from point A. Another way of transporting energy is with wave motion. Notice that motion from A is carried to B along the rope in the example which we have been studying. Figure 6-2-4 shows the motion at B imparted by the wave. Although there is no mechanism in our example for

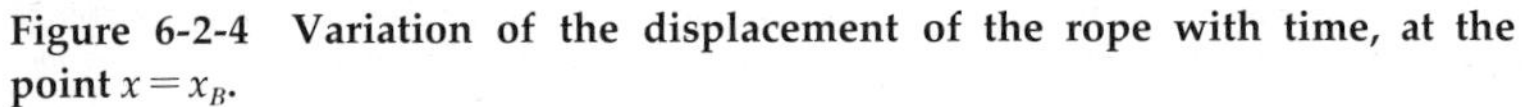

Figure 6-2-4 **Variation of the displacement of the rope with time, at the point** $x = x_B$.

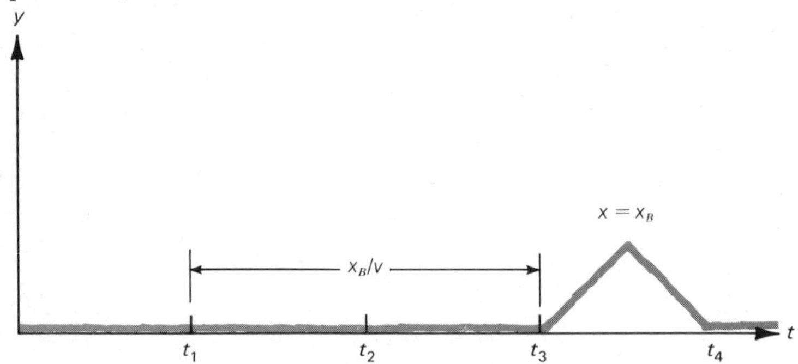

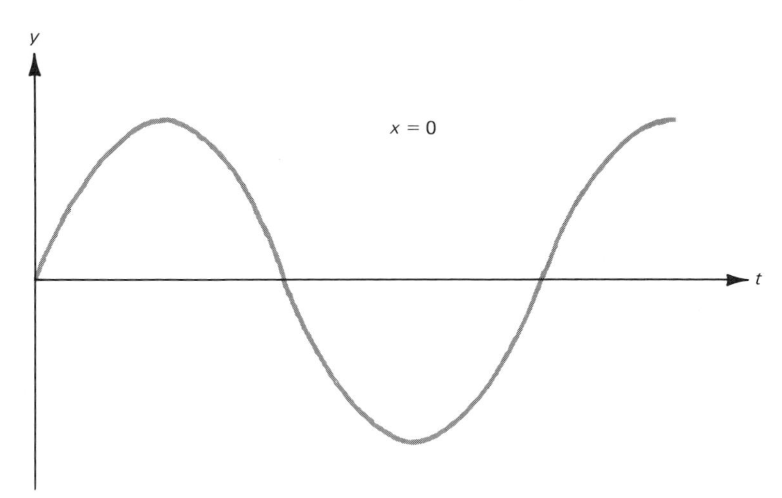

Figure 6-2-5 One end of the rope being wiggled with simple harmonic motion in the y direction.

absorbing the wave kinetic energy which arrives at B, this could be done.

Wave motion is a way of transporting energy which is very different from particle motion. Each part of the rope between points A and B moves in the y direction, at *right angles* to the direction of wave motion, when the wave passes it. No material object actually travels from A to B, only a wave which carries energy.

PROBLEM 3
Show, with a graph, how the velocity of a small piece of the rope at $x = x_B$ varies with time for the wave motion described by Fig. 6-2-4. How does the kinetic energy vary with time? If the wave shape were kept the same except that its height were doubled, by what factor would the kinetic energy increase?

Suppose now that point A moves along the y direction with simple harmonic motion, such that the motion of point A is as shown in Fig. 6-2-5. According to this figure, point A is at rest until time $t = 0$ when it begins to execute a motion in which the relation between y and t is a sine curve. Figure 6-2-6 shows y vs. x, the shape of the rope, at times t_1 and t_2. Once again, we have a wave motion traveling along the rope at velocity v. At time t_1 point B, at $x = x_B$, just begins to move. Figure 6-2-7 shows the motion at point B, and you can see from Figs. 6-2-6 and 6-2-7 that the time t_2 is that particular moment when point B has just finished one complete cycle of simple harmonic motion.

For a traveling sine wave such as that shown in Fig. 6-2-6 every point on the rope reached by the wave executes simple harmonic motion. Just as a single pulse can carry a burst of kinetic energy from A to B, a steady sine wave can continually transport energy in the direction the wave is traveling.

We have already discussed the amplitude and period of simple harmonic motion in Sec. 5-8, and these properties also refer to a

Figure 6-2-6 The rope at two times, t_1 and t_2, after the end at $x = 0$ has begun to move with simple harmonic motion.

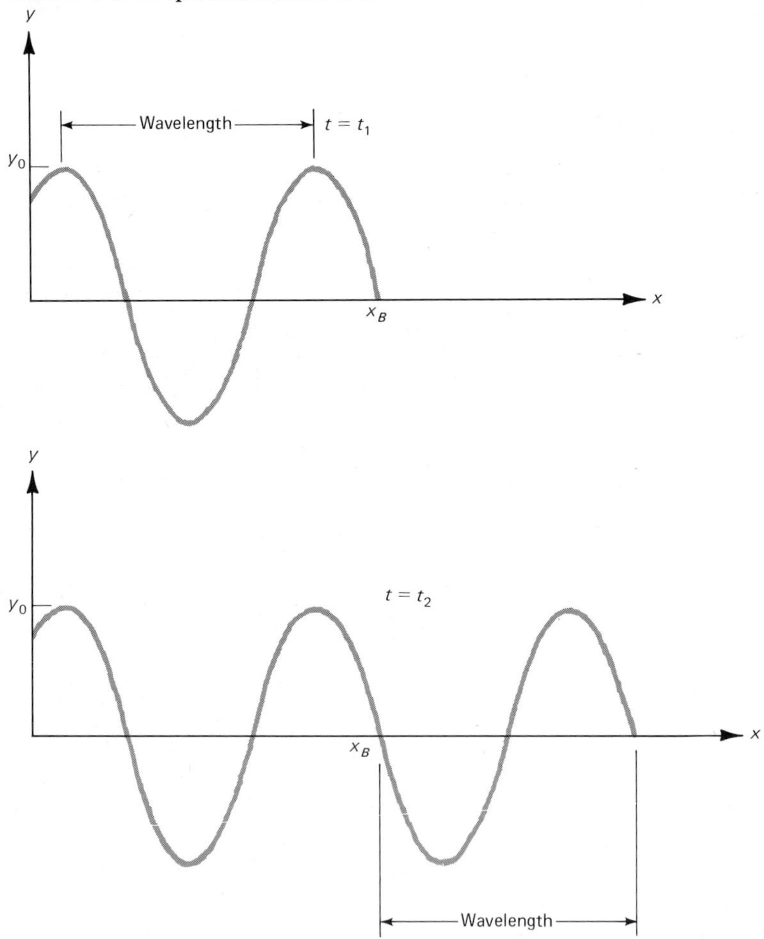

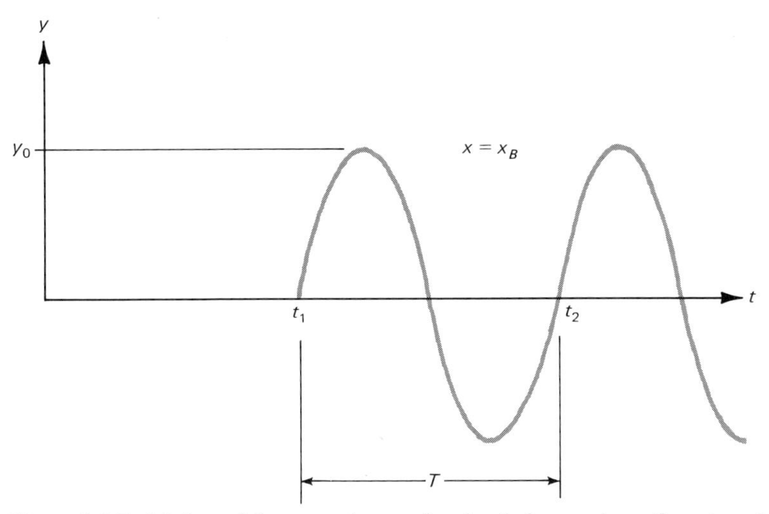

Figure 6-2-7 Motion of the rope at $x = x_B$ for simple harmonic motion at $x = 0$ beginning at $t = 0$. The amplitude y_0 and period T are shown.

traveling sine wave. In Figs. 6-2-6 and 6-2-7, the amplitude y_0 and period T are shown. In addition, a new quantity is indicated in Fig. 6-2-6, the *wavelength.* The wavelength of a traveling sine wave is the distance along the direction of motion for one complete cycle of the wave at one instant of time.

We can derive an important relationship by carefully comparing Figs. 6-2-6 and 6-2-7. Notice that the time difference $t_2 - t_1$ is equal to the period T, from Fig. 6-2-7. Also, note that during this time T the wave has traveled a distance equal to one wavelength, from Fig. 6-2-6. The symbol for wavelength is the Greek letter λ (lambda). If the speed of the wave is v, then we have found that

$$vT = \lambda \qquad \text{6-2-1}$$

The wavelength is the distance traveled by the wave in a time equal to one period.

So far we have used the example of waves on a rope to discuss wave motion in general and traveling sine waves in particular. Little has been said about the basic physics which leads to the possibility of wave propagation. One physical model of a rope, which can be used to show that waves propagate along it, is shown in Fig. 6-2-8. Although a rope doesn't stretch a great deal, unless it is made of rubber, it does stretch like a spring, by an amount proportional to the tension. In order to study wave motion

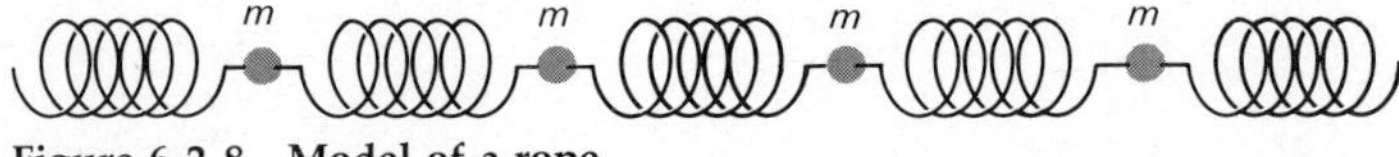

Figure 6-2-8 Model of a rope.

of the rope, its mass must not be neglected, and Fig. 6-2-8 is thus a model in which a rope is represented by a long spring with mass. For ease in calculating, the "springiness" and the "massiveness" are separated by assuming a line of massless springs connecting a line of point masses.

The physics for determining the motion of an array of many springs and many masses merely involves the use of $F = ma$ for each mass, with F found from the spring forces. But the resulting mathematics is too complex to go into here. The result is that the small motions of the individual parts of the rope, determined from Newton's mechanics, lead to a wave motion which propagates along the rope with a velocity given by $\sqrt{T/m_0}$, where m_0 is the mass per unit length and T is the tension (*not* the period). This is the correct velocity for waves on a rope, of the type which we have been studying.

The model in Fig. 6-2-8 suggests that there might be two kinds of wave which are possible, *transverse* and *longitudinal.* In Fig. 6-2-9, the line of masses is shown at one instant of time when there is a transverse wave, like the waves we have been considering,

Figure 6-2-9 Waves on a model rope, with (*a*) showing a transverse wave and (*b*) a longitudinal wave.

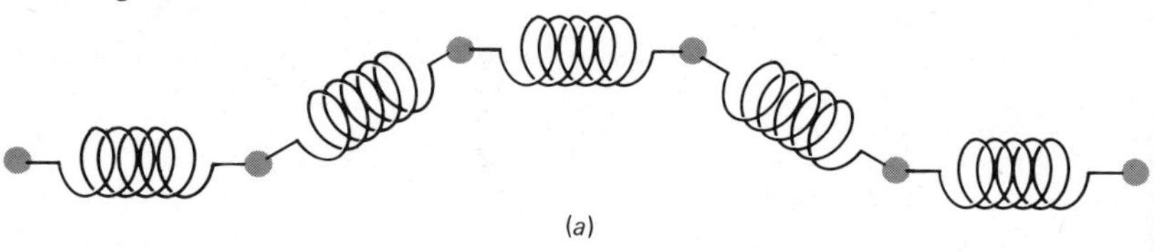

(*a*)

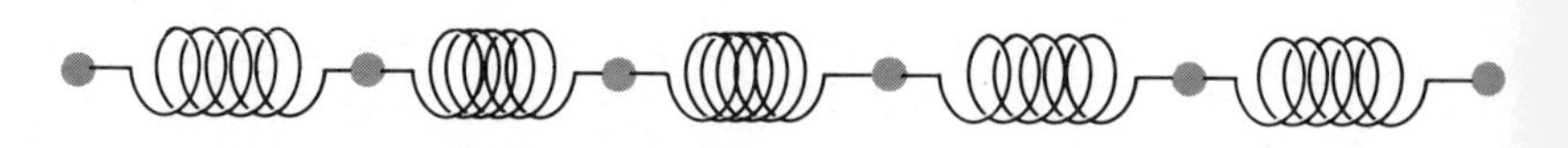

(*b*)

on the rope. This figure also shows a longitudinal wave at one instant of time. Here each mass moves a small distance *along the line of the rope* as the wave passes, instead of moving *perpendicular* to the rope as for a transverse wave. The result is a wave in which the line of masses remains straight but a *compressive* pulse propagates along it. At any instant of time this kind of pulse produces a pushing together of the masses at the location of the pulse.

For a rope the longitudinal waves are hard to observe without special instrumentation. The velocity of propagation given above, $v = \sqrt{T/m_0}$, applies only to the transverse waves, and longitudinal waves in general have a different velocity.

Basically, all matter is composed of atoms which are bound together by forces which allow a certain amount of springiness. Thus a model like that shown in Fig. 6-2-8 actually should apply to all matter, provided the masses and springs extend in three dimensions instead of along one line. We can think of the masses as individual atoms and the springs as the forces between them. The mathematics in three dimensions is even more complicated than for a single line of masses, but once again both longitudinal and transverse waves are found to be possible.

An everyday example of a longitudinal wave motion in three dimensions is the propagation of sound through air. Figure 6-2-10

Figure 6-2-10 A loudspeaker for generating sound waves and a microphone for detecting them.

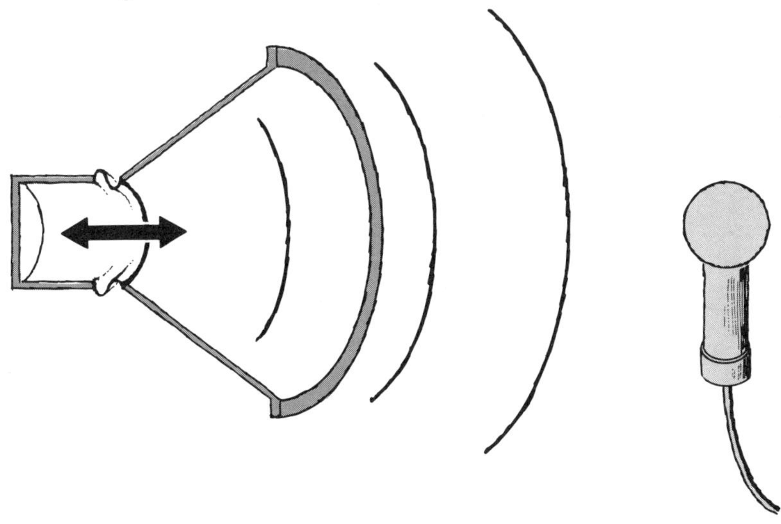

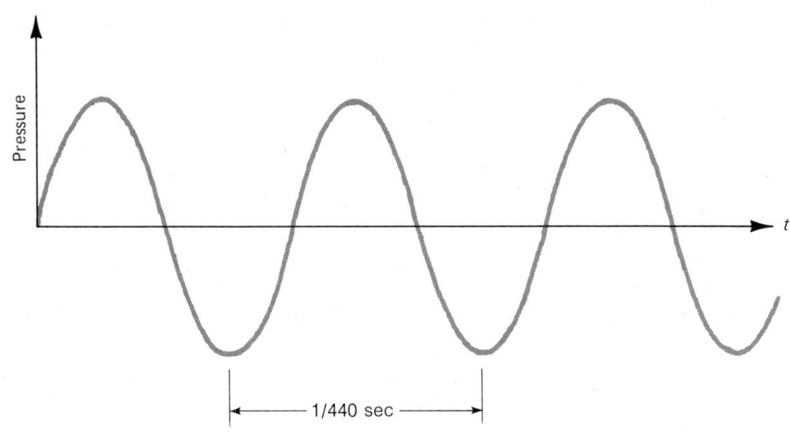

Figure 6-2-11 Pressure vs. time at the microphone, for a sound wave with a period of 1/440 sec.

shows a loudspeaker and a pressure gage which can respond to very small and rapid changes in air pressure. A microphone, or the human ear, is basically such a pressure gage. If the loudspeaker is broadcasting the sound of a tuning fork tuned to the note A above middle C on the piano, then a curve of pressure against time at the microphone will be as shown in Fig. 6-2-11. A tuning fork vibrated with an almost perfect sine wave, and the period for this note is $1/440$ sec.

For a note with a period of $1/440$ sec, the loudspeaker moves a sheet of material back and forth so as to compress the air next to it 440 times per second. This generates a longitudinal pressure wave which can be detected by an ear or a microphone. The wave travels out from the loudspeaker so that far away the places where the pressure is a maximum are along surfaces which are parts of a sphere. Such a wave is called a *spherical wave.*

Figure 6-2-12 shows a general picture for an ideal spherical wave motion. The waves propagate radially outward from a point source with velocity v. At any instant of time the points where the wave has maximum amplitude, the "wave crests," lie on spherical surfaces separated by one wavelength. Although spherical waves are not visible in an everyday context, circular waves on a water surface can easily be generated by throwing a rock into a pond, or even by dropping a small object into a full bathtub.

The only other kind of three-dimensional wave with which we will be concerned is called *a plane wave.* This is actually the

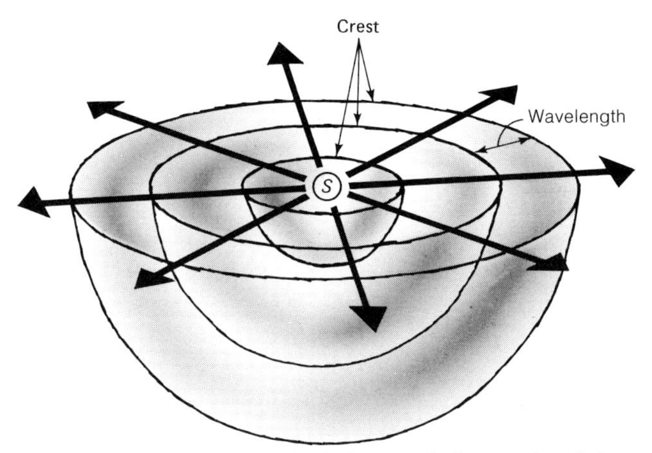

Figure 6-2-12 An ideal spherical wave being produced by a source S. The wave propagates radially outward, as shown by the arrows, and the wave crests move like expanding spheres.

simplest kind of three-dimensional wave, in which the wave crests at an instant of time lie on parallel planes a distance apart equal to the wavelength. Very far from a point source, a small area of spherical wave looks like a plane wave. Figure 6-2-13 shows a plane wave traveling in the $+x$ direction.

PROBLEM 4
The velocity of sound in air is about 1,000 ft/sec. What is the wavelength of the waves sent out by a tuning fork vibrating with a period of $1/440$ sec?

PROBLEM 5
The velocity of sound in water is much greater than in air. If the loudspeaker and microphone in Fig. 6-2-10 were operated underwater, explain whether the curve in Fig. 6-2-11 would differ. (Assume the apparatus involved is waterproof!)

6-3 INTERFERENCE

We have often utilized the principle of superposition for forces. That is, if two or more forces act on a body, the net force is found by vector addition of the individual forces. Each individual force acts as if the others were not present. A principle of superposition

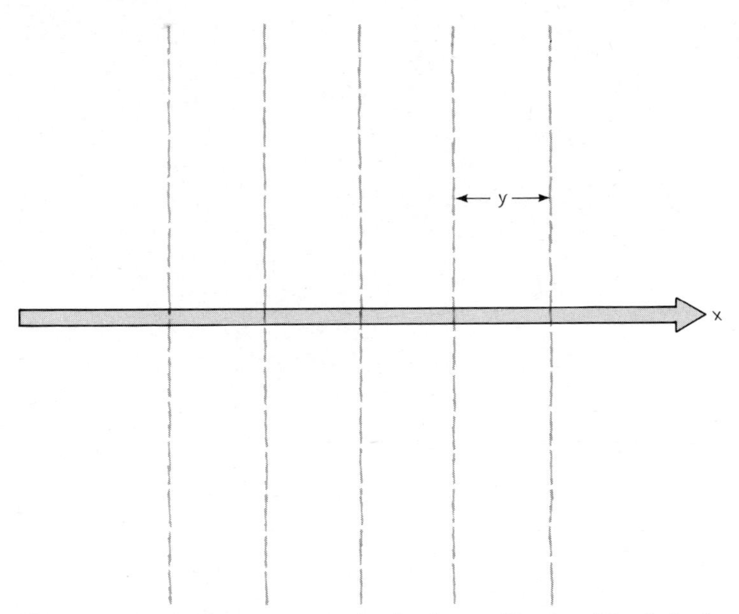

Figure 6-2-13 A plane wave moving in the $+x$ direction. The dashed lines show the wave crests.

also applies to wave motions. *If two or more waves are present, the net effect is calculated by combining the effects of all the waves, the effect of each wave being calculated as if the others were not present.*

A rather dramatic illustration of the principle of superposition for waves is shown in Fig. 6-3-1. This figure represents an example of two wave pulses traveling in opposite directions along a rope. In the example, the waves have identical shapes, but one produces a displacement in the y direction which is the opposite of the other. At time t_1 the waves are shown approaching each other; at time t_2 they are crossing; and at time t_3 they have crossed and are continuing on their ways.

There is a striking effect at time t_2 in Fig. 6-3-1. The dashed lines shown for this time show each wave as if the other were absent. According to the principle of superposition, the net displacement of the rope is found by summing the displacements given by the dashed curves. When the displacements are summed in this way, the solid curve results, and the effects of the two waves partially cancel. At one instant of time, shortly after t_2, the waves will cancel completely and the rope will look *completely straight*! Having chosen the two waves to be in a sense "equal and

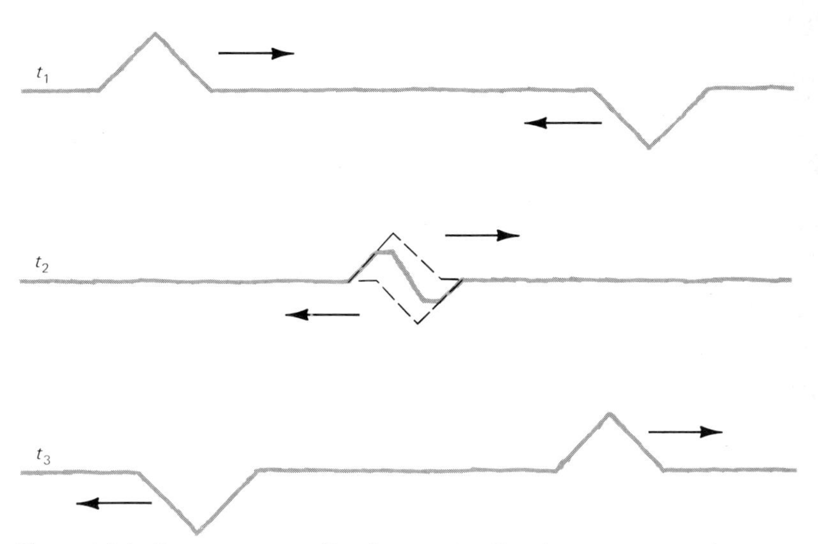

Figure 6-3-1 Two waves traveling in opposite directions on a rope, shown at three different times.

opposite" to each other, this complete cancellation is possible. When two waves meet, the net effect, found from the principle of superposition, is not generally zero. However, the net effect is always found by adding the effects of both waves, and it is possible for the sum to be zero.

At time t_3 in Fig. 6-3-1, after the waves have crossed, they are proceeding onward with their original shapes. Although the sum of the two waves does some strange things when the waves cross, neither wave actually disturbs the other. When the waves are crossing, they are said to *interfere* with each other, and the phenomenon of *destructive intereference*, in which two waves cancel, is one of the most striking features of wave motion. However, this interference is the result of superposition of two waves, *each unaffected by the other.* The waves interfere because the principle of superposition results in their effects cancelling, but they do not interfere in the colloquial sense of the word, which might imply that they change each other.

PROBLEM 6

Sketch the resultant of the two waves in Fig. 6-3-1 a short time after they cancel exactly. Let the time you take be as long after the instant when they cancel as t_2 is before it.

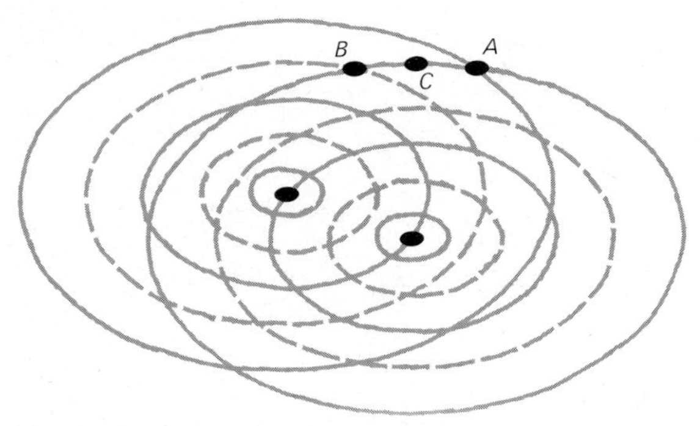

Figure 6-3-2 Top view of waves on a water surface generated by motion of two mechanical fingers. The solid circles represent crests and the dashed circles represent troughs, at one instant of time. The fingers are represented by the black dots at the centers of the circles.

Interference can also take place between sine waves. Figure 6-3-2 shows a top view of a water tank in which two mechanical fingers move up and down with simple harmonic motion. It is assumed that each finger moves in synchronism with the other, with the same amplitude and period. The fingers are sources for sine waves on the surface of the water. In the figure, which is like a snapshot at one instant of time, the solid lines show the crests and the dashed lines show the troughs of the sine waves.

In order to find the real wave pattern on the water surface, it is necessary to superpose the effects of the waves from both sources. Notice that at point A two crests overlap, while at point B a crest of one wave meets a trough of the other. At the instant of time represented by Fig. 6-3-2, the waves *interfere constructively* at A and *interfere destructively* at B.

Figure 6-3-3 shows the same picture as Fig. 6-3-2 at a time $T/2$ later, where T is the period of the waves. In a time T a wave travels one wavelength, so in this time interval the wave from each source has moved one-half a wavelength. Therefore, the crests in Fig. 6-3-3 are where the troughs were in Fig. 6-3-2. Similarly, the troughs in Fig. 6-3-3 are where the crests were in the preceding figure.

Once again, the two points A and B are shown at the same positions as in Fig. 6-3-2. There is still destructive interference at point B, where a crest and a trough again add. At point A, two troughs add to produce a bigger trough. This interference is still

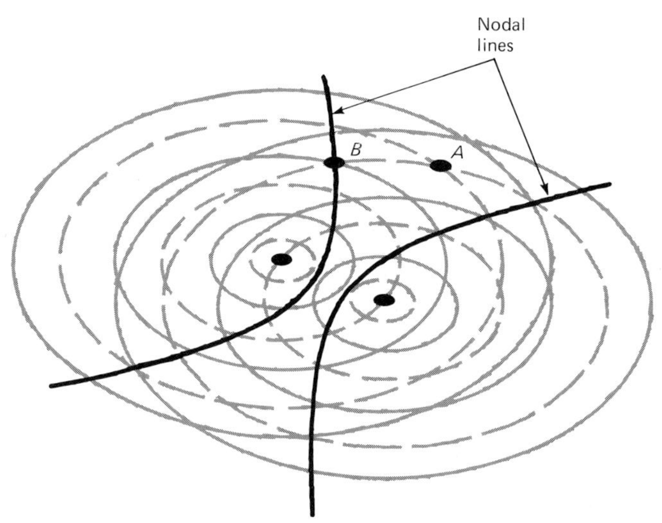

Figure 6-3-3 Same as Fig. 6-3-2, but at an instant of time $T/2$ later than that of the preceding figure. Two nodal lines are shown, along which the waves cancel at all times as a result of destructive interference.

constructive, in that the effects of the two waves combine to give a doubled displacement. Point B is called a *node*, where the two waves always destructively interfere leading to no motion. In Fig. 6-3-3 two nodal lines are shown, one passing through point B. There is destructive intereference all along these lines. Along the *antinode* line through point A there is constructive interference.

Another way of analyzing the way waves from the two sources interfere at points A and B is to draw curves of y vs. t at each point, where y is the height of the water measured from the height when the fingers are still and the water surface is flat. In Fig. 6-3-4, y vs. t is shown at the point B. The dashed curve labeled with the numeral 1 shows y vs. t for one source only, and the dashed curve numbered 2 shows y vs. t for the other source. The actual wave at point B when both sources are operating is the superposition of these two waves. They cancel exactly at every time, so at the point B the water never moves—a strange but true result.

Perhaps the variation of y with t at point A is now a foregone conclusion. However, in Fig. 6-3-5, for completeness, a dashed curve shows y vs. t for *either* source, and the solid curve shows

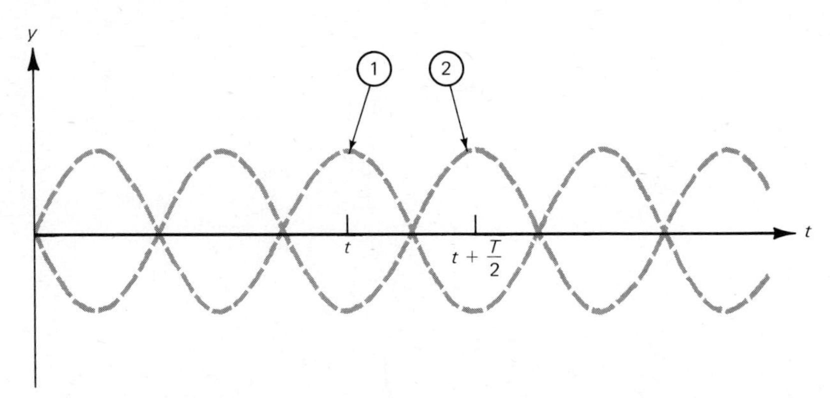

Figure 6-3-4 **Waves at the point B from the two sources shown in Figs. 6-3-2 and 6-3-3. The two waves, shown by curves numbered 1 and 2, interfere destructively to give zero displacement of the water surface at all times.**

a sine curve with twice the amplitude of the dashed curve, which results when the effects of both sources combine at this point.

PROBLEM 7

A line can be drawn on Fig. 6-3-2 which passes through point A and along which there is constructive interference at every point. Show, with a sketch, the relation of this line to point A and the two sources.

Figure 6-3-5 **Constructive interference at point A of Fig. 6-3-2 or 6-3-3. The dashed line represents the wave from either source.**

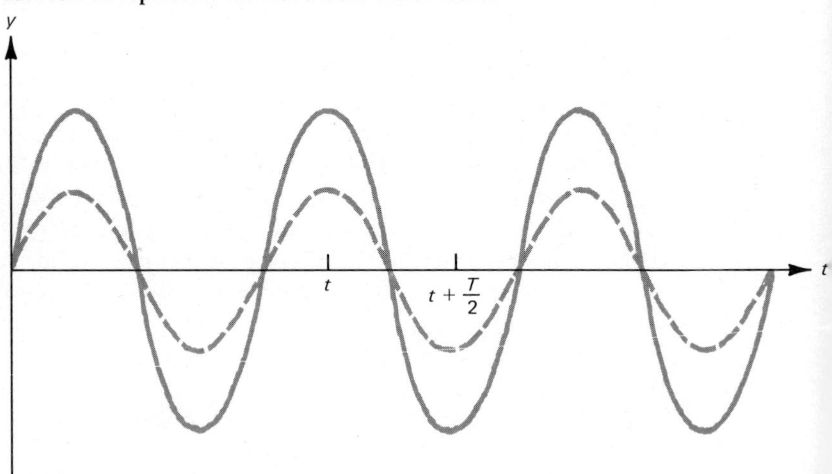

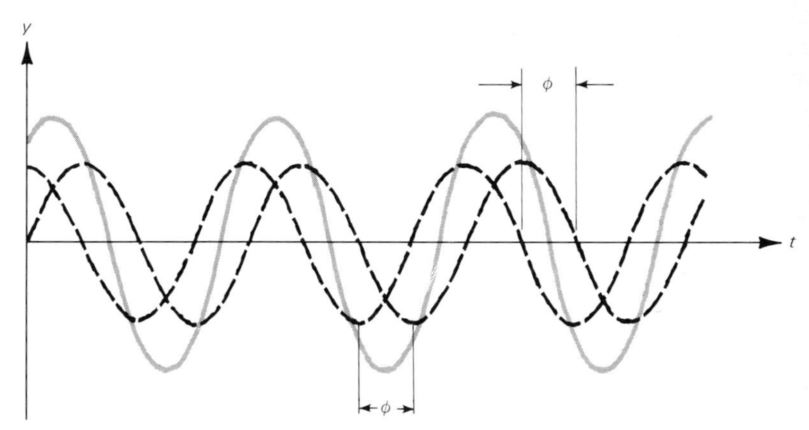

Figure 6-3-6 Two sine waves with equal amplitudes which have a phase difference, ϕ, equal to $T/4$. The solid curve shows the sum of the two waves.

When studying the interference of two waves with the same period, the concept of the *phase difference* between them is very convenient. For example, in Fig. 6-3-6, the dashed curves show two sine waves with equal amplitude which produce a displacement y at some point. For example, this figure could refer to point C in Fig. 6-3-2. The phase difference between these two waves is one-quarter of a period, $T/4$, and is labeled with the symbol ϕ (the Greek letter phi, pronounced "fee"). The phase difference is the time difference between corresponding points of two waves. You can see that ϕ is the same between two crests, two troughs, or any two similar points of the waves.

We will usually find it simplest to measure ϕ in periods or fractions of a period. Notice that in Fig. 6-3-4 the two waves which interfere destructively have a phase difference $T/2$, while in Fig. 6-3-5 the waves which interfere constructively have a phase difference 0. If we are given the amplitude of two waves, and their phase difference, we can then find out how they interfere.

Figure 6-3-6 also shows the sum of the two waves which have a phase difference of $T/4$. The sum is still a sine wave, and the sum of two or more sine waves with the same period and with any phase difference between them is still a sine wave.

PROBLEM 8

Make a sketch of y vs. t, like Fig. 6-3-6, for two sine waves with a phase difference of $(3/4)T$. Explain why the resultant wave has the same amplitude as it would have if the phase difference were $(1/4)T$.

6-4 LIGHT WAVES

Beginning with Huygens in the seventeenth century, there was a suspicion that light was in reality some kind of wave motion. However, it was not until the beginning of the nineteenth century that overwhelming experimental evidence supported this idea. One of the most clearcut experiments was performed in 1802 by an Englishman, Thomas Young, who found interference effects for light which were exactly like those for water waves in the example discussed in the previous section. Young was a brilliant medical student, with a special interest in sight and hearing, who made important contributions to the physics of sound and light, studied the tides, pioneered in the development of life-expectancy tables, and ultimately became an Egyptologist and deciphered the famous Rosetta stone.

The apparatus for performing Young's interference experiment is shown schematically in Fig. 6-4-1. Light from a small source is incident on two very narrow vertical slits. Only the light which passes through the slits is permitted to illuminate the screen. If the slits are sufficiently narrow, with widths of perhaps a few thousandths of an inch, then the light from each slit is found to spread out and illuminate a larger area of the screen than would be struck by particles traveling in straight lines from the very small "point" light source. This phenomenon, called *diffraction*, is characteristic of wave motion.

Figure 6-4-1 Schematic view of Young's apparatus. Light from the source which passed through the two narrow slits was observed on the screen. In the figure, the slit widths are much enlarged.

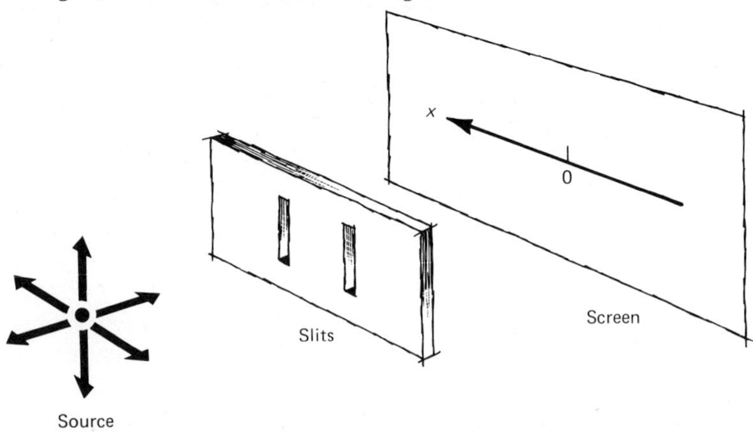

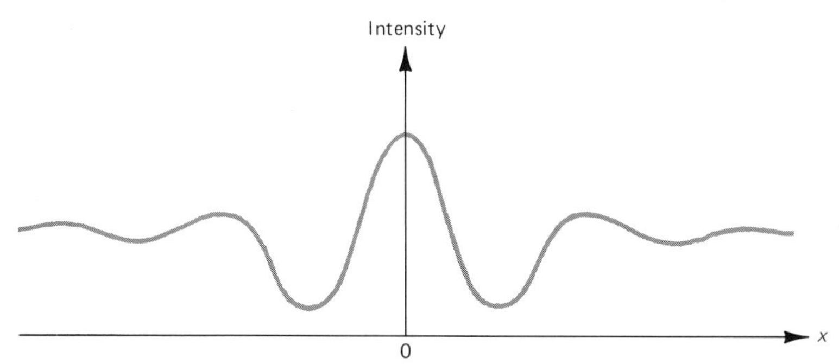

Figure 6-4-2 Slit widths are much enlarged. Intensity of light on the screen plotted against the x coordinate, for Young's experiment with white light.

The quantitative interpretation of diffraction is so complicated that we will not discuss it here. The observation of diffraction was, however, one of many relatively indirect pieces of evidence which supported the wave interpretation of light. In Young's apparatus, diffraction caused the light from the two narrow slits, which were very close together, to illuminate essentially the same area on a screen placed some feet away. Thus, Fig. 6-4-1 is certainly not drawn to scale.

In Fig. 6-4-1 an x axis is shown, with $x = 0$ at the center of the screen, and Fig. 6-4-2 shows how the light intensity on the screen is found to depend on x when white light passes through both slits. There is a bright region at the center, a dark region, and some further, less pronounced variations of intensity. However, if light of any one color shines on the two slits, from a special light source or from a white light source with a colored filter, then the intensity varies much more dramatically with x, as shown in Fig. 6-4-3.

For a source of one color, the intensity pattern can be clearly interpreted in terms of wave interference. Figure 6-4-4 shows the region of the apparatus which includes the double slit and the screen. Suppose the two slits are light sources which send out waves in synchronism, like the two mechanical fingers in the water tank discussed in the previous section. Since the distances to point *0* are equal, each wave takes the same time to travel from the slit to point *0*. Two waves crests sent out from the two slits at a given time arrive at point *0* together. Thus, if waves are sent out "in phase" (with zero phase difference) by both slits, then they are still in phase at point *0*.

Suppose now that s, the difference in the distances from the

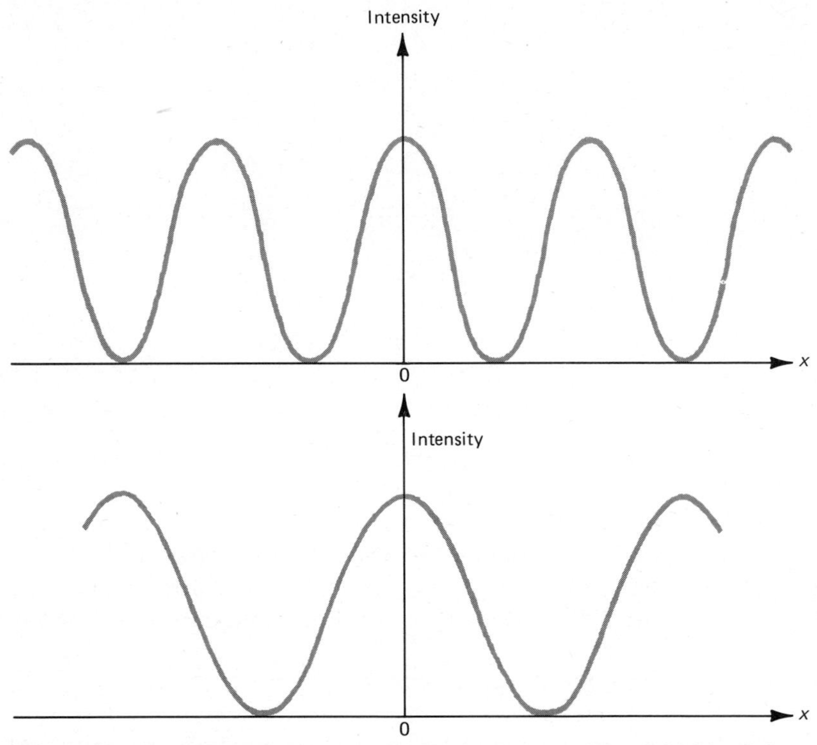

Figure 6-4-3 Variation of intensity with x for Young's experiment with blue light (top) or red light (bottom).

slits to point A, is such that the time for light waves to travel the distance s is $T/2$, half their period. Then waves which are emitted in phase at the slits have a phase difference of $T/2$ when they arrive at point A. A difference in transit times leads to a difference in phase. Figure 6-4-5 shows two waves at A with equal amplitudes and phase difference $T/2$ and, as we have seen before, the result is destructive interference so as to produce zero amplitude at A.

Figure 6-4-6 shows the slits and screen again, with two additional points B and C on the screen. The path-length difference s_B will cause a difference in transit times for the light from the two slits to reach point B. Suppose the time for light to travel the distance s_B is one full period T. In this case, the waves at point B will interfere constructively, since a phase difference of one period

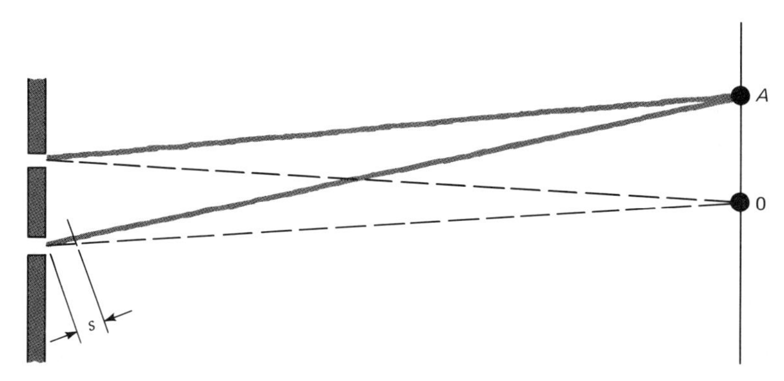

Figure 6-4-4 Enlarged view of part of Young's apparatus. The distance s is the difference in path lengths from the two slits to point A on the screen.

merely causes a wave crest emitted by the farther slit to arrive at B at the same time as the immediately following wave crest emitted by the nearer slit. For steady sine waves, a phase difference of any whole number of periods is the same as no phase difference at all.

Suppose point C is located so that the path-length difference from the two slits leads to a phase difference of $(3/2)T$. The difference $(3/2)T$ is equal to $T + (1/2)T$, and a phase difference of T is the same as no phase difference at all. Therefore, a phase differ-

Figure 6-4-5 Destructive interference at the point A of Fig. 6-4-4, assuming the path-length difference s causes a phase difference $T/2$ between the two waves at A.

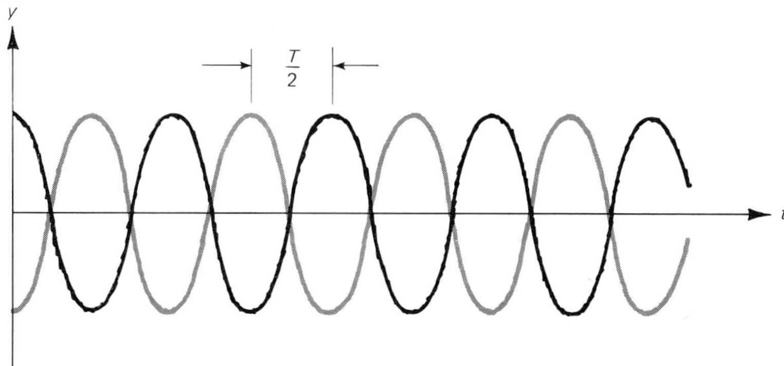

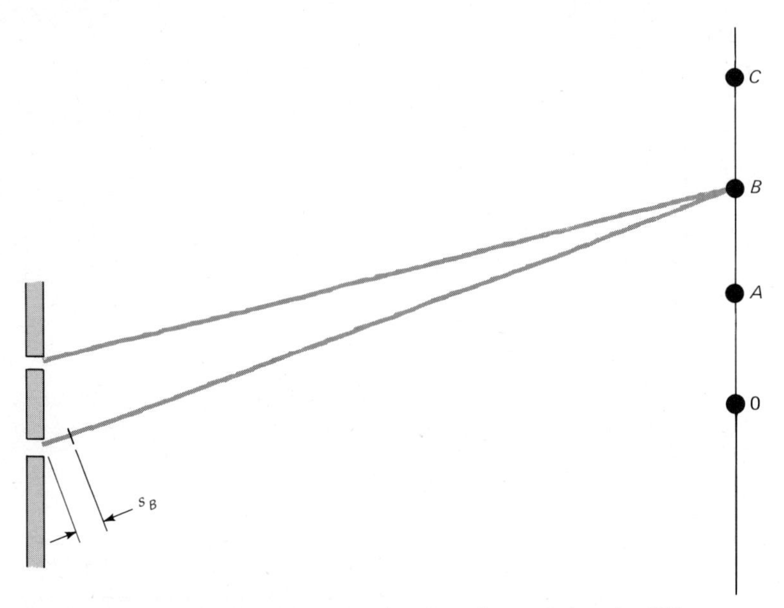

Figure 6-4-6 Young's apparatus, showing the path-length difference s_B for point B on the screen.

ence of $(3/2)T$ has the same effect as a phase difference of $T/2$, and thus, there is destructive interference at point C.

Summarizing, at a series of points along the screen, *0*, A, B, C, waves assumed to be emitted in phase at the two slits alternately interfere constructively and destructively. At the points in between, the waves can be combined, taking into account their phase difference, and the amplitude of the resultant wave can be calculated.

If the wave amplitude calculated for the light at each point on the screen in Young's experiment is squared, a curve like that shown in Fig. 6-4-7 results. This is exactly the shape of the experimentally observed light-intensity curves in Fig. 6-4-3. The energy carried by a mechanical wave, such as a wave on a rope or a sound wave in the air, is proportional to the square of its amplitude, just as is the total energy of a single particle executing simple harmonic motion. The *light intensity*, which is the energy per unit area carried by the incident light, was also found to be proportional to the square of a wave amplitude.

From Young's experiment, as well as others, the only natural interpretation was that light behaves like a wave motion. Furthermore, its wavelength could be found. The path-length difference s, appropriate to any given point on the screen in Young's experiment, can be calculated from a knowledge of the distance between the slits, the distance from the slits to the screen, and the location of the point on the screen. In particular, consider the path-length differences for two successive points which correspond to successive maxima in the light intensity, such as B and D in Fig. 6-4-7. If the path-length difference for one point leads to a phase difference T, as at point B in our discussion, then the next maximum at point D corresponds to a phase difference $2T$, greater by one period. The calculated distances s_B and s_D must therefore differ by one wavelength, since a wavelength is the distance traveled in a time T.

Thus, from the spacing between maxima (or minima) of the pattern on the screen of a two-slit light interference experiment, the wavelength of the light could be inferred. For various colors, the wavelength is different, which is why the patterns in Fig. 6-4-3 have a different spacing along the screen. A good average wavelength for visible light is about 5×10^{-7} m, which corresponds to a green color. Along the spectrum of colors made by the prism in Fig. 6-1-9, the wavelength varies from about 6.5×10^{-7} m for red light to about 4×10^{-7} m for violet light.

PROBLEM 9

For Young's apparatus, shown in Fig. 6-4-4, there is an approximate relationship,

Figure 6-4-7 Variation of the square of the amplitude of the resultant wave in Young's experiment with position x along the screen.

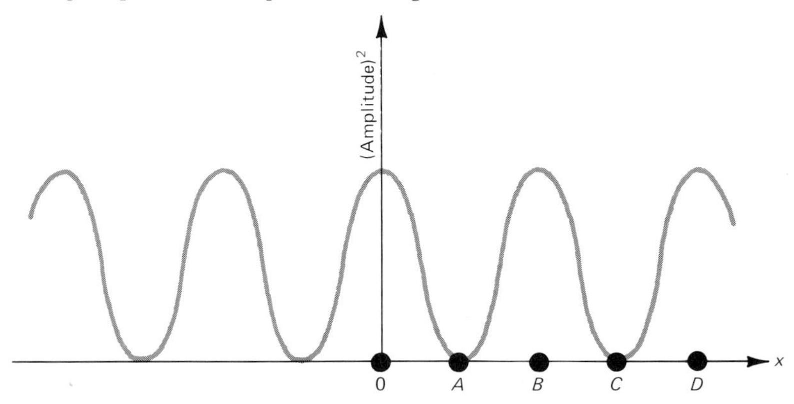

$$\frac{\overline{OA}}{D} = \frac{s}{d}$$

where $\overline{OA}$ is the distance from point O to point A, D is the distance from a point midway between the slits to point O, d is the distance between the centers of the slits, and s is the path-length difference for paths from either slit to point A. This equation is an excellent approximation if OA and d are much less than D. Verify this equation by making measurements on a drawing of the apparatus. Prove it if you can. It isn't easy, but a helpful hint might be to draw a line from A to the midpoint between the two slits and look for useful similar triangles.

PROBLEM 10

In Fig. 6-4-3, if the wavelength of the blue light is 4.5×10^{-7} m, estimate the wavelength of the red light, explaining your reasoning.

Since the velocity of light is known to be 3.0×10^8 m/sec, the period of light waves can be found from the wavelength λ by using the general relation for any wave, $vT = \lambda$. For light, $v = c$, so the period is given by

$$T = \frac{\lambda}{c} \qquad 6\text{-}4\text{-}1$$

For green light, with $\lambda = 5 \times 10^{-7}$ m, we find $T = 1.67 \times 10^{-15}$ sec. The small wavelength and extremely short period of light waves help explain why it took so long before experiments were done which unequivocally established that light behaved like a wave motion. However, the phenomenon of interference is so specifically "wavelike" that after Young's experiment any nonwave model for light was at an almost impossible disadvantage.

It should be emphasized that the straightforward interpretation of Young's experiment involves the assumption that waves interfere which are emitted from the two slits with no phase difference. For most light sources, the phase of the light has a well-defined value only for a period of time which is typically 10^{-9} sec, or less. With one source and two slits the changing phase doesn't matter, since the light waves from both slits change phase together, but with two ordinary independent light sources the phase difference changes so rapidly that interference phenomena cannot be directly observed.

The invention of the "laser" has just recently made it possible to perform interference experiments with two separate light sources. The phase of the light emitted by a laser is sufficiently stable that two of them can be made to interfere in a given way—for example, either constructively or destructively—for a time of order 10^{-3} sec.

PROBLEM 11

The x axis in Fig. 6-2-13 is a "ray" for the plane wave shown in that figure. The light rays shown in Fig. 6-1-4 can be interpreted as the directions of plane waves. Sketch some wave crests for a ray being refracted, like one of those in Fig. 6-1-4. Show from your sketch that successive crests are closer together in glass than in air. If this is true, in which medium is the velocity higher if the wave model is correct?

6-5 ELECTROMAGNETIC WAVES

For a long time after light was known to behave like a wave motion, little progress was made in understanding the nature of the waves. Sound waves and waves on a water surface can be understood in terms of motions of air or water. The sound waves are longitudinal compressions, and the water waves are transverse displacements. What were light waves?

Finally, in the mid-nineteenth century, James Clerk Maxwell, a British theoretical physicist, understood the nature of light waves. He was working with the theory of electromagnetism, the combined subject of electricity and magnetism, at a time when somewhat more was known about electromagnetism than we have so far discussed. We have primarily been concerned with electric and magnetic fields which don't change with time. In Maxwell's day, it was known that if a magnetic field changed with time, it gave rise to an electric field. Maxwell hypothesized that, similarly, a changing electric field would give rise to a magnetic field. He got this idea while trying to write down a more unified theory of electromagnetism. It led to an elegant and symmetric set of equations which described correctly all the then known electromagnetic phenomena and, in addition, led to a striking new prediction.

According to Maxwell's theory, transverse traveling electromagnetic waves were possible. A simple example of an electromagnetic wave is a plane sine wave, traveling, for example, in the

$+x$ direction. For such a plane wave, Fig. 6-5-1 shows the electric and magnetic fields at points along the x axis at one instant of time. For a plane wave, the fields along any other line parallel to the x axis vary with x in the same way. The wave is said to be transverse because the field directions are perpendicular to the direction of motion of the wave.

In Maxwell's theory, the velocity of electromagnetic waves could be predicted from constants which were already known. One was the constant in Coulomb's law, 9.0×10^9. The other constant had a similar role in the law which gives the magnetic field set up by a steady current. The predicted result for the velocity of the electromagnetic waves in vacuum was 3.0×10^8 m/sec, precisely the observed velocity of light!

The fact that Maxwell's theory predicted waves propagating in vacuum caused a great deal of consternation and controversy. All other known waves propagated in some medium, like air, water,

Figure 6-5-1 Electric and magnetic fields at one instant of time, for a plane electromagnetic wave traveling in the $+x$ direction. The fields are shown for points along the x axis. The E field is represented by dashed arrows and the B field by solid arrows. The fields are perpendicular to each other and to the direction of motion. The values of the fields vary with x like a sine wave.

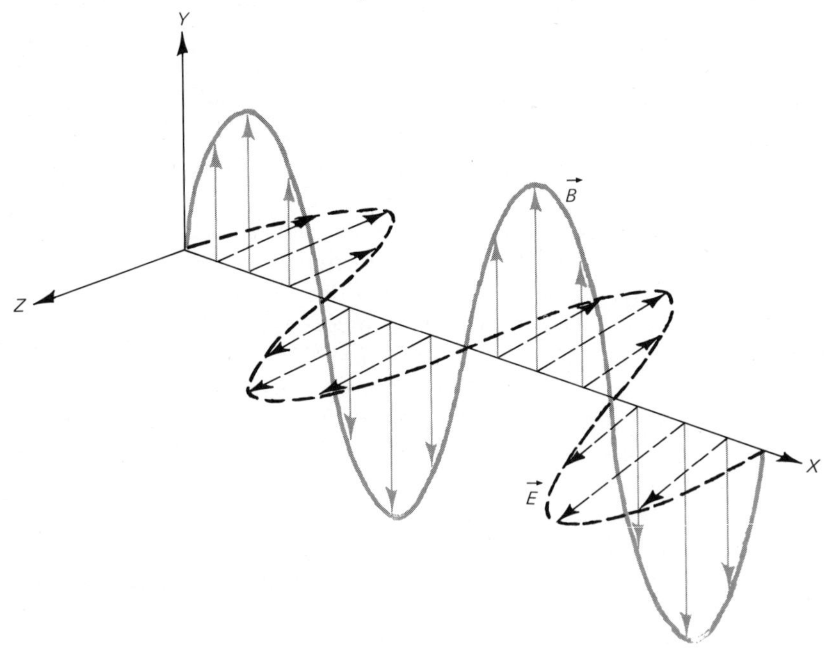

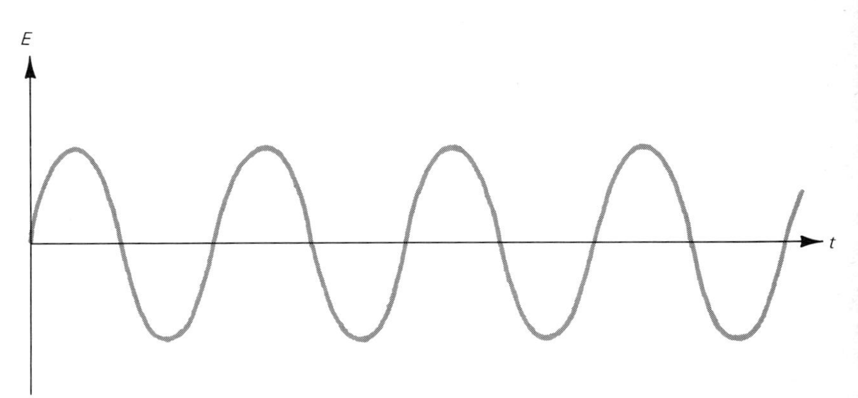

Figure 6-5-2 Variation of the electric field with time, at a given position, for the plane wave shown in Fig. 6-5-1.

or some other form of matter. However, Maxwell's waves needed no medium to propagate in. For a traveling electromagnetic sine wave, such as that illustrated at one instant of time in Fig. 6-5-1, the electric field at a given point would vary with time as shown in Fig. 6-5-2. The changing electric field gives rise to a magnetic field. In the same way, the changing magnetic field at any given point causes an electric field. At a given point, the two changing fields seem to cause each other. Spread out over space, the changing fields result in a traveling wave moving in vacuum.

The results of Maxwell's theory were quite shocking, but experiments soon supported the theory. Electromagnetic waves were created in the laboratory. Maxwell's work constitutes one of the giant steps in the development of physics, like Newton's discovery of the laws of motion and Einstein's development of the theory of relativity.

There was nothing in Maxwell's theory that specified a natural period for electromagnetic sine waves, but the theory in its more developed form predicted that an *antenna*, or source, for these waves would be most efficient for wavelengths of roughly the size of the antenna. Atoms tend to emit electromagnetic waves of very short wavelength, like visible light. Radio antennas generate electromagnetic waves with wavelengths ranging from meters to kilometers. And nuclei emit electromagnetic waves, called gamma rays, with wavelengths a millionth the wavelength of visible light. According to Maxwell's theory, all these electromagnetic waves are simply traveling transverse E and B fields.

With the full development of Maxwell's theory, and its verification in every detail, the wave theory of light seemed completely well established during the second half of the nineteenth century. However, beginning about 1900, a number of phenomena involving light were found to be difficult or impossible to explain using Maxwell's theory. The next section describes one such phenomenon.

6-6 THE PHOTOELECTRIC EFFECT

In 1887, a German physicist, Heinrich Hertz, became the first man to produce, in a laboratory experiment, the electromagnetic waves predicted by Maxwell. In the course of his research, he also discovered accidentally that when light shines on a metal surface, negatively charged particles are emitted from it. In 1899, J. J. Thomson performed an ingenious experiment, different from the one described in Chap. 4, to measure the value of q/m for these negatively charged particles. His measured value of q/m was however very nearly the same as that for the particles, christened *electrons*, which he had studied with the apparatus described in Chap. 4. Electrons were thus found to be emitted when light struck a metal surface, a phenomenon which became known as the *photoelectric effect.*

While a great many experiments were performed to study the photoelectric effect, technical difficulties often led to conflicting or uncertain results. None of the experimenters was fortunate enough to have hit upon the correct explanation of the data when Albert Einstein, in 1905, proposed a daring interpretation. Einstein's idea owed less to the experimental data than to a previous discovery by Planck, in 1900, of an example where the wave theory of light apparently had to be abandoned. Between them, Einstein and Planck reestablished a corpuscular theory of light. They each required that light energy be carried in particlelike bundles called *quanta*, or *photons.* The photoeffect, and Einstein's interpretation of it, is the most simple and direct evidence for the existence of light quanta.

Einstein's theory can be understood by means of the energy diagram in Fig. 6-6-1. The diagram represents the potential energy for an electron in the neighborhood of the surface of a metal plate. The x coordinate is in a direction perpendicular to the plate, and $x = 0$ corresponds to the surface of the plate. One electron in the metal might, for example, have total energy U_1, as shown in Fig. 6-6-1. Since we have chosen to define the potential energy V to

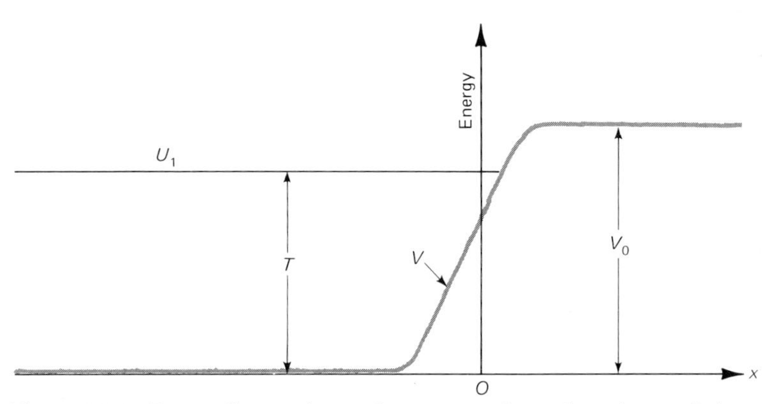

Figure 6-6-1 Energy diagram for an electron near the surface of a metal plate.

be zero well inside the metal, this total energy corresponds to an electron moving with kinetic energy equal to U_1 well inside the metal.

> PROBLEM 12
>
> The work done on a particle which moves a distance Δx is $F_x(\Delta x)$, where F_x is the component of force along the direction of motion. Assuming that energy is conserved, show that $F_x(\Delta x) = -\Delta V$, where ΔV is the change in potential energy when the particle moves Δx. If V varies with x as shown by the graph in Fig. 6-6-1, show on a graph how F_x varies with x. A qualitative picture without numerical values of F_x is all that is asked for.

According to the model represented by the energy diagram in Fig. 6-6-1 an electron with total energy U_1 moves perfectly freely, with no forces acting on it, except very near the surface of the metal. This model is reasonable in view of the experimental facts. Metals conduct electrical current so well that it is plausible to assume that electrons can move very freely within a metal. On the other hand, electrons are not continually escaping from metal conductors, so near the surface there must be a steep potential-energy "hill," which represents a strong inward force on electrons at the surface.

In the example shown in Fig. 6-6-1, an electron traveling in the $+x$ direction with total energy U_1 rapidly loses kinetic energy near the surface of the metal. Finally, at a point a little to the right of $x = 0$, where V is equal to U_1, its kinetic energy must be zero.

At this point it is stopped and then turns around and goes back in the $-x$ direction. When it has left the immediate neighborhood of the surface, its kinetic energy will again be equal to U_1, but it will be traveling in the opposite direction. In this model, the surfaces of a metal are perfect reflectors for electrons which move freely inside.

The interpretation of the photoelectric effect can now be reduced to understanding how light shining on the metal can raise the total energy of an electron from U_1 to some energy like U_2 in Fig. 6-6-2. An electron moving to the right inside the metal with total energy U_2 can escape, emerging with kinetic energy $(U_2 - V_0)$, as shown in Fig. 6-6-2. The data on the photoelectric effect available to Einstein suggested that, when light of a particular wavelength illuminated the surface of a metal, photoelectrons were emitted with kinetic energies that were always less than some maximum value which we will call T_{max}. Furthermore, if the intensity of the light was varied by an enormous factor, say 10^4, the maximum kinetic energy of the photoelectrons, T_{max}, remained the same.

The photoeffect data were very hard to understand in terms of light as an electromagnetic wave. Since such a wave involves an electric field which can exert a force on electrons, there is indeed some basis for the wave to give energy to electrons in a metal.

Figure 6-6-2 An electron with total energy U_2 can leave the metal with kinetic energy $U_2 - V_0$.

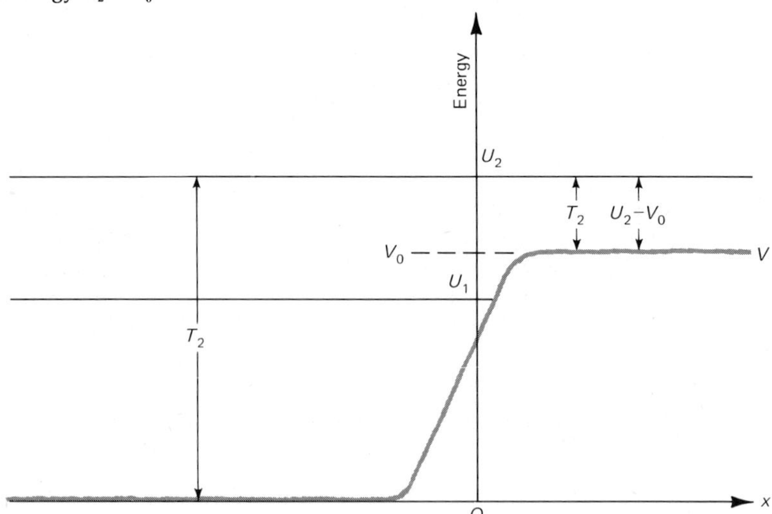

However, as the intensity of light increases, the electric field should also increase. If this is so, why should the kinetic energy T_{max} be independent of light intensity? Logically, it should *increase* for a stronger light source.

For some kinds of metal, T_{max} for the photoelectrons was measured for light of different colors, or wavelengths. This was an important clue to the explanation of the photoelectric effect. Given the velocity of light in air (which is essentially equal to c, the velocity in vacuum), the period T of each wave with a given wavelength λ can be found from Eq. 6-2-1, $T = \lambda/c$. Instead of working with the period, however, it will be more convenient now to use the frequency, which is defined as the number of periods or *cycles* per second, known as Hertz and abbreviated Hz. The frequency f is related to the period T by the simple equation

$$f = \frac{1}{T} \qquad 6\text{-}6\text{-}1$$

From Eq. 6-2-1 and Eq. 6-6-1,

$$f = \frac{c}{\lambda} \qquad 6\text{-}6\text{-}2$$

Figure P-6-13

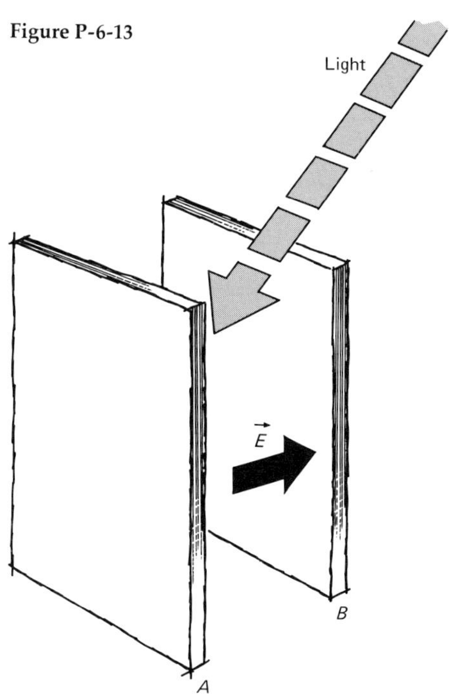

PROBLEM 13

Figure P-6-13 shows part of an apparatus for studying the photoelectric effect. There is a uniform electric field E between the plates A and B. This field tends to push photoelectrons emitted from A back toward that plate. As the magnitude of E is increased from zero, the number of photoelectrons from plate A which reach plate B decreases. Finally, at some value of E, say E_0, the flow of photoelectrons to plate B ceases. Show that

$$eE_0d = T_{\text{max}}$$

where d is the plate separation.

Figure 6-6-3 shows a typical graph of T_{max} vs. frequency for a good modern experiment in which a particular metal is illuminated with light of various colors. While T_{max} does not depend on the intensity of the light, it certainly does depend on the color, or frequency. In fact, below a certain frequency, f_0, no photoelectrons are emitted, while for frequencies above this value T_{max} follows a very simple straight line graph. The mathematical form of the graph shown in Fig. 6-6-3 is

$$T_{\text{max}} = k(f - f_0) = kf - kf_0 \qquad \text{6-6-3}$$

Figure 6-6-3 Variation of the maximum kinetic energy of photoelectrons, T_{max}, with frequency of the incident light.

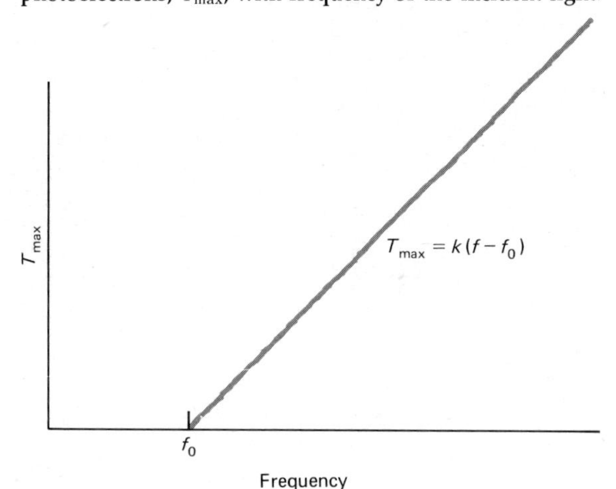

where k is a constant which can be found from the data. This equation applies for f greater than f_0. For f less than f_0 there are no photoelectrons.

Einstein's explanation of the photoelectric effect was inspired particularly by the fact that Eq. 6-6-3 fit the relatively poor data then available. He hypothesized that light was absorbed in the form of *quanta*, bundles of energy, and that for light of a given frequency the quanta always had a given energy. The energy of a single light *quantum*, or *photon*, was assumed to be given by the relation

$$E = hf \tag{6-6-4}$$

The quantity h is independent of f, and it has turned out to be one of the most important and fundamental physical constants. The value of h is

$$h = 6.625 \times 10^{-34} \text{ joule-sec} \tag{6-6-5}$$

(Like most fundamental constants, the value of h is now known to several more significant figures than are given above.)

PROBLEM 14

A radio station emits electromagnetic sine waves with a frequency of 10^5 Hz. The rate at which energy is emitted is 5×10^4 joules per second (50,000 watts). How many quanta are emitted per second, assuming Eq. 6-6-4 is true for all electromagnetic waves?

PROBLEM 15

The *electron volt* is a unit of energy defined to be equal to 1.60×10^{-19} joule. It is a very convenient unit for atomic physics. Suppose an electron is accelerated from rest in a uniform electric field of 1 N/C, for a distance of 1 m. Find its kinetic energy in joules and electron volts.

Recall that the potential energy of two point charges a distance r apart is given by

$$V = 9.0 \times 10^9 \frac{Q_1 Q_2}{r}$$

If Q_1 and Q_2 are both equal to the electron charge, e, and r is equal to 10^{-10} m, what is the value of V in electron volts?

A light quantum has a wavelength of 5×10^{-7} m. What is its energy in electron volts?

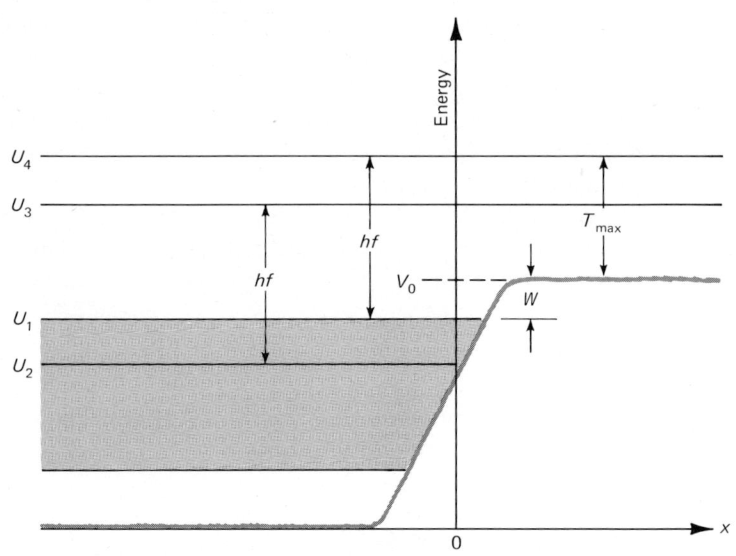

Figure 6-6-4 Energy diagram to illustrate Einstein's interpretation of the photoelectric effect. Free electrons in the metal have total energies in the shaded band. When light quanta of frequency f are absorbed, they raise the energy of one electron by an amount hf. Total energy U_4 is the maximum possible total energy for quanta with energy hf.

The interpretation of the photoelectric effect requires Eq. 6-6-4 and a model which takes account of all the "free" electrons in a metal. Figure 6-6-4 shows an energy diagram in the neighborhood of a metal surface, with a potential energy V which is the same as in Fig. 6-6-1. The shaded area represents the assumed range of total energies for the free electrons. The important feature of the model is that, while the electrons have a wide range of energies, there is a well-defined maximum energy U_1 above which there are essentially no electrons.

When light shines on the metal surface, it is assumed to be absorbed by the electrons just inside the metal. The absorption was assumed by Einstein to take place by light photons giving up quanta of energy to individual electrons, the amount of energy in every quantum being hf, as in Eq. 6-6-4. In Fig. 6-6-4 two energies, U_3 and U_4, are shown which result from absorption of quanta

hf by a "typical" electron with energy U_2 and by an electron with the maximum total energy U_1. If this model is correct, $T_{\max}$ is the kinetic energy, after it leaves the metal, of an electron which originally had maximum total energy U_1.

In terms of Einstein's picture, $T_{\max}$ is given by the equation

$$T_{\max} = hf - W \qquad 6\text{-}6\text{-}6$$

The energy W is shown in Fig. 6-6-4 and is the minimum photon energy which can possibly produce a photoelectron. This equation has exactly the same form as Eq. 6-6-3, which describes the experimental data shown in Fig. 6-6-3. The constant h in Eq. 6-6-6 is identified with the constant k in Eq. 6-6-3, and the energy W is equal to the energy hf_0.

PROBLEM 16

Photoelectrons are emitted from potassium metal when it is illuminated by light with a wavelength shorter than 6.0×10^{-7} m. For longer wavelengths, no photoelectrons are emitted. Find the energy W, in Eq. 6-6-6 for potassium. Find the maximum kinetic energy of electrons emitted by potassium when it is illuminated with light of wavelength 4.0×10^{-7} m.

To round out the experimental picture, the energy W is found to differ from one metal to another. Since it depends upon the surface forces which hold electrons in the metal, this is reasonable. The constant h is a universal constant for all metals and all frequencies of light.

While Einstein's explanation was daring, it was built on a foundation which included the striking theoretical result found by Max Planck about five years earlier. For the emission of light by an incandescent surface, Planck was the first person to succeed in making theoretical predictions which agreed with experiment. In order to do this, he was forced to assume that light was *emitted in quanta* of energy hf. The constant h is in fact called *Planck's constant.*

Complementing Planck's result, Einstein hypothesized that light was *absorbed in quanta* of energy hf. The numerical value of h required to fit the data on the photoelectric effect was the same as the value used by Planck to fit the data on emission of light. Since light appeared to be both emitted and absorbed in the form of quanta with energy hf, it was only natural to visualize the light as traveling in the form of such bundles of energy. Newton's corpuscles were back in style!

The work of Einstein and Planck resulted in a crisis in the theory of what had been called electromagnetic waves. The photoelectric experiments were not to be denied, nor were the interference experiments. Neither Maxwell's theory nor Einstein's seemed able to explain both types of phenomena, even after years of theoretical effort. Light appeared to present a particle-wave duality, to be unexplainable by either type of theory alone. In the next chapter, we shall see how this duality is finally embodied in quantum-mechanical theory.

SUMMARY

Light exhibits *refraction* and *color.* It is either a particle or wave motion with *finite velocity*

$$c = 3.00 \times 10^8 \text{ m/sec}$$

For a traveling sine wave with period T or frequency f ($f = 1/T$), the *wavelength* λ is related to the velocity and frequency by

$$v = \lambda f$$

The most important physical aspect of wave motion is the *principle of superposition*: Two or more waves combine by addition of the wave motions, taking each as if the other waves did not affect it. Application of this principle leads to *interference effects,* the most dramatic of which is the combination of two waves so that they exactly cancel, called *destructive interference.*

Observation of *interference for visible light* is strong evidence that it is a wave motion. Maxwell's equations, describing electric and magnetic phenomena, lead to the prediction of electromagnetic waves with velocity 3.00×10^8 m/sec, precisely the velocity of light. The conclusion is that *light is a form of electromagnetic wave.*

The *photoelectric effect* is an example of *particlelike behavior of light.* The equation developed by Einstein to explain the photo effect is

$$hf = T_{\text{max}} + W$$

with $h = 6.63 \times 10^{-34}$ joule-sec (Planck's constant). A particle or *photon* of light, with *energy* hf, leads to emission of photoelectrons with kinetic energies up to T_{max} from a material with *work function* W. The dual nature of light, as photon or wave, provided one of the most important foundations for the development of quantum mechanics.

QUANTUM MECHANICS

7-1 INTRODUCTION

In the preceding chapter the puzzling behavior of light, as both a particle and wave phenomenon, was discussed. The dilemma which this duality posed provided one important stimulus for the development of quantum mechanics. The second important stimulus was the realization that atoms could apparently exist only in certain *allowed states*, characterized by certain allowed values of total energy.

The simplest atom, hydrogen, consists of a single proton and a single electron. If they are a distance r apart, their potential energy, resulting from the electric attraction between them, is given by $V = 9.0 \times 10^9(e^2/r)$ (see Eq. 5-5-7). If they are moving, then their kinetic energies are added to this potential energy to find the total energy of the atom. On the basis of classical physics, the total energy of the hydrogen atom should be allowed to have any value within some range, limited solely by the requirement that the kinetic energy be low enough so that the proton and electron stay bound together.

Soon after the discovery of the nucleus by Rutherford, Geiger, and Marsden, a brilliant Danish theoretical physicist, Neils Bohr, invented a model of the hydrogen atom in which the total energy was only permitted to take on certain allowed, or *quantized,* values. As a result of this unconventional attribute of Bohr's model, many properties of hydrogen and ultimately of other atoms were satisfactorily predicted for the first time. However, the quantized energy values, which nature appeared to require, could only be put into the model by abandoning classical physics in an unsatisfying, ad hoc fashion.

From the invention of Bohr's model in 1913, until 1925, theoretical physicists grappled without much success with the problem of quantized energies, as well as with the dual nature of light. In 1924, Prince Louis de Broglie, an imaginative French physicist, suggested that the particle-wave duality might in fact be a universal attribute of nature: that just as light waves had a particle aspect, particles (like electrons) would be found to have a wave aspect. Then, in 1925 and 1926, Werner Heisenberg and Erwin Schrödinger independently invented two new theories, each of which rejected some of the most basic ideas of classical (newtonian) mechanics. Both theories predicted the quantized energies of Bohr and the universal particle-wave duality of de Broglie. After a short time, it was seen that the two theories were really different mathematical ways of expressing a single theory, now called *quantum mechanics.*

In this chapter, the basic ideas of quantum mechanics, and some applications of the theory, will be discussed. We will begin with the "matter waves" of de Broglie and then discuss the quantum theory mainly from the point of view adopted by Schrödinger. You may find the conceptual basis of quantum mechanics very difficult to believe at first. It appears to violate physical intuition based on everyday experience. Yet, not only does the theory predict correctly the behavior of atoms, it also works in the large-scale world. For large systems, in spite of its nonintuitive foundations, the theory predicts the same behavior as newtonian mechanics.

We can now think of newtonian mechanics as an approximation to the correct quantum mechanics, valid with very good accuracy for large-scale systems. In the realm of atomic physics, quantum mechanics provides our only successful basis for understanding the observed phenomena, and it has been verified in detail. We are forced, perhaps reluctantly, to regard quantum mechanics as a description of nature which is as correct as any other part of physics.

The basic equation written down by Schrödinger, which in essence replaces $\vec{F} = m\vec{a}$, is considerably more complicated mathematically. Obtaining solutions to this equation tends to require much more mathematical sophistication than is appropriate to this book. Therefore, our study of quantum mechanics will of necessity be more descriptive and less rigorous than was possible for classical mechanics. In this chapter, the essential features of quantum mechanics will be discussed by considering some very simple one-dimensional systems. In the next chapter we will discuss the quantum mechanics of real atoms.

7-2 MATTER WAVES

In 1924, Louis de Broglie suggested that if light waves appeared to behave like particles, perhaps particles, like electrons, should exhibit the characteristics of waves. In order to be more quantitative about the wavelike nature of particles, de Broglie was guided by a relationship between the wavelength and momentum of a light quantum.

When light is absorbed by an object, it gives up energy. (We have, for example, described the photoelectric effect in terms of energy absorption.) According to the electromagnetic-wave theory, light waves also carry momentum. In fact, the ratio of the energy carried by a light wave to the momentum is predicted to equal c, the velocity of light. Experiments verify this relationship between the energy and momentum of light, so that even if light really consists of quanta, their ratio of energy to momentum must also be equal to c. Using the photon energy relation $E = hf$, we can write the following equation:

$$\frac{E}{p} = \frac{hf}{p} = c \qquad \text{7-2-1}$$

For light considered as a wave phenomenon, $f\lambda = c$, where λ is the wavelength. Using this relation to substitute c/λ for f in Eq. 7-2-1,

$$\frac{hc}{\lambda p} = c \qquad \text{or} \qquad \lambda = \frac{h}{p} \qquad \text{7-2-2}$$

De Broglie proposed that Eq. 7-2-2 was valid for *any* particle with momentum p, as it is for light quanta. This conjecture did not create an immediate sensation because there was no direct experimental evidence to support it. However, a number of people who studied the de Broglie hypothesis concluded that the many experimental observations of particle motion did not in fact disprove it.

In all the particle experiments done up to the time of de Broglie's hypothesis, the wavelength predicted by Eq. 7-2-2 would be much smaller than the wavelength of light. As a result of the very short wavelengths and the nature of the experiments, interference and diffraction effects would not have been observable. While de Broglie and others proceeded with theoretical work based on the matter-wave hypothesis, at least one experimenter, G. P. Thomson (J. J. Thomson's son), set out to design an experiment specifically to observe interference effects for an electron beam.

PROBLEM 1

In an electron gun, electrons are accelerated from rest by a uniform electric field E equal to 10^6 N/C. In this gun, the electrons move a distance of 1 cm in this uniform field and then leave the gun and enter a region where E is zero. What is their momentum when they emerge from the gun? What is their wavelength? Compare this wavelength to the wavelength of green light.

To Thomson's chagrin, the first experimental verification of the wavelike behavior of electrons actually came accidentally when two Americans, Davisson and Germer, observed an interference effect which they correctly interpreted in terms of electron waves. Not long afterward, Thomson's experiment worked and also indicated that electrons were wavelike. In the next few years further experiments showed that other particles, for example, hydrogen ions and helium ions, also had wavelike properties, and in each experiment the de Broglie relation between wavelength and momentum, Eq. 7-2-2, was verified.

None of these experiments was as simple as Young's two-slit interference experiment with light. The wavelengths of the particles were so short that slit widths and slit spacings too small to be technically possible would have been required. Instead, interference effects were observed for particles scattered from crystals. In a crystal, atoms or molecules are arranged in a very regular three-dimensional array, and the scattering of waves from such an array can be understood by considering each atom to be the source of a scattered "wavelet." To illustrate this idea, Fig. 7-2-1 shows a line of atoms, with a plane wave incident from the left. Little scattered spherical wavelets are shown originating from each atom.

Figure 7-2-2 shows the same line of atoms as Fig. 7-2-1 and a point A at which the scattered intensity might be measured. Notice that the difference in path lengths from one atom to point A and from a neighboring atom to point A, labeled s on the figure, is essentially the same for all pairs of atoms if the distance to A is much greater than the spacing between atoms. If this distance s is an exact number of wavelengths, then the scattered waves will all interfere constructively at point A. The principle is very similar to that of Young's experiment, except that the distance between the "sources" whose waves interfere is the distance between two adjacent atoms. It is as if Young's experiment were performed with slits about 10^{-10} m apart, which is a typical atomic spacing in a crystal.

While Fig. 7-2-2 refers only to a single line of a few atoms, the principle of wave scattering from a real crystal is the same. However, because the atoms in a crystal are arrayed in three dimensions, the theory is more complicated and will not be discussed here. Instead, we will discuss, in the following section, a two-slit experiment for electrons which is like Young's experiment for

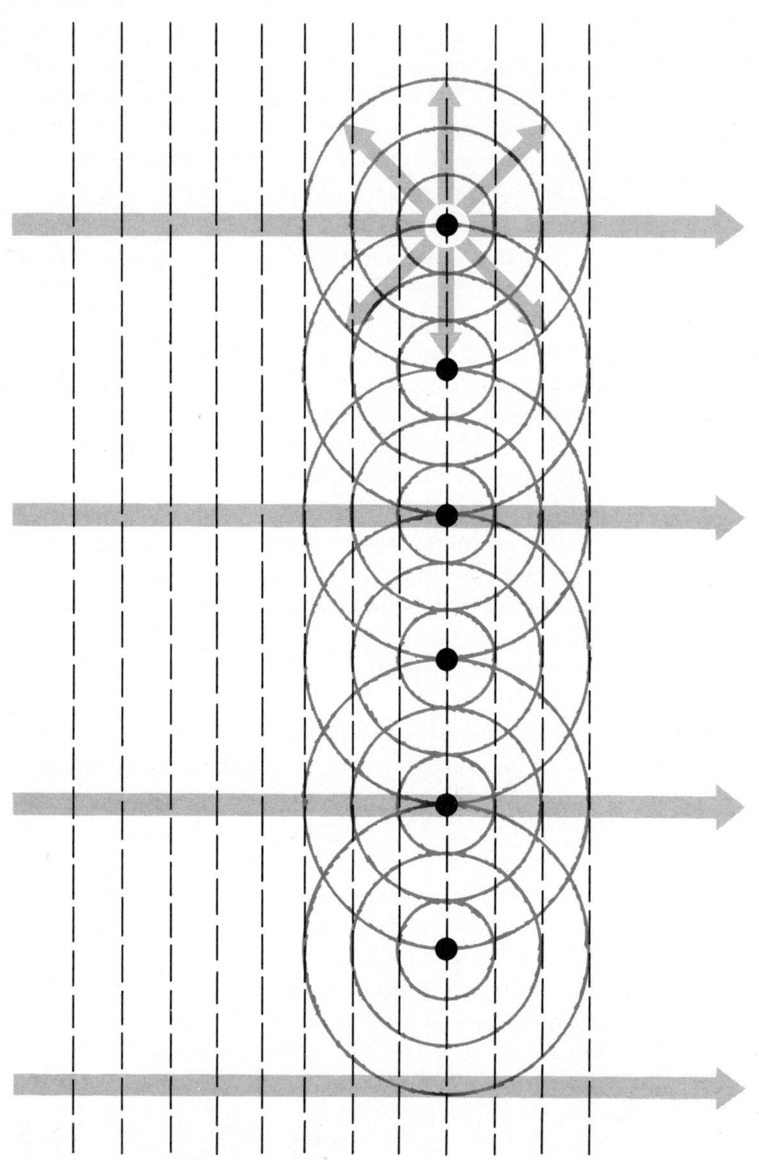

Figure 7-2-1 A plane wave traveling to the right, incident on a line of atoms. The wave crests of the plane wave are shown by dashed lines, and the wave crests of the scattered wavelets from the atoms are shown by solid lines.

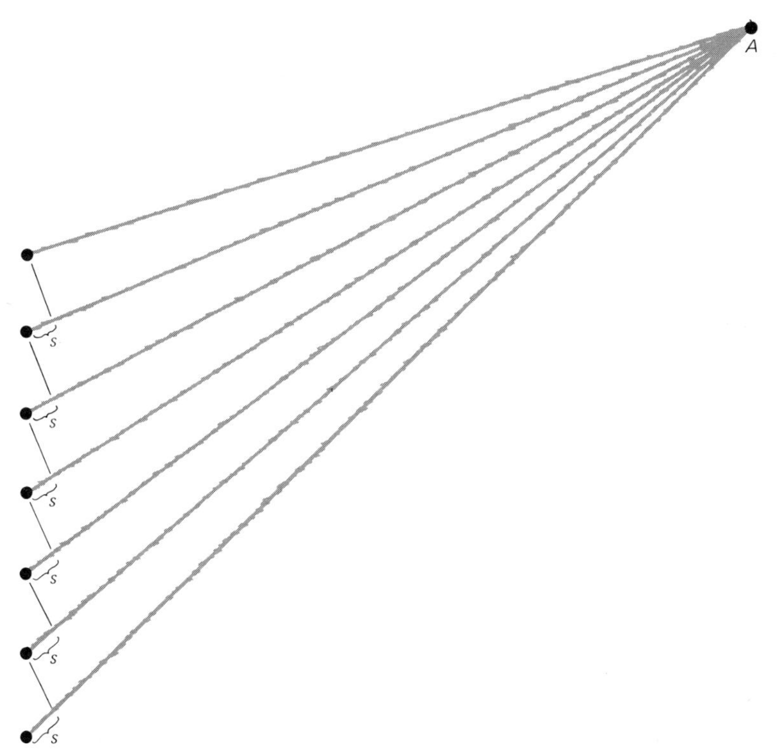

Figure 7-2-2 Path lengths from atoms in a line to a point A where the scattered wavelets shown in Fig. 7-1-1 are to be detected. If A is very far from the line of atoms, the path lengths will differ successively by the same distance s.

light. Except for technical difficulties such as experiment would have been performed in the 1920s, and in fact an essentially equivalent experiment was successfully performed in 1962.

7-3 PROBABILITY AMPLITUDES

Figure 7-3-1 shows a source S, assumed to emit electrons with a particular momentum, and therefore with a particular wavelength, according to Eq. 7-2-2. Because of the wavelike behavior of electrons, if only slit number 1 is open, the intensity of electrons at the screen is given by the curve labeled 1 in the figure. The electrons "diffract," so the beam spreads out to cover some area on the screen. If only slit number 2 is open, the intensity is given

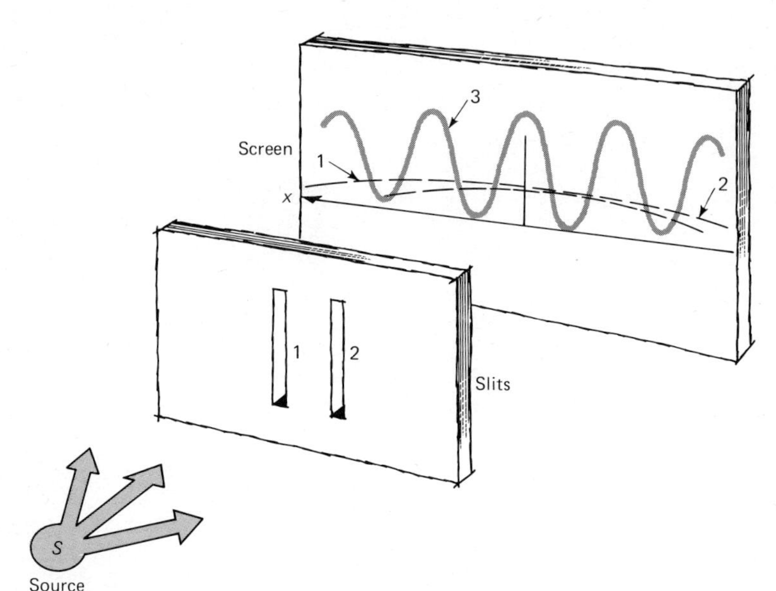

Figure 7-3-1 A hypothetical two-slit electron-interference experiment. Curves 1 and 2 show the electron intensity along the screen with either slip open and the other blocked. Curve 3 shows the intensity with both slits open.

by curve 2. However, if both slits are open, the intensity exhibits maxima and minima, shown by curve 3, as a result of wave interference.

Suppose the source intensity is so low that we can detect the arrival of electrons at the screen one at a time. If the screen is divided up into strips with a small width Δx, we can record, for each electron, which strip it landed in and plot a bar graph to keep track of the results. Figure 7-3-2 shows a series of such bar graphs for three successive times, t_1, t_2, and t_3.

At time t_1, only the first electron has arrived; by time t_2 a dozen or so have been detected; and by time t_3 a larger number of electrons have arrived. For a large number of electrons, the graph looks like the intensity distribution for a two-slit interference pattern, shown by a dashed line on the figure. The interpretation of the wave-particle duality for particles is that the dashed curve for time t_3, calculated from *wave interference*, gives the relative *probability* of finding *particles* at the screen.

The probability of finding an electron at a given place is found from a *probability amplitude*, which is found by combining waves according to the principle of superposition. Thus, interference effects occur. The actual probability is found from the probability amplitude by squaring it, just as light intensity is found from the square of the wave amplitude. Symbolically, the probability amplitude is almost always represented by the Greek letter ψ (pronounced "sigh" by physicists). The probability P for an electron to be found in a certain small strip of screen may be written

$$P = \psi^2 \qquad \text{7-3-1}$$

where ψ depends on the x coordinate of the strip.

The preceding interpretation of two-slit interference for elec-

Figure 7-3-2 Positions of individual electrons detected at the screen in a two-slit interference experiment. At times t_1, t_2, and t_3, an increasingly large number of electrons have been detected. The shaded areas represent a bar graph of number of electrons N in each interval Δx.

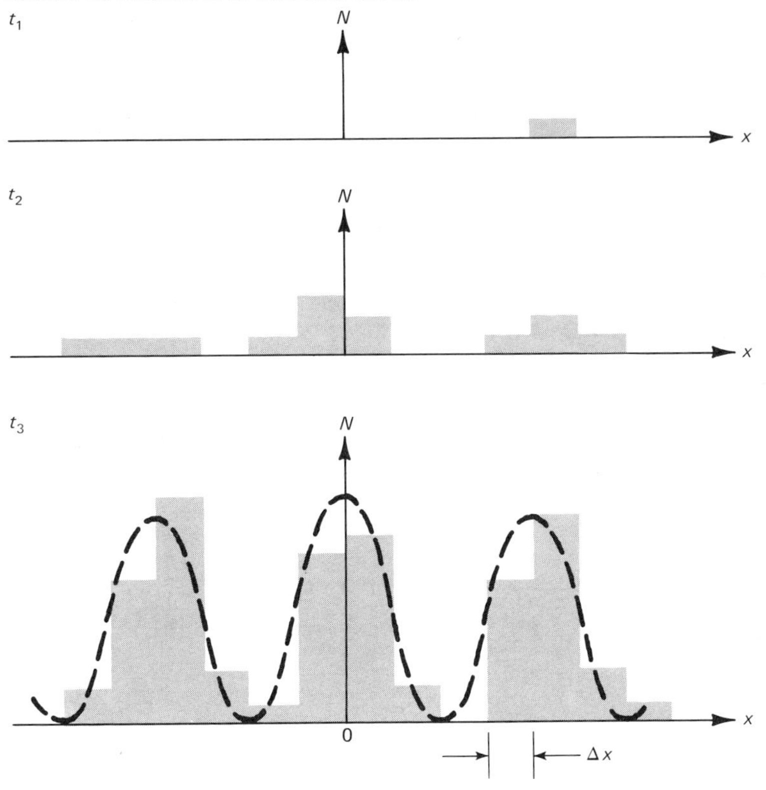

trons may seem quite strange. Each electron obviously proceeds from source to screen through *one slit or the other.* Yet, it never appears at the screen where there are minima of the *two-slit* interference pattern. It is as if each single electron were guided by the behavior of a wave passing through both slits. Moreover, this "guide wave," which exhibits interference effects at the screen, does not really determine where the electron goes, but only where it is likely or unlikely to go.

The wave interference is used to determine *probabilities*, such that a very large number of electrons will display a predictable pattern, although the behavior of each individual electron is not specified. To illustrate a bit more fully this probability idea, let us examine the flip of a coin. We cannot predict whether the result of an individual toss will be heads or tails. Yet, for an honest coin, we believe that a large number of tosses will tend to give equal numbers of heads and tails. The statement that "the probability of either a head or a tail is 50 percent" correctly describes the average result of a large number of tosses, without predicting the result of any one toss.

For an interference experiment with light quanta, the same interpretation can be followed as for electrons. The only difference is that the guide waves for light can be imagined to correspond to the electromagnetic fields of Maxwell's classical theory, which may make them seem a bit more concrete than de Broglie's waves. Thus, an interpretation which is the same for either so-called waves or so-called particles can be developed, one which embodies dual particle-wave behavior for either. The quantum mechanics of Schrödinger, based very much on de Broglie's idea, is a theory for calculating the guide waves, called probability amplitudes, which can be used to find probabilities for the behavior of particles or quanta.

A large number of physicists, including Einstein, found it difficult to accept the philosophy of the theory just outlined. If the probability interpretation truly describes nature, then, even if the present conditions are known perfectly, it is impossible to predict with certainty what will happen in the future. We will see this in more detail in the next section. For the example of two-slit interference, the probability amplitude can be used to accurately predict how a large number of electrons are distributed on the screen. But it is *impossible* to predict where any single electron will hit.

The idea that the world could only be described by probabilities evoked from Einstein a famous comment, that he could not believe that "God threw dice." In the years since the invention of quantum mechanics, many able people have tried to find an alternative, more deterministic way of describing the world. However, these efforts have not so far been successful.

7-4 THE UNCERTAINTY PRINCIPLE

Suppose there is an electron with momentum precisely equal to some value p_x, traveling in the $+x$ direction. Then according to de Broglie it behaves like a traveling wave with wavelength $\lambda = h/p_x$. The probability amplitude ψ at some instant of time looks like the wave shown in Fig. 7-4-1*a*, which extends over all values of x.

Figure 7-4-1 (*a*) The probability amplitude at one instant of time for an electron of momentum Px traveling in the $+x$ direction. (*b*) The variation with time of the probability amplitude at a fixed value of x, as well as ψ^2 and ψ^2_{av}.

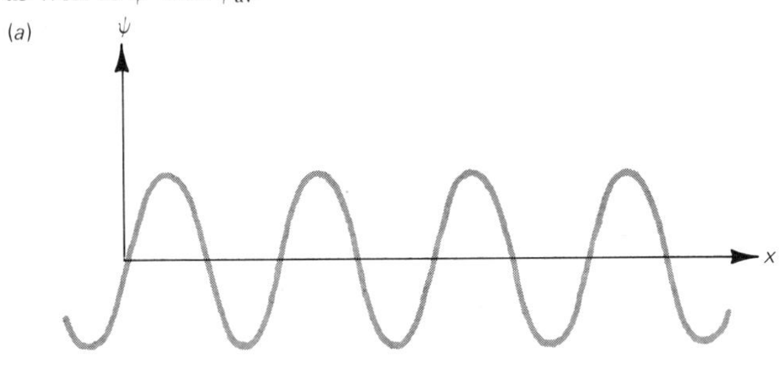

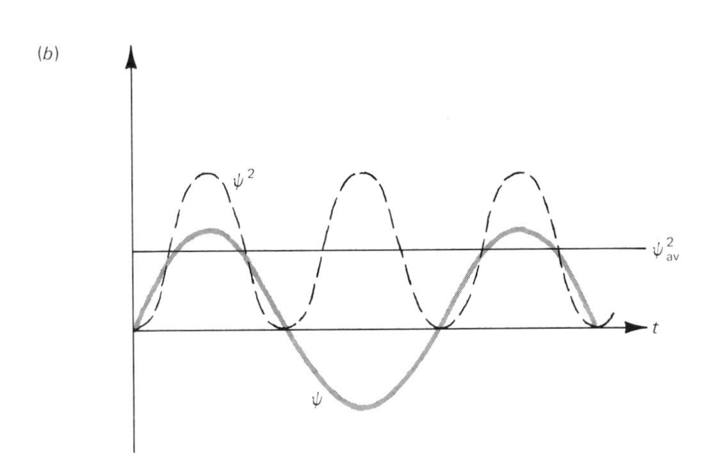

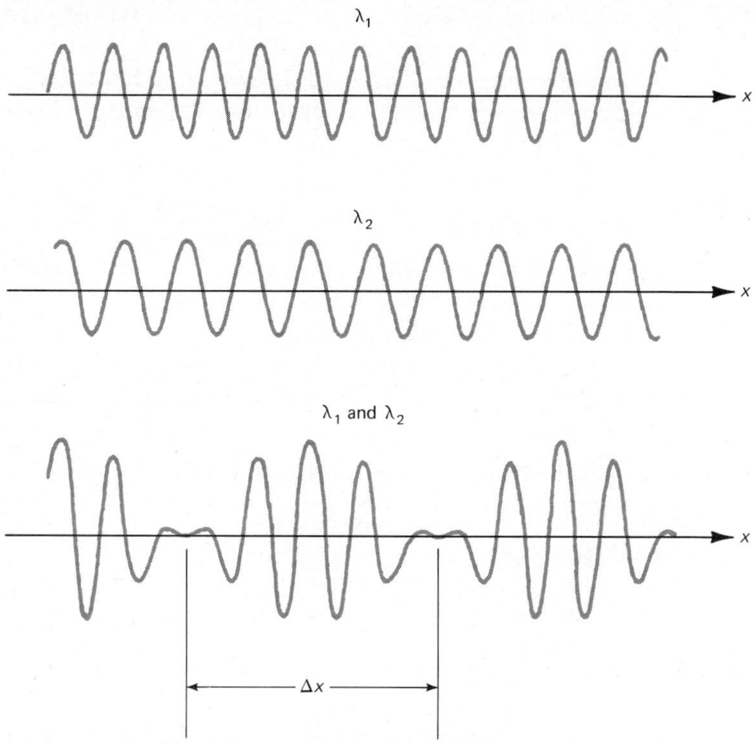

Figure 7-4-2 Interference, at one instant of time, for two waves with equal amplitudes and slightly different wavelengths.

The variation of ψ with time, at a particular value of x, is shown in Fig. 7-4-1*b*, as well as the variation of ψ^2. The average value of ψ^2, labeled ψ^2_{av}, gives the average probability of finding the electron at the position x. Notice that the value of ψ^2_{av} will be the same for any value of x. Thus, if we represent ψ as a traveling sine wave, as in Fig. 7-4-1, the probability of finding the electron is the same *anywhere* in space.

The probability amplitude shown in Fig. 7-4-1, spread out over all values of x, seems hardly appropriate for describing a real electron whose location is known to some reasonable accuracy. For example, if an electron is shot from an electron gun at time $t = 0$ with a velocity v in the x direction, we can predict its motion at later times quite well with classical physics. At some time t after the electron has left the gun we can say very accurately where it is. How can this be reconciled with de Broglie's hypothesis, which seems to say that we know nothing of the electron's x coordinate?

The first clue to understanding the connection between the de Broglie waves and the classical motion of an electron is to consider the superposition of two waves which have a small difference in wavelength. Figure 7-4-2 shows two waves with wavelengths λ_1 and λ_2 which differ by about 20 percent. (We will eventually assume the difference is much less than this, but the figure would then be too difficult to draw and examine.)

In Fig. 7-4-2 we see the result of interference between two waves with different wavelengths. They interfere alternately constructively and destructively along the x axis. The amplitude of the resultant wave *varies* with x, and its wavelength is equal to the average of the wavelengths of the two interfering waves. Ultimately, we will identify these waves with de Broglie waves, and we will also be concerned with their velocity, which has not yet been discussed. For the moment, we will continue to superpose waves of various wavelengths and observe the resultant wave at one instant of time.

Suppose a third wave, with wavelength λ_3, is added to the re-

Figure 7-4-3 Interference of a third wave with the resultant of the two waves shown in Fig. 7-4-2. This wave interferes destructively with alternate maxima of the sum of the two waves in Fig. 7-4-2.

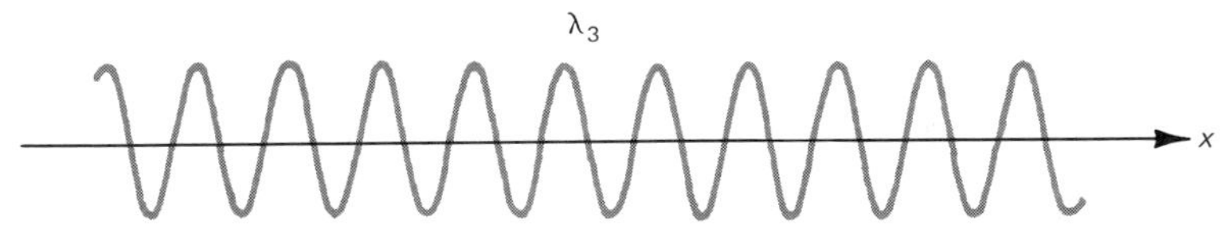

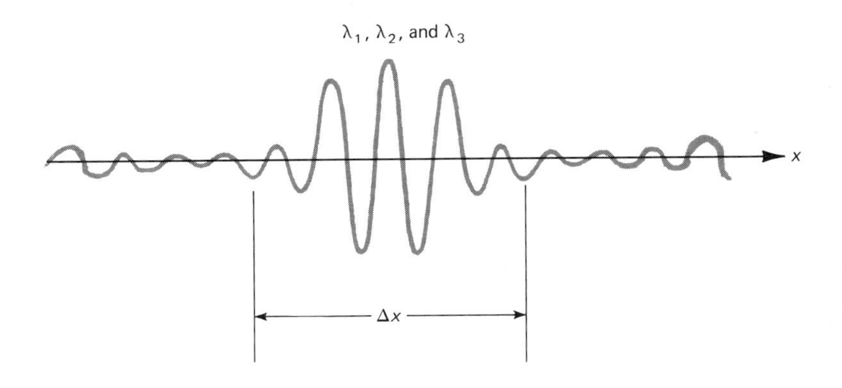

sultant of the first two. This wave and the superposition of the three waves with different wavelengths are both shown in Fig. 7-4-3. By using the third wave we have reinforced the resultant at some places and canceled it at others. Furthermore, by adding together an *infinite* number of waves, it is possible to obtain a resultant which is small everywhere except in a length Δx, as shown in Fig. 7-4-4.

In order to obtain a resultant wave like that shown in Fig. 7-4-4, the relative amplitudes of the superposed waves, and their frequencies, are described by a curve like that shown in Fig. 7-4-5. A range of wavelengths, mainly extending from λ_1 to λ_2, leads to a wave localized in a region of length Δx. There is a relation between the spread in wavelengths, $\lambda_2 - \lambda_1$, which we can call $\Delta\lambda$, and the spread in x, Δx, which we can understand by considering Fig. 7-4-2.

In Fig. 7-4-2 there is destructive interference at the beginning and end of the interval labeled Δx and constructive interference at the center. A relationship between $\Delta\lambda$, the average wavelength λ, and the number of wavelengths N in the length Δx follows from this. We will assume that $\Delta\lambda$ is much less than λ. As x changes by an amount λ, the two waves shift relative to each other by an amount $\Delta\lambda$. However, in the whole interval, Δx, the waves must shift by one full wavelength λ. This is necessary if the destructive interference at the beginning of the interval is to be repeated at the end. Thus,

$$N(\Delta\lambda) = \lambda \qquad 7\text{-}4\text{-}1$$

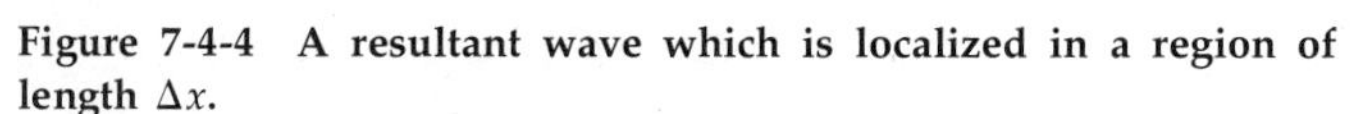

Figure 7-4-4 A resultant wave which is localized in a region of length Δx.

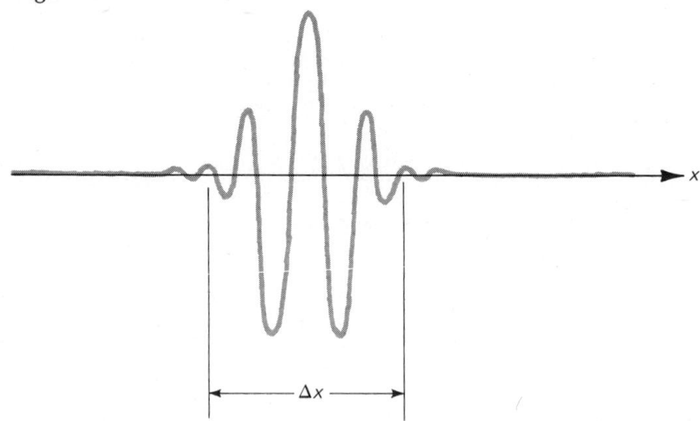

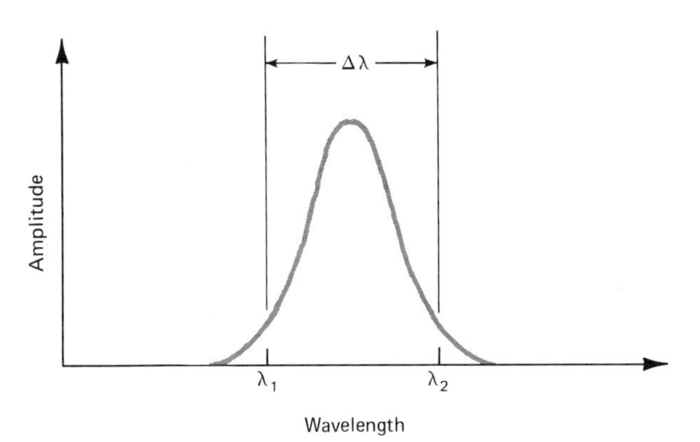

Figure 7-4-5 The relation of amplitude and wavelength for the infinite number of waves which must be combined to give the wave shown in Fig. 7-4-4.

The quantity N in Eq. 7-4-1 is just $\Delta x/\lambda$, so that substituting this value for N, Eq. 7-4-1 may be written

$$\frac{\Delta x}{\lambda}\Delta\lambda = \lambda \qquad \text{or} \qquad \Delta x = \frac{\lambda^2}{\Delta\lambda} \qquad \text{7-4-2}$$

This is the relation which connects Δx and $\Delta\lambda$ in Figs. 7-4-4 and 7-4-5. The reason is that when we add a third wave to help cancel some of the maxima, as in Fig. 7-4-3, its wavelength must be between the wavelengths of the original pair, λ_1 and λ_2. Succeeding waves, for canceling all the maxima except the one remaining in Fig. 7-4-4, also have wavelengths between the two extremes λ_1 and λ_2.

Now, suppose the waves we've been studying are interpreted as de Broglie waves, or probability amplitudes. The spread in wavelength $\Delta\lambda$ can be related to a spread in momentum Δp. For de Broglie waves of wavelengths λ_1 and λ_2,

$$p_1 = \frac{h}{\lambda_1} \qquad p_2 = \frac{h}{\lambda_2} \qquad \text{7-4-3}$$

The difference $p_1 - p_2$, which will be called Δp, is then given by

$$\Delta p = \frac{h}{\lambda_1} - \frac{h}{\lambda_2} = h\left(\frac{1}{\lambda_1} - \frac{1}{\lambda_2}\right) \qquad \text{7-4-4}$$

Putting the terms in parentheses over a common denominator,

$$\Delta p = h\frac{\lambda_2 - \lambda_1}{\lambda_1\lambda_2} \qquad \text{7-4-5}$$

The difference $\lambda_2 - \lambda_1$ is $\Delta\lambda$, and for small $\Delta\lambda$ the product $\lambda_1\lambda_2$ can be written as λ^2, where λ is the average wavelength. The final result is then

$$\Delta p = \frac{h(\Delta\lambda)}{\lambda^2} \qquad 7\text{-}4\text{-}6$$

From Eq. 7-4-6, $\lambda^2/\Delta\lambda = h/\Delta p$. Substituting this value of $\lambda^2/\Delta\lambda$ into Eq. 7-4-2,

$$\Delta x = \frac{h}{\Delta p} \qquad 7\text{-}4\text{-}7$$

Thus, according to the de Broglie wave picture, the region Δx in which a particle can be localized is related inversely to the precision Δp with which its momentum is known. Notice first that Δx is proportional to Planck's constant, which is a very small number. In particle beam experiments like J. J. Thomson's, the value of Δp is typically enormous compared with h. As a result, an electron "wave packet," like that shown in Fig. 7-4-4, can be localized in a region of x which is completely unmeasurable. Wavelike electrons can thus be localized almost like classical particles, provided they have a large enough spread in momentum.

We will not go into it in detail here, but the velocity calculated for a "particle" of mass m, represented by a wave packet like that

Figure 7-4-6 The probability amplitude which results from a wave packet of de Broglie waves.

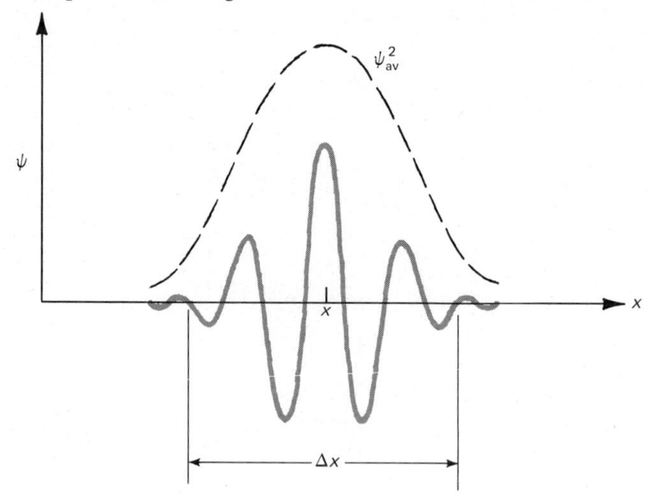

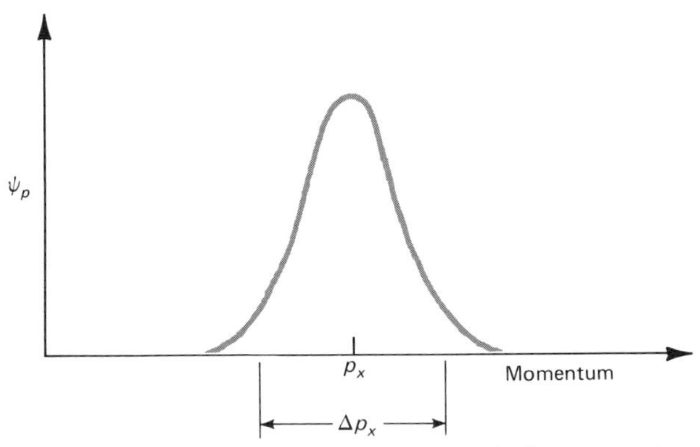

Figure 7-4-7 The mean momentum and the momentum spread for the de Broglie waves used to form the wave packet in Fig. 7-4-6. The curve shows how the amplitudes of the de Broglie waves vary with momentum and is to be interpreted as the probability amplitude for a single electron, described by the wave packet, to be observed to have various momenta.

shown in Fig. 7-4-4, turns out to be just the correct classical velocity p/m. The velocity of an individual de Broglie wave for a particular momentum p is *not* p/m, but depends on p. However, the wave packet moves at the speed p/m. This is not a quantum-mechanical phenomenon but just the result of ordinary superposition of waves with different wavelengths and velocities. If each of two waves has a different velocity, then the resultant wave has yet another velocity.

Figures 7-4-6 and 7-4-7 are just like Figs. 7-4-4 and 7-4-5, except that they are labeled as graphs of probability amplitudes. The wave packet localizes a particle in a region of width Δx because at a given instant in time the probability amplitude for finding the particle at various values of x looks like the curve in Fig. 7-4-6. The probability ψ^2, averaged over time, has the smooth appearance of the dashed curve of Fig. 7-4-6. The curve in Fig. 7-4-4 is interpreted as the momentum-probability amplitude, whose square gives the probability for the electron to have various momenta. The probability is a maximum at the classical value of p_x but includes a range of momenta (Δp_x). According to quantum mechanics, Eq. 7-4-7 tells us the best that nature will allow in the way of knowing both p_x and x. It is *not* that each electron has a known momentum and position and that for many there must be a spread. Even for a single electron, the momentum and position can only be known with accuracies Δp and Δx related by Eq. 7-4-7. In fact, it is possible to superpose waves in a less efficient way, so that for a given Δx, Δp_x can be larger, or for a given Δp_x, Δx can be larger. To express this, Eq. 7-4-7 can be rearranged and rewritten with the symbol $\geq$, which means "greater than or equal to," in the form

$$(\Delta x)(\Delta p_x) \geq h \qquad 7\text{-}4\text{-}8$$

Equation 7-4-8 is a famous relation, known as the *Heisenberg uncertainty principle.* It states that for any particle, the x coordinate and the x component of momentum *cannot simultaneously be precisely known.* It is the most concise statement of the indeterminacy inherent in quantum theory. For example, if we know a particle's momentum extremely well, then we also know its velocity, so we can predict very well how it moves in time. However, in this case Eq. 7-4-8 states that we are not permitted to know its position well at all, so the idea of a classical path is not allowed. Conversely, if the position of a particle is known very well, its momentum is required to be very uncertain, and therefore also its velocity. Again, we cannot trace its motion along a classical, determined path.

The uncertainty principle goes against our experience with particles like bullets or baseballs, which clearly follow determined paths. However, for large-scale motion, even of atomic particles, the uncertainties in momentum and position required to fulfill Heisenberg's equation introduce such small effects that classical physics is an excellent approximation to the motion of a wave packet. It is when we limit Δx to the size of an atom that a relatively enormous Δp results, ruling out the idea of a classical trajectory for an electron in an atom.

Until now, we have expressed the uncertainty principle in a form which resulted from consideration of one-dimensional motion. Heisenberg and others have shown that three uncertainty equations, like Eq. 7-4-8, apply *separately* to each component of position and momentum,

$$\begin{aligned}(\Delta x)(\Delta p_x) &\geq h \\ (\Delta y)(\Delta p_y) &\geq h \\ (\Delta z)(\Delta p_z) &\geq h\end{aligned} \qquad 7\text{-}4\text{-}9$$

These uncertainty relations often express, with little mathematical complication, the essence of detailed results which can only be obtained with great effort from the quantum theory. For many problems a detailed calculation is impossibly difficult, and the uncertainty relations provide general principles, almost like conservation laws, for helping to guess something about the answer without knowing the details.

PROBLEM 2

An electron is traveling in the $+x$ direction, with a velocity of 3×10^7 m/sec. It passes through a slit of width 10^{-6} m and strikes a screen a distance D behind the slit. Call the slit width Δy, and calculate the minimum allowed value of Δp_y, the uncertainty in the component of the electron's momentum in the y direction. From symmetry, what do you expect the average value of p_y to be for many such electrons? From Δp_y and p_x, find the uncertainty in the y coordinate of such an electron when it arrives at the screen, $(\Delta y)_{sc}$, if the distance D is 2 m. Use the relation

$$\frac{(\Delta y)_{sc}}{D} = \frac{\Delta p_y}{p_x}$$

Try to derive this equation, using similar triangles.

7-5 QUANTUM MECHANICS

As was brought out in the discussion of the Heisenberg uncertainty principle, the wavelike behavior of particles makes it necessary to abandon classical, or newtonian, mechanics. The basic idea of classical mechanics is to predict the motion of a body —the path it takes and the velocity it moves with—from knowledge of its position and velocity at a given time, and from the forces which act on it. Newton's second law is used to find the acceleration, and from the acceleration, given the starting point and the starting velocity, the position and velocity at all times can also be found.

If particles are wavelike, then they do not have a precisely defined motion, in the classical sense. A probability amplitude can be used to describe where the particle is likely to be at a given time, and what its momentum—and thus, its velocity—is likely to be. But, in contrast to classical mechanics, the position and velocity cannot both be simultaneously known with perfect accuracy.

In this section we will apply Schrödinger's quantum mechanics to a simple example, called the motion of a particle in a box. To keep the mathematics as uncomplicated as possible, we will only consider one-dimensional motion. Figure 7-5-1 shows an energy diagram for our example, where V is the potential energy of a particle of mass m moving along the x axis. Notice that V is constant for x between $-a$ and $+a$, and its value there has been taken equal to zero. At $x = -a$ and $x = +a$ the potential energy rises steeply to the value V_0. From the shape of the curve for V, we see that the only place where forces act on a particle (resulting in a change of V) will be near $x = -a$ and $x = +a$.

Figure 7-5-1 also shows the total energy U for the particle, and as long as U is less than V_0, the particle moves back and forth between points A and B. We say it is *bound* in a *potential energy well.* Its kinetic energy $U - V$ is constant except near $x = -a$ and $x = +a$, where the particle's direction of motion changes. The reason this problem is called motion of a particle in a box is that it describes motion like that of a marble rolling back and forth on the bottom of a flat-bottomed box, with walls which are *perfectly elastic.* The term perfectly elastic means that when the marble hits a wall, it bounces off without losing any kinetic energy.

PROBLEM 3

Sketch a graph which shows qualitatively how the force F_x varies with x if the potential energy is as shown in Fig. 7-5-1.

In order to further simplify the calculation, it is convenient to assume an idealized case where the potential energy V changes

Figure 7-5-1 Energy diagram for a particle in a box.

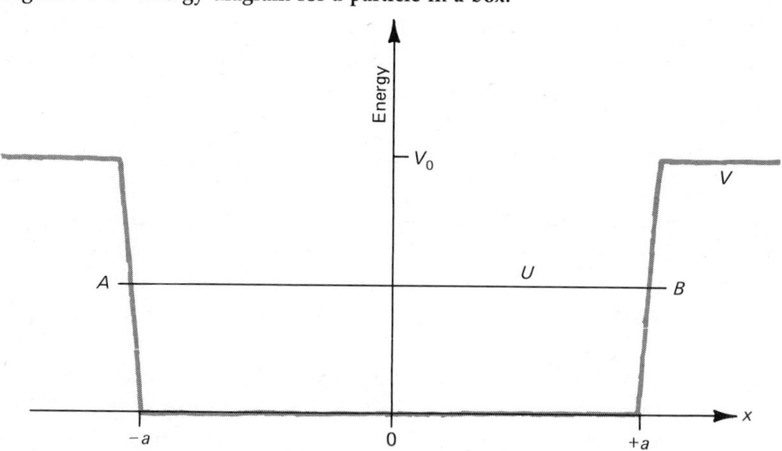

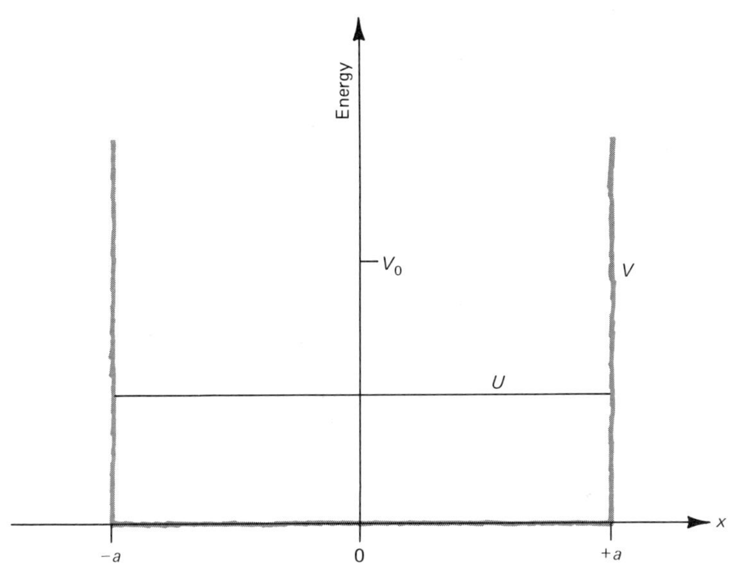

Figure 7-5-2 Simplified energy diagram for a particle in a box.

abruptly to an infinite value at $x = -a$ and $x = +a$. This is shown in Fig. 7-5-2 and corresponds to reflection of the particle at the walls of the box by an infinitely strong force.

The Schrödinger equation has a role in quantum mechanics somewhat analogous to the role played by Newton's second law in classical mechanics. Given the potential energy and the total energy, as in Fig. 7-5-2, the Schrödinger equation enables us to find the probability amplitude which describes the likelihood of finding the particle at various values of x. The force acting on the particle is introduced by means of the potential energy, not directly as in the second law. The probability amplitude is found instead of the path of the particle, because that is all nature permits us to know.

For one-dimensional motion of a particle in a box, with total energy U, the Schrödinger equation takes the form

$$\frac{\Delta(\Delta\psi/\Delta x)}{\Delta x} = -\frac{8\pi^2 m}{h^2}(U - V)\psi \qquad 7\text{-}5\text{-}1$$

where m is the mass of the particle and h is Planck's constant. The left-hand side of the equation is the rate of change with respect to x of the quantity in parentheses, $(\Delta\psi/\Delta x)$. This quantity is itself the rate of change of ψ with respect to x. Thus, we have a rate of change of a rate of change. Perhaps this concept will be a bit less formidable if we look at the familiar quantity, acceleration. Acceleration is the rate of change with respect to time of velocity. But velocity is the rate of change with respect to time of position. Thus, acceleration is also a rate of change of a rate of change:

$$a = \frac{\Delta(\Delta x/\Delta t)}{\Delta t}$$

Equation 7-5-1 is called a *differential equation*, which can be used to determine how ψ varies with x. If we look at the right side of Eq. 7-5-1 for x between $-a$ and a, where V is equal to 0, the right-hand side is a constant times ψ, where the constant is $-(8\pi^2 mU/h^2)$. Moreover, for x less than $-a$ or greater than $+a$, where V is assumed to be infinitely large, the right-hand side is essentially equal to an infinitely large constant, $8\pi^2 mV/h^2$, times ψ. If ψ is to be a physically reasonable quantity, it is not permitted to have an infinite rate of change of its rate of change. In the region where V is infinite this would result, unless the right-hand side were made to equal zero by setting ψ equal to zero. Thus, part of the solution for the probability amplitude ψ is that it is zero outside the well. This fits our intuition, that the particle should only be found inside the potential-energy well, but, as we shall see later, this result is only strictly true for the special case where V is infinitely large outside the well.

Inside the well, we have seen that the Schrödinger equation requires the rate of change of the rate of change of ψ be equal to a negative constant times itself. In Fig. 7-5-3 a sine curve is drawn along with its rate of change with respect to x, $\Delta\psi/\Delta x$, and the rate of change of its rate of change. The curve in Figure 7-5-3*c* is equal to $-(2\pi/\lambda)^2$ times ψ. Thus, if ψ is a sine curve, the rate of change with respect to x of $\Delta\psi/\Delta x$ is indeed found to be a negative constant times ψ itself. Such a variation of ψ with x can then satisfy Eq. 7-5-1 for x between $-a$ and a, provided the constant $(2\pi/\lambda)^2$ is equal to the constant on the right-hand side, $(8\pi^2 mU/h^2)$. We have thus found solutions to Schrödinger's equation for all values of x, which can be summarized below:

$x < a$	$\psi = 0$	
$-a < x < +a$	$\psi =$ a sine wave, written mathematically $\sin(2\pi x/\lambda)$, with $(2\pi/\lambda)^2 = 8\pi^2 mU/h^2$	7-5-2
$x > a$	$\psi = 0$	

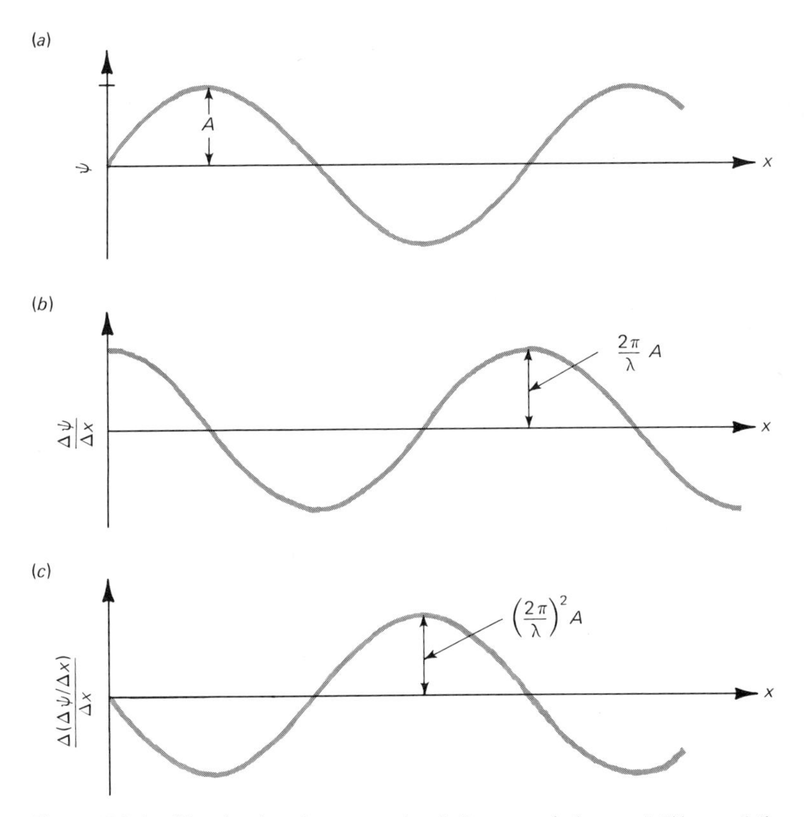

Figure 7-5-3 **Sketch of a sine curve for ψ, its rate of change $\Delta\psi/\Delta x$, and the rate of change of $\Delta\psi/\Delta x$. The curve in (c) is seen to be $-(2\pi/\lambda)^2$ times ψ.**

In order for the solution of the Schrödinger equation to be acceptable, we will now demand that the probability amplitude does not change instantaneously from zero outside the box to a finite value just inside the box. One rationale for this requirement is that nature does not exhibit instantaneously rapid (or "discontinuous") changes; another is that imposing this condition gives a result which agrees with experimental facts. Given a choice between rigorous mathematics and agreement with experiment, the physicist will seldom hold out for rigor.

In Fig. 7-5-4, parts of three sine curves are shown in the region of the left side of the well. The requirement that ψ not change abruptly at $x = -a$ rules out all but the one curve which starts from $\psi = 0$ at $x = -a$. The same requirement at the other side of the well must also be met, so we are therefore limited to sine curves which go through 0 both at $x = -a$ and at $x = +a$. The three longest

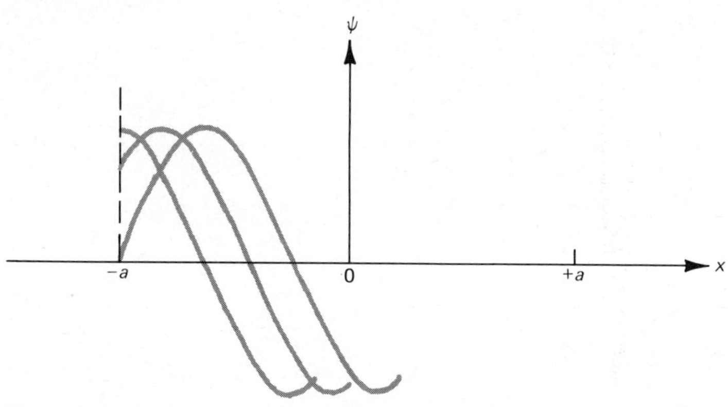

Figure 7-5-4 Parts of three possible sine curves for ψ, showing their behavior at $x = -a$.

wavelength curves which meet this requirement are shown in Fig. 7-5-5.

Summarizing, solutions of the Schrödinger equation have now been found which also satisfy the requirement that the probability amplitude does not jump suddenly from a value of zero at the walls of our potential-energy well. These solutions are restricted to certain allowed values of wavelength, such that the sine curves are 0 at $x = -a$ and $+a$. Examining Fig. 7-5-5, the allowed wavelengths can be seen to be such that an integral number of half-wavelengths add up to the width $2a$ of the potential-energy well, or

$$n \frac{\lambda}{2} = 2a \qquad n = 1, 2, 3, \ldots \tag{7-5-3}$$

The fact that only certain sine curves are acceptable solutions of the Schrödinger equation has striking physical consequences, which will be discussed in the following section. The probabilities found from the squares of the amplitudes in Fig. 7-5-5 will also be discussed and compared with what might have been expected classically.

7-6 QUANTIZED ENERGY STATES

For a particle in a box, we have found certain allowed results for ψ. A particle described by any one of these probability amplitudes is said to be in a *state* characterized by a particular *wave function* ψ. The "state" of a particle, or of a system of particles, is a shortened term for "physical state."

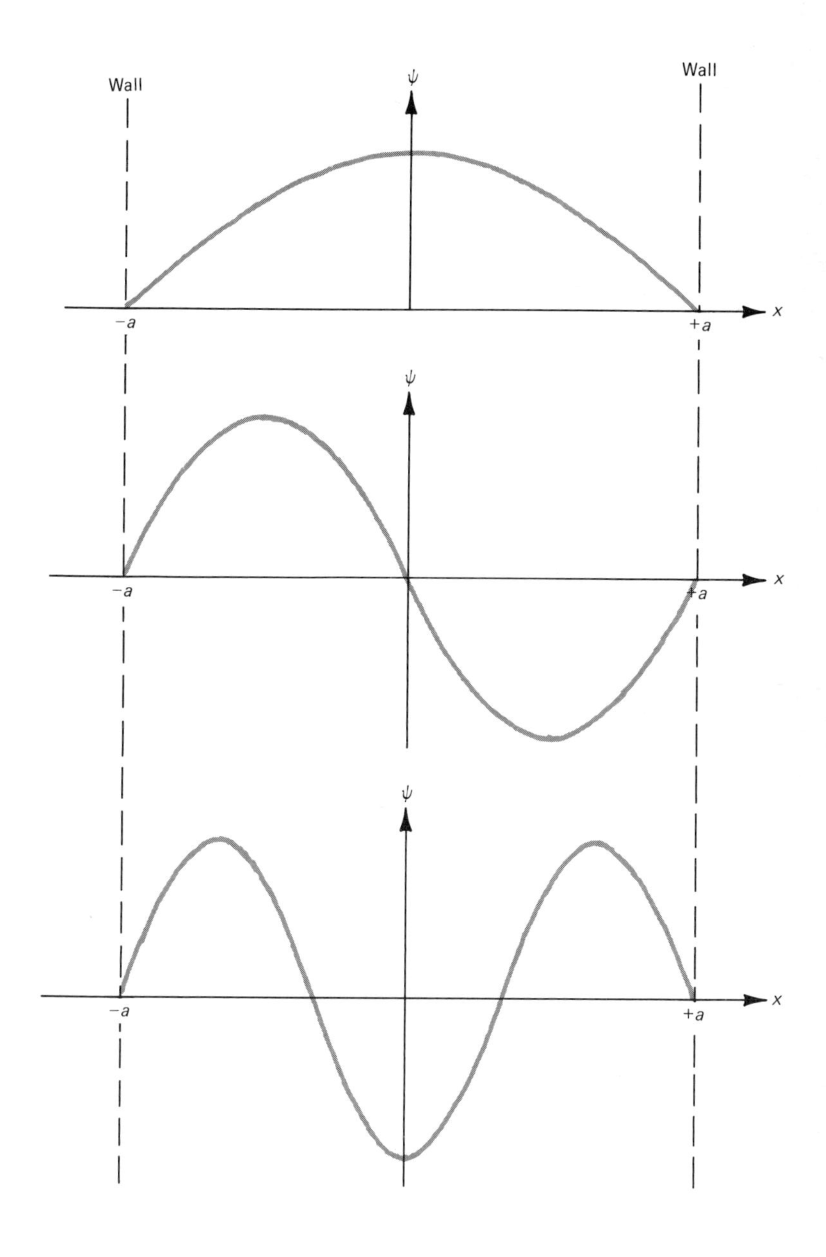

Figure 7-5-5 Three solutions for ψ for the particle in a box with infinitely high walls. These are the three longest wavelength sine waves which give the value $\psi = 0$ at both walls.

Classically, the state of a particle is described by the variation of its velocity and position with time. Quantum mechanically, a state is described by giving the *wave function*, the probability amplitude for the particle to be found at any position. From the wave function, the energy and the probability amplitude for finding the particle with various momenta can also be found. We have not described the procedure for finding the momentum amplitude in the case of a particle in a box, but in Sec. 7-4 we saw, for example, that a wave function which localized a particle in a region Δx also gave rise to a probability amplitude for the particle to have various momenta.

The particle in a box has been found to have a number of allowed states, with wave functions ψ_1, ψ_2, ψ_3, and so on. Figure 7-6-1 shows the first three allowed wave functions, as well as curves proportional to their squares, which give the probability of finding the particle at various values of x. The curves labeled P are interpreted as giving the probability per unit length along the x axis. This means that the probability for finding the particle in a region of length Δx, at some value of x, is given by $P = (\psi_x)^2(\Delta x)$, where ψ_x is the value of the wave function at the point x.

PROBLEM 4

Sketch curves like Fig. 7-6-1 for the wave functions ψ_4 and ψ_5. Also, show curves of P for these wave functions, which describe the relative probability of finding the particle at various values of x.

There is a simple physical requirement that the probability of finding the particle *somewhere* must be equal to 1. The particle exists, and therefore at any time it must be someplace. The wave function ψ must be adjusted in size, or *normalized*, so that the sum of the probabilities for all the little intervals Δx which make up the whole x axis is equal to 1. This can always be done by simply multiplying a curve with the proper shape, found from the Schrödinger equation, by some appropriate number. In general we will not need to worry about this normalization, and when it matters we will assume that it has been taken care of.

Before examining the probabilities shown in Fig. 7-6-1 in more detail, it is worthwhile first to explore the consequence of two results obtained in the preceding section. In order to satisfy the Schrödinger equation, a relation between the wavelength of the sine curve for ψ and the total energy U was obtained, as given in

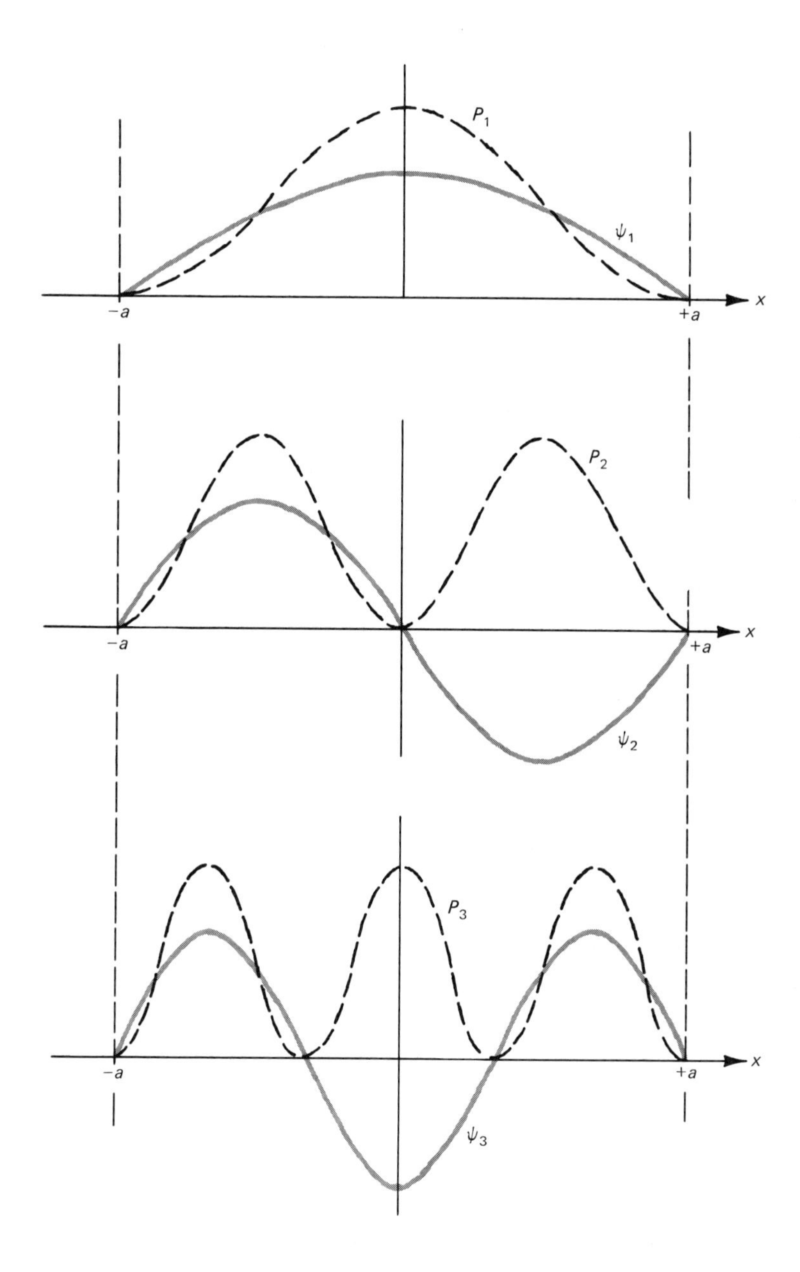

Figure 7-6-1 The three longest-wavelength allowed wave functions for the particle in a box with infinite walls. The probability curves, which give the variation with x of the probability of finding the particle, are also shown.

Eq. 7-5-2:

$$\left(\frac{2\pi}{\lambda}\right)^2 = \frac{8\pi^2 m}{h^2}U \qquad 7\text{-}6\text{-}1$$

Also, in order for the sine curves to go through zero at the edges of the well, a second equation involving the wavelength resulted, Eq. 7-5-3,

$$n\frac{\lambda}{2} = 2a \qquad n = 1, 2, 3, \ldots \qquad 7\text{-}6\text{-}2$$

By combining the two equations above, we can now obtain a new and interesting equation for the total energy U. Rearranging Eq. 7-6-1 so as to solve for U,

$$U = \frac{h^2}{8\pi^2 m}\left(\frac{2\pi}{\lambda}\right)^2 \qquad 7\text{-}6\text{-}3$$

Then, solving Eq. 7-6-2 for the wavelength,

$$\lambda = \frac{4a}{n} \qquad 7\text{-}6\text{-}4$$

Substituting for λ in Eq. 7-6-3 the value obtained in Eq. 7-6-4, we finally find

$$U_n = \frac{h^2}{32ma^2}n^2 \qquad 7\text{-}6\text{-}5$$

where the subscript n is given to U to indicate that U_n is the total energy for a given value of the integer n.

We have now found that, according to quantum mechanics, there is a series of allowed states for the particle in a box, characterized by wave functions ψ_1, ψ_2, ψ_3, . . . , and that each has a definite energy U_1, U_2, U_3, ... Only certain discrete, or *quantized*, values of energy are allowed. This is certainly different from the classical motion of a particle in a box, where any speed, and therefore any energy, is allowed. It is this attribute of the quantum theory, the existence of *discrete states with quantized values of energy*, from which the name "quantum mechanics" is derived.

PROBLEM 5

Suppose an electron is in a one-dimensional box, like that shown in Fig. 7-5-2. Let the width of the box be 10^{-10} m, approximately the size of an atom. Find the lowest allowed energy U_1 in joules and electron volts.

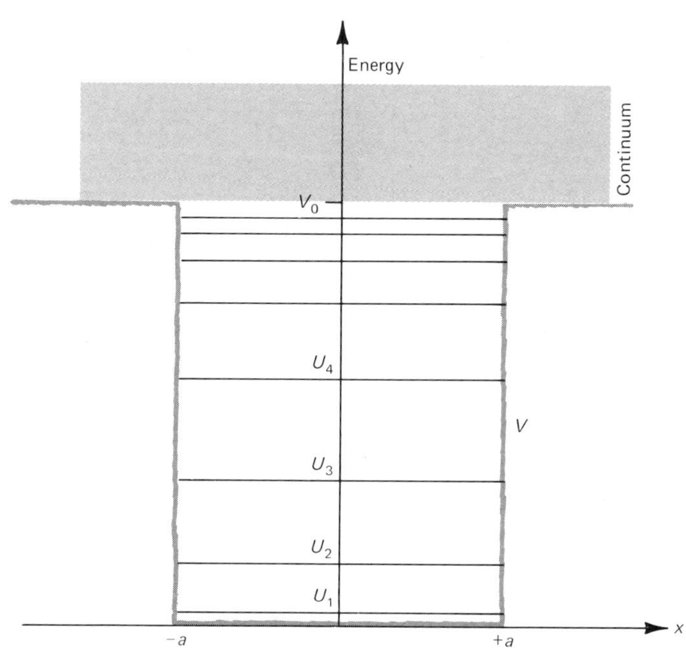

Figure 7-6-2 Energy diagram showing the allowed energies for a particle in a box. Energies less than V_0 are quantized, whereas any value of energy greater than V_0 is allowed. (For energies greater than V_0, the particle is of course no longer bound in the box.)

The quantum theory predicts discrete states for any bound-state problem in which a particle is confined to some region of space, not just for the particle in a box. However, when the particle is no longer bound, all values of energy are allowed. In the case of the particle in a box, if V_0 is taken to be large, but not infinite, we can make an energy diagram like that shown in Fig. 7-6-2. Here the quantized values of total energy U are shown for U less than V_0, and a shaded *continuum* of allowed energies is also shown for U greater than V_0.

In discussing wave packets and the uncertainty principle, we have seen that classical mechanics is an excellent approximation for the motion of large-scale objects, where the effects of the uncertainty principle are unmeasurable. Similarly, we may ask whether the classical motion of a particle in a box is a good approximation to the truth for motion on a large scale. For laboratory scale particles in boxes of visible dimensions with reasonable energies, the value of n in Eq. 7-6-5 turns out to be enormous. It is left for a problem to verify this for a specific example. As a consequence, the change in U from one allowed state to the next is a very small fraction of U, so the energy appears to be able to take on essentially any value.

Furthermore, the probability distribution ψ_n^2 for a very large value of n has a very large number of equal-sized maxima extending across the box. The change from $(\psi_1)^2$ to $(\psi_3)^2$ in Fig. 7-6-1 shows the beginning of this trend. For very large n the spacing between maxima of ψ_n becomes so small as to be unobservable, so the value of $(\psi_n)^2$ will appear in the laboratory to be uniform across the width of the box. We thus find that for large values of n the quantum-mechanical results approach those of classical mechanics for both the energy states and probability distribution.

PROBLEM 6

Consider a small particle in a small box, where "small" refers to laboratory-sized units of mass and length. Let the mass of the particle be 10^{-6} kg and the width of the box be 10^{-5} m. Find the minimum energy U_1 for this particle, from the quantum theory of a particle in a box. Since we've taken $V = 0$ inside the box, the total energy U_1 may be identified with the kinetic energy. What velocity does this value of kinetic energy correspond to?

Suppose the particle discussed above moves with a velocity of 10^{-2} m/sec. Find its kinetic energy and find a value for n if this energy is identified with U_n in the quantum-mechanical problem. For this value of n estimate the percent difference between U_n and U_{n+1}.

Neils Bohr, the Danish physicist who was so intimately involved in the development of quantum mechanics, invented the name *correspondence principle* for the requirement that the results of quantum mechanics be identical with the results of newtonian mechanics when motion on a large scale is considered. The correspondence principle was an important guide in the early development of quantum theory, and it has been shown in detail that it is always satisfied.

7-7 BARRIER PENETRATION

In the problem of the particle in a box, we introduced an important simplification by assuming that the potential energy was in-

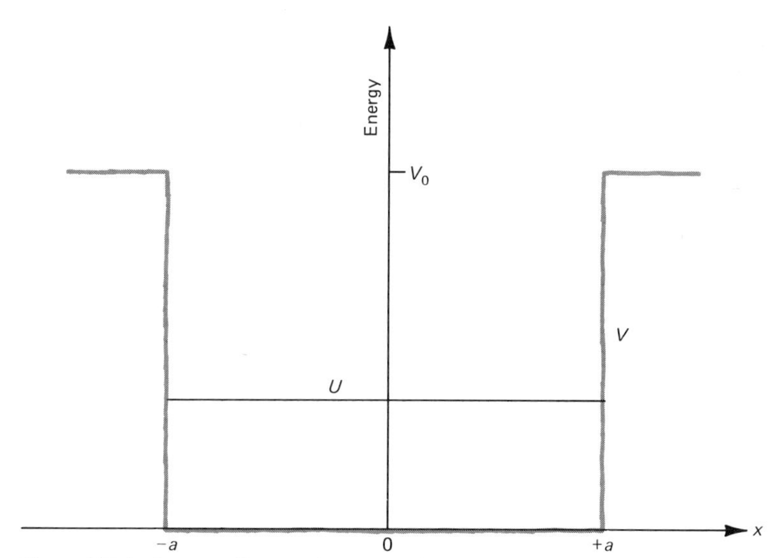

Figure 7-7-1 Energy diagram for a particle in a box, with V_0 not infinite.

finite outside the box. Now we will investigate the result if this assumption is not made. Figure 7-7-1 shows an energy diagram for the problem, with V_0 not a great deal bigger than U. Classically, the regions outside the box are completely forbidden, whether V_0 is a lot greater than U or just a little greater. However, we will find that the quantum-mechanical result depends upon V_0.

The Schrödinger equation, Eq. 7-5-1, for $x > +a$, may be written

$$\frac{\Delta(\Delta\psi/\Delta x)}{\Delta x} = -\frac{8\pi^2 m}{h^2}(U - V_0)\psi \qquad \text{7-7-1}$$

Since V_0 is greater than U, the right side of the equation is equal to a *positive* number times ψ. In Sec. 7-5 we found that inside the box, when the right-hand side was a *negative* constant times ψ, the solution was a sine curve. However, changing to the region outside the box, the constant on the right changes sign, and this eliminates the possibility of a sine wave solution. Instead, the solution is a curve called an *exponential.* In Fig. 7-7-2, a particular exponential curve is shown, which decreases toward the value $\psi = 0$ at large values of x. In the figure the curves for $\Delta\psi/\Delta x$ and for $\frac{\Delta(\Delta\psi/\Delta x)}{\Delta x}$ are also shown. The distance labeled α, in which the

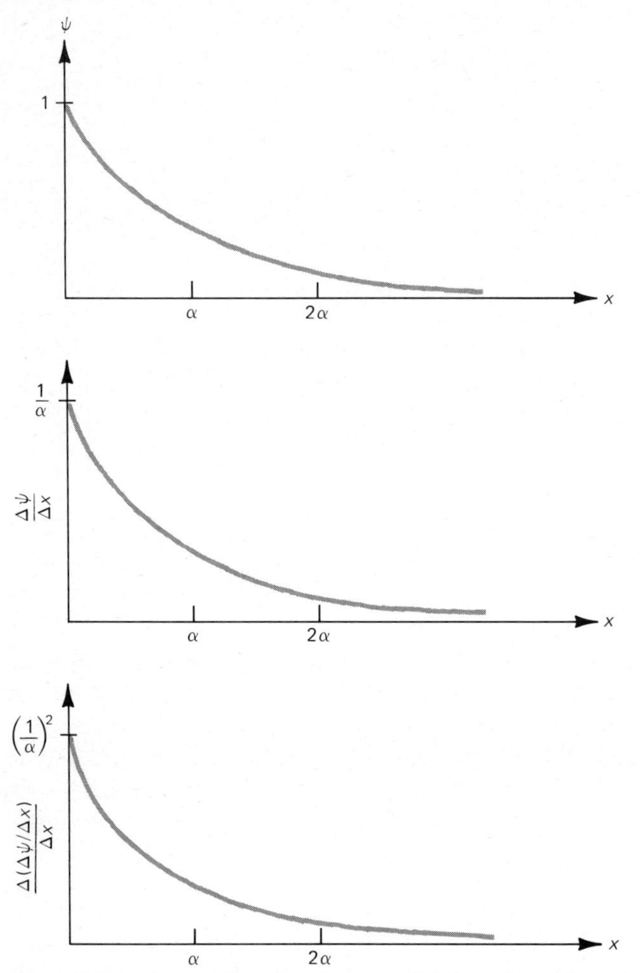

Figure 7-7-2 A wave function ψ represented by a decreasing exponential, along with curves of $\Delta\psi/\Delta x$ and of the rate of change of $\Delta\psi/\Delta x$.

curve drops to about one-third its value at $x = 0$, plays a role analogous to the wavelength for the sine wave. Equation 7-7-1 is solved, provided

$$\left(\frac{1}{\alpha}\right)^2 = \frac{8\pi^2 m}{h^2}(V_0 - U) \qquad \text{7-7-2}$$

In order to find solutions to the particle-in-a-box problem when V_0 is finite, we now combine sine wave solutions inside the box with exponentials outside, subject to the condition that the curve for ψ remains smooth at the walls, where the two kinds of curves are joined. We can attempt to justify this rather arbitrary condition as a manifestation of an inherent smoothness with which nature changes, but its main justification is that it leads to results in accord with experiment. Figure 7-7-3 shows the resulting wave

Figure 7-7-3 The first three allowed wave functions for a particle in a box with finite walls.

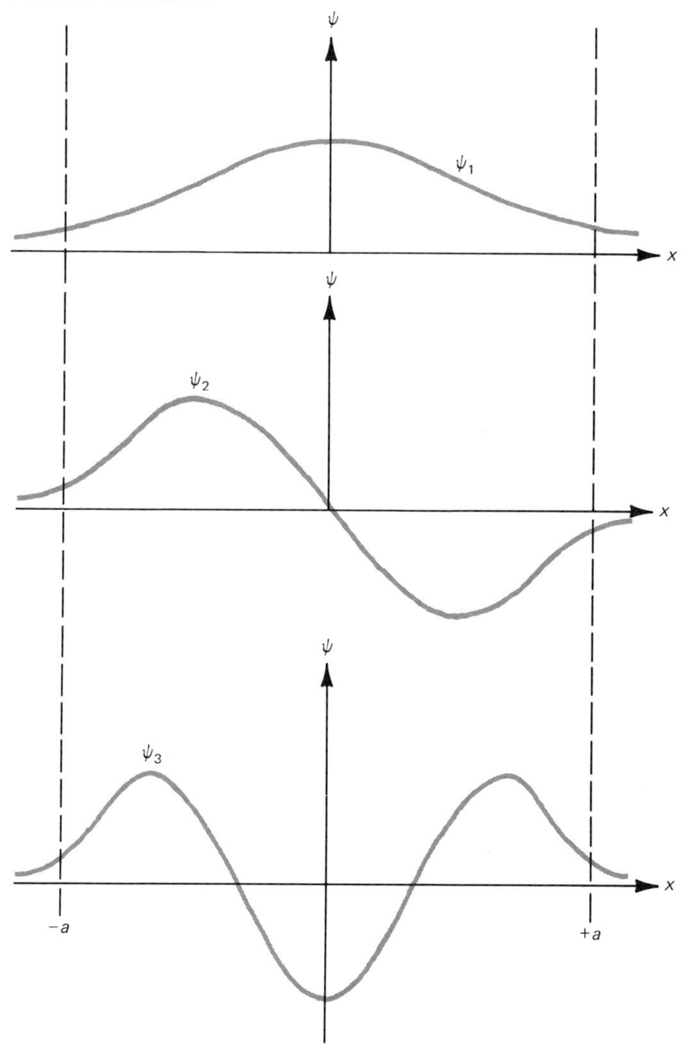

functions ψ_1, ψ_2, and ψ_3 for the particle in a box with finite V_0. Again, only certain wave functions are allowed, reminiscent of those for the potential well with V_0 infinite.

A very nonclassical result is implied by the exponential "tails" of the wave functions shown in Fig. 7-7-3. There is a relatively small but nonzero probability for finding the particle *outside* the well, in regions of space where its total energy is *less* than its potential energy. These regions are classically absolutely forbidden since they correspond to negative values of kinetic energy. This purely quantum-mechanical effect, the *tunneling* of a particle into classically forbidden regions, is called *barrier penetration.*

The distance of penetration is characterized by the value of the constant α which characterizes the exponential curves. Even for very small values of the energy difference $V_0 - U$ the value of α is only of order of atomic dimensions. As a consequence, barrier penetration tends to produce significant effects only in the atomic realm and is usually unobservable for laboratory-sized physical systems. In some special instances, however, barrier penetration is directly observable in the laboratory.

PROBLEM 7

Explain why the energies U_1, U_2, and U_3 for a particle in a box like that shown in Fig. 7-7-1 are all lower than the corresponding energies for the case when V_0 is infinitely large. Do this by comparing the wavelengths of the wave functions in the two cases.

PROBLEM 8

Suppose an electron with energy equal to 1.60×10^{-19} joules is in the box shown in Fig. 7-7-1. If V_0 is equal to 8.0×10^{-19} joules, find the characteristic length for the exponential parts of its wave function.

PROBLEM 9

Explain why the exponential tails of the wave function ψ_2 in Fig. 7-7-3 fall off less rapidly than for ψ_1.

The type of potential-energy curve shown in Fig. 7-7-4 leads to an interesting form of barrier penetration. Here the potential-energy "walls" are not only of finite height but of finite thickness. If a particle is initially inside the box, between $-a$ and $+a$, its wave function has tails which extend through the walls and lead to some probability for the particle to be in the regions $x < -b$ or $x > +b$. In these regions, the total energy U is greater than V, so

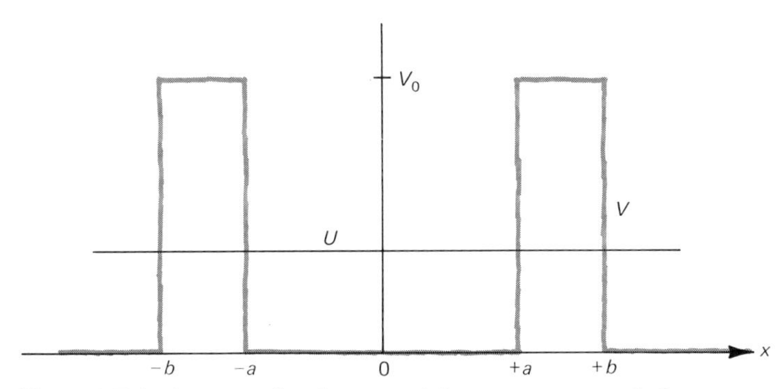

Figure 7-7-4 An example of a potential-energy curve such that a particle of energy U, initially in a box, can escape by barrier penetration.

that the wave function becomes a sine wave, as shown in Fig. 7-7-5.

The situation shown in Fig. 7-7-5 does not remain unchanged with time. Since the particle can escape from the well completely if it tunnels through the walls, it will gradually "leak out." As times goes on the wave function inside the box will steadily decrease and there will be a greater and greater likelihood of finding the particle somewhere outside the box. If we place detectors around the outside of the box, we can't predict exactly when the particle will be observed. However, if we wait long enough, the particle will eventually be detected. This is precisely the situation for emission of alpha particles from radioactive nuclei. They "tunnel" through a potential-energy barrier and escape from the nucleus in which they were held.

Figure 7-7-5 A wave function for the potential-energy curve shown in Fig. 7-7-4, illustrating the escape of a particle by barrier penetration. A wave function like ψ_3 for the particle in a box has been chosen as an example.

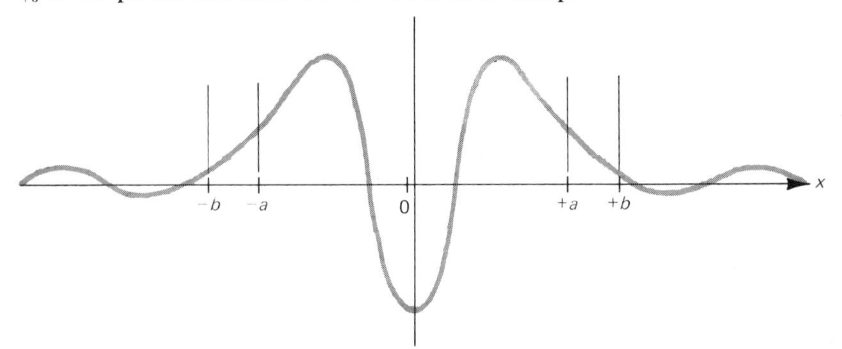

The simplest one-dimensional bound-state problems have now been discussed, and a large part of the new physics which comes out of the Schrödinger equation has been introduced. When we consider potential-energy curves less simple than the box, the results change in detail but remain qualitatively the same. For example, the *quantized harmonic oscillator* refers to a particle in the potential-energy well

$$V = \frac{1}{2}kx^2$$

shown in Fig. 7-7-6. The allowed wave functions for this particle look quite similar to those that are shown in Fig. 7-7-3. The allowed energies in this case are

$$U_n = \left(n + \frac{1}{2}\right)hf \qquad \text{7-7-3}$$

where $n = 0, 1, 2, \ldots$ and f is the frequency which would be calculated for the classical simple harmonic motion of the particle. Just as for the square well, we find a set of allowed energies, spaced differently here because of the different shape of the potential-energy curve.

Figure 7-7-6 The harmonic oscillator potential well, with the lowest allowed states of total energy shown. Note the spacing hf between adjacent energy levels.

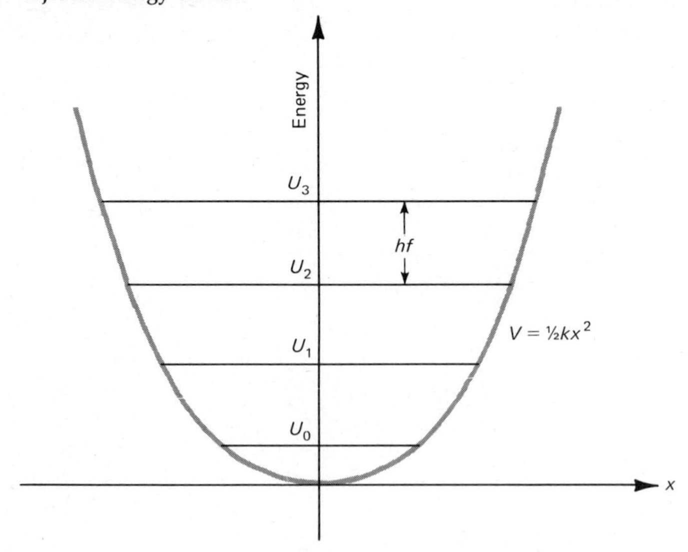

PROBLEM 10
Consider two identical traveling waves, with the same wavelength, velocity, and amplitude, traveling in opposite directions. Explain the appearance of "standing waves" when they cross each other, i.e., a resultant wave motion with no motion at some points due to destructive interference. Show that the nodal points, where the waves destructively interfere, are spaced a half-wavelength apart.
From your investigation of standing waves, above, interpret the wave functions in Fig. 7-6-1 in terms of standing de Broglie waves. Compare the quantity $p^2/2m$ for these waves (which would classically be the total energy for this problem) with the energies found from Eq. 7-6-5.

When we consider three-dimensional problems, and potential-energy curves less simple than for the box, the mathematics of Schrödinger's equation becomes quite formidable. For all except the most simple physical systems it is in fact impossible to find exact wave functions. However, quantum mechanics is extremely well verified by experiment for those physical systems, like the hydrogen atom, where essentially exact calculations can be made.

In recent years, accurate results have been obtained for some of the more complex problems through the use of electronic computers. Approximate methods, which can in principle lead to accurate answers, are known but require staggering amounts of arithmetic. Few such computations are possible for human beings using desk calculators, but the fast electronic computer, performing a million arithmetic operations per second, has widened the range of accurate quantum-mechanical answers. Even for problems which can not yet be accurately solved, crude calculations are valuable and can shed light on phenomena which would remain completely mysterious without quantum mechanics.

SUMMARY

Objects, such as electrons, originally conceived solely as particles, exhibit wavelike behavior. A two-slit interference experiment with an electron beam leads to results like those for light. The *particle wavelength* is associated with its momentum according to the *de Broglie relation*:

$$\lambda = \frac{h}{p}$$

The particle-wave duality is incorporated in the physical theory, called quantum mechanics, by means of *wavelike probability amplitudes.* The probability amplitude, usually called ψ, is calculated instead of the particle motion. Then, the *probability* of finding a particle at a given location in space and time is given by the *square of the probability amplitude*:

$$P = \psi^2$$

The uncertainty principle is a direct consequence of the particle-wave duality. A particle localized in a small region of space has a probability amplitude given by a superposition of waves with a range of wavelengths. The de Broglie relation relates such a wavelength range to a momentum range. Therefore, an expression can be found relating the *spreads in likely values of position and momentum,*

$$(\Delta x)(\Delta p_x) \geq h$$

Similar equations also hold for the y and z components of momentum. According to the uncertainty principle it is *impossible to precisely specify, simultaneously, both the position and momentum of a particle.*

The motion of a particle in a box is an example of a simple one-dimensional problem which is easy to solve with the quantum-mechanical theory. *The Schrödinger equation,* plus physically plausible conditions on the wave function, give certain *allowed solutions* for the wave function, ψ_1, ψ_2, ψ_3, etc.

For each allowed solution for ψ there is a corresponding total energy for the particle. The *allowed energies* for one-dimensional motion of a particle in a box with infinitely rigid walls are

$$U_n = \frac{h^2}{32ma^2} n^2$$

for $n = 1, 2, 3, \ldots$ and a box width of $2a$.

For all *bound-state* problems, a limited number of *discrete states* are allowed quantum mechanically, while in classical physics an infinite number of states are allowed. The discrete states have specific energies, so that *energies are said to be quantized,* meaning that the energy can only vary by "jumps" rather than continuously.

When the Schrödinger equation is applied to bound-state problems, *barrier penetration* is usually found to occur, which means that the probability for finding a particle outside the allowed region for classical motion is usually not zero. The particle is sometimes said to *tunnel* into, or through, the classically forbidden region.

ATOMIC PHYSICS

8-1 SPECTROSCOPY

In the last chapter the discussion of quantum mechanics was mainly based upon the particle-wave duality. The apparent existence of quantized energies for atoms was mentioned as a second major influence on the development of quantum mechanics, but was not discussed in detail. We have seen that quantum-mechanical theory does indeed lead to quantized energy states, but so far we have discussed them only in the context of simple one-dimensional problems. In this chapter quantum mechanics will be applied to atoms. We will begin with a description of the phenomena which stimulated Bohr to invent his early quantized atomic model, discuss that model, and then go on to use the ideas of quantum theory to understand a variety of the most basic atomic properties.

Although there is a great variety of observable atomic phenomena, all in principle describable with quantum mechanics, the study of *atomic spectra* was the initial testing ground of the quantum theory and is still a very important part of atomic physics. The *spectrum* of an atom is a description of the wavelengths of light which it emits.

The simplest spectroscope consists of a light source, a slit, a prism, and a screen, as shown schematically in Fig. 8-1-1. In the absence of the prism, the lens would focus the light coming from the slit at the point *0* in Fig. 8-1-1. However, the prism bends the light rays so that for light of one color or wavelength the image

Figure 8-1-1 Simplified sketch of a spectroscope. Rays are shown for a particular wavelength of light, focused at point *A*. Point *B* would be the focus for a different wavelength.

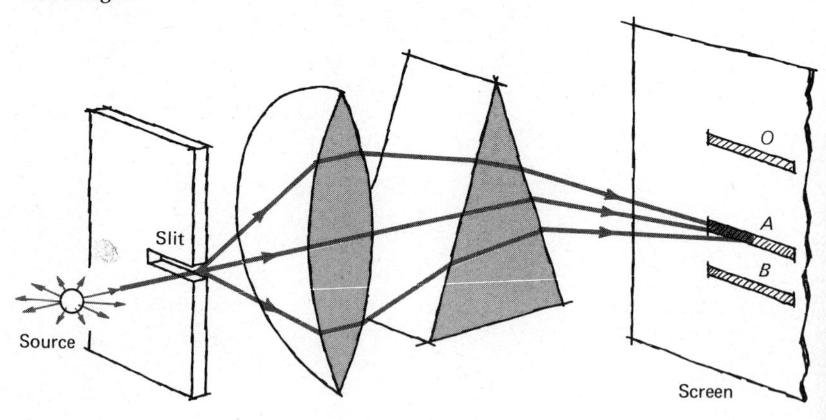

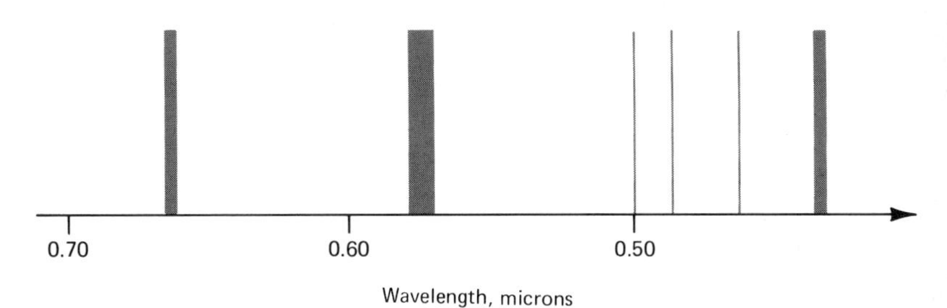

Figure 8-1-2 Some prominent lines of the helium spectrum.

might be at the point A. The prism bends light of different wavelengths by different amounts, so a different wavelength might result in an image of the slit at point B. The slit must be narrow, in order for images at A and B to be distinct from each other.

One of the first important discoveries in atomic physics was that the light emitted by "excited" atoms often consists of a spectrum of *discrete wavelengths*, as opposed to the continuous spectrum of a rainbow. Atoms can be most easily excited by passing an electric current through a gas (as in a neon sign) or by heating in a flame. Figure 8-1-2 shows some of the prominent wavelengths of visible light emitted by helium. The thicknesses of the vertical lines are meant to show qualitatively how strongly light of each wavelength is emitted. The unit of wavelength used in Fig. 8-1-2 is the *micron* (abbreviated μ), equal to 10^{-6} m. The range of wavelengths for visible light extends from about 0.40 microns (violet) to about 0.65 microns (red).

The wavelengths in Fig. 8-1-2 are often called *spectral lines.* This is because in experimental spectroscopy, as in Fig. 8-1-1, the image of the slit formed by light of each single wavelength is a line on the screen. Photographs of spectra are taken by replacing the screen of Fig. 8-1-1 with a photographic emulsion coated on a glass plate. The photographic image shows the lines very similarly to Fig. 8-1-2.

Early investigators discovered that each chemical element emitted its own characteristic pattern of spectral lines. The spectrum of an element might vary somewhat depending upon how the atoms were excited, mainly by changes in the relative intensities of the spectral lines, but not in their wavelengths. It was recognized that atomic spectra might provide clues to atomic structure, and very careful spectroscopic studies were made for various atoms under different conditions. Furthermore, in addition to atomic spectra, there are molecular spectra, which are wavelengths emitted specifically by given molecular combinations of atoms. This adds another dimension to spectroscopy, the study of chemical binding and the structure of molecules.

From the enormous variety of spectroscopic observations we will choose one particularly important example for detailed discussion: the spectrum of hydrogen, the lightest atom. Perhaps predictably, the spectrum of the lightest and simplest element is the least complicated. By 1885, detailed observations had been made of the hydrogen spectrum, and a startling empirical law was then discovered by a Swiss teacher named Johann Balmer. Balmer discovered that the wavelengths of the known hydrogen lines could be found from the following relation:

$$\lambda = b\frac{n^2}{n^2 - 2^2} \qquad 8\text{-}1\text{-}1$$

Each known hydrogen line corresponded to an integer value for n, greater than 2, in the equation above ($n = 3, 4, 5$), and b was a constant, equal to 0.365 microns. Balmer had not the slightest physical basis for his formula, but it fit the experimental data to an accuracy of 0.1 percent or better!

Only four hydrogen lines were known to Balmer when he invented his formula, the lines corresponding to $n = 3, 4, 5$, and 6, but he soon found out that others had been seen, which corresponded to $n = 7, 8$, etc. A few years after Balmer proposed his formula, 35 of his predicted series of lines were seen in the spectrum of hydrogen emitted by the sun. Figure 8-1-3 shows the *Balmer series spectral lines*, extending from about 0.65 μ, for $n = 3$, to 0.37 μ for n enormously large. The wavelength range of the series almost perfectly matches the range for visible light, a fortunate circumstance which facilitated experimental measurements.

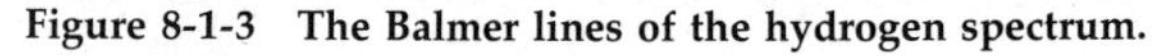

Figure 8-1-3 The Balmer lines of the hydrogen spectrum.

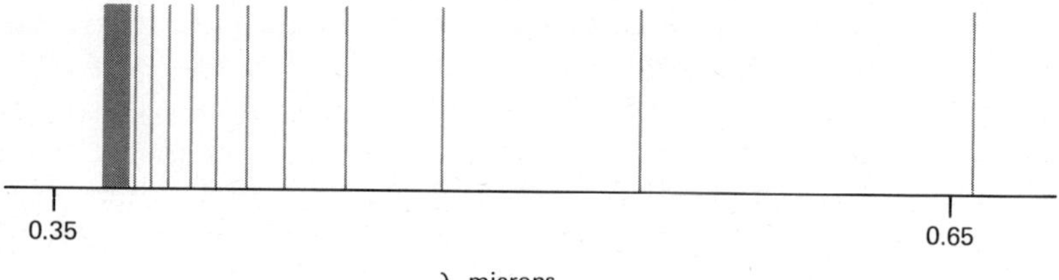

PROBLEM 1
What are the longest and shortest wavelengths in the Balmer series?

Balmer had the imagination to guess at a possible generalization of the formula in Eq. 8-1-1, to cases where the term 2^2 in the denominator might be 1^2, 3^2, 4^2, etc. However, the new wavelengths found in this way were not in the range of visible light. As a result, spectroscopists did not observe these lines until more than 20 years later, but the observations then completely confirmed Balmer's formula in a more general form,

$$\frac{1}{\lambda} = B\left(\frac{1}{n_2^2} - \frac{1}{n_1^2}\right) \qquad \text{8-1-2}$$

PROBLEM 2
The relation between the wavelengths makes it possible to identify the Balmer series in light from a distant galaxy even though all the wavelengths are about 10 percent longer than in the laboratory on earth. Can you explain the wavelength shift in terms of a relative velocity of the galaxy either toward or away from the earth? What is the velocity?

In Eq. 8-1-2, if n_2 is taken to be 2, and n_1 taken to be 3, 4, 5, etc., the result is the same as Eq. 8-1-1, provided the constant B is equal to 10.97 for λ in microns. The Balmer series thus corresponds to $n_2 = 2$, and n_1 any integer greater than 2. Of the other series predicted by Eq. 8-1-2, one, with $n_2 = 1$, $n_1 = 2, 3, 4$, etc., has wavelengths shorter than that of the shortest visible wavelength, which is violet. These lines, which are said to be in the *ultraviolet*, were discovered about 1910.

If we take $n_2 = 3$, $n_1 = 4, 5, 6$, etc., in Eq. 8-2-1, wavelengths are found which are longer than that of the longest visible wavelength, which is red. This series of lines, said to be in the *infrared*, was also discovered about 1910. The complete success of Balmer's formula hinted strongly at some underlying order within the hydrogen atom, which, if it could be discovered, might be the key to understanding the structure of atoms. However, it took many years before anyone succeeded in establishing a physical basis for Balmer's formula. In 1913, Bohr made a major breakthrough, which will be discussed in the following section.

8-2 THE BOHR ATOM

In 1913, soon after Rutherford discovered the nucleus, Neils Bohr constructed a theory which was a combination of classical physics, the quantum theory of light, and a daring new quantum postulate. The Bohr theory embodies a number of the most important ideas of quantum mechanics, and it signaled the beginning of a 15-year period during which the modern quantum theory was developed. For some important cases, Bohr's theory can be used to get very simply the same answers as the more correct quantum mechanics of Schrödinger and Heisenberg.

Following Rutherford's work, Bohr was led to visualize the hydrogen atom as shown in Fig. 8-2-1. The atom was imagined to consist of one electron moving in a circular orbit about the hydrogen nucleus, a proton. It was assumed that the electron was held in its circular orbit by the electrical force between it and the proton, calculated according to Coulomb's law for point charges. We will further assume that the proton remains fixed and that only the electron moves. Because the proton is nearly 2,000 times as massive as the electron, this is a good approximation.

A major difficulty with Bohr's picture was that it violated the laws of classical electromagnetism, which require that an accelerated charge radiate electromagnetic waves. For example, waves broadcast by radio transmitters are created by accelerating charges in the antennas. Bohr simply assumed that inside atoms, for some unknown reason, electrons could travel around circular orbits *without radiating.* This drastic measure seemed unavoidable, because otherwise energy would be carried off in electromagnetic waves and the electron would spiral in toward the nucleus. A calculation based on classical electromagnetism leads

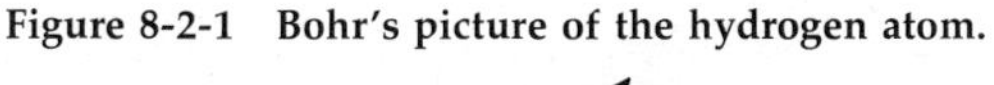

Figure 8-2-1 Bohr's picture of the hydrogen atom.

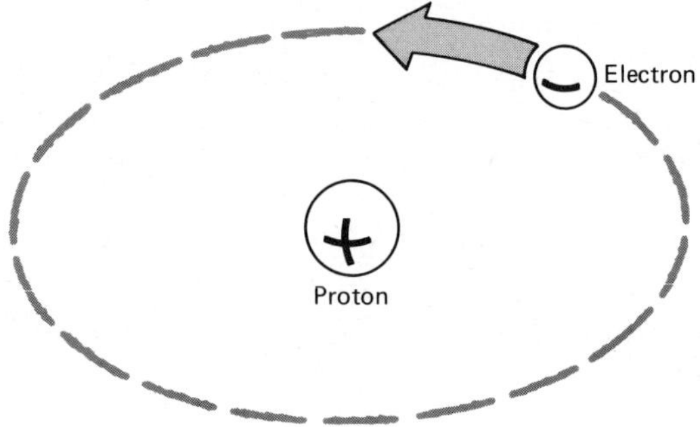

to the result that a hydrogen atom, with an electron at a radius of about 10^{-10} m, would collapse in a time of about 10^{-10} sec.

Having decided that electrons could be allowed to move in circular orbits, Bohr then made two postulates which led to a predicted hydrogen spectrum in very good agreement with experiments. These were:

1 Only orbits with certain radii were allowed.
2 When an electron moved from a large orbit to a smaller one, its decrease in total energy appeared in the form of a single light quantum which was emitted by the atom.

Bohr discovered that allowed radii could be specified which led to quanta with frequencies in agreement with spectroscopic evidence, provided that only orbits corresponding to specific values of *angular momentum* were allowed. Although we have discussed linear momentum, this is the first instance where it is necessary to introduce the idea of angular momentum. The angular momentum, called L, of a particle moving in a circular orbit of radius r is given by the product of r times the linear momentum,

$$L = mvr \qquad \text{8-2-1}$$

Angular momentum has a relationship to rotational motion similar to that of linear momentum to straight-line motion.

Bohr's first postulate was that the only orbits allowed for the electron in the hydrogen atom were circular orbits with certain *quantized values of angular momentum L,* such that

$$L = n\frac{h}{2\pi} \qquad \text{8-2-2}$$

where h is Planck's constant and n is any integer 1, 2, 3, ... From Eq. 8-2-2 and ordinary classical mechanics, we can find the radii of the allowed orbits. For a circular orbit of radius r, the angular momentum is given by $L = mvr$, as in Eq. 8-2-1. Combining this equation with the allowed values of L according to Eq. 8-2-2,

$$mvr = \frac{nh}{2\pi} \qquad \text{8-2-3}$$

In addition to this equation, there is another relation, obtained by applying Newton's second law to the circular motion,

$$\frac{mv^2}{r} = F = (9.0)(10^9)\frac{e^2}{r^2} \qquad \text{8-2-4}$$

This equation says that the centripetal force is given by the coulomb force between two point charges $+e$ and $-e$, the charges of the proton and the electron. Equations 8-2-3 and 8-2-4 may be used to solve for the radius r. First, solving Eq. 8-2-3 for v,

$$v = \frac{nh}{2\pi mr} \qquad \text{8-2-5}$$

Then, substituting this value of v into Eq. 8-2-4,

$$\frac{mv^2}{r} = \frac{m}{r}\left(\frac{nh}{2\pi mr}\right)^2 = \frac{n^2h^2}{4\pi^2mr^3} = (9.0)(10^9)\frac{e^2}{r^2} \qquad \text{8-2-6}$$

Equation 8-2-6 can now be solved for the radius r,

$$r = \left(\frac{h^2}{4\pi^2m(9.0)(10^9)e^2}\right)n^2 \qquad \text{8-2-7}$$

The numerical value of the collection of constants preceding the

Figure 8-2-2 The electrostatic potential energy for two oppositely charged point charges with magnitudes Q_1 and Q_2.

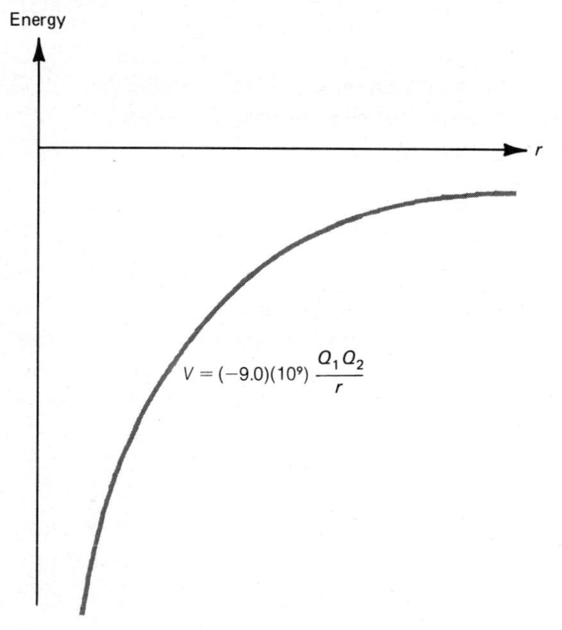

factor n^2 is equal to 0.529×10^{-10} m, so Eq. 8-2-7 may be written in the form

$$r = a_1 n^2 \qquad 8\text{-}2\text{-}8$$

with $a_1 = 0.529 \times 10^{-10}$ m. According to Bohr's quantization condition on the angular momentum, the symbol n stands for any integer 1, 2, 3, . . . , so the constant a_1 in Eq. 8-2-8, known as the *first Bohr radius*, is the smallest allowed value of r. For this value of r, the diameter of the electron orbit is close to 10^{-10} m, which is the typical size of an atom according to a variety of estimates.

In order to predict the spectrum of atomic hydrogen, from Bohr's second postulate, it is now necessary to find the total energy of the electron when it is in any allowed circular orbit. The total energy U is equal to the sum of the kinetic energy T and the potential energy V. The potential energy of two point charges with opposite signs when they are a distance r apart is shown in the graph of Fig. 8-2-2. In Chap. 5 (Sec. 5-5) we have discussed a similar potential-energy curve for the repulsion of like charges. Here, because the charges attract each other, the potential energy decreases as r gets smaller. For the Bohr atom, with charges $+e$ and $-e$, we can write

$$V = -(9.0)(10^9)\frac{e^2}{r} \qquad 8\text{-}2\text{-}9$$

The kinetic energy can be found from Eq. 8-2-4, the centripetal force equation. If we multiply each side of that equation by $r/2$, an expression for the kinetic energy results,

$$T = \frac{1}{2}mv^2 = (9.0)(10^9)\frac{e^2}{2r} \qquad 8\text{-}2\text{-}10$$

The magnitude of T turns out to be just half the magnitude of V. The total energy can be found by adding the potential energy from Eq. 8-2-9 to the kinetic energy from Eq. 8-2-10,

$$U = -\frac{(9.0)(10^9)e^2}{2r} \qquad 8\text{-}2\text{-}11$$

Equation 8-2-11 gives the total energy for an electron in a circular orbit of radius r, but only certain values of r are allowed in the Bohr model. Substituting for r from Eq. 8-2-8,

$$U = -\frac{(9.0)(10^9)e^2}{2a_1}\frac{1}{n^2} = -R\left(\frac{1}{n^2}\right) \qquad 8\text{-}2\text{-}12$$

All the constants have been combined and represented by a single constant R, whose value is given by

$$R = 21.7 \times 10^{-19} \text{ joules} \qquad \text{8-2-13}$$

Thus, according to Bohr's model, the electron in the hydrogen atom has only certain allowed, or *quantized, energies* U_n given by

$$U_n = -\frac{R}{n^2} \qquad \text{8-2-14}$$

An energy diagram for the hydrogen atom, drawn according to Eq. 8-2-14, is shown in Fig. 8-2-3. In this figure, all the energies are negative, because the potential energy is negative and its magnitude is larger than the magnitude of the kinetic energy. Zero energy corresponds to an infinitely large value for n, representing an infinitely large orbit, corresponding to an essentially free, unbound electron.

PROBLEM 3
Find, for the first Bohr orbit, the de Broglie wavelength (h/p) of the electron and compare it with the circumference of the orbit. Can you find a general relation for all Bohr orbits? The connection between de Broglie wavelength and orbit circumference provides a rather tenuous link between the Bohr theory and wave mechanics—interesting but not to be taken too seriously.

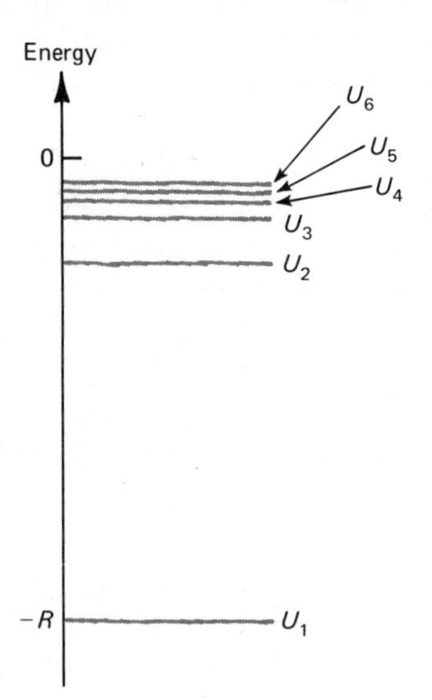

Figure 8-2-3 Energy-level diagram for the hydrogen atom, according to the Bohr model.

PROBLEM 4

A negative mu-meson is a cosmic ray particle which has all the same attributes as the electron but is about 210 times as massive. Suppose a mu-meson and a proton form an atom. Compare the first Bohr radius a_1 and the energy R for mu-mesonic hydrogen with their values for ordinary hydrogen.

The value of the constant R, which is typical of atomic energies, is seen to be a very small fraction of a joule. As a consequence, atomic physicists have found it convenient to adopt an energy unit called the *electron volt*, abbreviated eV, of more convenient size:

$$1 \text{ eV} = 1.60 \times 10^{-19} \text{ joule} \qquad \text{8-2-15}$$

This is the energy that would be gained by an electron released from rest in an electric field of one newton/coulomb after traveling one meter. In terms of this more natural unit,

$$R = 13.6 \text{ eV} \qquad \text{8-2-16}$$

Remember that the absolute value of the total energy depends upon an arbitrary choice of the zero of potential energy and hence has no physical significance by itself. It is *energy differences* which really matter. For example, the lowest energy, U_1, in Fig. 8-2-3 is $-R$ or -13.6 eV. The highest energy is zero and corresponds to U_n for an electron and proton so far apart that they are "free" and not bound together in a hydrogen atom. Thus, if we want to *ionize* a hydrogen atom, which means to separate the electron from the proton, we must supply a minimum of 13.6 eV of energy if the atom has energy U_1 initially. This is the *energy difference* between U_n, for n very large, and U_1.

Total energies less than zero correspond to bound states of the hydrogen atom, circular orbits in Bohr's model. A total energy of zero corresponds to a free proton and a free electron at rest, and total energies greater than zero correspond to a free electron and a free proton with kinetic energy. The bound-state problem, corresponding to the hydrogen atom, is the one of interest here.

According to Bohr's second postulate, emission of light from the hydrogen atom takes place when the total energy of the electron decreases, for example from U_3 to U_2. The difference in energy between these two states is equal to the energy of the light quantum emitted,

$$hf = U_3 - U_2 \qquad \text{8-2-17}$$

Written in a slightly different way, this equation can be seen to be simply a statement of energy conservation,

$$hf + U_2 = U_3 \qquad \text{8-2-18}$$

In Eq. 8-2-18, the final energy of atom + radiation, $hf + U_2$, is set equal to the initial atomic energy, U_3. In general, a hydrogen atom emits light if it makes a transition from an initial state of energy U_i to a final state of energy U_f, with U_f less than U_i. (Subscripts i and f stand for "initial" and "final.") In order to conserve energy, the light quantum must then have energy hf given by

$$hf = U_i - U_f \qquad \text{8-2-19}$$

The energies U_i and U_f in Eq. 8-2-19 must be allowed energies, given by Eq. 8-2-14, with values of n which we can call n_i and n_f. Thus, Eq. 8-2-19 can be written

$$hf = -\frac{R}{n_i^2} + \frac{R}{n_f^2} = R\left(\frac{1}{n_f^2} - \frac{1}{n_i^2}\right) \qquad \text{8-2-20}$$

In order for the initial state to have a higher energy than the final state, so that hf is a positive energy, n_i must be greater than n_f. Equation 8-2-20 is in complete agreement with the empirical relation, Eq. 8-1-2, discovered by the spectroscopists. Since $f = c/\lambda$, Eq. 8-2-20 can be rearranged in the form

$$\frac{1}{\lambda} = \frac{R}{hc}\left(\frac{1}{n_f^2} - \frac{1}{n_i^2}\right) \qquad \text{8-2-21}$$

which is identical to Eq. 8-1-2 provided R/hc is equal to the empirical constant B. In fact, the value of R/hc predicted by the Bohr model is in excellent agreement with B.

By combining the quantum theory of radiation with the ad hoc condition that angular momentum was quantized, Bohr arrived at the first real physical understanding of atomic spectra. His model of hydrogen was extended to other atoms, and the picture of atoms as "miniature solar systems," with electrons revolving around the nucleus like planets around the sun, was a very appealing one. Ultimately, such orbits were later found to be incompatible with the wave-particle duality, or the uncertainty principle. However, while the fully developed quantum mechanics rejected Bohr's orbits, it confirmed from the first principles of either the Heisenberg or Schrödinger theories, his quantum condition for angular momentum.

PROBLEM 5

An uncertainty relation connects the uncertainty in r with the uncertainty in the radial component of momentum p_r:

$$(\Delta r)(\Delta p_r) \geq h$$

Find the uncertainty in p_r if (Δr) is about one-tenth of the radius of the first Bohr orbit. Compare with the momentum of the electron in the first Bohr orbit. Is motion in the first Bohr orbit compatible with the uncertainty principle?

The idea of quantized energies, introduced by Bohr, turned out to be a fundamental aspect of quantum mechanics. At least for hydrogen, the Bohr model even gave an easy way to calculate the correct energies. Through Bohr's second postulate, epitomized by Eq. 8-2-19, the observed spectral lines were linked to transitions of the atom from one quantized energy state to another of lower energy. In Fig. 8-2-4, arrows on an energy diagram indicate

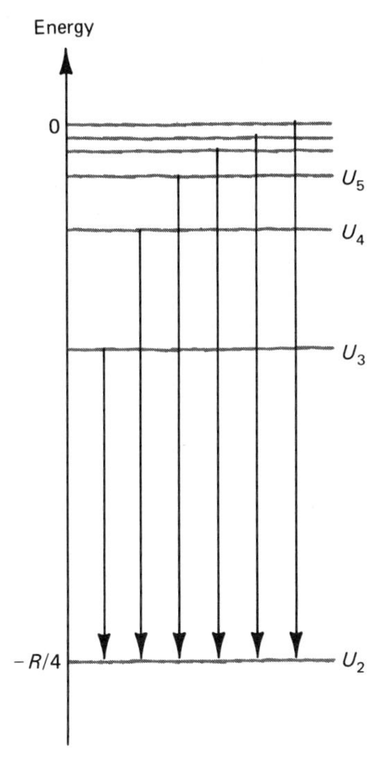

Figure 8-2-4 Hydrogen-atom transitions which give rise to the spectral lines of the Balmer series.

the transitions which lead to the Balmer series. This diagram is simply a pictorial representation of Eq. 8-2-20, for $n_f = 2$.

Ordinary hydrogen atoms are in their lowest possible energy state, with total energy U_1, called the *ground state.* In a gas discharge tube, or in a flame, or when bombarded by high energy electrons, hydrogen atoms can be given the energy required to raise them to the states with energies U_2, U_3, etc., called *excited states.* Typically, an atom remains in an excited state for a very short time, of order 10^{-9} sec or less, after which it makes a transition to a lower energy state and emits a quantum of radiation.

PROBLEM 6

Singly ionized helium has one electron and a nucleus with charge $+2e$. Predict the wavelengths of some spectral lines of singly ionized helium in the visible range, from about 0.35 to 0.65 microns.

PROBLEM 7

For a heavy atom with atomic number Z the energy required to remove one of the two innermost electrons—to knock it completely out of the atom—is approximately given by $E \approx Z^2R$, where R is the same constant as in the Balmer and Bohr formulas for hydrogen, equal to 13.6 eV. Explain this formula in terms of an approximate picture and the Bohr model.

When an inner electron is knocked out, an electron with $n = 2$ can make a transition to the vacant $n = 1$ state. This results in emission of an x-ray quantum. Estimate the wavelength of x-rays from iron ($Z = 26$).

8-3 THE HYDROGEN ATOM

In this section we will see the results of applying the more correct Schrödinger quantum theory to the hydrogen atom. While the mathematics involved when the Schrödinger equation is applied to the hydrogen atom is much too complicated to include here, many of the results can be qualitatively understood in terms of the simple examples of the preceding chapter. To begin a quantum-mechanical study of the hydrogen atom, consider first states of the electron which correspond to purely radial classical motion. In this type of motion, the electron moves back and forth *through* the proton, much as a simple pendulum can be made to swing back and forth through a point directly under its point of support. The idea that the electron can move right through the

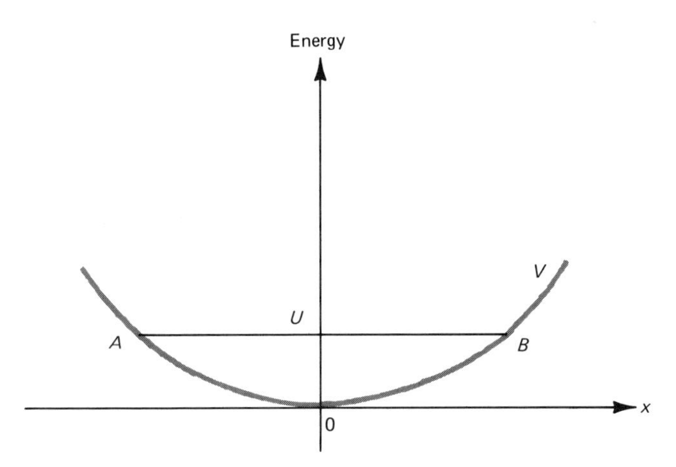

Figure 8-3-1 Energy diagram for motion of a classical pendulum with total energy U.

location of the proton is rather shocking. We will first discuss the quantum-mechanical problem and return to this question after having seen the results.

Figure 8-3-1 is an energy diagram for pendulum motion, while Fig. 8-3-2 is an energy diagram for motion of an electron in a hydrogen atom along an x axis passing through the proton at $x = 0$. The potential energy in Fig. 8-3-2 corresponds to the effect of an attractive Coulomb force between the proton and electron and looks quite different from the potential-energy curve for the pendulum. In each figure, a total energy U is shown which corresponds to a bound state, either classically or quantum mechanically. The points labeled A and B show the positions at which the classical motion would reverse, where the kinetic energy $U - V$ is zero. With reference to motion of a particle in a box, points A and B are like the points where the particle hits the walls.

In order to find the electron wave function, it is necessary to solve the Schrödinger equation for three-dimensional motion. While Fig. 8-3-2 can give you a feeling for the motion we are discussing, the electron is really not physically constrained to move along a given line, like the x axis in Fig. 8-3-2. The problem which must really be discussed is motion of the electron in the *three-dimensional region* surrounding a proton, but such that its wave function depends only upon the distance r from the proton. This restriction leads to the simplest wave functions for the hydrogen atom.

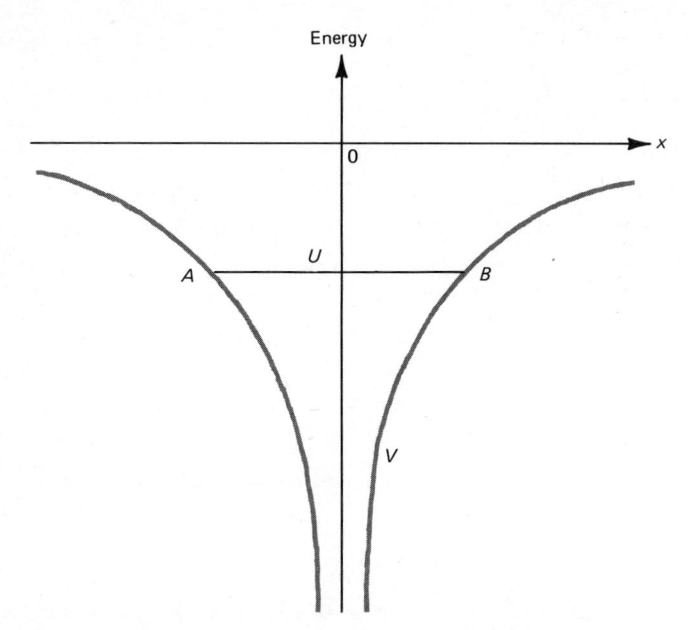

Figure 8-3-2 Energy diagram for motion of an electron with total energy U along a line through a positive point charge.

The solution of the Schrödinger equation in three dimensions is too complex to discuss here in detail. However, as with the particle in a box, the Schrödinger equation plus the requirement that the wave function be "well behaved" leads to certain allowed wave functions. The phrase "well behaved" means that both the wave function and its slope are not allowed to change discontinuously, and that the wave function is not permitted to get infinitely large outside the classically allowed region of electron motion. Recall that these are the conditions, for the particle in a box, which led to certain allowed sine waves inside the box and decreasing exponentials outside.

Figure 8-3-3 shows the first three allowed wave functions for the hydrogen atom, plotted against r, with the first three Bohr radii shown on the r axis for comparison. Each wave function tails off like a decreasing exponential for large values of r, and each of these allowed wave functions corresponds to a particular value of the total energy of the electron, U. The wave function ψ_1 does not cross the r axis, while ψ_2 crosses the r axis once, ψ_3 crosses the axis twice, and so on. The situation is at least reminiscent of the

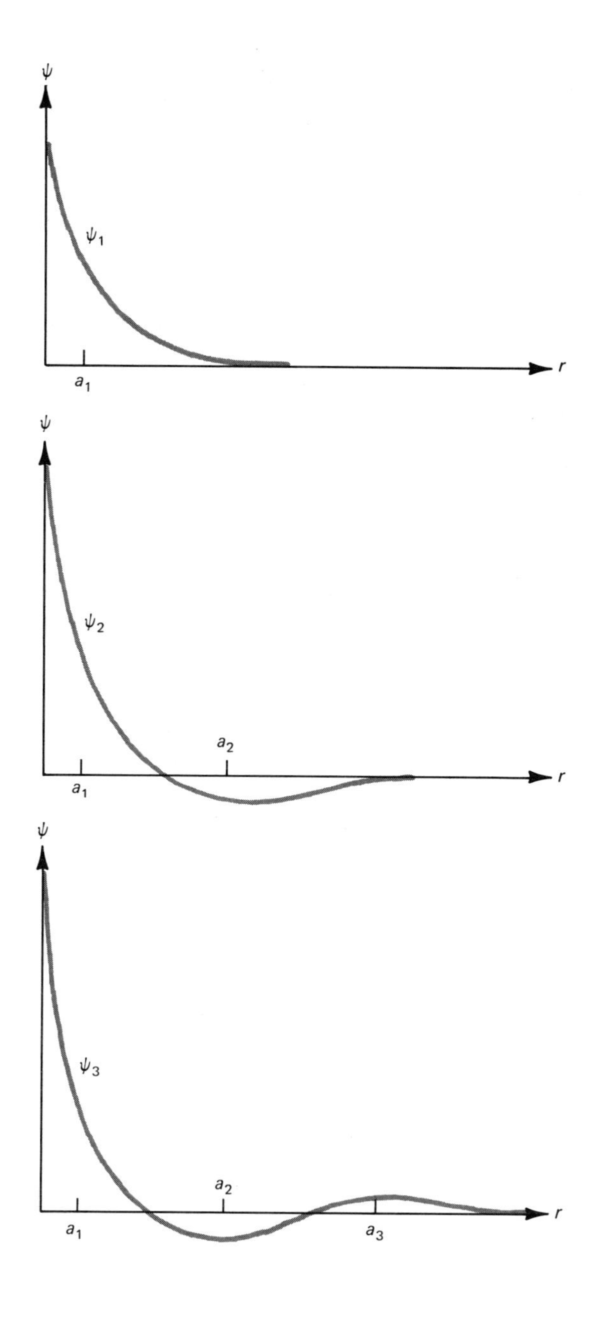

Figure 8-3-3 Wave functions for the three lowest energy states of hydrogen.

wave functions for the one-dimensional motion of a particle in a box.

From the Schrödinger equation, the allowed energies are found to be given by the formula

$$U_n = -R\frac{1}{n^2} \qquad \text{8-3-1}$$

where R is the same constant which appeared in the Bohr model, and this equation is exactly the same as that obtained by Bohr. The energy differences are therefore again in excellent agreement with the quantum energies for the observed spectrum of hydrogen.

The old Bohr orbits are replaced by probability amplitudes ψ_n. While the correct quantum theory predicts precisely the same quantized energies as the Bohr model, it is based on a very different physical picture. In particular, the states of the hydrogen atom described by the wave functions presented so far have *zero* angular momentum and correspond to classical motion of the electron back and forth through the proton rather than circularly around it.

Since the hydrogen atom represents a three-dimensional problem, the interpretation of ψ^2 is that it is the probability *per unit volume* for finding the electron at various places. Graphs of $(\psi_1)^2$ and $(\psi_2)^2$ are shown in Fig. 8-3-4. Note that the Bohr radii a_1 and a_2, which correspond to these two states in the Bohr model, are the radii within which the electron is quite likely to be found. The ground-state wave function has reasonably large values in a sphere about 10^{-10} m in diameter, which corresponds to the size of the hydrogen atom. The proton diameter is now known to be about 10^{-15} m, and so the proton is hardly in the way when the electron motion is smeared out by quantum mechanics. Even in the unlikely event that the electron is inside the proton, it is found that the proton behaves like a transparent glob of charge as far as electrons are concerned.

The quantum theory is not limited to states with zero angular momentum. We have merely considered them first. The theory predicts, from the Schrödinger equation and from requirements that the wave function be well behaved, that the angular momentum L is quantized according to the relation

$$L = \sqrt{l(l+1)}\,\frac{h}{2\pi} \qquad l = 0, 1, 2, \ldots, n-1 \qquad \text{8-3-2}$$

The quantity l is known as the *angular momentum quantum num-*

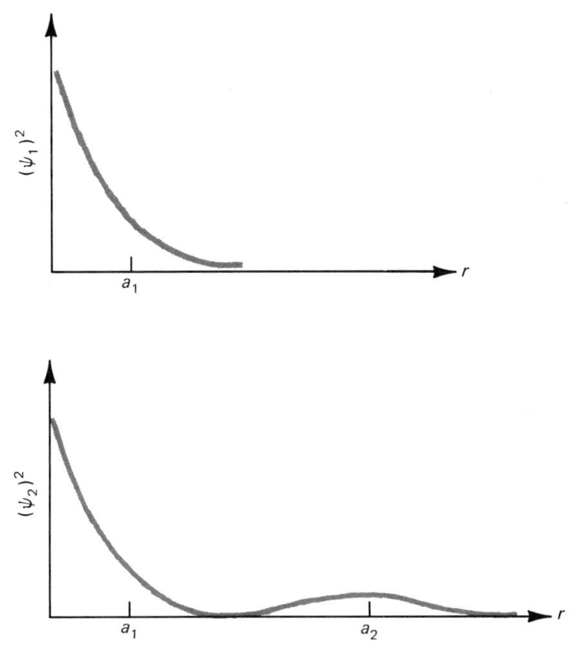

Figure 8-3-4 Distributions of probability per unit volume, ψ^2, for the two lowest energy states of hydrogen.

ber and may be any integer less than n, including zero. For large values of l, Eq. 8-3-2 is almost identical to the Bohr condition, while for small values, especially $l = 0$ (not allowed in the Bohr theory), there is a significant difference.

We have now seen the appearance of two separate quantum numbers in the Schrödinger theory, a *principal quantum number n,* which determines the energy of a state, through Eq. 8-3-1, and an *angular momentum quantum number l.* The maximum value of l is limited to $n - 1$. The energy is found not to depend upon l, but the wave functions differ for different choices of l. For $l = 0$ we have the states in which ψ depends only upon r. We say these wave functions are spherically symmetric. For other values of l, the wave functions vary in space in a way which corresponds very roughly to elliptical orbits of the electron about the nucleus. For l different from 0, ψ is equal to zero at the nucleus, unlike the wave functions in Fig. 8-3-3. For the maximum value of l, the wave function corresponds roughly to Bohr's circular orbit.

The table below summarizes the possible values taken on by the two quantum numbers for states with various quantized energies:

ENERGY	ALLOWED ANGULAR MOMENTUM QUANTUM NUMBERS, l	PRINCIPAL QUANTUM NUMBER, n
U_1	0	1
U_2	0, 1	2
U_3	0, 1, 2	3
U_n	$0, 1, 2, \ldots, n-1$	n

Summarizing, in this section we have seen some of the results of the correct quantum theory of the hydrogen atom. The square of an electron wave function, ψ^2, is interpreted as the probability per unit volume for finding the electron at various locations. There are no classical orbits. However, the quantized energies found from the correct calculation are the same as in the Bohr theory. Angular momentum is also found to be quantized, in a somewhat different way from that described by Bohr's ad hoc assumption. For the hydrogen atom, states with a given principal quantum number may have various values of angular momentum, but the energy of the state depends only upon the principal quantum number.

PROBLEM 8

A hydrogen atom is in a state with principal quantum number 4. It emits three quanta in succession to reach the ground state. What are the quantum energies in electron volts?

PROBLEM 9

The probability per unit volume, $(\psi_2)^2$, in Fig. 8-3-4 peaks inside the first Bohr radius which is at $r = a_1$. However, we might hope that the electron in this state is most likely to be found near the second Bohr radius at $r = 4a_1$. Find the volume of a sphere with radius a_1 and compare it with that of a spherical shell from $r = 3a_1$ to $r = 5a_1$. By estimating averages for the probability density $(\psi_2)^2$ in the two volumes, find out where the electron is most likely to be found.

PROBLEM 10

If the nucleus is a point charge, the quantum mechanics for hydrogen gives the same energy U_2 for $l = 0$ or 1. Which energy might be most affected by a finite nuclear size? How would it change? Explain your reasoning.

8-4 THE ELEMENTS

Suppose we now ask about the results of the quantum theory when it is applied to atoms more complicated than hydrogen. The obvious sequence would be to study atoms with two electrons, three electrons, and so on. For each atom the nucleus would be assumed to have a number of protons equal to the number of electrons, so that the atom would have no net electric charge. This, incidentally, cannot be the entire picture, since the atomic weights are such that the nuclei contain more mass than that of the needed number of protons. There are massive electrically neutral particles in the nucleus as well as protons, which exert so little force upon the electrons they can be neglected in calculations of electron wave functions.

Actually, it is impossible to calculate exact wave functions for even the two-electron atom, and it is extremely difficult to calculate even approximately the wave functions of complicated atoms. The difficulties lie in the mathematics, not the physics. However, it is possible to understand qualitatively a great deal about the more complicated atoms, and in this section we will see how the chemical properties of the elements can be understood.

Figure 8-4-1 shows two potential-energy graphs, one for an

Figure 8-4-1 Potential-energy graphs for an electron in the field of a positive point charge, with and without shielding.

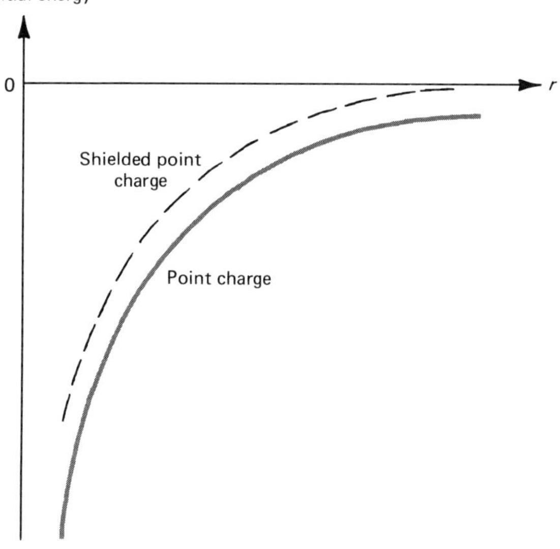

electron moving in the electric field of a positive point charge located at $r = 0$, and another for an electron moving in the electric field of a positive point charge surrounded by a cloud of negative charge. For a many-electron atom, the cloud of negative charge represents all the electrons other than the one we are considering. Even for a two-electron atom the other electron can be represented as a cloud of negative charge, distributed as determined by its wave function. The main difference between the two curves is that the cloud of negative charge has the effect of *shielding* the positive charge, or reducing its effect at the larger values of r.

Now consider the quantized total energies for an electron whose potential energy arises from the electric field of a point positive charge surrounded by a cloud of negative charge. The main new result, in comparison with the hydrogen atom, is that the energies depend upon the angular momentum quantum number as well as the principal quantum number. Figure 8-4-2 shows qualitatively the allowed values of U for the first three principal quantum numbers, where U has two subscripts, the first for the principal quantum number and the second for the angular mo-

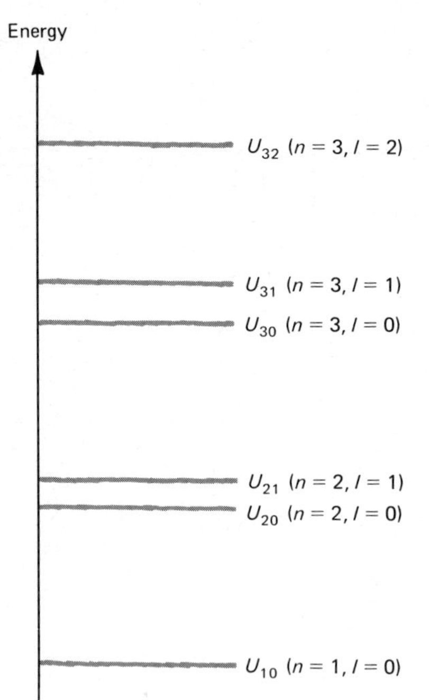

Figure 8-4-2 Qualitative variation of electron energy with principal quantum number for the potential given by a shielded point charge.

mentum quantum number. For example, U_{21} ("U sub-two-one") corresponds to the principal quantum number 2 and the angular momentum quantum number 1.

To get a physical feeling for the ordering of energies in Fig. 8-4-2, the most useful idea is that higher energies mean the electron is less strongly bound in the atom. An electron with a high energy needs less energy to be removed from the atom. For a given value of the principal quantum number, states with higher angular momentum have higher energies and are less strongly bound. This is because the wave functions of the high angular momentum states are large in regions of space further away from the nucleus. For zero angular momentum, we have seen that the wave functions peak right at the nucleus, and these are the most strongly bound states.

The lowest energy state of a given atom, which is its normal state or ground state, is found by putting electrons in states so that their total energy is as small as possible. The most significant discovery for being able to understand the chemical properties of the elements was made by Wolfgang Pauli soon after the invention of the quantum theory. It is that, at most, two electrons may be in a given state. This requirement is called the *Pauli exclusion principle*, because it excludes any other electrons from being in a state already occupied by two. The many electrons in complex atoms must therefore occupy a variety of states. They cannot all be in the lowest energy state because of the exclusion principle.

In order to find out how the electrons occupy various states in different atoms, there is one other result, from the Schrödinger equation, that is needed. For a given value of the angular momentum quantum number there are a number of states which correspond, classically, to the same elliptical orbit oriented in different ways in space. For example, the orbit might lie either in a vertical plane, in a horizontal plane, or in a plane at some angle to the horizontal. The quantum theory leads to a surprising result, which corresponds to a classical orbit being restricted to certain *quantized orientations.* For a given angular momentum quantum number l, there are $2l + 1$ allowed orientations. Each corresponds to a *separate state* from the standpoint of the Pauli exclusion principle and can be occupied by two electrons at most.

The table below lists the number of states, and the number of electrons which may occupy them, for each value of the angular momentum quantum number. We are now ready to find out the

ANGULAR MOMENTUM QUANTUM NUMBER, l	NUMBER OF STATES, $2l + 1$	NUMBER OF ALLOWED ELECTRONS, $2(2l + 1)$
0	1	2
1	3	6
2	5	10
3	7	14
4	9	18

way the electrons are arranged for the elements, restricting the discussion, for simplicity, to the first twenty.

Table 8-4-1 lists the first 20 elements, those with 20 or fewer electrons, and gives the number of electrons in the states corresponding to the various values of the n and l quantum numbers. Hydrogen has one electron with $n = 1, l = 0$, while helium has two such electrons. The next element, lithium, must have its third electron in the next higher energy state with $n = 2, l = 0$, as a result of the exclusion principle.

Notice, from Fig. 8-4-2, that there is a big difference in energy from $n = 1$, $l = 0$ to $n = 2$, $l = 0$, and this means that the two *inner* $n = 1$ electrons in lithium are tightly bound while the single *outer* $n = 2$ electron is very loosely bound. Small principal quantum numbers correspond to wave functions which are large close to the nucleus and larger principal quantum numbers to wave functions which are large further from the nucleus. The terms *inner* and *outer* electrons are often applied to those with small or large principal quantum numbers, respectively.

Chemically, lithium is a very reactive metal, which reflects the fact that its loosely bound electron causes it to combine easily with certain other elements which would like to gain an extra electron. On the other hand, helium has two tightly bound electrons and is an example of an *inert gas*, an element which refuses to enter into chemical combination under almost any circumstances.

PROBLEM 11

What are the quantum numbers for the first excited state above the ground state for lithium?

The next element, beryllium, has four electrons, two tightly bound and two less tightly bound. It is a metal, with a chemical *valence* of two. The valence is a number which is used in chemistry

Table 8-4-1

ATOMIC NO.	NAME	NUMBER OF ELECTRONS WITH:					
		$n=1$	$n=2$		$n=3$		$n=4$
		$l=0$	$l=0$	$l=1$	$l=0$	$l=1$	$l=0$
1	Hydrogen	1					
2	Helium	2					
3	Lithium	2	1				
4	Beryllium	2	2				
5	Boron	2	2	1			
6	Carbon	2	2	2			
7	Nitrogen	2	2	3			
8	Oxygen	2	2	4			
9	Fluorine	2	2	5			
10	Neon	2	2	6			
11	Sodium	2	2	6	1		
12	Magnesium	2	2	6	2		
13	Aluminum	2	2	6	2	1	
14	Silicon	2	2	6	2	2	
15	Phosphorous	2	2	6	2	3	
16	Sulfur	2	2	6	2	4	
17	Chlorine	2	2	6	2	5	
18	Argon	2	2	6	2	6	
19	Potassium	2	2	6	2	6	1
20	Calcium	2	2	6	2	6	2

to describe the number of electrons an atom may like to contribute —or sometimes to gain—by entering into chemical combination with atoms of other elements. Incidentally, metals, which like to contribute electrons when they form chemical compounds, are good electrical conductors because some of the loosely bound electrons become essentially free charge carriers in a metallic solid.

The next six elements correspond to increasing numbers of electrons with principal quantum number 2 and angular momentum quantum number 1. Boron, carbon, and nitrogen have rather complicated chemical properties, participating in chemical compounds which involve various kinds of electron-sharing. Oxygen, however, is just two electrons short of filling up all the states with $n = 2$, $l = 1$. It would bind these two electrons rather strongly and typically forms compounds in which two electrons are supplied by other atoms. Since beryllium has two relatively loosely bound electrons, you might expect the compound BeO, beryllium oxide, to exist, and it does.

The next element, fluorine, is just one electron away from filling up all the states with principal quantum number 2. It very actively reacts chemically with elements which can supply one electron, and is an example of a chemical element known as a halogen. Lithium fluoride is an example of a compound whose existence we can correctly predict from the electron structures of fluorine and lithium. After fluorine comes neon, which has 10 electrons.

For neon, the eight electrons with $n = 2$ are said to constitute a *closed shell.* All the allowed states with $n = 2$ are occupied by these electrons, and this results in a particularly tightly bound structure. Closed-shell atoms do not like to give up electrons in chemical reactions. They also do not like to add an electron. For neon, an eleventh electron would have to go to a state with $n = 3$, outside the closed-shell configuration. The electric field seen by such an added electron is so weak it is not bound at all. Therefore, neon, like helium, is an inert gas, an element which does not combine chemically in any way with other atoms.

Just as neon behaves like helium, the next element, sodium, with one loosely bound electron, behaves like lithium. The chemical properties of the atoms are beginning to repeat, because they depend only on the least tightly bound *outer,* or *valence,* electrons. The chemical properties of all the elements from sodium to argon are very similar to the properties of the elements from lithium to neon. There is a relatively large energy difference between the states $n = 3, l = 1$ and $n = 3, l = 2$ (see Fig. 8-4-2) which causes argon to be an inert gas, with a closed shell of $n = 3$, $l = 1$ electrons.

A Russian chemist, Dmitri Mendeleev, published a paper in 1869 in which he introduced the idea of a periodic repetition of the properties of the elements. The *periodic table* of the elements soon became established as an important empirical tool in chemistry. Undiscovered elements were predicted from gaps in the table, and much of the chemistry of poorly studied elements could be inferred from knowledge of similar ones in the same family. In the 1930s quantum mechanics made this empirical chemistry understandable, and the modern chemist is now much more involved than the physicist in quantum mechanical calculations of atomic and molecular wave functions.

As far as we have gone, electron states with a given value of n are seen to be filled in order of increasing l, and then states of the next larger value of n are similarly filled. However, beginning with potassium, the shielding effect begins to complicate matters. The inner electron structure of potassium, which is essentially the closed-shell structure of argon, binds an electron with $n = 3$, $l = 2$ less strongly than one with $n = 4$, $l = 0$. The high angular momentum electron sees a weaker attractive electric field than one with larger principle quantum number but zero angular momentum. The reason is that the $l = 0$ electron penetrates the argon core as it moves back and forth through the center of the atom.

Following argon, potassium and calcium have one and two electrons, respectively, with $n = 4$, $l = 0$. Then, 10 succeeding elements, all metals, finally result from the filling of states with $n = 3$, $l = 2$. Their chemical properties are mainly determined by the two $n = 4$, $l = 0$ electrons. Many of them, such as manganese, iron, cobalt, and nickel, have very similar chemical behavior. The addition of six ($n = 4$, $l = 2$) electrons completes a period of 18 elements and leads to another inert gas, krypton.

PROBLEM 12

Both magnesium and silicon have two electrons in their highest energy group of states. Yet one has valence +2 and the other +4. Which is which? Explain why. (Hint: Look at the next lower energy states.)

PROBLEM 13

Predict the atomic number of the next element heavier than calcium which has very similar chemical properties.

The chemistry of elements heavier than krypton can also be understood in terms of quantum mechanics, with the competition between the filling of inner and outer shells leading to considerable complication. An obvious question which arises is whether quantum mechanics can predict any end to the number of elements. The answer is no; yet there are in fact no known natural elements with more than 92 electrons. The reason for this is that it appears to be impossible to form a stable nucleus containing more than 92 protons.

8-5 CHEMICAL BINDING

In the previous section atoms of various chemical elements have been described in terms of their individual electron structures.

For each element, the inner electrons, with small values of the principal quantum number, are tightly bound to the nucleus and are not involved in chemical reactions. The outer electrons, with the largest values of the principal quantum number, determine the chemical properties of a particular element. For example, the *alkali metals*, lithium, sodium, potassium, etc., all have a single loosely bound outer electron which determines their similar chemical properties.

An alkali metal combines with a halogen by means of the simplest type of chemical bond. A halogen is an element which has so many outer electrons with the same quantum numbers that the addition of one more would result in a filled shell. Since a filled shell results in very tight binding of all the electrons in it, halogens are eager to capture an extra electron. The binding between an alkali metal and halogen results from a simple transfer of one electron, forming a closed-shell positive ion from the alkali metal atom which loses the electron, and a closed-shell negative ion from the halogen atom which gains it. Sodium chloride, bound in this way, is shown schematically in Fig. 8-5-1. This is called *ionic* binding, because the constituents of the molecule are ions, which attract each other by electrostatic forces. The ionic binding described above is characteristic of chemical compounds formed from elements with valences +1 and −1. If the two ions of sodium chloride are widely separated, they attract each other almost like point charges, with a force that is approximately inversely proportional to the square of the distance between their centers. However, if the ions get so close that their electron clouds overlap, then the electrostatic repulsive force between the positive nuclei pushes them apart. This situation can be represented by a graph of electrostatic potential energy as a function of distance between

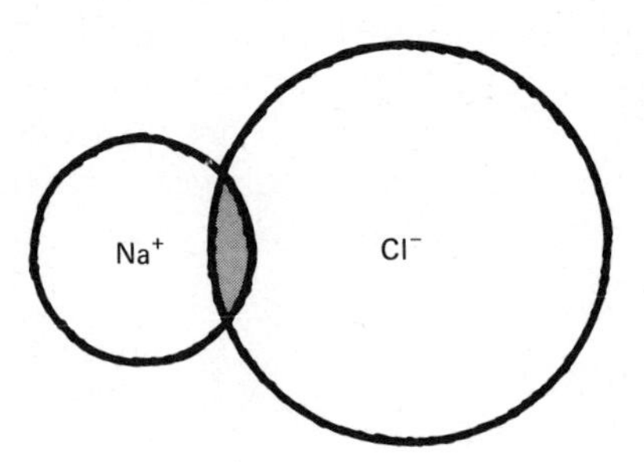

Figure 8-5-1 Ionic binding of the sodium chloride molecule. The ions attract each other until they slightly "penetrate."

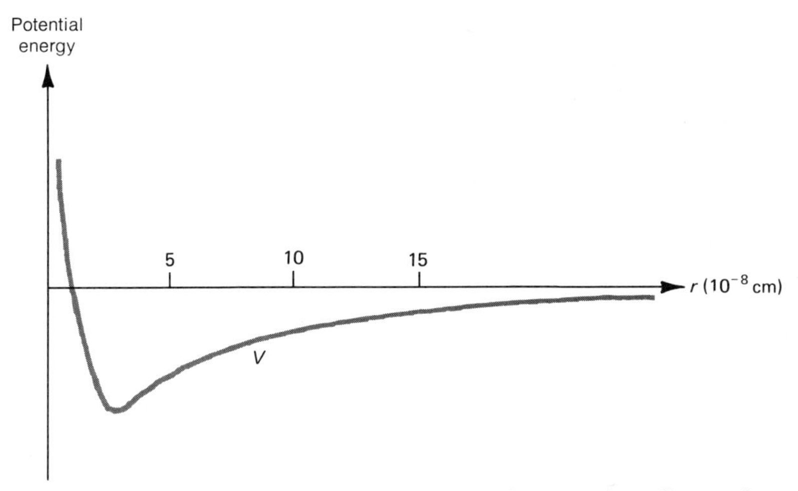

Figure 8-5-2 Variation of the electrostatic potential energy of a sodium and a chlorine ion with their distance of separation.

the sodium and chlorine ions, as in Fig. 8-5-2. The equilibrium position of the two ions is that with the lowest potential energy, when the distance between them is about 3×10^{-8} cm.

PROBLEM 14

Estimate the energy needed to unbind the ions of the NaCl molecule, assuming they behave like point charges located at the centers of the ions and are 3×10^{-8} cm apart in the molecule. This establishes a scale of energies for quantized vibrational energy states of the molecule. Compare with the similar quantity R for the hydrogen energy levels. Compare with the energy of a visible light quantum.

Figure P-8-15

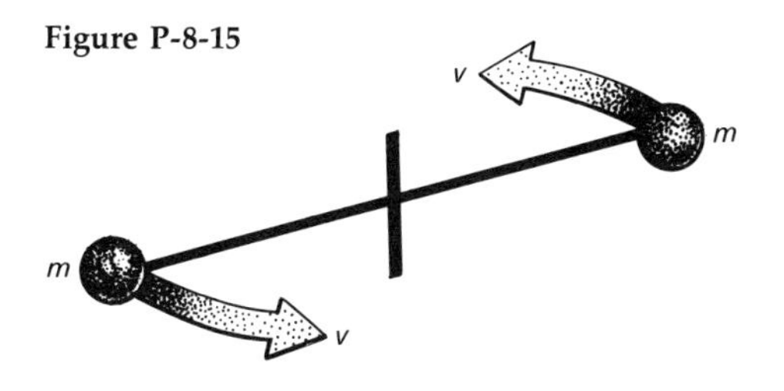

PROBLEM 15
A diatomic molecule like NaCl has most of its mass in the nuclei. It can spin about an axis through the line connecting the nuclei. A simple case, for equal mass nuclei is shown in Fig. P-8-15. The angular momentum about the center of rotation must be quantized in units of $h/2\pi$, however. As a result the kinetic energy is quantized. For an oxygen molecule, O_2, with the nuclei at a fixed distance apart of about 4×10^{-8} cm, find the energy in electron volts for $L = h/2\pi$, which is called the first excited rotational state.

For many chemical compounds, the bond between atoms is more subtle than the ionic bond described above. The bond arises from a *sharing* of outer electrons between atoms such that the valence electrons no longer belong exclusively to a single atom. The simplest example of such a *covalent bond* is that of the hydrogen molecule. Hydrogen atoms unite in pairs to form hydrogen molecules, and hydrogen gas is molecular hydrogen, not atomic hydrogen. This type of binding can be made plausible by the following example.

Consider, as in Fig. 8-5-3, two hydrogen nuclei (protons, with charge $+e$) located a distance d on either side of an electron. It is easy to calculate from the inverse square law that the electrostatic force of attraction between each proton and the electron is four times as great as the repulsive force between the protons. Thus, if there were a bit more than $(1/4)e$ of negative charge midway between the protons, there would be a net attractive force acting on them.

Figure 8-5-4 shows two hydrogen atoms separated from each other and then bound together in a hydrogen molecule. Their electron clouds are shown, and in the case of the molecule, there

Figure 8-5-3 Idealized bonding of two protons by an electron between them. The forces on the right-hand proton due to the other proton and the electron are shown.

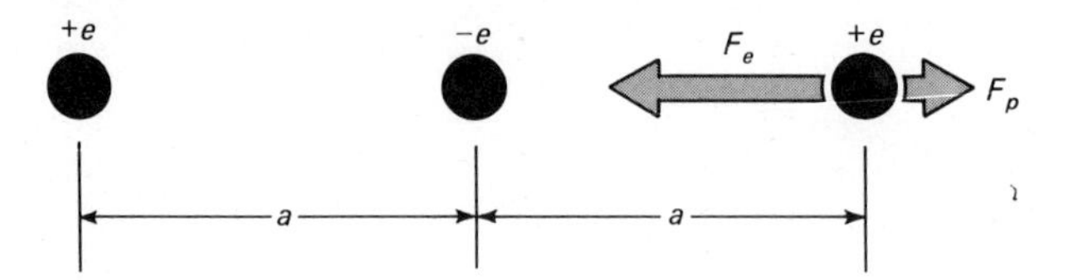

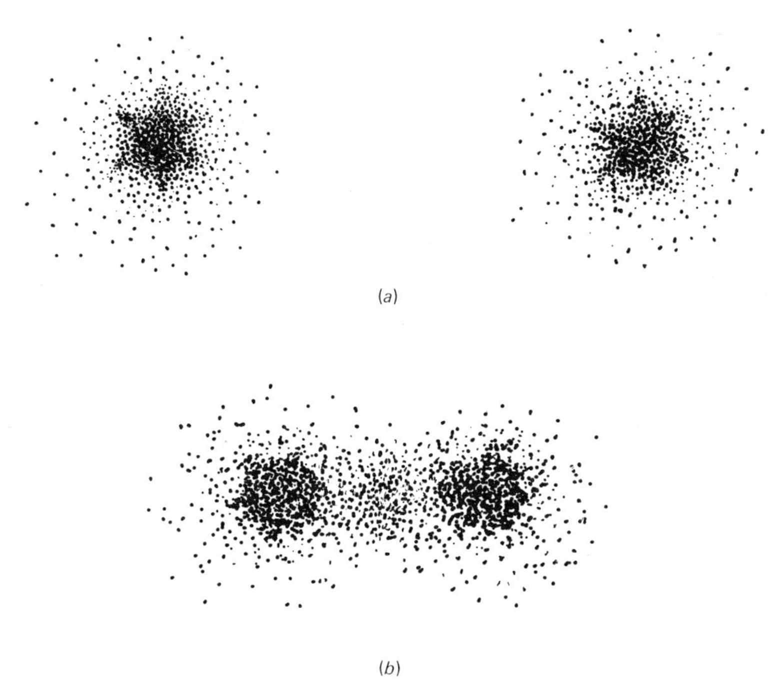

Figure 8-5-4 Two hydrogen atoms, widely separated and covalently bonded.

is indeed an excess of negative charge between them, which results in the type of binding illustrated by the example discussed above. The molecular electron wave functions can be found, at least approximately, from a quantum-mechanical calculation for two electrons shared by two protons. The sharing produces the tendency for the electrons to be found between the protons, which in turn results in the attractive force which binds the protons into a molecule.

In many chemical compounds the binding between atoms is mainly covalent, particularly in the case of compounds of carbon, hydrogen, and oxygen, which are called *organic compounds* because of their central importance in biochemistry. A simple organic compound, ethyl alcohol, is shown diagramatically in Fig. 8-5-5. The four valence electrons of the carbon atom at the left end of the molecule are shown to form covalent bonds with the valence electrons of three hydrogen atoms and with one valence electron of the other carbon atom.

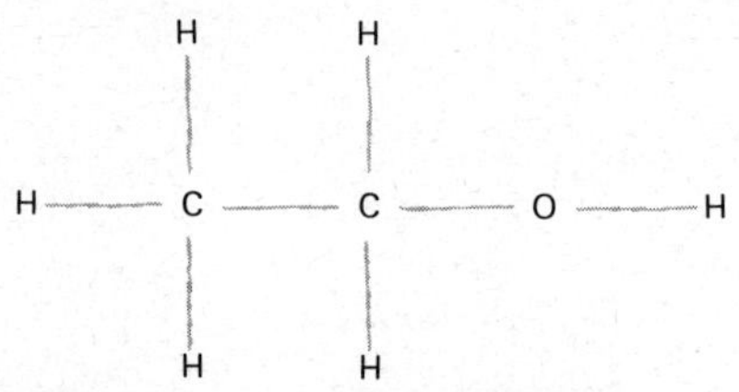

Figure 8-5-5 Covalent bonds of a simple organic chemical, C_2H_5OH, or ethyl alcohol.

Covalent bonds between hydrogen and carbon lead to the possibility of long chain molecules. Figure 8-5-6*a* shows diagramatically the structure of ethylene, which includes a "double bond" involving two electrons from each of two carbon atoms. Under the proper conditions a great many ethylene molecules can *polymerize*, which means that they can combine into a long chain, as shown in Fig. 8-5-6*b*. This chain, called polyethylene, is a common plastic,

Figure 8-5-6 Ethylene and polyethylene.

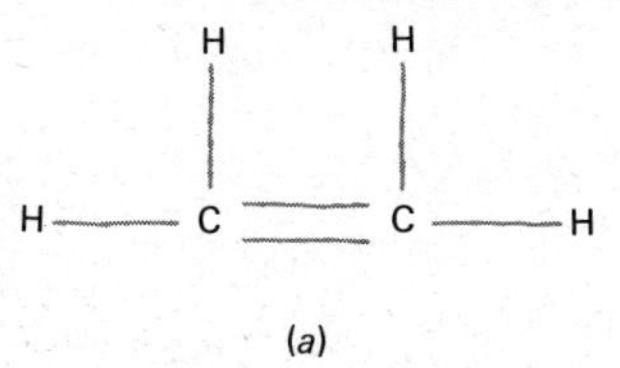

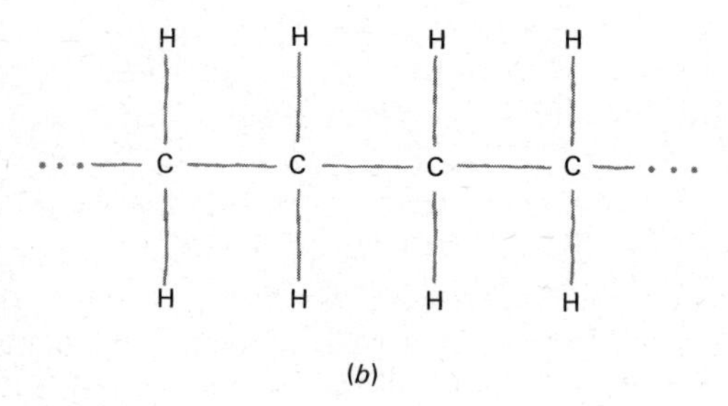

and the covalent carbon bonds make it possible. Other common polymers are the hard transparent plastics like lucite and polystyrene, synthetic rubbers, oleo-margarines, etc.

PROBLEM 16

Estimate the formulas for chemical compounds of the following pairs of elements, utilizing your knowledge of their electron structure and of the relative energies shown in Fig. 8-4-2.

Potassium and chlorine

Carbon and chlorine

Magnesium and oxygen

Boron and fluorine

While it is easiest to discuss bonds which may be characterized as purely ionic or purely covalent, there are many chemical compounds in which both aspects of bonding are simultaneously important. Also, there are bonds which are too weak to be important for the formation of single molecules but which are important in holding atoms or molecules together in bulk matter. The most important is the metallic bond, which is a kind of covalent bond in which a valence electron from each atom is shared with all the others in a solid. This implies that valence electrons in bulk metals are not constrained to stay near any given atom but are in a sense free to move through the material. This picture is consistent with the physical properties of metals, in particular with their high electrical conductivity.

SUMMARY

Bohr assumed that the hydrogen atom consisted of an electron in a circular orbit around a proton. By *quantizing the allowed values of angular momentum*, he arrived at a discrete set of allowed states of the atom, characterized by certain orbital radii:

$$r_n = a_1 n^2 \qquad n = 1, 2, 3, \ldots$$

where the constant a_1, the radius of the smallest allowed Bohr orbit, is

$$a_1 = 0.529 \times 10^{-10} \text{ m}$$

The total energy of the hydrogen-atom electron was then found also to be limited to certain allowed values, corresponding to the allowed radii. The *quantized energies* are given by

$$U_n = -\frac{R}{n^2}$$

where the constant R is given by

$$R = 21.7 \times 10^{-19} \text{ joule} = 13.6 \text{ eV}$$

The hydrogen spectrum is then understood as photons emitted when the atom makes transitions between states with energy U_i and U_f:

$$hf = U_i - U_f$$

The Bohr model predicted photon energies given by the formula as follows:

$$hf = R\left(\frac{1}{n_f^2} - \frac{1}{n_i^2}\right)$$

where n_i and n_f are the quantum numbers of the initial and final states of an atom which emits the photon with energy $E = hf$.

Bohr orbits are not compatible with the more correct modern quantum theory. Instead, the electron is described by a *wave function,* which gives the probabilities for finding the electron at various locations. However, this theory leads to the same quantized energies as the Bohr theory, given by a *principal quantum number n.* In addition, for a given value of n there are a number of states with different values of the *angular momentum quantum number l,* with l equal to

$$0, 1, 2, \ldots, n - 1$$

For atoms with more than one electron, the quantized energies depend upon *both* quantum numbers of the electrons, n and l. An atom with atomic number Z has Z electrons arranged in the lowest energy arrangement. The *exclusion principle* permits only *two electrons per state,* however, where $(2l + 1)$ *states* correspond to one value of n and l.

The outer electrons determine the chemical properties of the element, and because of the arrangement of allowed states, there is a *periodic variation* of chemical properties as atomic number increases. The *valence* summarizes the outer electron structure of an element.

Atoms of different elements are combined into molecules by the interaction of their outer electrons. Depending upon how the electrons interact, the bonding can be *ionic*, *covalent*, or *metallic*.

NUCLEAR PHYSICS

9-1 THE NUCLEUS

In Rutherford's model, the atom consisted of a central nucleus, positively charged, surrounded by negatively charged electrons. From the fact that alpha particles were scattered by the nucleus as if it were a point charge, it was inferred that the nucleus, if not actually a point charge, was smaller than the distance of closest approach of the alpha particle. Figure 9-1-1 shows the scattering of a "point" alpha particle by a nucleus represented as a sphere of positive charge. If the size of the nucleus is less than the distance of closest approach, as in Fig. 9-1-1*a*, then the scattering angle θ for a given impact parameter b is the same as would be calculated for a point-charge nucleus. If, however, the alpha particle penetrates the nuclear charge, as in Fig. 9-1-1*b*, then the angle θ is smaller than would be calculated for point-charge scattering.

It is easy to understand qualitatively why the scattering angle is smaller if the alpha particle penetrates the nuclear charge. In such a case, some of the charge of the nucleus will exert a force on the alpha particle in a direction such as to reduce the scattering angle. For the early scattering experiments the distance of closest approach to a point-charge nucleus could be calculated to be less than 10^{-14} m. Since the point-charge theory worked in such cases, the nuclear radius was inferred to be less than 10^{-14} m, or less than 10^{-4} of the atomic radius.

PROBLEM 1

Calculate the distance of closest approach for a 4 MeV alpha particle which collides head on with an iron nucleus with $Z = 26$. Use conservation of energy, realizing that at "closest approach" the kinetic energy is zero and the potential energy must be 4 MeV. This neglects the motion of the heavy iron nucleus, which is a reasonable approximation. Estimate the potential energy as if the alpha particle and the nucleus were point charges.

For describing the electrons of the atom according to the quantum theory it could be assumed that the nucleus was simply a massive positive point charge. From the electron structure it was possible to understand atomic spectra, chemical binding, electrical conductivity, and almost all the attributes of atoms. The electrons gave the atom its size and also determined the interactions of two or more atoms, therefore explaining many properties of molecules

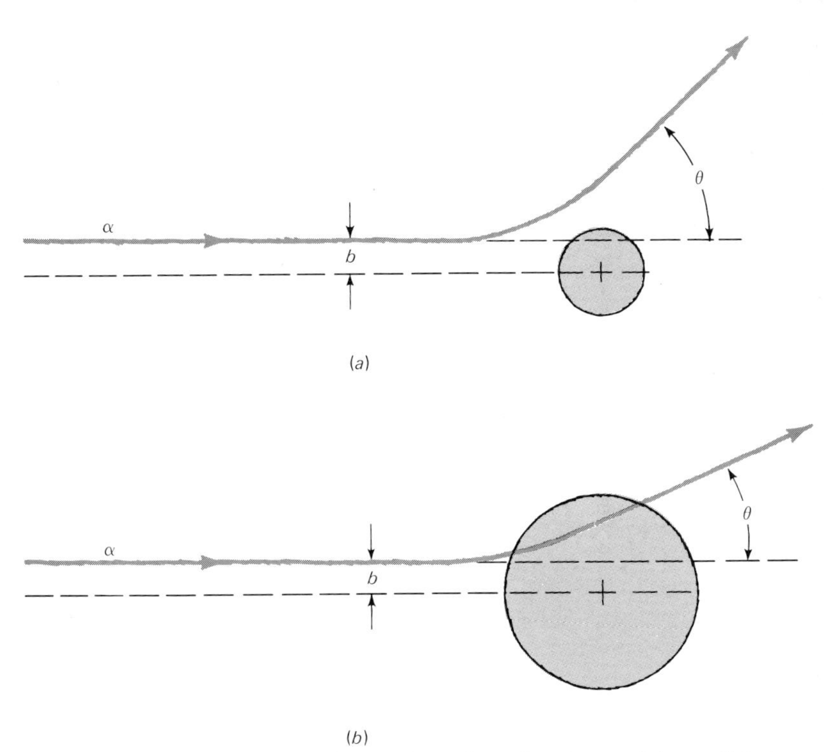

Figure 9-1-1 Schematic view of scattering of a point-charge alpha particle by a nucleus of finite size. In (*a*) the nuclear size is not detectable; in (*b*) it is.

and of bulk matter. Detailed understanding of the nucleus wasn't needed, and *atomic physics* came to mean the physics involving the electrons in atoms. Experimentally, and theoretically, nuclear physics developed more slowly than atomic physics, and as a separate subject.

The one major inference about the nucleus which came from atomic physics was that the different elements were understood to have 1 to 92 atomic electrons, so that (for the net charge to be zero) the number of protons in the nucleus must also range from 1 to 92. However, except for hydrogen, the atomic masses were two to three times as great as the masses of the requisite number of protons. Thus, the typical nucleus was known to contain at least half of its mass in the form of matter with no net charge.

In the 1920s, light elements were bombarded with high-energy alpha particles, and for the first time *nuclear reactions* were found to occur. A nuclear reaction, like a chemical reaction, can be conveniently represented as a symbolic equation. For example, one of the first reactions to be studied was

$$\alpha + \mathrm{N} \rightarrow \mathrm{O} + p \qquad 9\text{-}1\text{-}1$$

The equation describes the combination of a nitrogen nucleus and an alpha particle to produce an oxygen nucleus and a proton. The dream of the ancient alchemists, the transmutation of one element into another, had finally been achieved. Even gold was produced in some cases (though hardly in bulk quantity).

A reaction such as that of Eq. 9-1-1 could be interpreted as resulting from the passage of a fast alpha particle through a nucleus. Sometimes, the alpha particle might collide with a proton of the nucleus in such a way as to cause it to be ejected and the alpha particle to be captured. There were also nuclear reactions in which it appeared that fast neutral particles were ejected from the nucleus, and over a period of many years experimenters attempted to establish the identity of these particles.

Figure 9-1-2 shows schematically a nuclear reaction experiment in which a beryllium target was bombarded with alpha particles. A neutral particle could sometimes be inferred to travel along the path shown by the dashed line because it caused a fast recoil proton to emerge from a slab of paraffin. The fast proton, having a net electric charge, was capable of being detected by the disruption it produced in matter. In general, when a fast charged particle travels through matter, it can collide with atomic electrons and give them sufficient energy to escape from the atoms in

Figure 9-1-2 Schematic view of an experiment in which uncharged particles from the interaction of alpha particles with beryllium are detected by protons which they eject from a paraffin block.

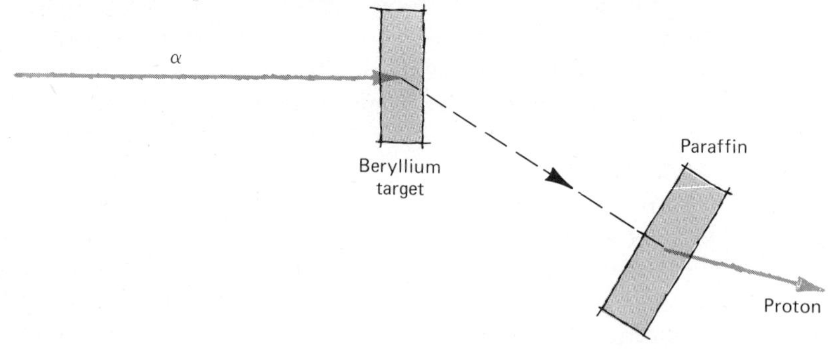

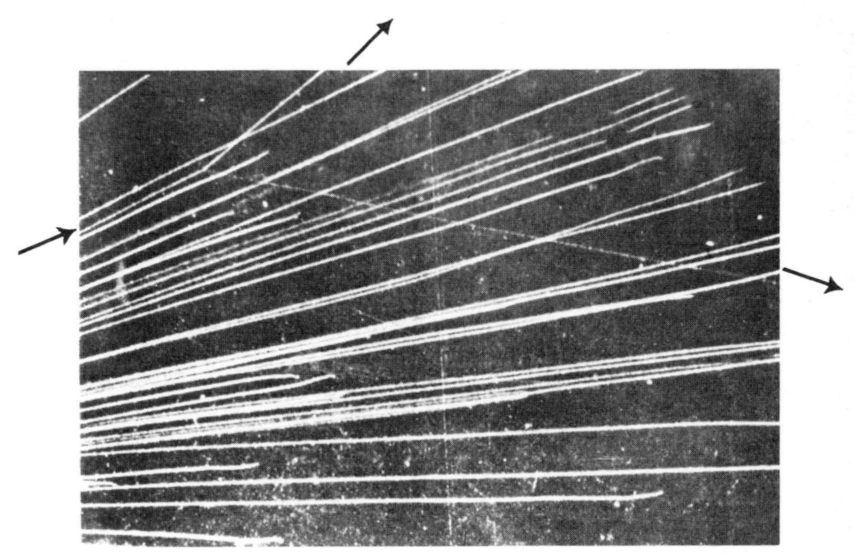

Figure 9-1-3 Collision of an alpha particle with a proton. Alpha particles from a radioactive source enter the gas of a hydrogen-filled cloud chamber from the left. One of them, indicated by an arrow, is seen to scatter. The scattered alpha particle and recoil proton travel in directions indicated by arrows. *(Courtesy of Prof. P. M. S. Blackett.)*

which they are bound. In this case, free electrons are produced, as well as positive ions (the atoms minus one electron). This process, called *ionization,* makes it possible to detect otherwise invisible charged particles.

The ionization along the path of a charged particle can actually be used to make the path visible, in a variety of ways. The earliest technique, which was discovered in the 1920s, was the *cloud chamber.* If a gas is so saturated with water vapor that fog is about to form, then droplets of water will form preferentially around ions in the gas. As a result, the path of an ionizing particle can be seen as a line of minute water droplets and can be photographed by shining a bright light through the cloud chamber.

Figure 9-1-3 shows an early picture of alpha particles traversing a cloud chamber. In one case, indicated by arrows, an alpha particle collides with a nucleus of the gas (hydrogen for this picture) and creates a fast recoil proton. The charge of the proton is half that of the alpha particle. As a result, the proton creates fewer ions in traveling a given distance, so its track has fewer water droplets per unit length and appears thinner. Fast charged particles gradually lose kinetic energy as a result of creating ions, and this is often the main process by which they are stopped in matter.

Were neutral particles to pass through the cloud chamber shown in Fig. 9-1-3, they would create no tracks themselves, but a recoil proton resulting from a collision would be visible, just as the proton from an alpha-proton collision is visible in Fig. 9-1-3. Chadwick, who was a collaborator of Rutherford's, realized in 1932 that the mass of the neutral particles emitted from nuclear reactions could be inferred from the angles and kinetic energies of the recoil protons they produced, purely from classical mechanics. In fact, from previously measured data he could infer that each neutral nuclear particle had a mass very nearly equal to the mass of the proton. In this way, he discovered the uncharged constituent of nuclei and named it the *neutron.* The neutron and proton masses are now known very accurately and only differ by about 0.14 percent.

PROBLEM 2

For a head-on collision of a particle of mass m_1 with a stationary particle of mass m_2, conservation of energy and momentum requires the following relation:

$$E_2' = \frac{4m_1m_2E_1}{(m_1 + m_2)^2}$$

The energy E_2' is the kinetic energy of the particle with mass m_2 after it has been struck. The energy E_1 is the kinetic energy of the incident particle with mass m_1.

Suppose, in Fig. 9-1-2, that the neutral particles produce recoil protons with maximum kinetic energy 5 MeV. Predict the maximum kinetic energy for recoil nitrogen nuclei, assuming the mass of the neutral particles is half the proton mass, equal to the proton mass, or twice the proton mass. In fact, data for nitrogen recoils were used to estimate the neutron mass.

If you want to test your algebraic ability, try to derive the formula above.

In Chap. 4, it was mentioned that in measuring the ratio of charge to mass for ions, *isotopes*, which are atoms with identical chemical properties and different masses, were discovered. One example was chlorine, with atomic number 17, which therefore has 17 atomic electrons and 17 protons in its nucleus. The two common isotopes have atomic masses close to 35 and 37 times the proton mass. We can therefore infer that the lighter isotope has 18 neutrons in the nucleus and the heavier one has 20. The standard symbols for the two chlorine isotopes are:

$_{17}Cl^{35}$ $\quad$ $_{17}Cl^{37}$

The letters Cl stand for chlorine; the number 17 stands for its atomic number, equal to the number of protons in the nucleus; and the numbers 35 and 37 are the total numbers of neutrons plus protons in the nucleus for each isotope. To find the number of neutrons, it is necessary to subtract 17 from 35 or 17 from 37.

9-2 NUCLEAR STRUCTURE

A nucleus contains many protons and neutrons in a very small volume of space. Even though the electric force between the protons tends to push them apart, there is a *nuclear force* which is so strong that it can hold the protons in the nucleus. The most pressing question in nuclear physics since its birth in the 1920s has been the nature of this nuclear force. At present, the question remains unanswered, although some knowledge of the force has been accumulated. The main known facts are that the range of the force is very short and that it is the same for protons and neutrons. Since charge doesn't matter in determining the nuclear force, protons and neutrons are often both called by the same name, *nucleons*, in nuclear physics. The short range of the nuclear force means that it is unimportant between nucleons more than about 10^{-15} m apart. At distances shorter than this the force rapidly becomes very strong.

Knowledge of the nuclear force has been accumulated by scattering experiments, as well as in a variety of other ways. It is now known that the force depends upon factors other than the relative positions of the nucleons, such as their velocities. Therefore, it *cannot* be accurately represented by a potential-energy curve. Various hypotheses have been made about the force, but in no case has it been possible to precisely predict nuclear properties, or even the force between two nucleons. Most physicists expect that ultimately the basic nature of the nuclear force will be known, and it will then be possible to calculate the properties of nuclei as accurately as the properties of atoms. Perhaps quantum mechanics will work, once the correct representation of the force is discovered. However, there is a real possibility that ordinary quantum mechanics will prove to be only an approximation, good for the atom and unsatisfactory for the nucleus.

In spite of the lack of really fundamental knowledge, much is understood about nuclear phenomena. For nuclei which contain more than a few neutrons and protons a very simple model, in which the affect of the nuclear force is approximated with a potential-energy curve, is a fairly good approximation to the facts. Figure 9-2-1 shows a potential-energy diagram for the neutrons in the nucleus according to this model, along with the energies U_1, U_2, etc., of a few allowed neutron states.

The potential-energy curve of Fig. 9-2-1 is similar to the one we studied in detail in Chap. 7 for the particle in a box. The difference is that now it refers to the variation of potential energy with r, the radius measured from the center of the nucleus. The quantum-mechanical problem is three-dimensional, like that of a particle moving in a spherical can with gravity turned off. The exact solution can be found but involves more mathematics that we want to go into here. As usual, there are quantized energy levels, symbolized by the values of total energy shown in Fig. 9-2-1.

It should be emphasized that Fig. 9-2-1 represents an approximate potential energy for *one* neutron moving under the influence of the short-range forces of *all the other nucleons* in the nucleus. At the edge of the nucleus, the curve of V vs. r has a steep slope, representing a strong attractive force. Inside the nucleus, the po-

Figure 9-2-1 A nuclear "square well" potential-energy curve for neutrons, with the energies of some allowed states indicated schematically.

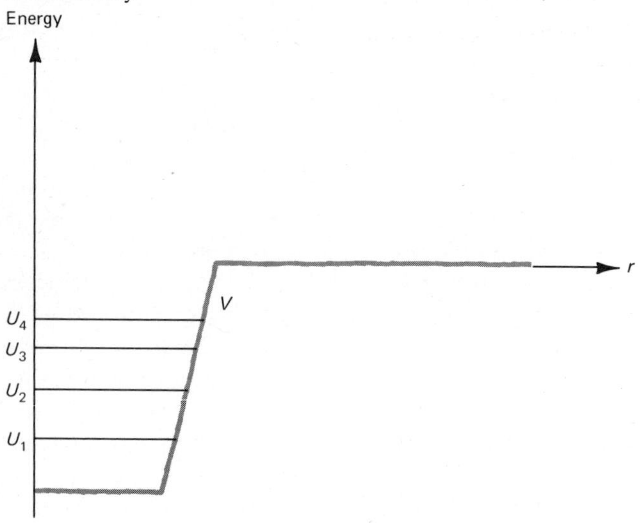

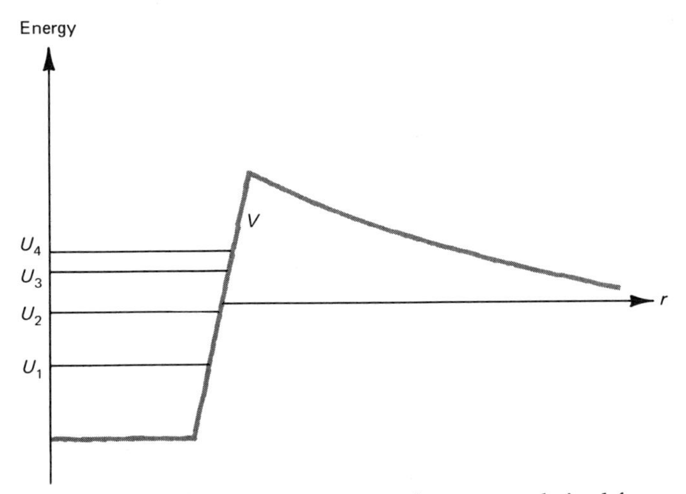

Figure 9-2-2 A potential-energy curve for protons, derived from a combination of a "square well" nuclear potential energy with an electrostatic repulsion. A few proton energy levels are schematically indicated.

tential energy is flat, corresponding to essentially no net force. (When the neutron is surrounded by nucleons, there tends to be no net force on it.) There is also no force when the neutron is more than about 10^{-15} m outside the edge of the nucleus.

Figure 9-2-2 shows the potential energy for a proton, analogous to that for the neutron in Fig. 9-2-1. The potential-energy curve includes the long-range repulsive effect of the Coulomb force. The nuclear part of the potential-energy curve, the same as for neutrons, is superimposed on the potential-energy curve which represents the electric repulsion. At the nuclear radius, and inside the nucleus, the nuclear force completely dominates, and the potential energy from the electric force has been neglected in this region. However, outside the range of the nuclear force the electric force is dominant and leads to the potential energy shown in the figure. Again, the protons may only have certain quantized energies, determined by a quantum-mechanical calculation involving this potential-energy curve, and a few are shown in Fig. 9-2-2.

An actual nucleus consists of protons and neutrons, and in Fig. 9-2-3 the two potential-energy curves of Figs. 9-2-1 and 9-2-2 are shown side by side on the same graph for convenience. The Pauli exclusion principle, which limits the number of electrons per atomic state, also applies to neutrons and protons. It limits the number of neutrons or protons per nuclear state to two. As a result, a typical nucleus can be envisioned as having neutrons and protons filling up the lowest energy states, in analogy to the behavior of electrons in an atom. In Fig. 9-2-3 the neutrons and protons which occupy the various states are symbolized by black circles, with a maximum of two per state.

In Fig. 9-2-3 the state of highest energy occupied by a proton has an energy closely equal to that of the state of highest energy occupied by a neutron. As a result of the shift in potential energy for the protons, caused by the Coulomb force, there are fewer protons than neutrons in a nucleus when the maximum energies of each type of particle are roughly the same. If these energies are more than a little different, the nucleus is unstable, so that Fig. 9-2-3 is the correct qualitative picture of the structure of a typical stable nucleus.

Figure 9-2-3 Schematic picture of the energy levels occupied by neutrons and protons in a typical nucleus. Unless the maximum energies of neutrons and protons are similar, the nucleus will undergo radioactive decay. For similar maximum energies and stable nuclei, there is an excess of neutrons over protons.

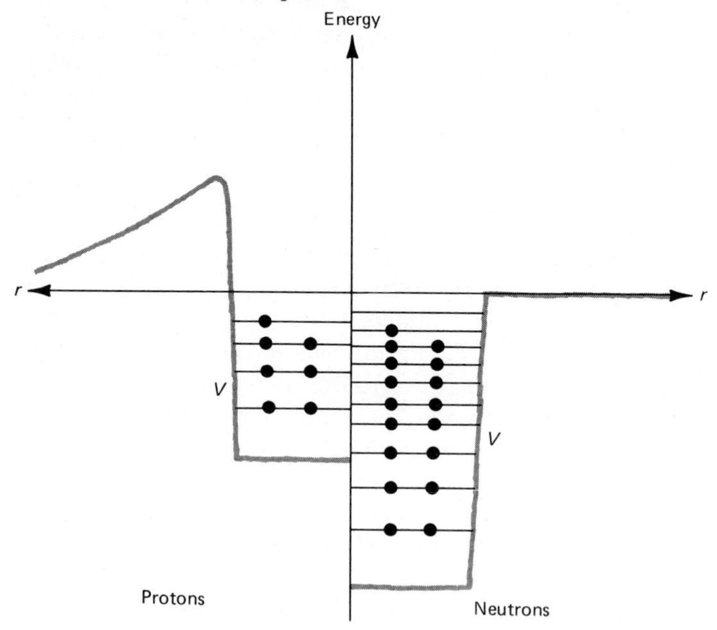

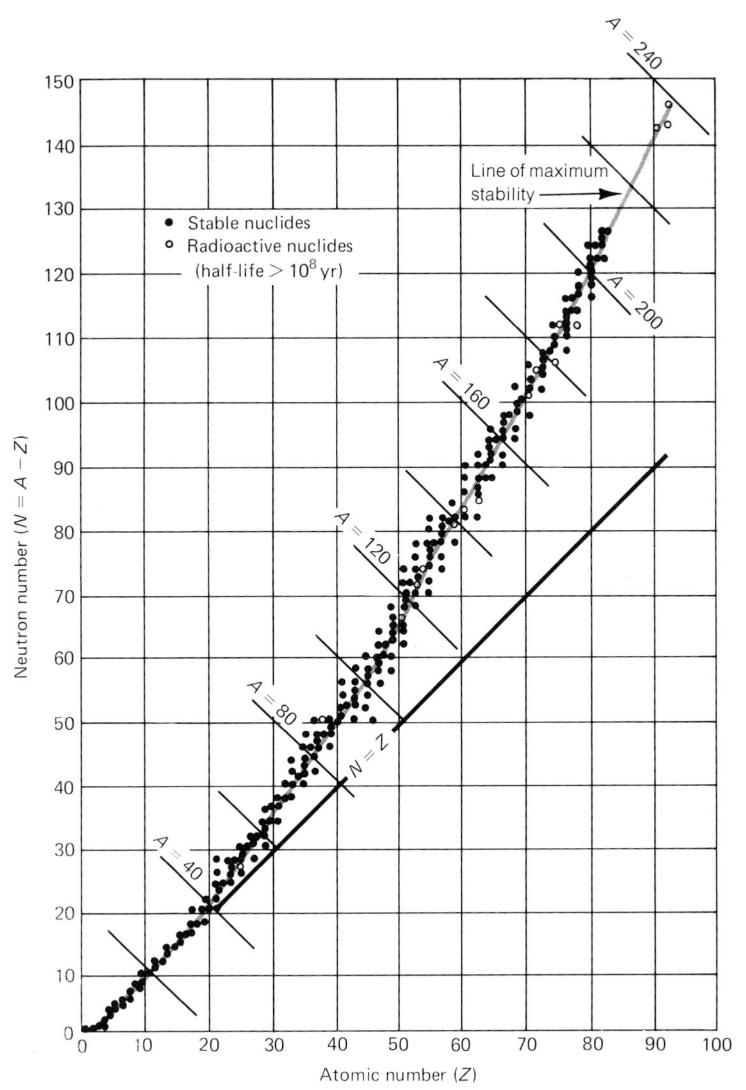

Figure 9-2-4 A graph showing the makeup of the stable nuclides.

Figure 9-2-4 shows all the known stable *nuclides.* The term nuclide is sometimes used to specify any particular kind of nucleus, characterized by a specific number of neutrons and protons. Notice that for low atomic number, when the Coulomb force is less important because the nuclear charge is small, the number of neutrons is roughly equal to the number of protons. As the atomic number increases, the number of neutrons exceeds the number of protons by a larger and larger amount. In the figure, three common symbols are used, defined as follows:

Z = *atomic number*, the number of protons in the nucleus, or the number of electrons in the atom.

N = *neutron number*, the number of neutrons in the nucleus.

$A = N + Z$, the total number of nucleons (neutrons plus protons) in the nucleus, often called the *mass number*, and approximately equal to the atomic mass.

In terms of the quantities listed above, the symbol for the chlorine isotope, ${}_{17}Cl^{35}$, is a way of specifying a nucleus in the form ${}_{Z}(\text{name of element})^{A}$. Notice that all the elements with Z greater than 83 are unstable, except that elements 90 and 92 (thorium and uranium) are so weakly radioactive that they still exist in nature.

Figure 9-2-5 again shows the approximate nuclear potential-energy curve and the distance R, called the nuclear radius, from the center of the nucleus to the center of its edge region. Experiments such as alpha-particle scattering by nuclei, or electron scattering by nuclei, can be used to determine the radius R. The results of such experiments are summarized in the formula

$$R^3 = [(1.20)(10^{-15})]^3 A \text{ cubic meters} \qquad 9\text{-}2\text{-}1$$

where A is the total number of nucleons in the nucleus. From this relationship it is possible to find that the volume occupied per nucleon is the same in all nuclei, or that the density of all nuclei is the same. While the nuclear force between two nucleons

Figure 9-2-5 The nuclear radius R, defined by the distance to the midpoint of the nuclear "edge."

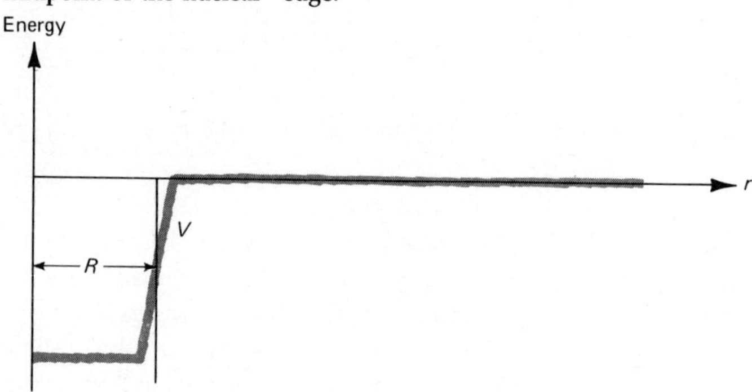

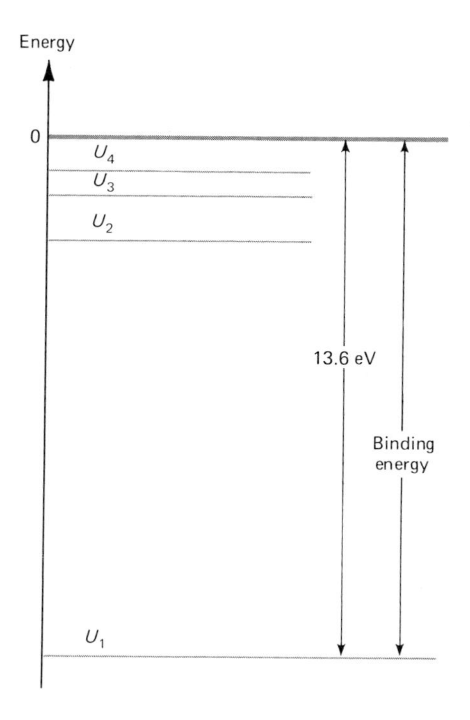

Figure 9-3-1 The concept of binding energy, illustrated for the hydrogen atom. An energy of 13.6 eV is needed to separate the electron and proton of the atom. This is the binding energy.

is strongly attractive at a distance of about 10^{-15} m, it becomes repulsive at distances which are significantly smaller than this. As a result, nucleons are no more tightly packed in nuclei with large numbers of nucleons, like lead, than in nuclei with small numbers of nucleons, like carbon. Nucleons seem to behave somewhat like hard spheres coated with "stickum."

PROBLEM 3
From Eq. 9-2-1 find the volume per nucleon and the density of nuclear matter in tons per cubic centimeter.

9-3 NUCLEAR BINDING ENERGIES

For the hydrogen atom, the electron energy-level diagram looks like that shown in Fig. 9-3-1. The ground state of hydrogen, with energy U_1, has a total energy 13.6 eV lower than that of a separate electron and proton at rest a long distace apart. This quantity, 13.6 eV, is the *binding energy* of the hydrogen atom. To "unbind" the atom, and free the electron, an energy of 13.6 eV must be supplied.

For nuclei, the binding energies of the neutrons and protons are of the order of a million times greater than the binding energies of electrons in atoms. A graph which summarizes the systematic variation of nuclear binding energies is shown in Fig. 9-3-2. The convenient unit of energy in nuclear physics, used for this graph, is one million eV, abbreviated MeV. Since the graph gives the average binding energy per nucleon, the total binding energy of a given nucleus is found by multiplying this average energy by the mass number A, which is the total number of nucleons.

The curve shown in Fig. 9-3-2 summarizes experimental data which cannot be accurately calculated from theory. Nuclei with mass numbers near 60 or 70, like iron (whose chemical symbol is Fe), are the most tightly bound. Heavier nuclei have more weakly bound nucleons, presumably because of the Coulomb repulsive force between the protons. Lighter nuclei are more weakly bound because the nuclear force between many nucleons is a bit stronger than between few. This is particularly striking for very light nuclei. The isotope of hydrogen, $_1H^2$, which is sometimes called *deuterium*, consists of only two nucleons, one proton and one neutron. Its total binding energy is 2.2 MeV, about 1 MeV

Figure 9-3-2 The variation of nuclear binding energy with mass number. The average binding energy *per nucleon* is given, equal to the total binding energy of the nucleus divided by the sum of the number of neutrons and protons.

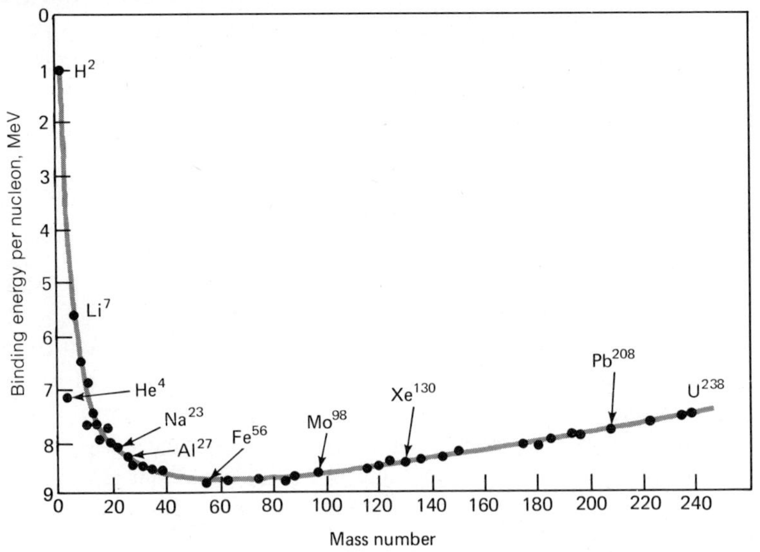

per nucleon. The most common isotope of lithium, ${}_3Li^7$, has a binding energy of about 40 MeV, more than 5.5 MeV per nucleon.

The curve shown in Fig. 9-3-2 is drawn so that binding energy is measured downward, with lower points for the most tightly bound nuclei. This is to match the curve to energy-level diagrams, where the tightest binding corresponds to the lowest total energy. Although a smooth curve is drawn, the actual binding energies of nuclei scatter around the curve. The most striking variation from the smooth curve is that for He^4, the alpha particle, which is a very strongly bound combination of two neutrons and two protons.

Experimentally, the data shown in Fig. 9-3-2 might in principle be obtained by breaking up nuclei into individual nucleons and carefully keeping track of the net energy required. This would be difficult, except for the lightest nuclei. The data in Fig. 9-3-2 are actually obtained from Einstein's famous equation

$$E = mc^2 \qquad 9\text{-}3\text{-}1$$

This equation is a deduction from the special theory of relativity, and defines a *mass energy*, mc^2, for a body at rest with mass m. The proportionality factor c^2, the square of the velocity of light (in m/sec), relates mass in kilograms to energy in joules.

According to the theory of relativity, the total energy U of a particle with mass m is given by the equation

$$U = T + V + mc^2 \qquad 9\text{-}3\text{-}2$$

The law of conservation of energy applies to this total energy U, which includes the mass energy. If m does not change, then the sum $T + V$ (kinetic plus potential energy) is conserved, as assumed in classical mechanics. We may look upon mc^2 as a kind of potential energy proportional to the mass. It is called either the *rest energy* or *mass energy*.

Consider the formation of a hydrogen atom in its ground state, from a free electron and a free proton, at rest and far apart. The total energy in the initial state, U_i, according to Eq. 9-3-2, is the sum of the rest energies

$$U_i = m_p c^2 + m_e c^2 \qquad 9\text{-}3\text{-}3$$

In the final state, we have a hydrogen atom *plus* light quanta, which are emitted when the hydrogen atom drops to its ground state from the high-energy state at which it is formed. The total energy emitted in the form of quanta must be equal to the binding energy, 13.6 eV, which we shall call E_b (where the subscript b stands for "binding"). Suppose the hydrogen atom is at rest in the final state, with mass m_h (where the subscript h stands for hydrogen). Then the total energy of the hydrogen atom plus the quanta which constitute the final state is:

$$U_f = m_h c^2 + E_b \tag{9-3-4}$$

Since total energy is conserved, U_i, from Eq. 9-3-3, is equal to U_f, from Eq. 9-3-4. Therefore, equating the right-hand sides of the two equations,

$$m_p c^2 + m_e c^2 = m_h c^2 + E_b \tag{9-3-5}$$

Rearranging Eq. 9-3-5,

$$E_b = c^2[(m_p + m_e) - m_h] \tag{9-3-6}$$

Thus, the quantity given by c^2 times the mass difference $[(m_p + m_e) - m_h]$ is predicted to equal the binding energy of the hydrogen atom. *Because the hydrogen atom is a bound structure of an electron and proton its mass is less than the individual masses of its constituents.*

Because it is often useful to know mass energies, rather than masses, in nuclear physics, we often specify a mass by giving mc^2 in MeV. For example, the mass of the proton is 1.672×10^{-27} kg. Therefore,

$$m_p c^2 = 1.50 \times 10^{-10} \text{ joules} = 938 \text{ MeV} \tag{9-3-7}$$

Also,

$$m_e c^2 = 0.511 \text{ MeV} \tag{9-3-8}$$

The binding energy of hydrogen, 13.6 eV, is 1.45×10^{-8} times $m_p c^2 + m_e c^2$. Thus, if we could determine the mass m_h and the mass m_p to an accuracy better than 1 part in 10^8, Eq. 9-3-6 could be used to determine E_b. For the hydrogen atom, it is much easier to determine E_b by measuring either the minimum energy needed to ionize the atom or the energy emitted in the form of quanta when the atom is formed.

Since nuclear binding energies are much greater than for atoms, mass determinations are useful for determining these binding energies. Consider the *deuteron*, the nucleus of a hydrogen isotope with one proton and one neutron. The equation which corresponds to Eq. 9-3-6 is:

$$E_b = c^2[(m_p + m_n) - m_d] \qquad 9\text{-}3\text{-}9$$

The masses m_p, m_n (the neutron mass), and m_d (the deuteron mass) can be directly measured to an accuracy of 1 part in 10^4 or better, and the mass energies are

$$m_p c^2 = 938.2 \text{ MeV}$$

$$m_n c^2 = 939.5 \text{ MeV}$$

$$m_d c^2 = 1875.5 \text{ MeV}$$

Substituting these values in Eq. 9-3-9,

$$E_b = 938.2 + 939.5 - 1875.5 = 2.2 \text{ MeV} \qquad 9\text{-}3\text{-}10$$

When a neutron and a proton combine to form a deuteron in its ground state, a single electromagnetic quantum, or gamma ray, is emitted. The gamma-ray energy can be accurately determined to be 2.2 MeV, the same as predicted by Eq. 9-3-10 from the relation $E = mc^2$.

Since the deuteron has two nucleons, its *binding energy per nucleon* is (2.2)/2, or 1.1 MeV, as shown for the isotope ${}_1H^2$ in Fig. 9-3-2. Nuclear mass determinations to an accuracy of 1 part in 10^4 or better can be made for other nuclei and are in fact used to determine total binding energies. From these the average binding energies per nucleon, shown in Fig. 9-3-2, are found.

PROBLEM 4

From the masses given, determine whether the following two nuclei are stable or unstable with respect to alpha decay:

${}_{13}Al^{27}$ and ${}_{88}Ra^{226}$

Masses:

${}_{13}Al^{27}$	26.9901
${}_{11}Na^{23}$	22.9971
${}_{88}Ra^{226}$	226.0960
${}_{86}Rn^{222}$	222.0869
${}_{2}He^{4}$	4.0039

If alpha decay takes place, what is the alpha-particle kinetic energy?

9-4 RADIOACTIVITY

In this section and the next we will discuss aspects of nuclear physics which mainly involve nuclear energy. We will begin with the type of radioactivity which has been introduced in previous chapters, alpha emission. At any given instant there exist "virtual alpha particles" within any nucleus. The alpha particle is so tightly bound that it tends to form easily from the population of neutrons and protons. An analogy is a dance attended by men and women singly, where couples form, split up, and re-form out of the crowd. The alpha particles may be thought of as quartets of nucleons which exist temporarily inside the nucleus like the couples at the dance.

When an alpha particle is formed, its motion may be analyzed by an energy diagram like that shown in Fig. 9-4-1, considering the alpha particle as a separate entity moving in a potential established by the rest of the nucleons. If the alpha particle has an energy U_1, as shown in Fig. 9-4-1, then the tails of its wave function will extend through the potential-energy "barrier" between points A and B on the figure. As in the barrier penetration problem discussed in Sec. 7-7, there will be a small but significant probability for the alpha particle to be found outside the nucleus, beyond point B. Here its energy U_1 is greater than V, and it may legally exist as a free particle and escape from the nucleus forever. This qualitative picture describes alpha-particle radioactivity, and quantitative calculations made using this model are in good agreement with experiment.

Notice that if the alpha-particle energy is relatively low, like U_2 in Fig. 9-4-1, it can never escape. There is no region outside the nucleus for classically allowed motion, and the potential-energy barrier extends to all radii. Heavy nuclei tend to undergo alpha decay because the nucleons are so weakly bound that formation of an alpha particle, which is more tightly bound, leads to a high energy, like U_1 in Fig. 9-4-1. For light nuclei, the alpha-particle energy typically is like U_2 in the figure, and such nuclei are never alpha-radioactive.

The fact that the wave function for an alpha particle with energy equal to U_1 in Fig. 9-4-1 has a small but finite value outside the nucleus leads to a certain *decay rate D* for the nucleus, defined as the probability of alpha emission during a time interval of one second. To make an analogy to a kind of game, suppose you had

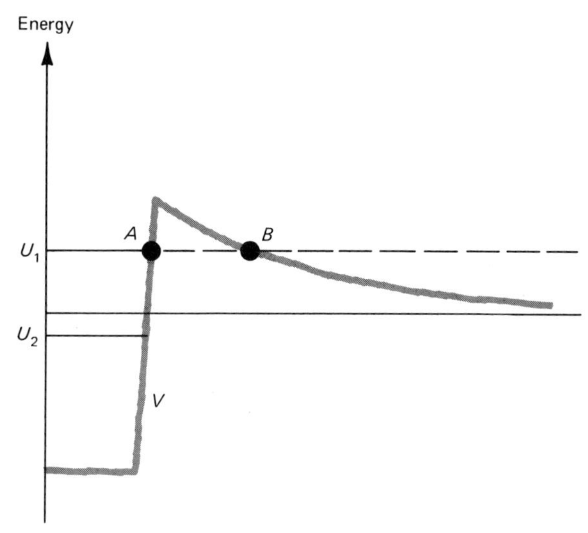

Figure 9-4-1 Energy diagram for an alpha particle in a nucleus. For the energy level U_1, radioactive decay can take place. For energy U_2, the nucleus is stable.

a bowl with 99 black marbles and 1 white one and you picked out 1 per second, without looking. Suppose further that if you picked a black marble you replaced it in the bowl and stirred the marbles, and that if the white marble is picked it is removed and the game is over.

The bowl of marbles is analogous to a single nucleus which can emit an alpha particle, symbolized by the white marble. Each second there is a 1 percent chance of picking the white marble, corresponding to the chance of the α particle being found outside the nucleus, and giving a decay rate D equal to 0.01. Once the white marble is picked, or the alpha particle is emitted, the nucleus is changed, and the game is over.

A sample of a radioactive substance usually contains a large number of identical radioactive nuclei, all "trying" to decay with a given decay rate D. After each decay, a residual nucleus is left behind, which we shall assume for simplicity is stable and does not decay radioactively. The original nucleus in such problems is often called the parent nucleus and the residual one the daughter nucleus. Suppose at some time which we can call $t=0$ there are N_0 parent alpha-radioactive nuclei. If the decay rate is D, then, at $t = 0$, N_0D nuclei decay per second, emitting N_0D alpha particles per second. Once a nucleus decays it does not decay again, so after a while there are fewer parent nuclei and the rate of emission of alpha particles is less than N_0D.

PROBLEM 5
Suppose an alpha particle is released from rest at the surface of a uranium nucleus. Find its kinetic energy far from the nucleus, in MeV. (Use conservation of energy and the radius of the uranium nucleus given by Eq. 9-2-1.) Alpha particles are observed from the radioactive decay of uranium isotopes with energies around 4 MeV. Compare this with the result of your calculation, and with your expectations if the "tunneling" picture of Fig. 9-4-1 is correct.

PROBLEM 6
The standard unit of activity for radioactive sources is the curie, defined by

$$1 \text{ curie} = 3.7 \times 10^{10} \text{ disintegrations per second}$$

How much power, in watts, is given off by a 1 curie source of 4 MeV alpha particles?

The mathematical curve which describes the change in the number of parent nuclei with time is a *decreasing exponential*, shown in Fig. 9-4-2. After a time $t_{1/2}$, called the *half-life*, the number of parent nuclei is half the original number. After a time $2(t_{1/2})$ the number of parent nuclei is one-fourth the original number. Beginning at any time, the number of parent nuclei will decrease by a factor of two after a time $t_{1/2}$. Clearly, a large value of decay constant D leads to a short time $t_{1/2}$, and a small decay constant leads to a long half-life. The mathematical relation between $t_{1/2}$ and D is simple,

$$t_{1/2} = \frac{0.69}{D} \qquad 9\text{-}4\text{-}1$$

Throughout the discussion above we have neglected chance fluctuations in the overall decay rate of the radioactive nuclei in the sample. If the number of nuclei is very large, as it usually is, the chance fluctuations are unimportant.

Suppose now that the daughter nucleus, remaining after radioactive decay, is not in its ground state. For example ${}_{92}U^{238}$ is alpha-radioactive and usually decays to a daughter nucleus, ${}_{90}Th^{234}$, in an excited state. The daughter nucleus, like an excited

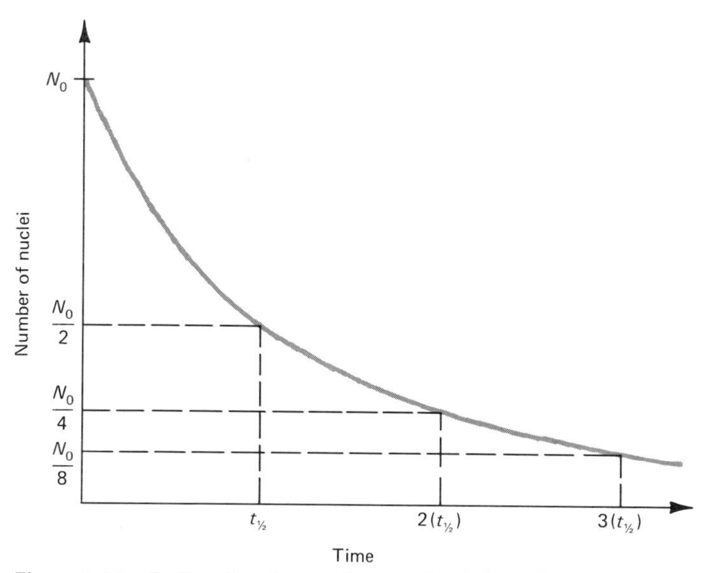

Figure 9-4-2 Radioactive decay of a sample of N_0 nuclei at time $t = 0$. The number of original nuclei decreases according to an exponential curve, with a half-life indicated as $t_{1/2}$.

atom, emits quanta of electromagnetic energy as it makes one or more transitions to go from the excited state to the ground state. However, the quantum energies hf are now typically 10^6 eV instead of the few eV typical of light quanta emitted by atoms. Quanta emitted by nuclei are still, like light, electromagnetic wave packets, but with wavelengths a million times less than the wavelength of visible light. Physicists who first studied radioactivity observed very penetrating radiation which, unlike alpha particles, could pass through inches of materials like wood or iron. They found the radiation to be uncharged and named it *gamma radiation.* These high-energy "light" quanta are the *gamma rays* they discovered.

PROBLEM 7

A naturally occurring isotope of uranium, ${}_{92}U^{238}$, decays with a half-life of about 4.5×10^9 years. This decay leads to the nucleus ${}_{90}Th^{234}$ which is also radioactive, and in fact a chain of more than a dozen radioactive decays takes place, finally ending up with a stable nucleus of lead, ${}_{82}Pb^{206}$. This is one of several stable lead isotopes.

All the decays following the uranium decay have half-lives short compared to that of the uranium. Suppose you are a geologist and you discover a rock with only the one lead isotope Pb^{206} present in it and that the number of these lead nuclei was equal to the number of U^{238} nuclei present. Estimate the age of the rock.

A third form of radioactivity is *beta radiation*, which is the most complicated to understand. The simplest example of beta decay is the decay of the free neutron:

$$\underset{\text{Neutron}}{{}_0n^1} \rightarrow \underset{\text{Proton}}{{}_1H^1} + e^- + \text{neutrino} \qquad 9\text{-}4\text{-}2$$

Although neutrons exist indefinitely inside stable nuclei, a lone neutron spontaneously decays into three particles, a proton, an electron, and a *neutrino.* The half-life for the decay, in which the electron and neutrino are *created* when the neutron changes into a proton, is almost 10 minutes. The neutron is *not* a bound state of proton, electron, and neutrino. It is a particle with rest energy 1.3 MeV greater than that of the proton, and this energy is released in the decay described by Eq. 9-4-2. The energy appears in two forms, the kinetic energies of the electron and neutrino, and their mass energy. The electron has a mass energy m_ec^2 equal to 0.511 MeV, and the neutrino, a particle which we have not previously discussed, has zero mass energy.

The neutrino interacts with matter so weakly that neutrinos can travel *through the diameter of the earth* with only a small probability of interacting with any atom or nucleus. However, neutrinos carry energy and momentum, like light quanta, which also have no mass or charge. The neutrino was "invented" by Wolfgang Pauli to account for the apparently missing energy and momentum when only the proton and electron were observed in beta decay experiments. It was not until the 1960s that neutrinos were directly detected by means of their very weak interactions with matter. Figure 9-4-3 shows Pauli's way of announcing his idea. The name "neutron," proposed by Pauli was changed to "neutrino," a diminutive proposed by Fermi, when Chadwick discovered the heavier particle we now call the neutron.

Purely on the basis of symmetry, we could write a beta decay reaction for the proton, too;

$$1.8\ \text{MeV} + {}_1H^1 \rightarrow {}_0n^1 + e^+ + \text{neutrino} \qquad 9\text{-}4\text{-}3$$

This reaction assumes the existence of a *positron* or positive electron. There is also a need for 1.8 MeV of added energy on the left-hand side to supply added rest mass on the right-hand side. The

Zurich, December 4, 1930

Dear radioactive ladies and gentlemen,

I beg you to most favorably listen to the carrier of this letter. He will tell you that, in view of the "wrong" statistics of the N and Li^6 nuclei and of the continuous beta spectrum, I have hit upon a desperate remedy to save the laws of conservation of energy and statistics. This is the possibility that electrically neutral particles exist which I will call neutrons, which exist in nuclei, which have a spin 1/2 and obey the exclusion principle, and which differ from the photons also in that they do not move with the velocity of light. The mass of the neutrons should be of the same order as those of the electrons and should in no case exceed 0.01 proton masses. The continuous beta spectrum would then be understandable if one assumes that during beta decay with each electron a neutron is emitted in such a way that the sum of the energies of neutron and electron is constant. . . .

I admit that my remedy may look very unlikely, because one would have seen these neutrons long ago if they really were to exist. But only he who dares wins and the seriousness of the situation caused by the continuous beta spectrum is illuminated by a remark of my honored predecessor, Mr. Debye, who recently said to me in Brussels: 'O, it is best not to think at all, just as with the new taxes.' Hence one should seriously discuss every possible path to rescue. So, dear radioactive people, examine and judge. Unfortunately I will not be able to appear in Tübingen personally, because I am indispensable here due to a ball which will take place in Zürich during the night from December 6 to 7.

Your most obedient servant,

W. Pauli

Figure 9-4-3 Part of a letter sent by Wolfgang Pauli to Hans Geiger and Lise Meitner, who were attending a conference in Tubingen.

positron was in fact discovered in 1933 and identified with the *antielectron* which had appeared in a theory of the electron developed by Dirac some years earlier. This was the first known *antiparticle*, so called because several of its properties, such as its charge, were opposite to those of its corresponding particle, the electron. We now believe that for every particle there is an antiparticle.

When nuclei beta decay, either Eq. 9-4-2 or 9-4-3 describes the transformation of one nucleon of the nucleus. The 1.8 MeV of energy needed for positron emission can be supplied if the binding energy of the daughter nucleus exceeds that of the parent nucleus by more than 1.8 MeV.

PROBLEM 8

Some stable cobalt, ${}_{27}Co^{59}$, is placed in a nuclear reactor, where it captures neutrons. Afterward, it is found to emit fast electrons and gamma rays. Write down the nuclear reactions which might explain the observations.

PROBLEM 9

For the following radioactive decays, complete the right sides of the reactions:

$${}_{90}Th^{234} \rightarrow {}_{91}Pa^{234} +$$
$${}_{11}Na^{22} \rightarrow e^{+} + \quad +$$
$${}_{28}Ni^{60} \rightarrow \gamma +$$
$${}_{84}Po^{212} \rightarrow {}_{82}Pb^{208} +$$

PROBLEM 10

Complete the following radioactive decay reaction, which produces a thorium isotope:

$${}_{92}U^{238} \rightarrow \alpha + \cdots$$

The alpha particle energy is 4.2 MeV. The mass of the uranium nucleus is 235.988 times the mass of the proton. The mass of the alpha particle is 3.971 times the proton mass. What is the mass of the thorium nucleus?

Actually, sometimes alpha particles with energy of 2.7 MeV are also observed from the decay of this uranium nucleus. Explain why gamma rays are emitted very soon after these low-energy alpha particles and give their energy.

9-5 NUCLEAR ENERGY

In radioactive decay, a nucleus is transformed into a more tightly bound collection of nucleons, and the binding energy in the final state is greater than that in the initial state. As a result, energy is available, which can be in the form of kinetic energy of the particles emitted in the radioactive decay, added mass energy of the decay particles, or gamma-ray quanta.

Energy from ordinary radioactive decay is at present used for small, lightweight sources of heat and electricity, which are important for certain special applications (as in space vehicles). However, the cost of the radioactive isotopes, and of the conversion apparatus, makes this form of nuclear energy much more expensive than the energy available from *nuclear fission.* Fission was discovered quite by accident in the late 1930s by Otto Hahn and Fritz Strassman.

The binding-energy curve, Fig. 9-3-2, shows that the binding energy per nucleon is about 1 MeV less for the heavy elements than for the most tightly bound moderate-weight nuclei. Thus, if a very heavy nucleus with mass number about 200 were to split into two pieces of roughly equal mass, about 200 MeV of energy would be available. For some isotopes, like ${}_{92}U^{235}$, fission can be triggered by absorption of a slow neutron, and splitting into roughly equal-mass pieces takes place almost immediately afterward. Slow neutrons can easily reach a nucleus and be absorbed, since there is no repulsive electric force acting to keep them away.

When slow neutrons are absorbed by ${}_{92}U^{235}$ and fission takes place, a typical reaction might be

$$ {}_{92}U^{235} + {}_{0}n^{1} \rightarrow {}_{54}Xe^{140} + {}_{38}Sr^{94} + {}_{0}n^{1} + {}_{0}n^{1} + 200 \text{ MeV} \qquad 9\text{-}5\text{-}1 $$

A variety of possible nuclear fragments can result, and the nuclei of xenon and strontium in Eq. 9-5-1 represent just one possibility. Also, various numbers of free neutrons can be produced, but two is a typical number. The 200 MeV of energy liberated comes off mainly in the form of kinetic energy of the neutrons and fission fragments.

PROBLEM 11

Are the two "fission fragment" nuclei in Eq. 9-5-1 stable? If not, how do they decay? Taking 5 MeV as a typical energy release for a radioactive decay, estimate the energy given off by radioactive decays which ultimately lead to stable products. (Hint: See Fig. 9-2-4.)

Usually, slow-neutron capture by a nucleus produces either a stable isotope or one which undergoes beta decay. Fission is a rare mode of nuclear decay and is exhibited by only a few heavy nuclei. (The isotope ${}_{92}U^{235}$ occurs in nature, but accounts for less than 1 percent of natural uranium.) A most important attribute of the fission reaction, Eq. 9-5-1, is that it produces neutrons which can in turn produce more fissions. In a nuclear reactor, adjustable neutron absorbers are used to control the probability that a fission-produced neutron produces another fission. This probability is adjusted so that on the average each fission reaction produces just one succeeding one, resulting in a steady reaction rate appropriate for the reactor design and power demand.

Fission was discovered in Germany, just before the start of World War II, and some physicists realized that the fission reaction could possibly lead to the creation of a nuclear-energy bomb. In both Germany and the United States massive research programs were undertaken to develop nuclear weapons. The effort in the United States succeeded some time after the war with Germany ended. It was possible to make a fast, explosive "chain reaction" in which on the average the neutrons from one fission would lead to more than one fission. Suppose, for example, that each fission led to two more after a time of about 10^{-6} sec. Then these two would lead to four, again in a very short time; these four would lead to eight, and so on. The mathematics of such growth leads to 2^n fissions in a time $n \times 10^{-6}$ sec. In less than a thousandth of a second, fission energy equal to thousands of tons of TNT could be generated in such a runaway reaction. This was found to occur, and terrifying bombs were made.

The development of nuclear weapons resulted in an unprecedented involvement of scientists in politics and government. When the weapons were successfully made, it was the physicists and chemists who worked on them who first appreciated some of the possible consequences. Many of them became involved in political activity in an attempt to control the effects of nuclear energy and to avoid what many of them saw as the possible end of our civilization in nuclear warfare.

In either a bomb or in a reactor, the fission energy ultimately generates heat, explosively in one and in a controllable fashion in the other. In nuclear power plants the heat generated in reactors is presently used to make steam for use in conventional electric power generating equipment. The cost is competitive with the cost of coal or oil as a fuel for making steam. Furthermore, the reactor can also be used to make more fissionable fuel by capturing some of the surplus neutrons in material which, after neutron capture, yields fissionable nuclei other than ${}_{92}U^{235}$.

Like fission, fusion of light nuclei into heavier ones also leads to an increase in the average binding energy per nucleon. There is therefore also a consequent release of energy, in the form of gamma rays and fast charged particles. Unlike the discovery of fission energy, fusion energy was discovered purposefully, in an attempt by nuclear physicists to understand the source of energy in the sun and other stars. Hans Bethe, in 1939, proposed a series of reactions which, in effect, are equivalent to

$$4({}_1H^1) \rightarrow {}_2He^4 + 2e^+ + 2\nu + 24 \text{ MeV} \qquad 9\text{-}5\text{-}2$$

According to Bethe's theory, which has since been confirmed in laboratory experiments, four protons could combine to give one helium nucleus, two positrons, two neutrinos, and 24 MeV of energy. For the reactions represented by Eq. 9-5-2 to take place, nuclei must collide so that they approach to within about 10^{-15} m of each other. Because of the repulsive Coulomb force between positively charged nuclei, the colliding nuclei must have high kinetic energies in order to approach this closely. Such kinetic energies are the normal consequence of the enormously high temperatures which occur at the center of a star.

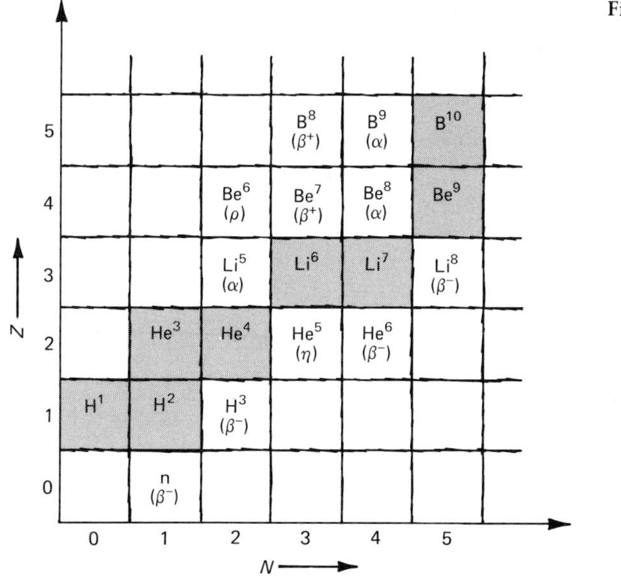

Figure P-9-12

PROBLEM 12

After a star has "burned hydrogen" for a long time, according to the reaction of Eq. 9-5-2, it will become rich in helium. Can it then continue to radiate energy by "helium burning"? The low-mass stable and unstable nuclei are shown in Fig. P-9-12. Hypothesize helium-burning reactions if you can. (Stable nuclei are shaded.)

The only earthbound apparatus which has so far been successful in producing large amounts of energy from fusion reactions is a uranium fission bomb, which can produce temperatures high enough to "ignite" an explosive fusion reaction. A hydrogen bomb results, with explosive energy up to 1,000 times that of the early fission weapons. Great efforts are now being made to produce controlled fusion reactors capable of producing electric power or heat more economically than fission reactors. Such efforts have not yet been successful, although encouraging progress has been made. The central problem is to contain a sufficiently hot gas of hydrogen isotopes in such a way that most of them undergo fusion reactions before either escaping from the container or dissipating their energy in a useless way.

PROBLEM 13

A fusion reaction is

$${}_3\text{Li}^7 + {}_1\text{H}^1 \rightarrow {}_2\text{He}^4 + {}_2\text{He}^4 + Q$$

From the curve of binding energy per nucleon, estimate the total binding energies of the initial and final nuclei and the energy given off in the reaction, Q.

PROBLEM 14

The fusion reaction above requires that a proton comes close to "touching" a lithium nucleus, in that the distance between their centers is about equal to the sum of their radii. Estimate the distance from Eq. 9-2-1 and find the kinetic energy required for them to approach this close in a head-on collision. At what absolute temperature T does the thermal energy, $(3/2)kT$, equal this energy? (k, Boltzmann's constant, is equal to 1.38×10^{-23} joules/degree.)

SUMMARY

Atomic nuclei are formed from *neutrons* and *protons*, together called *nucleons*. The nuclei are held together by the strong nuclear force, which acts when nucleons are about 10^{-15} m apart. Nuclear radii vary from about 1 to 5×10^{-15} m. *Isotopes* of a given chemical element have nuclei with a given number of protons and various numbers of neutrons.

The nucleons in a nucleus have *quantized states*, characterized by specific *energy levels*, analogous to the states of atomic electrons. The *exclusion principle* applies to neutrons or protons as it does to electrons.

The energy differences between nuclear states are typically measured in MeV (million electron volts) and are roughly a million times larger than energy differences between atomic electron states. Nuclear masses can be used to find their total energies and their *binding energies* by use of the Einstein relation: $E = mc^2$.

Medium-weight nuclei are found to be more tightly bound than either heavier or lighter nuclei. As a result, energy can be obtained by splitting a heavy nucleus apart (*fission*) or by joining together two light nuclei (*fusion*). Fusion reactions produce the energy radiated by our sun and other stars.

Three types of naturally occuring radiation are called α, β, and γ rays. Such radiation occurs when, by *radioactive decay*, a nucleus can reach a lower energy state. The decay occurs randomly, leading to an *exponential decrease* in the number of radioactive nuclei in a given radioactive sample. The *half-life* can be used to characterize the decay rate.

Alpha decay takes place by tunneling of an alpha particle inside the nucleus through a potential energy barrier at the nuclear surface. *Beta decay* occurs when a neutron changes to a proton in a nucleus by emission of an electron, or when a proton changes to a neutron by emission of a positron. *Neutrinos* or *antineutrinos* are also emitted in beta decay. After either alpha or beta decay, the residual nucleus may be in an excited state, in which case it reaches its ground state by emission of electromagnetic radiation, called *gamma rays,* which are very high-energy (MeV) photons.

ELECTRONS IN SOLIDS

10-1 CRYSTALS

The preceding chapters have mainly emphasized the physics of individual atoms, without shedding much light on the properties of bulk matter. Since even a relatively small quantity of matter may contain of the order of 10^{23} atoms or molecules, understanding bulk matter on the basis of the properties of individual atoms is not a small undertaking. Progress has been made, however, particularly for matter in the gaseous state and in the simplest forms of the solid state. For gases, the understanding is based on the theory called *statistical mechanics*, which applies to large numbers of identical particles moving chaotically. Paradoxically, the utter chaos of a gas is helpful in simplifying the problem. For solids, our understanding is reasonably complete only for especially simple solids called *crystals*. In a crystal the enormous numbers of atoms are arranged in a regular, repetitive pattern.

This chapter discusses some topics which are part of the field known as *solid-state physics*. They are mainly topics that relate to electrical conduction in solids. We will tend to discuss the solid state as if solids were perfect crystals, which isn't generally true. However, there are important solids which are nearly perfect crystals, such as the germanium and silicon from which transistors and other electronic devices are made. In addition, there are many solids which are *polycrystalline*, consisting of many small crystals bound together in a more or less chaotic arrangement. Common metals and alloys, such as steel, are polycrystalline. While some of their properties are determined by the way in which the constituent crystals are "glued" together, other properties, like the electrical conductivity, are closely related to the behavior of a single large crystal.

One of the most easily accessible crystalline solids is ordinary table salt. With a good hand magnifier or a low-powered microscope, the individual grains may easily be seen to be in most cases well-formed little cubes. The external shape reflects the regular ordering of the constituent atoms in what is called a *cubic lattice*. Figure 10-1-1 shows the basic pattern of sodium and chlorine atoms which is repeated over and over again to form a salt crystal. Actually, the crystal is formed from sodium and chlorine *ions* which look more like the picture in Fig. 10-1-2. Sodium chloride is an example of an *ionic crystal*, held together by electrostatic forces between the constituent ions. From the standpoint of electrical conduction, ionic crystals are quite simple: The electrons are very tightly bound in closed shells of the ions which form the crystal, and the crystals are therefore electrically insulating.

PROBLEM 1

From the electrostatic force binding sodium and chlorine atoms together, make an order of magnitude estimate of the stress in newtons per cm^2 needed to pull apart a perfect sodium chloride crystal. Assume the spacing between ions in about 3×10^{-8} cm. (Very strong steels have a strength of about 10^5 N/cm^2. Comment on a comparison of your answer with this value.)

An example of a *covalent crystal* is the diamond, which is a form of carbon. The carbon atom has four valence electrons, and the diamond crystal structure, as shown in Fig. 10-1-3, is such that each carbon atom has four nearest neighbors with which it forms

Figure 10-1-1 The structure of sodium chloride. Shaded and unshaded circles represent the locations of sodium and chlorine atoms.

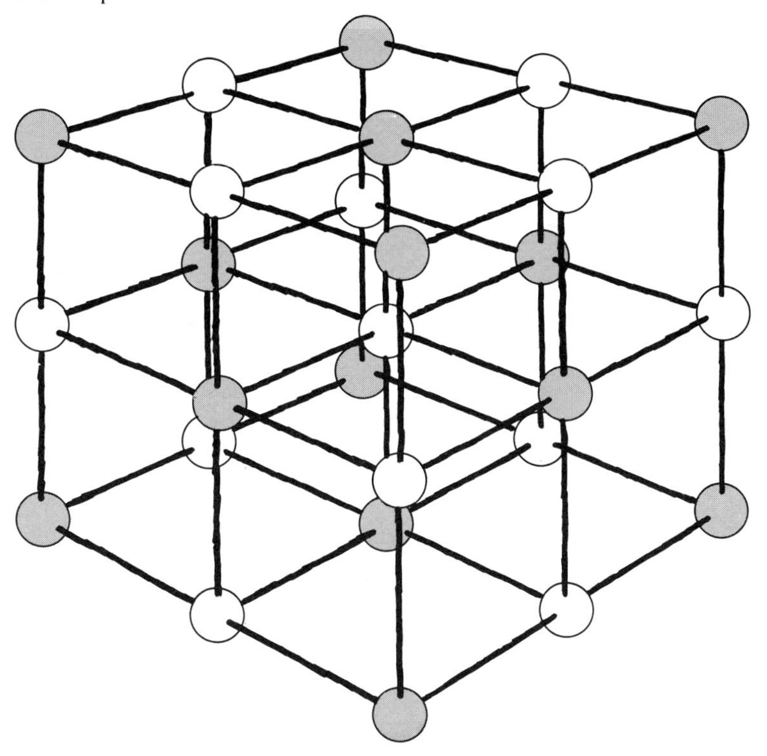

covalent bonds. All the valence electrons are utilized in the covalent bonds, and poor electrical conductivity results from the fact that the covalent bonds tend to firmly anchor the electrons. However, in covalent crystals of germanium and silicon the bonds are weak enough so that thermal vibrations of the crystal can occasionally free an electron. These elements are observed to conduct electric currents much more readily than good insulators. Such materials, called *semiconductors*, are of great practical importance and will be discussed in detail later in this chapter.

PROBLEM 2
Use the structure of the diamond lattice to guess what the methane molecule, CH_4, looks like. Show your result in a perspective drawing.

Figure 10-1-2 One plane of a sodium chloride crystal, showing sodium ions (Na^+) and chlorine ions(Cl^-).

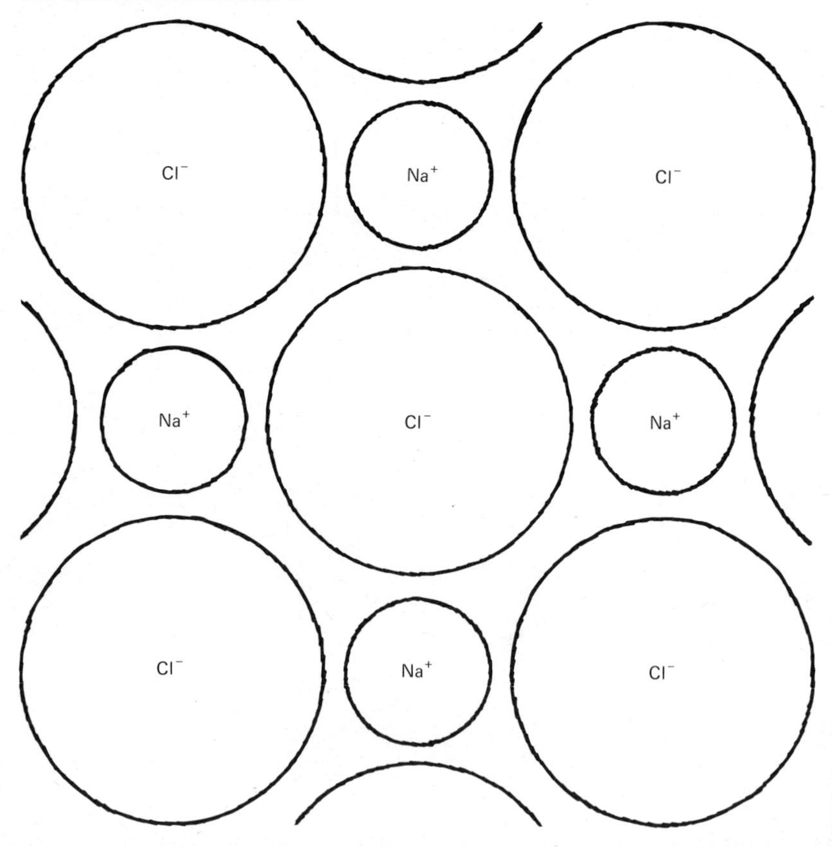

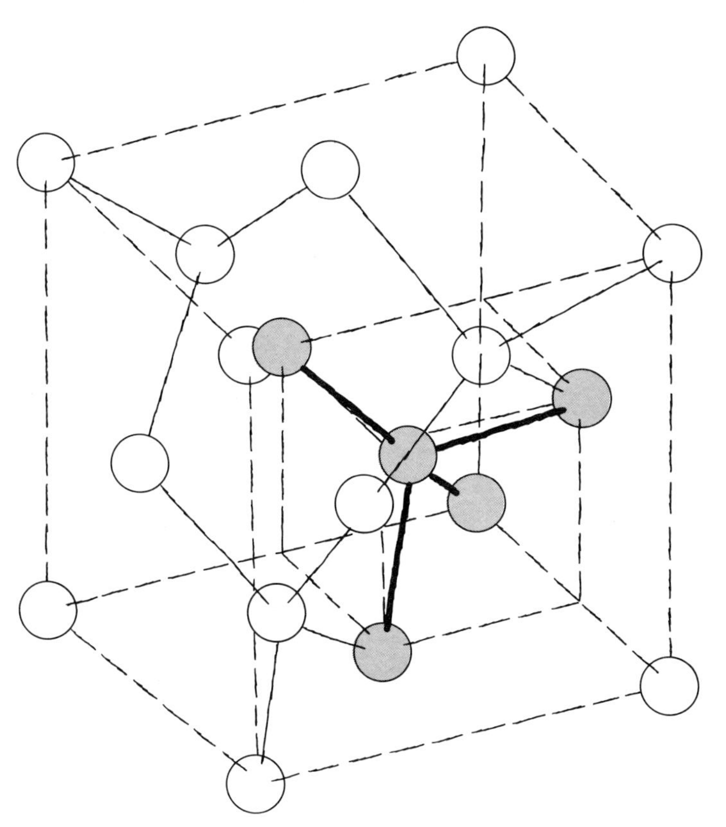

Figure 10-1-3 The structure of diamond. The larger cube indicated by dashed lines is an arrangement which repeats to form the crystal. The smaller cube contains one carbon atom at its center, covalently bonded to its four nearest neighbors, with all five atoms shaded to identify them. Each atom has four bonds, although all of them are not shown for all the atoms in the figure.

An example of a *metallic crystal*, which is a good electrical conductor, is sodium, shown in Fig. 10-1-4. This crystal, whose structural arrangement is called *body-centered cubic*, consists of ions at the corners of a series of cubes and also at the centers of the cubes. No specific bonding is shown between the ions, but minus signs indicate that the charge of the valence electrons is distributed throughout the crystal. Since each ion has only a single valence electron, there is no way in which we can form a three-dimensional crystal structure with covalent bonds linking neighboring atoms. (For carbon, with more valence electrons, this was possible.)

For sodium, and other metals, one or more valence electrons

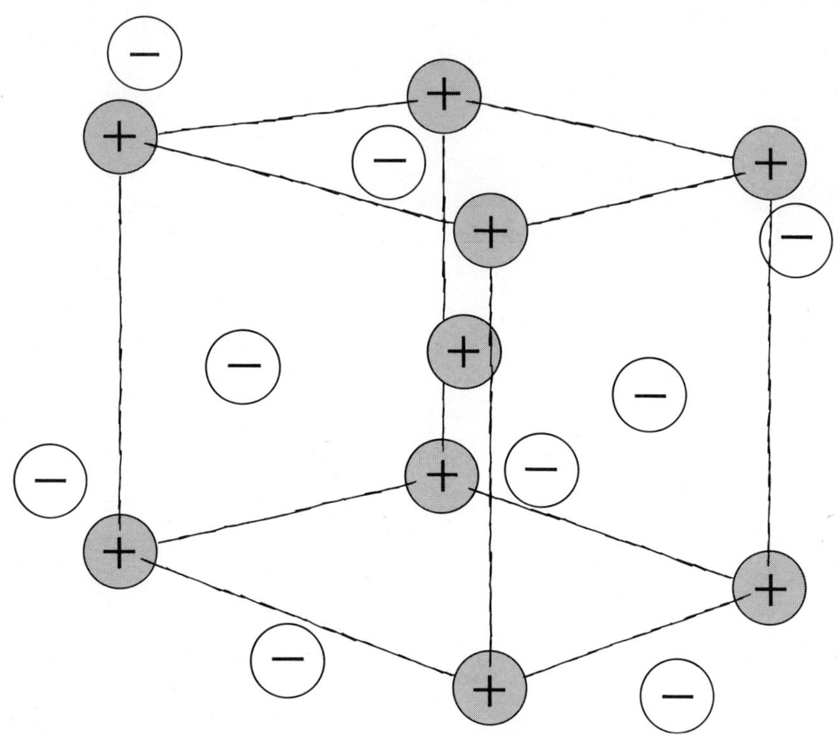

Figure 10-1-4 Structure of sodium metal. The sodium ions, in a body-centered cubic array, are indicated by + signs, and the electrons distributed throughout the lattice are indicated by − signs.

per atom form a sea of negative charge, which helps to bind the ions together in a way which is somewhat like that for the covalent bond. However, these valence electrons don't belong to specific pairs of atoms, as in the carbon crystal, and as a result they can move freely through the crystal. They are called *conduction electrons* and make possible the easy flow of electric current through metals.

10-2 THE FREE-ELECTRON MODEL

In the early twentieth century the idea that the conduction electrons in a metal could move freely was idealized by considering these electrons to be like gas molecules in a closed container. According to this *free-electron model*, the conduction electrons were in chaotic motion with an average kinetic energy determined by the temperature of the metal, and they "rattled around" inside a piece of metal, bouncing elastically off the walls of their container, which in this case were the surfaces of the metal. When the photoelectric effect was discussed in Chap. 6, this idea of the electrons in a metal was used.

The free-electron model was quite successful in describing some of the phenomena associated with metals, but unsuccessful in other instances. With the advent of quantum mechanics, the model was modified, and, in conjunction with the Pauli exclusion principle, the modified model can be used to understand a wide variety of metallic phenomena. We will mainly be interested in using the model to understand electrical conduction.

In applying quantum mechanics to the free-electron model, we consider the electrons to be described by the theory of a particle in a box. In Chap. 7, this problem was discussed in detail, assuming that the particle motion was *one-dimensional*. For a box with walls at $x = +a$ and $x = -a$, we found that the wave functions allowed for the particle were sine waves with particular values of wavelength.

The wavelength of a free electron, λ, can be connected to its momentum, p, by the de Broglie relation: $\lambda = h/p$. The allowed wave functions for the particle in a box represent traveling waves moving in the $+x$ and $-x$ directions with particular wavelengths, and therefore with particular momenta according to the de Broglie relation. The kinetic energy for a given momentum is given by the relation KE $= p^2/2m$. Following the procedure in Chap. 7 we can take the potential energy to be zero for an electron inside the metal, and its total energy U is then given by the kinetic energy. The table below summarizes the results for the allowed states of a particle in a one-dimensional box of width $2a$, as previously found in Chap. 7. The last row of the table gives the general form of the result for any value of n, where n must however be an integer.

Table 10-2-1 States for a particle in a one-dimensional box.

STATE	WAVELENGTH, λ	MOMENTUM, p	ENERGY, $p^2/2m$
ψ_1	$4a$	$\frac{h}{4a}$	$\frac{h^2}{32ma^2}$
ψ_2	$\frac{4a}{2}$	$\frac{h}{2a}$	$\frac{h^2}{8ma^2}$
ψ_n	$\frac{4a}{n}$	$\frac{nh}{4a}$	$\frac{n^2h^2}{32ma^2}$

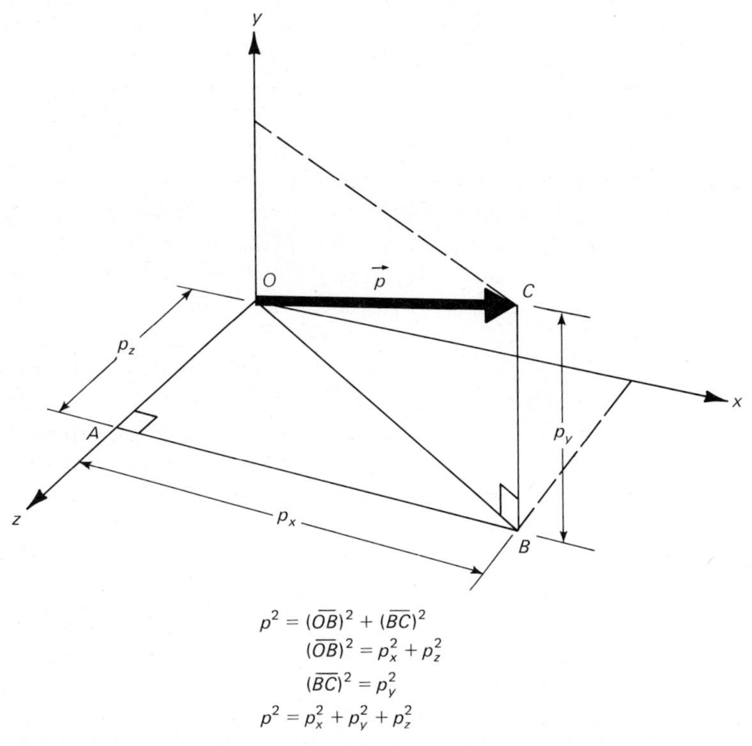

Figure 10-2-1 The pythagorean theorem in three dimensions for the square of a momentum vector. Right triangle OBC **is used to find** p^2**, and right triangle** AOB **is used to find** OB.

In order to discuss free electrons in a real metal, the results for one-dimensional motion of a particle in a box must be extended to three dimensions. The box can be taken to be a cube with each side of length $2a$. The result for this three-dimensional motion is that each *component* of momentum has the same set of allowed values as for the one-dimensional problem. To find the allowed values of energy, we first write the expression for p^2 in terms of the components p_x^2, p_y^2, and p_z^2:

$$p^2 = p_x^2 + p_y^2 + p_z^2 \qquad 10\text{-}2\text{-}1$$

This relation is easy to derive from the pythagorean theorem of geometry for a right triangle. Figure 10-2-1 illustrates the derivation. The quantum-mechanical result for each component of momentum may be written from the values given in Table 10-2-1:

$$p_x = \pm n_x \frac{h}{4a} \qquad n_x = 1, 2, 3, \ldots$$
$$p_y = \pm n_y \frac{h}{4a} \qquad n_y = 1, 2, 3, \ldots$$
$$p_z = \pm n_z \frac{h}{4a} \qquad n_z = 1, 2, 3, \ldots$$

The particle's energy U is given by the kinetic energy $p^2/2m$, with p^2 given by Eq. 10-2-1, so that we find finally

$$U = \frac{p^2}{2m} = \frac{h^2}{32ma^2}(n_x^2 + n_y^2 + n_z^2) \qquad 10\text{-}2\text{-}2$$

where the three *quantum numbers* n_x, n_y, and n_z must be integers.

Consider a cube of metal 1 cm on a side, with atoms about 10^{-8} cm apart. The cube then contains about 10^{24} atoms and about 10^{24} conduction electrons. The electrons will tend to arrange themselves in the lowest possible energy states, corresponding to the 10^{24} states with smallest values for the sum $n_x^2 + n_y^2 + n_z^2$. However, the Pauli exclusion principle *limits the number of electrons to two* for each particular choice of the three quantum numbers. This is because each choice specifies a particular quantum-mechanical state, and the exclusion principle states that at most there can be two electrons per state. Just as the exclusion principle has a profound influence on the structure of individual atoms, it controls the behavior of electrons in solids.

An enormous number of states must be occupied by the electrons in any macroscopic piece of metal. Fortunately, a great deal can be learned about the behavior of metals, and solids in general, without specifying the individual electron states. Instead, we combine the states into groups characterized by certain ranges of electron energies. The *band theory*, discussed in the following section, is formulated on this basis.

In this section we have considered a cube of metal because it is a simple three-dimensional shape. From the quantum mechanics of a particle in a box and from the exclusion principle, we have arrived at the result that the conduction electrons occupy an enormous number of states, each with a specific quantized energy which may be found from Eq. 10-2-2. We may ask, What if the metal is in the form of a "glob" instead of a cube? It requires some mathematical manipulation to show it, but it can be shown that once again the result is that a band of quantized states are occupied by the electrons. There is nothing special about the cubical shape except a degree of mathematical simplicity.

PROBLEM 3

In Eq. 10-2-2, let the integers n_x, n_y, and n_z be allowed to range from 1 to some maximum value N. Discuss why this gives N^3 different energy states, with maximum energy proportional to $(N/a)^2$.

Show that the maximum energy is the same for any sized cube, provided the number of states up to the maximum energy is proportional to the volume of the cube.

10-3 THE BAND THEORY OF SOLIDS

According to the band theory, the electron energy levels for all the electrons of the metal cube discussed in the previous section would be described by an energy diagram like the one in Fig. 10-3-1*a*. The most tightly bound inner electrons of the atoms have the lowest total energies. Each of these electrons is completely anchored to its own atom, and, for identical atoms, they all have essentially the same energy. For these electrons, all the states are represented by the lowest energy line in Fig. 10-3-1*a*.

For the less tightly bound atomic electrons, the close spacing of the atoms gives rise to quantum-mechanical tunneling phenomena, similar to that discussed in Chap. 7. Figure 10-3-1*b* shows very schematically the potential-energy diagram for a line of atoms, and three values of the total energy for electrons of a single atom are also indicated. The energy U_1 corresponds to tightly bound electrons, with negligible probability of tunneling from one atom to an adjacent one. However, for energies U_2 and U_3 such tunneling becomes more likely.

U_3 might represent the energy of a valence electron. These electrons are somewhat like the free electrons in a box, inasmuch as they can move through the entire crystal by tunneling. The detailed quantum-mechanical calculation of the energy levels for valence electrons leads to a band of energies, such as the valence band in Fig. 10-3-1*a*. The number of allowed states in the band is limited, however, unlike the states for a particle in a box. If there are N atoms in the crystal, each discrete state for an isolated atom becomes part of a band with N states in the crystal.

The conduction electrons in a solid are much like free particles in a box. However, the fact that the atoms of a crystal are arranged in a regular structure modifies the wave functions somewhat and introduces successive bands of conduction electron energy states. Just as there is a valence band, we thus have a conduction band, as shown in Fig. 10-3-1*a*, formed from the lowest energy band of these electron states.

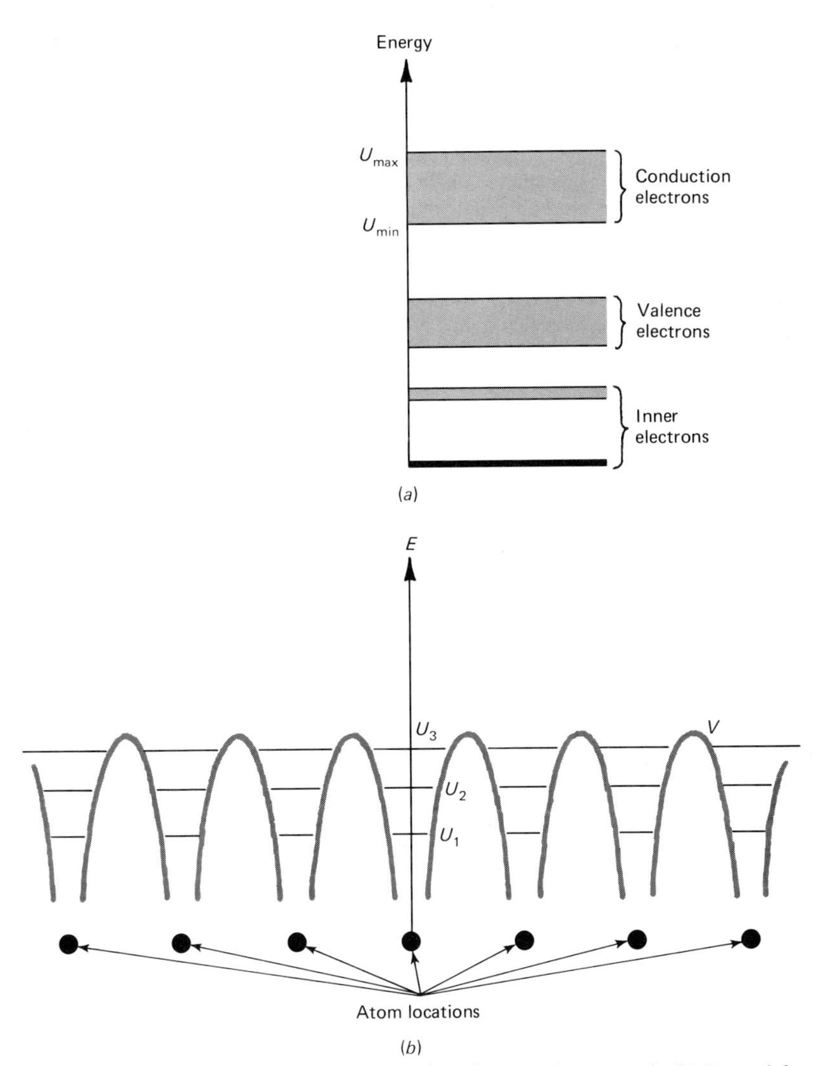

Figure 10-3-1 (*a*) Energy-level diagram for electrons in a metal. (*b*) Potential-energy curve for the electrons, showing energies U_2 and U_3 where tunneling from one atom to the next is significant.

PROBLEM 4

In Fig. 10-3-1, let U_{min}, the energy at the bottom of the conduction band, be that given by Eq. 10-2-2 with $n_x = n_y = n_z = 1$. Utilizing the result of the preceding problem, show that U_{max} is the same for a given number of conduction electrons per atom, regardless of the total volume of material. Energies associated with the band structure of solids are always independent of the actual size of the sample, as in this particular case.

The attribute of a solid which is most important in determining its electrical properties is the extent to which the conduction and valence bands are filled. Ideally, the valence bands are all exactly full, there being exactly as many valence electrons as there are allowed states. However, there are two important factors which can modify this conclusion. First, a valence electron can be broken loose from its normal place in the crystal by giving it enough energy to raise it to the conduction band. Second, in a real crystal there can be certain kinds of imperfections which give rise to vacant states in the valence band. For good conductors, the number of states in the lowest conduction band exceeds the number of conduction electrons, and the band is then only partially filled, unlike the valence bands. This fact is essential in understanding the high electrical conductivity of metals.

Suppose an electric field exists inside a metal. On the classical picture, the field would produce a force which would cause the free electrons to be accelerated in a particular direction (opposite to the field direction, since they have negative charge). This force would thus produce a net motion of all the free electrons, which would constitute an electric current in the metal. From a quantum-mechanical point of view, the acceleration of the electrons by the field requires that some of them occupy higher energy states than they would if the field were absent. In order for this to take place, such states must be available, which is not the case for a filled valence band. However, the highest energy electrons in an unfilled conduction band will be able to move into vacant states just above them in energy, and as a result electric current can flow.

We can now characterize solids as either electrical conductors or insulators depending upon whether they have a partly filled conduction band. Figure 10-3-2 shows schematically the valence and conduction bands for a metal and an insulator. For the insulator, there is an energy gap between the highest occupied valence state and the lowest unoccupied state, which is in the conduction band. Typically, the gap-energy difference E_g, shown on the figure, is several electron volts. The energy which thermal vibration of the crystal is likely to be able to transfer to a valence electron is only a fraction of an electron volt, which is insufficient to raise it to the conduction band. This band is therefore normally so nearly empty that the electrical conductivity may be as small as 10^{-20} times that of a typical metal.

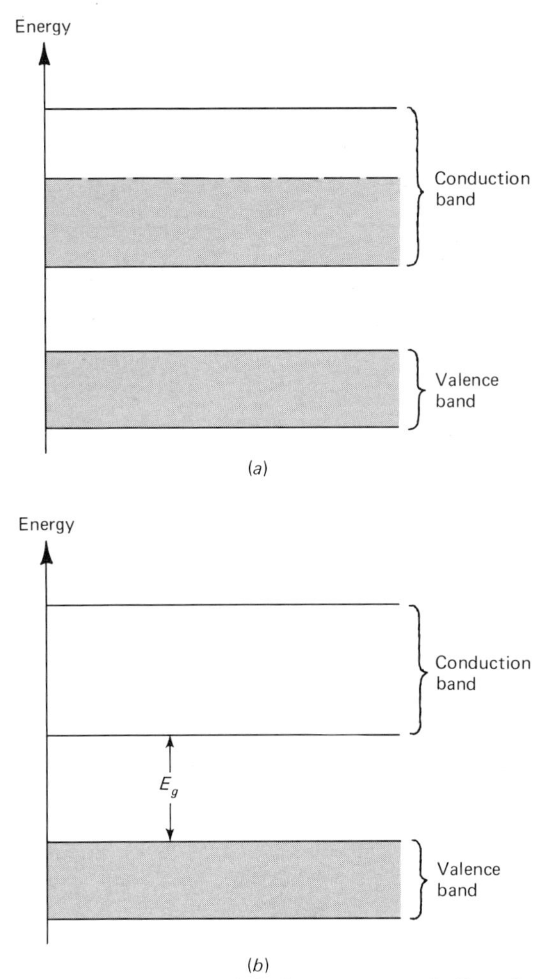

Figure 10-3-2 Occupied electron states, indicated by shading, for a metal (*a*) and for an insulator (*b*).

We might ask if there are substances for which the gap energy E_g is not so high as to make them really good insulators? There are, and such substances are known as semiconductors. For silicon, a common semiconductor, the gap energy is 1.1 eV, still much greater than the typical thermal vibration energy. However, there is a small probability that at room temperature a valence electron can actually receive sufficient energy from thermal vibrations to raise it to the conduction band. As a result, very pure silicon crystals at room temperature exhibit electrical conductivity which is much better than that of good insulators. However, the conductivity of silicon may still be a factor 10^{10} worse than for a good metallic conductor.

PROBLEM 5

Nearly perfect crystals of insulators are transparent to visible light, while conductors are opaque. Semiconductors like germanium and silicon are transparent in the infrared but opaque in the visible. Explain, using the band theory of solids.

Conduction in a semiconductor takes place because of electron motion in *both* the conduction and the valence band. If an electron is raised from the valence band to the conduction band, it becomes a current carrier, in an obvious way. However, we must also note that there is now a "hole" or vacancy in the valence band. This hole can move and actually acts like a mobile positive charge.

To visualize conduction by electrons and holes in a semiconductor, it is helpful to consider diagrams like Fig. 10-3-3. In the figure, part of a silicon crystal is shown schematically in a two-dimensional sketch, although real silicon is a crystal like diamond, whose four covalent bonds per atom do not lie in a plane. In Fig. 10-3-3, the covalent bonds are indicated by double lines connecting neighboring silicon atoms, and in Fig. 10-3-3*a* all the atoms are connected by normal bonds. In Fig. 10-3-3*b* a photon is shown breaking a covalent bond by giving its energy to a valence electron, which is then raised to the conduction band and is represented by a circle around a minus sign. The vacant site in the valence bonds is represented by a circle and a plus sign. The photon has created a *hole-electron pair*, as could other sources of energy such as vibrational heat energy.

The main point of Fig. 10-3-3 is to illustrate how the conduction electron and the hole in the valence bonds move under the influence of an applied electric field. In Fig. 10-3-3*c* we see the same pair as in the previous sketch, but imagine that an electric field has just been turned on, as shown by the E vector. A short time after the field is turned on, the situation will be as shown in Fig. 10-3-3*d*. The electric field exerts a force to the right on the conduction electron, and it is moving to the right with velocity v_- in this figure. The field also exerts a force to the right on the valence electrons, which in most cases cannot move because there is no vacant state in the valence energy band. However, the hole in Fig. 10-3-3*b* represents a vacancy in the valence structure which,

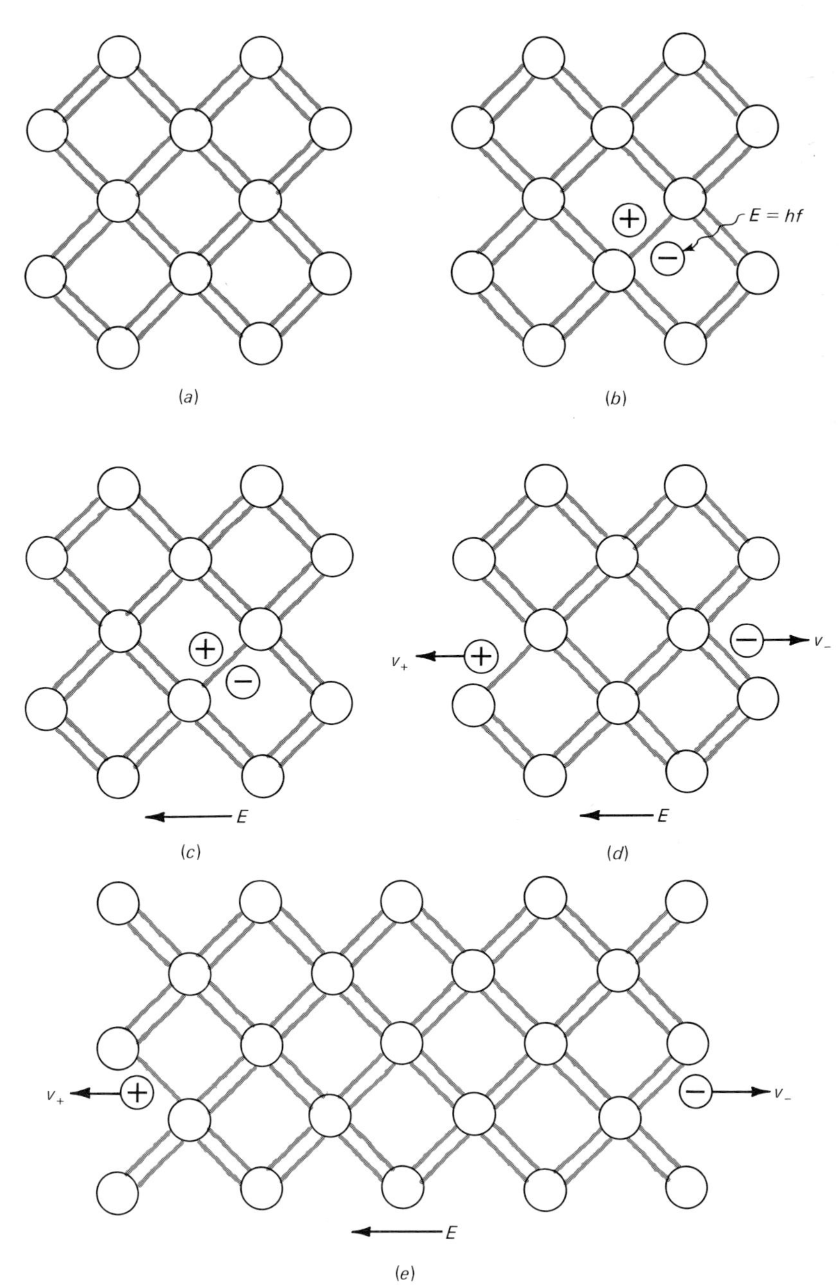

Figure 10-3-3. Motion of a hole-electron pair under the influence of an electric field.

as a result of the applied electric field, is most likely to be filled by an electron to its left tunneling into it. The hole then appears to move to the left, and it is in a different place in Fig. 10-3-3*d* than was in Fig. 10-3-3*c*. A velocity v_+ is shown for the hole to emphasize that it appears to continually drift to the left as time goes on because of the motion of valence electrons to the right under the influence of the E field.

The final figure, Fig. 10-3-2*e*, shows the electron-hole pair at a later time than Fig. 10-3-3*d*. The electron and the hole have moved farther from the place where the pair was originally formed, the electron moving to the right and the hole in the opposite direction. The hole has behaved exactly like a positive charge, moving in the direction of the field E. Furthermore, the crystal structure in the neighborhood of the hole actually has an extra positive charge. Without the hole the normal number of electrons just balances the positive charges of the silicon nuclei. With the hole, there is one negative charge missing, resulting in a net positive charge $+e$. A hole thus moves through the crystal very much like a positive electron!

PROBLEM 6

A conduction-band electron need not behave as if it has the mass of a really free electron, because it interacts with the crystal lattice as it moves. An important tool for determining the effective mass is called *cyclotron resonance* which is observed when the conduction electrons move in circular orbits because of an applied magnetic field. Show that as a result of the magnetic force always at right angles to its velocity the electron moves in a circular orbit whose radius is determined by the equation:

$$\frac{mv^2}{R} = evB$$

Find the revolution period for this motion and show that it is independent of v but depends upon the mass. The period can be measured, and the mass inferred, by observing that the electrons absorb energy from an oscillating electromagnetic field when the period of the field equals the cyclotron period. Consider a case when the oscillating electric field is along a line in the plane of the circular electron orbit, and explain how the field accelerates the electron.

Throughout the problem, assume the electron can move completely freely, as in a vacuum. In a metal, the conduction electrons scatter so rarely in moving through the crystal that this is actually a valid approximation.

10-4 DOPED SEMICONDUCTORS

In the previous section, holes and electrons in a pure semiconductor were discussed, and in such a material they are made in pairs and exist in equal numbers. In this section we will consider the consequences of *doping* a semiconductor crystal, by adding small quantities of particular kinds of impurity atoms. The object of doping is to produce semiconductor materials which have either a surplus of holes or of conduction electrons, and such materials make possible a variety of important electronic devices. In discussing doped semiconductors, electrons and holes produced in pairs, as discussed in the previous section, will generally be ignored. The doping effects are typically so much more important that this is a valid approximation.

In order to understand how a doped semiconductor behaves, consider the example of a silicon crystal which is "perfect" except that phosphorous atoms, which have five valence electrons instead of four, are occasionally substituted for silicon atoms throughout the crystal. The material in the neighborhood of one of these phosphorous atoms is schematically shown in Fig. 10-4-1. The phosphorous atom is labeled P, and the surrounding silicon atoms are labeled S. The covalent bond valence electrons are again represented by pairs of lines, as in Fig. 10-3-3.

Figure 10-4-1. The region of a silicon crystal in the vicinity of a phosphorous donor ion.

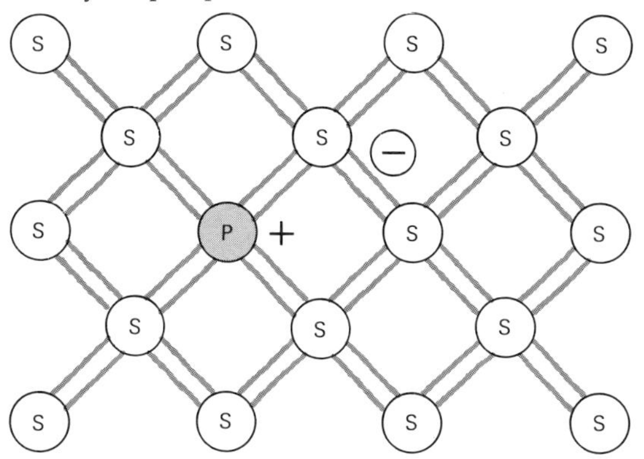

The phosphorous atom is seen to participate normally in the valence bonding of the crystal, with four of its valence electrons forming covalent bonds with electrons from four neighboring silicon atoms. The atomic structure of phosphorous is exactly the same as that of silicon, except that its nucleus has one more positively charged proton and it therefore has one more outer electron. This extra electron is shown schematically in Fig. 10-4-1 as a circle around a minus sign, quite far away from the phosphorous atom. The atom has a plus sign next to it to signify the one extra positive charge of the nucleus if only the four covalently bonded electrons are considered.

The situation shown in Fig. 10-4-1 can be described as a singly charged phosphorous ion bound in the crystal as if it were a silicon atom and an electron relatively loosely bound to this phosphorous positive ion. This electron is very easy to remove from the neighborhood of the ion, the energy required being only about 0.01 eV. If this much energy is given to the electron, it is raised from a bound state near its phosphorous atom to the conduction band of the crystal as a whole. Therefore, the phosphorous atom is called a *donor* because it can easily donate a conduction electron to the crystal.

The positive charge that balances the charge of the conduction electron is the extra proton in the phosphorous nucleus. This charge cannot move and in no way behaves like a hole. Thus, the addition of a donor such as phosphorous to a silicon crystal produces conduction electrons without corresponding holes. The energy needed to excite these electrons to the conduction band is less than the thermal vibration energy characteristic of room temperature. Therefore, when phosphorous is used to dope silicon, there is approximately one conduction electron per phosphorous atom at room temperature. This type of silicon is called *n type* where *n* stands for negative charge carriers, meaning conduction electrons.

In discussing semiconductor devices it is convenient to use energy diagrams like the one shown in Fig. 10-4-2. This figure represents a piece of *n*-type semiconductor. The x axis represents one direction in space, and in this figure there is no variation of the properties of the semiconductor with x. However, this will not always be the case. The band structure of the pure semiconductor material is represented by the valence and conduction bands shown in the figure with shading. The plus signs with circles around them signify the presence of positively charged

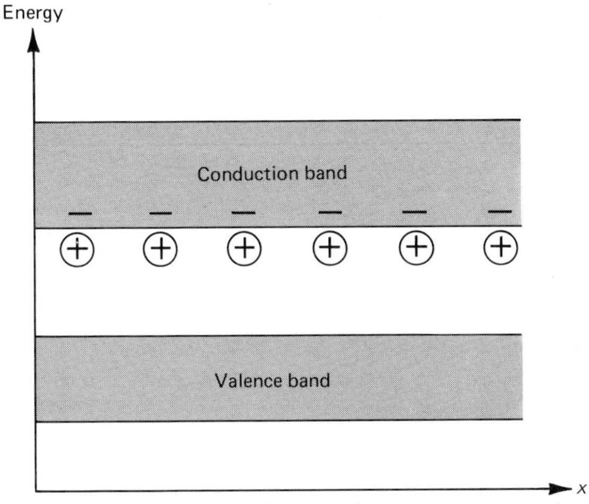

Figure 10-4-2 Energy diagram for an n-type semiconductor. The bound donor ions are symbolized by +signs, and the conduction electrons by −signs.

donor ions, immobile and bound in the crystal structure. The minus signs without circles around them represent electrons with energies near the bottom of the conduction band, freed from the donors, and able to move through the crystal. The temperature is assumed high enough so that all the donor atoms are ionized, with one electron from each occupying a state in the conduction band.

Note that the net charge of the n-type material is zero. For each conduction electron there is a positive ion with equal and opposite charge. The significant property of the n-type material is that the negative charges are free to move while the positive charges are not. Circles around the + charge, and not around the − charge, are used to signify this distinction.

If silicon is doped with aluminum instead of phosphorous, a semiconductor is produced in which there are free holes and bound negative ions. Aluminum is the element with one fewer proton in its nucleus than silicon, and one less outer electron. When it is bound in a silicon crystal, it creates a hole because of the missing valence electron. This hole need not be at the aluminum atom, but may be bound in its vicinity, as shown in Fig. 10-4-3. However, it takes very little energy to unbind the hole from the negative ion, about 0.01 eV, roughly the same as that required to separate an electron from a phosphorous donor.

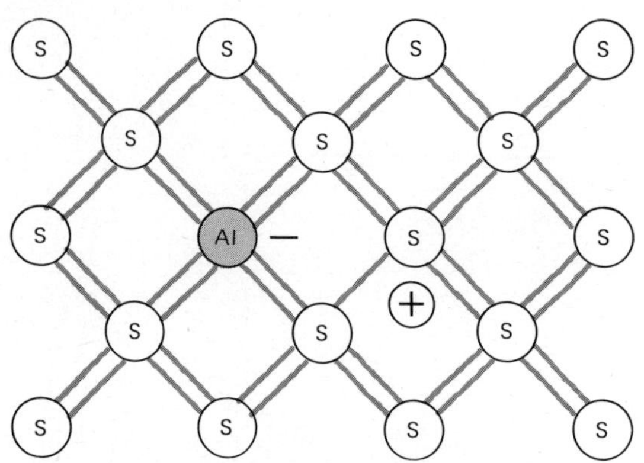

Figure 10-4-3 The region of a silicon crystal in the vicinity of an aluminum acceptor ion.

At room temperature, the holes are almost all freed from their aluminum *acceptor* ions by thermal energy. The acceptor ions are called by that name because they have accepted valence electrons from the other atoms of the crystal. In this way, they create holes, which are free charge carriers, and negative ions at the sites of the acceptor atoms which are not free to move. Semiconductor material which has been doped with acceptors is called *p type* because the mobile charge carriers are holes, which conduct electric current like positive charges.

Figure 10-4-4 Energy diagram for a p-type semiconductor. The bound acceptor ions are shown by − signs, and the free holes by + signs.

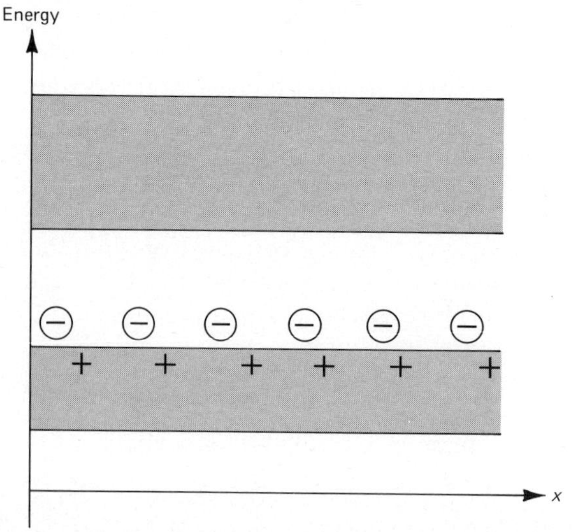

An energy diagram appropriate to a piece of p-type semiconductor is shown in Fig. 10-4-4. The minus signs with circles around them represent the immobile negative acceptor ions, while the plus signs represent free holes near the top of the valence band. The energy diagram is drawn so that electron energies increase upward. Thus, electrons tend to make downward transitions. As a result, a hole near the bottom of the valence band would be filled by a valence electron dropping from above. The hole would thus rise. On this type of energy diagram, holes therefore tend to float to the top of the valence band, like bubbles in water, while electrons tend to sink to the bottom of the conduction band.

PROBLEM 7

Gallium has a valence of 3 and arsenic has a valence of 5. Which is used to produce n-type semiconductors and which to produce p-type? The alloy gallium arsenide has become important as a semiconductor, with a diamond crystal structure and properties like germanium or silicon. Explain.

PROBLEM 8

Electrons moving through a wire in a direction perpendicular to a magnetic field are acted upon by a force $\vec{F}_{mag}$ in the direction shown

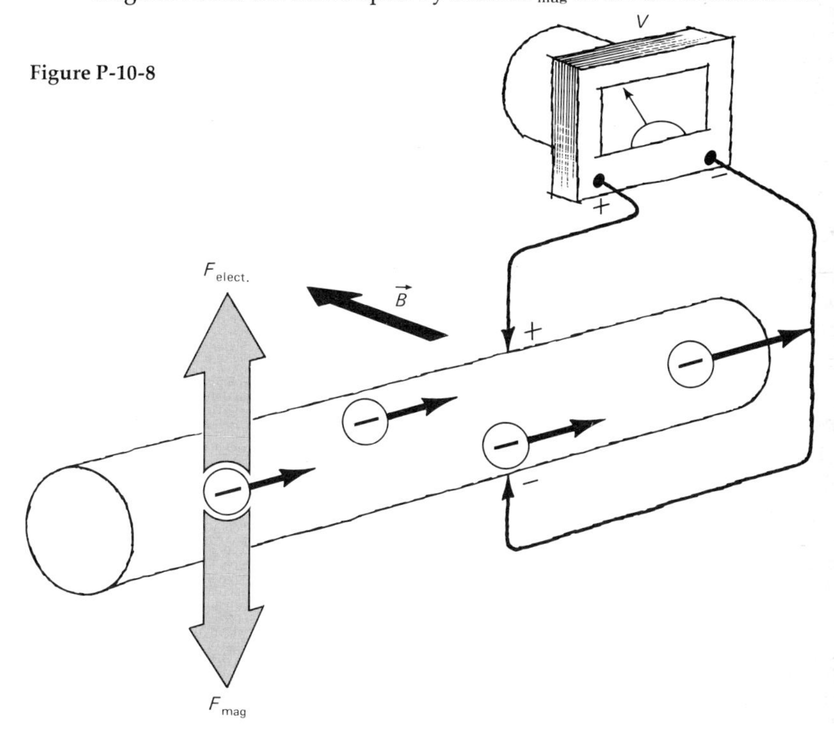

Figure P-10-8

in Fig. P-10-8. As a result a surplus of electrons accumulates on one side of the wire until an electric field is set up whose force F_{elect} balances the magnetic force. The presence of this field can be detected by measuring the voltage between the two sides of the wire, called the *Hall voltage* after the man who discovered this effect. Discuss whether this Hall effect can be used to tell whether current flow is due to holes or electrons. (Hint: For a given direction of electron motion, compare the electric field directions for free electron motion and hole motion.)

10-5 THE p-n JUNCTION

Conduction electrons and holes in semiconductors can move under the influence of applied electric fields to give rise to electrical conductivity. However, when there is no electric field applied, the charge carriers are still in random motion, occasionally being scattered when they interact with the crystal lattice. The crystal lattice at a given temperature T may be thought of as vibrating chaotically, with each atom of the lattice jiggling around with mean kinetic energy $^3/_2kT$ (where k is the Boltzmann constant). The interaction of electrons and holes with the lattice also leads to their possessing an average kinetic energy of $^3/_2kT$.

The random motion of electrons and holes leads to important consequences when we consider a sample of semiconductor material which is not uniformly doped with either donors or acceptors. If, for example, we try to establish a local region with a high density of conduction electrons, then the random electron

Figure 10-5-1 Formation of a *p-n* junction in a bar of doped semiconductor.

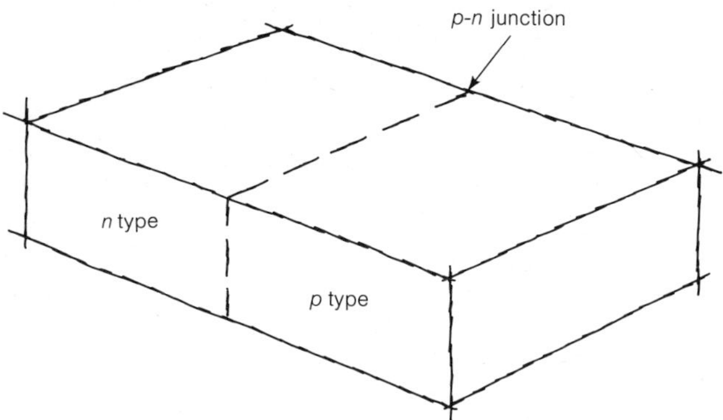

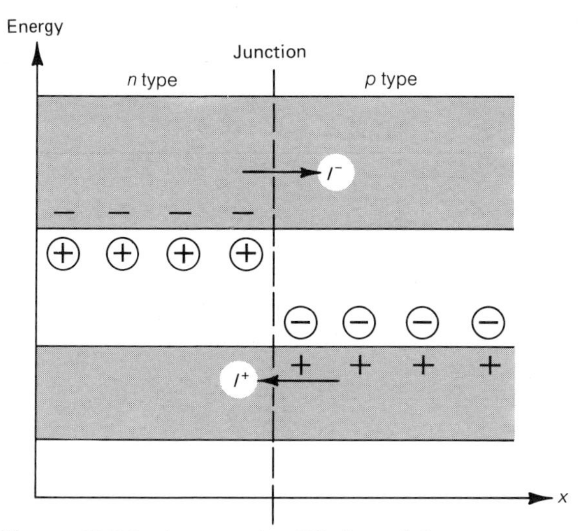

Figure 10-5-2 An oversimplified, and incorrect, energy diagram at a *p-n* junction. The diffusion currents I^+ and I^- will act so as to produce an energy diagram with a step at the junction, shown in Fig. 10-5-3.

motion will tend to result in electrons moving away from their donor ions and spreading through the semiconductor. This process of transport by random motion is called *diffusion* and is not solely important for the motion of charges in semiconductors. For instance, perfume molecules travel by diffusion through the air.

A single piece of semiconductor can be doped with donors and acceptors so that part of it is p type and part n type. Figure 10-5-1 shows schematically a bar of material which has a region called a *p-n junction,* in which there is an abrupt transition from n type to p type. Figure 10-5-2 shows an *incorrect* energy level diagram for the region near the junction, with the n-type and p-type regions each shown as in Figs. 10-4-2 and 10-4-4. The reason the picture shown in Fig. 10-5-2 is incorrect is that it ignores the process of diffusion. To the left of the junction in the figure there is a high density of conduction electrons. To the right there are essentially none. By diffusion, we expect that electrons will flow from left to right, as shown by a current arrow labeled I^- in the figure. Similarly, holes will diffuse in the opposite direction, as shown by the arrow labeled I^+.

If we consider the electric effects of the diffusion currents shown in Fig. 10-5-2, both of them will lead to a net positive charge in the *n*-type region to the left of the junction. Conduction electron flow will leave a surplus of positive donor ions, while hole flow will lead to a surplus of positive holes. Similarly, the diffusion currents will lead to a net negative charge to the right of the junction. The fields due to these net charges act so as to oppose the flow of diffusion current, so that ultimately there will be an equilibrium condition with electrons kept in the *n* region and holes in the *p* region as a result of electric fields. Figure 10-5-3*a* shows the correct energy diagram for a *p-n* junction, which shows the effect of the electric field set up near the junction by a shift in conduction-band and valence-band energy levels.

It is easiest to first discuss Fig. 10-5-3 from the standpoint of the motion of conduction electrons. The upward slope in the conduction band at the junction corresponds to an increase in potential energy for an electron moving from left to right across the junction, because there is an opposing force due to an electric field. The origin of this field is the charge distribution near the junction shown in Fig. 10-5-3*b*. As we shall see, there are donor and acceptor ions in the junction region but few electrons and holes, and the charges of the ions give rise to the charge distribution.

An electron moving to the right of the junction, in "hole country," leads a precarious existence. Ultimately, it drops from the conduction band into one of the holes in the valence band. The energy it gives off can be in the form of light or can be transmitted to the lattice as vibrational energy. Such a transition from the conduction to the valence band causes both an electron and a hole to disappear, or "annihilate" each other. This is precisely the reverse of the creation of an electron-hole pair by absorption of a photon or of thermal vibration energy.

Holes moving to the left across the junction are also annihilated by the surplus of electrons in the *n*-type material. Thus, a *depletion layer*, containing few charge carriers but a normal population of donor and acceptor ions, is formed near the junction. The resulting charge distribution in the depletion region is that shown in Fig. 10-5-3*b*, and the field arising from this charge is shown in Fig. 10-5-3*c*. This field exerts a force to the left on conduction electrons and leads to the potential-energy change shown in Fig. 10-5-3*a*.

In order to see how equilibrium is reached, with no net current

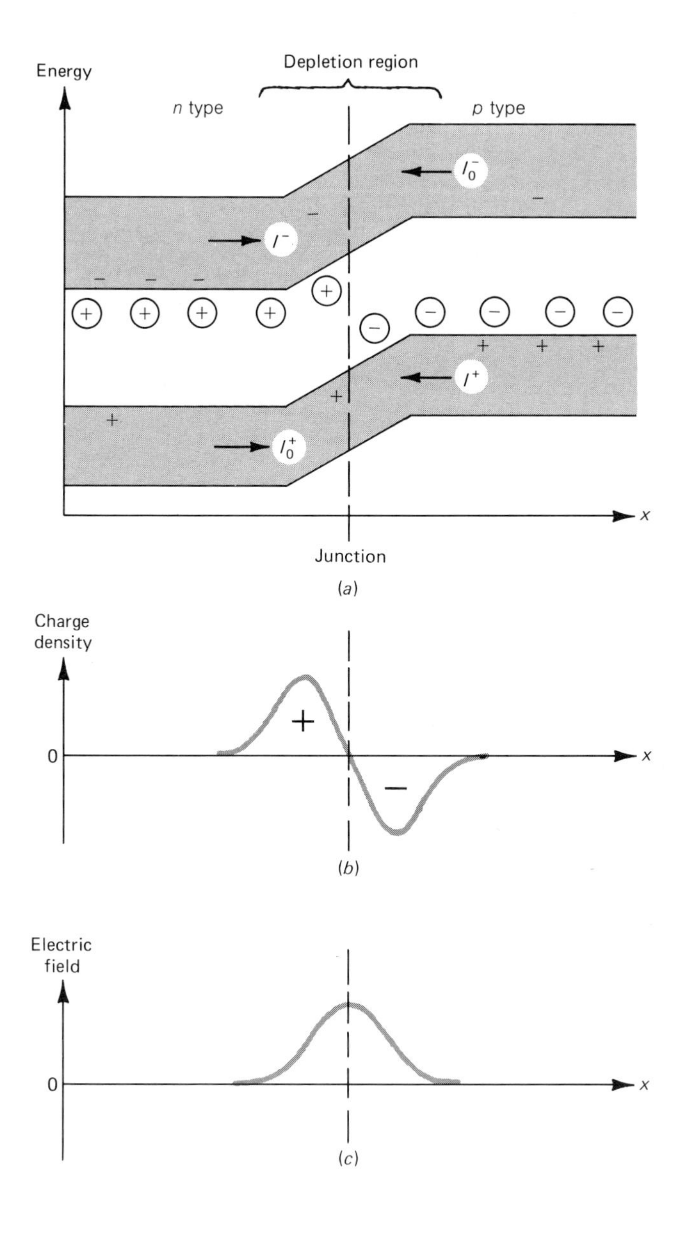

Figure 10-5-3 Energy diagram, charge distribution, and electric field in the vicinity of a *p-n* junction. The currents I^+ and I^- are due to diffusion of majority carriers, while I_0^+ and I_0^- are from diffusion of minority carriers. The current arrows point in the true direction of motion of the charge carriers.

flow across the junction, it is now necessary to also consider currents which arise from the relatively small number of thermally generated hole-electron pairs in the semiconductor. In both the n and p regions *minority carriers* arise from this pair-creation process, holes in the n-type material and electrons in the p-type materials. The oppositely charged *majority carriers* can be ignored, being negligible in number compared to the similar carriers from donors and acceptors. The minority carriers are of particular interest because when they diffuse to the junction, the electric field *attracts them across.*

The currents I_0^- and I_0^+ in Fig. 10-5-3a represent the flow of thermally generated minority carriers. In terms of the four currents shown in Fig. 10-5-3a, the net current I across the junction from right to left (p region to n region) is given by

$$I = I^- + I^+ - I_0^- - I_0^+ \tag{10-5-1}$$

The currents in Eq. 10-5-1 are positive for the directions shown in Fig. 10-5-3a, i.e., electron flow to the right and hole flow to the left both contribute to net conventional *positive* current flow to the left.

At equilibrium, an isolated p-n junction must have $I = 0$, so that the minority carrier currents I_0^- and I_0^+, in Eq. 10-5-1, must balance the majority carrier currents I^- and I^+. There is a particular height of potential barrier at the junction which reduces the majority carrier currents so that this balance is achieved. Since most of the conduction electrons are near the bottom of the conduction band, as the potential-energy barrier is increased, the flow of majority carrier electrons from the n region to the p region is very sharply reduced. Similarly, the flow of holes from p region to n region is also greatly reduced. Thus, the electric field barrier reduces the majority carrier flow so as to exactly balance the very small current of minority carriers.

When a p-n junction is used as a circuit element, it is called a *junction diode.* Its very asymmetrical voltage-current relation is shown in Fig. 10-5-4. The inset in the figure shows the diode symbol, which has an arrow pointing in the direction of "easy" current flow. As the current-voltage curve indicates, when the applied voltage is zero there is no current flow, which corresponds to the equilibrium discussed above. For a *positive voltage,* a voltage applied so as to cause current flow from p region to n region, the current is seen to rapidly increase with voltage, while for a negative voltage the current is small and relatively insensitive to voltage.

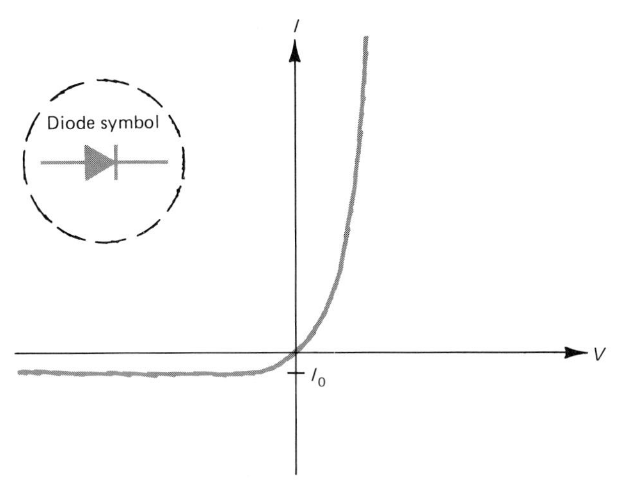

Figure 10-5-4 Current-voltage relationship for a *p-n* junction diode.

The reason for the current-voltage curve shown in Fig. 10-5-4 is that a positive voltage lowers the potential barrier at the junction while a negative voltage raises it. The positive voltage "opens the floodgates" for many more majority carriers, such as electrons with lower conduction-band energies, to cross the junction. An applied negative voltage, on the other hand, cuts off the majority carrier current essentially completely and leaves the small minority carrier current which is independent of voltage. The junction diode is thus a circuit element with a *preferred direction* of current flow, which leads to important applications in electronic circuitry.

PROBLEM 9
Light-emitting diodes are made from gallium arsenide *p-n* junctions. If the light is in the visible, what is the minimum energy gap you would expect for the semiconductor, in electron volts? (Hint: The light is emitted when the diode conducts a forward current. It comes from electron-hole annihilations.)

10-6 THE TRANSISTOR

The junction transistor is, in essence, a piece of semiconductor with two *p-n* junctions which are close together. The bar shown in Fig. 10-6-1 is a simple example, which we will find can be under-

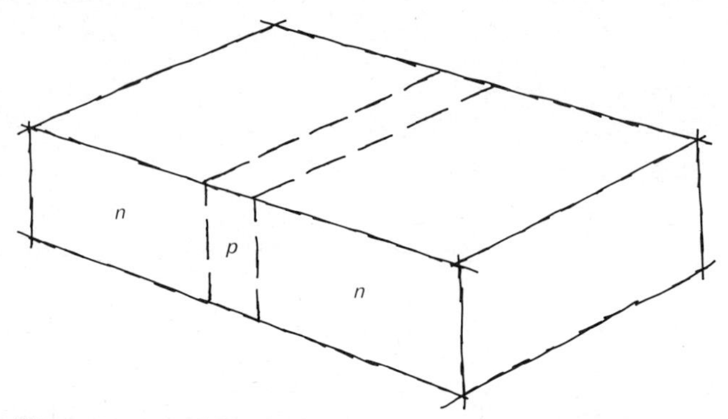

Figure 10-6-1 **Schematic view of a bar of semiconductor doped so as to form two *p-n* junctions. This arrangement constitutes an *n-p-n* transistor.**

stood easily on the basis of the preceding treatment of the junction diode. In normal operation, a voltage is applied across one junction in the *forward* direction, the direction of easy current flow, and across the other junction in the *reverse* direction. Figure 10-6-2 shows the forward voltage supplied by a battery with voltage V_1 and the reverse voltage supplied by the battery V_2. The diode characteristics shown in Fig. 10-5-4 are redrawn in Fig. 10-6-2, with voltages V_1 and V_2 shown on the graph.

The terminals marked e and b in Fig. 10-6-2 stand for *emitter* and *base* of the transistor. The voltage V_1 causes a relatively large current to flow across one of the two junctions, called the *emitter-base junction* of the transistor.

The forward current in the emitter-base junction has one dominant component, I_e^- in Fig. 10-6-2, which consists of electrons flowing from the n-type emitter region to the p-type base region. The forward hole current, from base to emitter, is kept comparatively small by only lightly doping the base region. Electrons which go from emitter to base become minority carriers in a p region, and they may suffer one of three fates:

1 They can annihilate with a hole.
2 They can flow to the wire which constitutes the base lead of the transistor.
3 They can flow across the base to the other junction of the transistor.

In a good transistor, the dominant process is number 3 above. While flow to the base lead is caused by an applied electric field due to the battery voltage V_1, flow to the other junction takes place by diffusion. In order for this flow to dominate, the base region is made very thin (often 0.001 in or less).

When an electron diffuses across the base region and reaches the other junction, it is "collected." This junction, called the *collector-base junction*, has a reverse voltage applied to it by battery V_2. As a result, minority carriers which reach the junction are

Figure 10-6-2 Voltages applied to junctions for transistor operation. The current I_e^- is the main current of the emitter-base junction. The majority of the electrons which constitute this current diffuse to the collector-base junction.

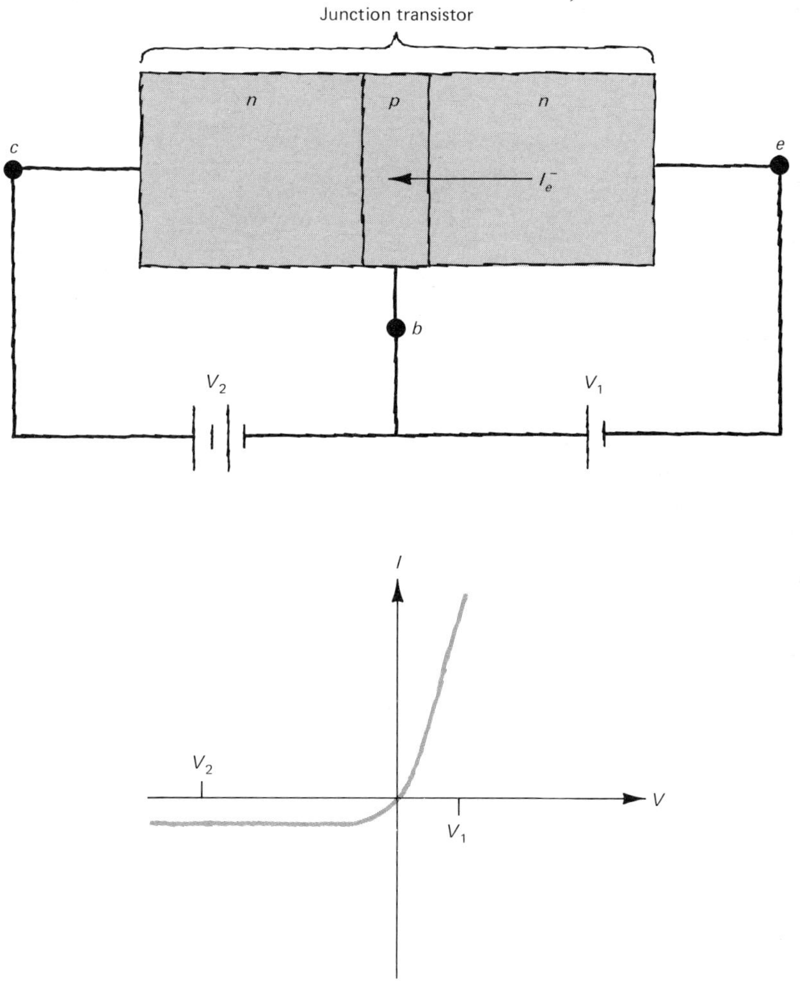

accelerated across it. Except for the very small current arising from thermally generated carriers, the minority carrier current is mainly from the electrons which diffuse across the base region from the emitter-base junction. These electrons ultimately flow to the collector lead of the transistor, labeled *c* in Fig. 10-6-2. Thus, the electron current I_e^-, from the emitter, mainly flows to the collector of the transmitter. Just a small fraction of the current goes to the base terminal.

The electrical behavior of a transistor is often summarized by a set of curves such as those shown in Fig. 10-6-3. Except for very low values of the collector-emitter voltage these curves tell a very simple story: A given collector-emitter current, say 10 mA, corresponds to a much smaller base-emitter current, say 0.1 mA. In fact, there is a simple approximate relation:

$$I_{ce} \approx (\beta)(I_{be}) \qquad 10\text{-}6\text{-}1$$

where β is approximately a constant, equal to about 100 for the example of Fig. 10-6-3.

Figure 10-6-3 Characteristic curves for a junction transistor. Each of the four curves refers to a specific value of base-emitter current, I_{be}, given in milliamperes (10^{-3} A). As long as V_{ce} is $\geq$ 2V, $I_{ce} \approx (100)I_{be}$.

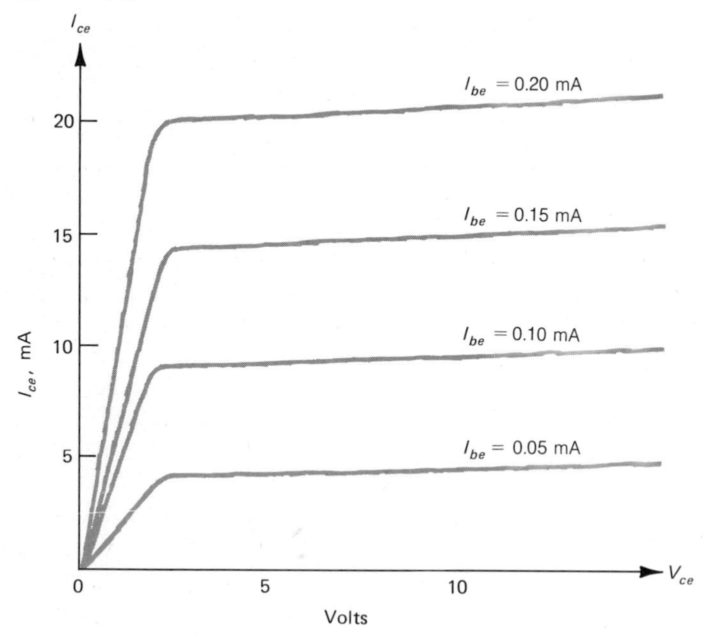

Equation 10-6-1 indicates the main importance of the transistor in electronic technology: The collector-emitter current is many times larger than the base-emitter current. In electronic circuits, this makes the transistor useful as a *current amplifier*. The physical reason for the relation shown in Eq. 10-6-1 is that because of diffusion almost all the current at the emitter-base junction flows to the collector-base junction.

When we speak of an amplifier in electronics, we mean that a small amount of input energy can produce a large change in output energy. An analogous device is the fire hydrant, which is an amplifier in much the same sense that a transistor is. The valve of the fire hydrant can be opened and closed with relatively little energy, thereby controlling a very powerful flow of water. We do not obtain free energy with this "amplifier": The energy of the flood of water which we can unleash is provided by a pumping station or by the gravitational potential energy of a raised water reservoir. Our efforts are amplified in the sense that we can *control* a large amount of energy with a small energy expenditure.

Similarly, the transistor amplifier, which often is connected as shown in Fig. 10-6-4, controls a large current I_{ce} at high voltage V_{ce} by means of a small current I_{be} at low voltage B_{be}. The batteries shown in this figure produce the requisite forward and reverse voltages across the junctions in a slightly different way than was shown in Fig. 10-6-2, more appropriate for the operation of practical electronic circuits. The expenditure of a small amount of electrical energy in the *base circuit* to increase the current I_{be} and voltage V_{be} slightly can produce a large change in the collector current I_{ce} and in the energy delivered by the battery with voltage V_c to an electrical device, schematically indicated by the box labeled "load." The collector circuit battery provides the main energy source just as the water pump does for the hydrant. This energy is controlled by small changes in the energy delivered to the base circuit, which is like the valve of the hydrant.

PROBLEM 10

Electric power is equal to the product current × voltage. Assume V_{be} is about 1 V and V_c is 10 V for the circuit in Fig. 10-6-4. Assume further that I_{be} is about 0.1 mA and that the transistor is described by the curves in Fig. 10-6-3. What is the ratio of the power used from the battery connected to the collector to the power used from the battery connected to the base?

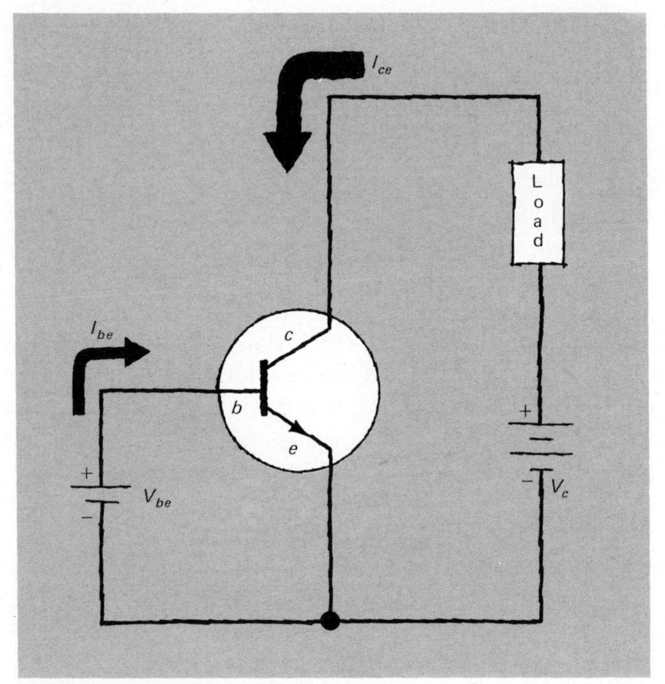

Figure 10-6-4 Schematic of a transistor amplifier circuit. The standard transistor symbol is shown, with emitter, base, and collector leads labeled *e*, *b*, and *c*. The small current I_{be} determines the large current I_{ce} which transfers energy from the battery V_c to the load.

PROBLEM 11

Figure 10-6-4 shows an *n-p-n* transistor. Sketch a circuit which functions the same way using a *p-n-p* transistor, which has the emitter arrow pointing the other way in its symbol.

SUMMARY

The conduction electrons in a metal can be described approximately by the *free-electron model,* in which they occupy the quantized states for particles in a three-dimensional box. The *exclusion principle* is operative, so that the electrons must occupy states covering a *band* of energies.

In a real crystal, because electrons can tunnel from atom to atom, there are valence bands as well as conduction bands. The electrical conductivity of a material is high if it has a partly filled conduction band. If the valence band is filled and separated by an *energy gap* from an *empty* conduction band, the material is an insulator for a large gap, and a semiconductor if the gap is relatively small. In a *pure* semiconductor, current flows as a result of motion of both *conduction electrons* and *holes*, generated in *pairs.*

A semiconductor can be *doped* with either *donor* or *acceptor* impurity atoms, which contribute either free electrons or free holes, alone, yielding *n-type* or *p-type* materials, respectively. If an abrupt *p-n junction* is formed, an energy barrier exists at the junction, such that an applied voltage which lowers the barrier causes a great increase in current while a voltage in the opposite direction does little. This is called a semiconductor *junction diode.*

If two *p-n* junctions are separated by a thin *base* region, a *junction transistor* results. Most of the emitter current diffuses across the base and arrives at the collector terminal. However this *collector-base* current is controlled by the *small emitter-base current* which flows, leading to the possibility of *amplification.*

ELECTRIC CIRCUITS

11-1 INTRODUCTION

The subject of this chapter is the applied physics and technology of electric circuits and electronic devices. The next chapter, on optics, can also be described as "applied" rather than "pure" science. The distinction between science and technology can be emphasized by assigning to the scientist the role of discovering the basic laws of nature, and to the technologist or engineer the role of producing useful applications of the scientists' basic knowledge. There is truly a real distinction, but it should not be oversimplified or reduced to snobbery. The interdependence of science and technology is reciprocal; each derives much of its strength from the other.

The modern engineer is often aptly called an applied scientist. He may work in areas so close to the frontiers of basic knowledge that he does research which is hardly distinguishable from that of his "pure" colleagues. Similarly, the pure scientist, particularly the experimenter, finds himself deeply involved with the use and development of technology. For example, particle accelerators are engineered to a large extent by the elementary particle physicists who use them. In another instance, the need for intense light sources can convert the physicist into a laser technologist. However, he may also find that the applied scientist has already created the research tool he needs.

The progress of pure science has been intimately connected with the development of technology. The invention of the telescope influenced must of the work of Galileo and Newton. In the nineteenth century, basic experiments which ultimately led to an understanding of the atom depended more on a developing technology for pumping a vacuum than on any other factor. Today, pure research takes place in a style that would be inconceivable without powerful electronic computers and other examples of the applied science and technology of solid state electronics. The interdependence of pure and applied science is greater now than ever before, and applied science encompasses many areas which are interesting and rich in intellectual challenge, as well as potentially useful.

11-2 A SERIES CIRCUIT

Figure 11-2-1 shows an electric circuit which includes a source of *electromotive force*, abbreviated *emf*, and three *circuit elements* labeled Z_1, Z_2, and Z_3. The source of emf may be thought of as a

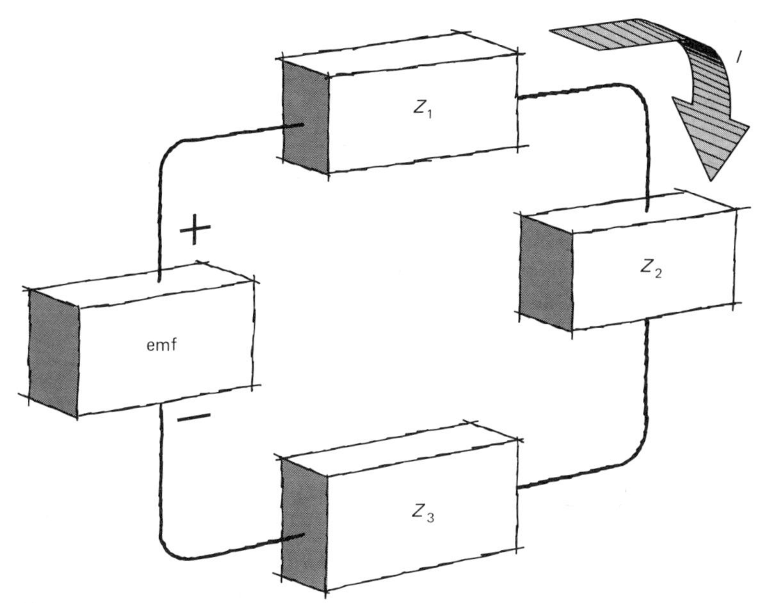

Figure 11-2-1 **A simple series circuit. The emf (electromotive force) causes a current I to flow through three circuit elements, Z_1, Z_2, and Z_3.**

kind of pump, which forces charge around the circuit, through the circuit elements. The various parts of the circuit are connected with wires to form a single closed loop. An arrow is shown, labeled I, the symbol for electric current, to indicate the flow of electric current around the loop, from the "+" terminal of the emf to the "−" terminal. The circuit elements can be light bulbs, motors, electric heaters, etc., and a circuit can involve either one or many circuit elements.

One of the fundamental laws of nature—which appears to be obeyed unconditionally—is that net quantities of electric charge never appear or disappear. A body can only be given a net charge if an equal and opposite charge is left somewhere else, and the net charge of an isolated system cannot change. This law, called *conservation of charge*, has an important consequence in electric circuits: Equal currents must enter and leave each point of a circuit, so that the net current into any point must be zero. Otherwise, net charge would have to appear or disappear at the point. In a circuit which forms a single loop, like that in Fig. 11-2-1, conservation of charge requires that the current be unchanged everywhere around the loop, for at each point of the circuit the ingoing current must equal the outgoing current as in Fig. 11-2-2*a*.

When circuit elements are connected one after the other in a circuit, as in Fig. 11-2-1, they are said to be in *series*, or the circuit is called a *series circuit*. The main attribute of such a circuit is that the current through each element is the same as that through the others. Just as charge conservation holds along the wire, it holds through the circuit elements. A flashlight is a concrete example of a simple series circuit and is shown schematically in Fig. 11-2-3. The figure shows symbolically a battery, a switch, and a bulb connected in series. The switch is "open" as shown in the circuit and "closed" as shown in the inset of the figure. An open switch doesn't conduct a current, while a closed switch has a conducting path through it which passes current as easily as a wire. Since this is a series circuit, when the current through the switch is zero, the current through the bulb is also zero, so the light is off. Upon closing the switch, current flows and causes the thin metal filament of the bulb to glow white hot.

When the flashlight is lit, the bulb gives off energy in the form of light and heat. The battery supplies this energy from chemical potential energy, which is converted to electrical energy by chemical reactions within it. Then, the electrical energy is converted to light and heat in the bulb. The overall function of the circuit is to utilize chemical energy in a controlled and convenient fashion. One main use of electric circuitry is to effect such energy transformations, with electrical energy as an intermediary. A candle converts chemical potential energy into light

Figure 11-2-2 Charge conservation at a point in a circuit. In *(a)* and *(b)* the current leaving the point must equal the current entering the point, leading to the current equations shown on the figure.

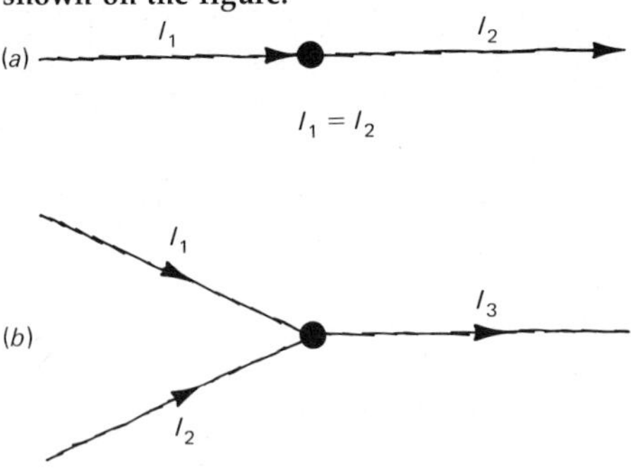

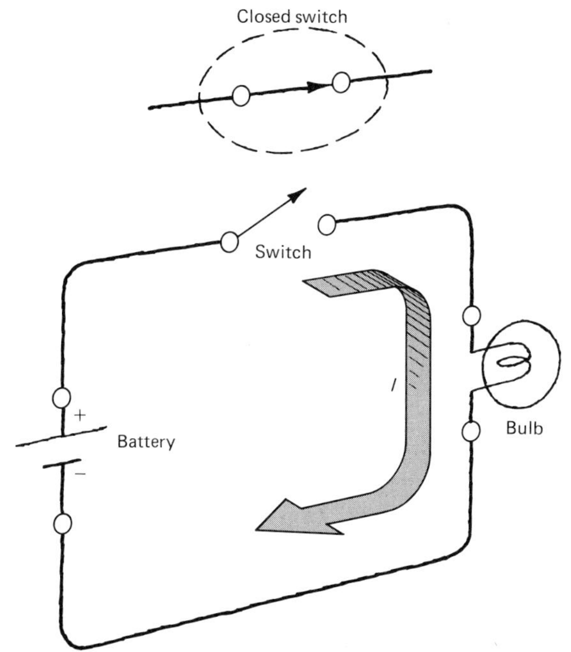

Figure 11-2-3 Schematic circuit of a flashlight. The current I flows only if the switch is closed.

and heat in a more direct, but much less convenient, way than a flashlight.

PROBLEM 1

Suppose you have a battery and a light bulb and two switches. Devise a circuit that enables the light to be lit only if both switches are closed. Devise one that will light the light if either switch is closed.

PROBLEM 2

Suppose you have a battery, a light bulb, and two *double throw switches*, of the type diagrammed below:

Devise a circuit so that either switch can control the light bulb—i.e., determine whether it is on or off—regardless of the position of the other. This circuit is useful for controlling a light from either end of a hallway or staircase.

11-3 ELECTRIC CURRENT

The movement of charge through a circuit, which we call the flow of electric current, can take place by motion of either electrons or ions. Atomic ions consist of single atoms with either an excess or a deficiency of electrons, and molecular ions are groups of atoms, bound together to form a molecule, which also have an excess or deficiency of electrons. Some possible charge carriers, represented symbolically, are

e^- H^+ Cl^- and O_2^-

The superscripts indicate the net charge in units of the size of the electron charge, −1 for the electron, the chlorine ion, and the oxygen molecular ion, and +1 for the hydrogen ion.

In gases and liquids ionized atoms and molecules are able to move, and conduction can therefore involve motion of ions. In solids, however, the atoms are bound together too tightly to permit motion, and current flow takes place entirely by electron motion. Electrons can move freely in metallic solids, which are called conductors, but not in other solids, like sulfur or polystyrene, which are called insulators. A relatively small number of solids conduct much more readily than insulators and much less well than metals. These are called semiconductors and are very important for the construction of modern electronic devices. They are discussed in the preceding chapter.

Free-electron motion in liquids is inhibited by the fact that electrons tend to be captured by atoms to form negative ions. Conduction in liquids is thus essentially entirely ionic, the opposite of the situation in solids, while in gases conduction can be either ionic or by electron motion, or both, depending upon the conditions. The table below summarizes the kinds of charge carriers which are important for current flow in various states of matter:

STATE	CHARGE CARRIER	SIGN OF CHARGE CARRIERS
Gas	Electrons and ions	+ and −
Liquid	Ions	+ and −
Solid	Electrons	−

From a purely electrical point of view, the motion of negative charge in one direction is indistinguishable from the motion of positive charge in the opposite direction. Figure 11-3-1 shows

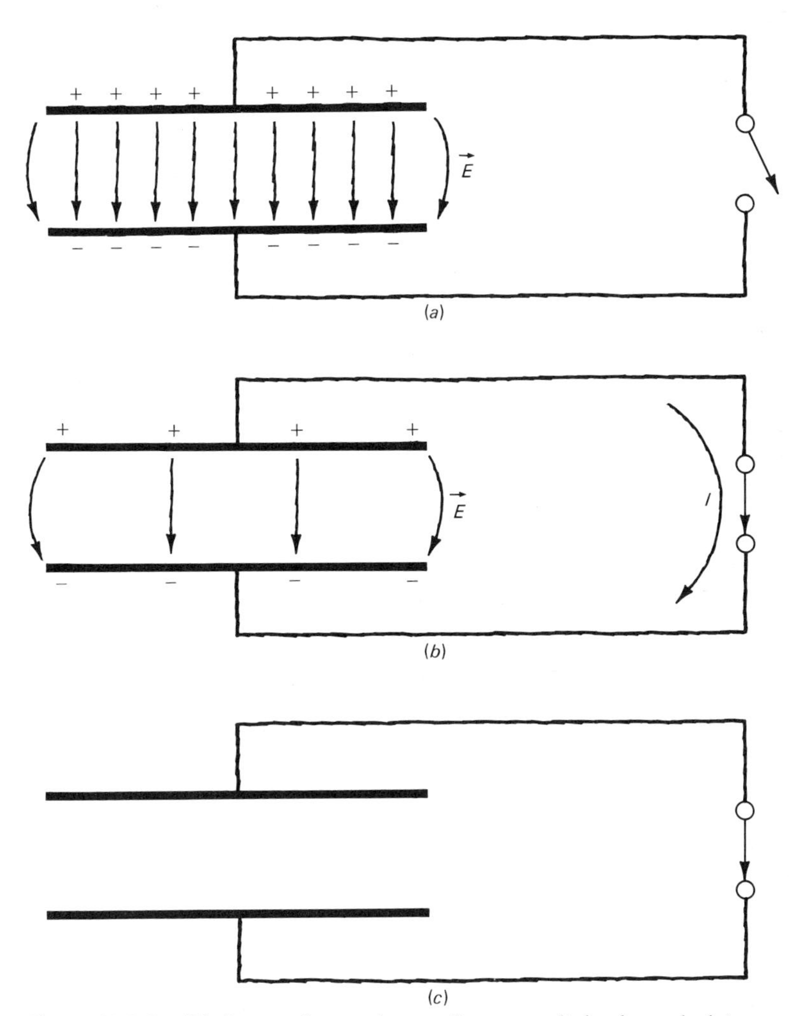

Figure 11-3-1 Discharge of a condenser. Two oppositely charged plates in *(a)* are connected by a wire when the switch is closed in *(b)*. Soon afterward, the flow of current has allowed the net charges on each plate to combine, and the electric field E is zero between the plates, as in *(c)*.

an illustrative example in which two oppositely charged plates become uncharged after a switch is closed in an electric circuit. The electrical effect of closing the switch is that a strong electric field between the plates disappears, which could equally well result from the flow of electrons from the bottom plate to the top plate, or of positive charges from the top plate to the bottom plate.

Conventionally, we analyze the circuit of Fig. 11-3-1 as if positive charge moved, as shown by the current arrow. If the circuit consists of metallic conductors, we ignore the known fact that electrons are actually the charge carriers. Figure 11-3-2 shows a hypothetical series circuit in which current flows through various kinds of material as a result of motion of different charge carriers. Although we may understand the properties of each individual circuit element on the basis of the detailed motion of the true charge carriers, we consider the electric effects in the circuit as a whole as if positive charge were moving. For the circuit of Fig. 11-3-2, we assume that a *conventional current*, indicated by the arrow labeled I, flows *everywhere* in the circuit.

PROBLEM 3

If one can detect very small mass changes, it might be possible to determine whether electrons or positive ions flowed from one

Figure 11-3-2 Current flow around a series circuit, as a result of motion of ions and electrons. The conventional current I is clockwise everywhere in the circuit, in the direction of motion of positive charges and opposite to the direction of motion of negative charges. Ions are shown with the charges circled; electrons are shown as uncircled minus signs.

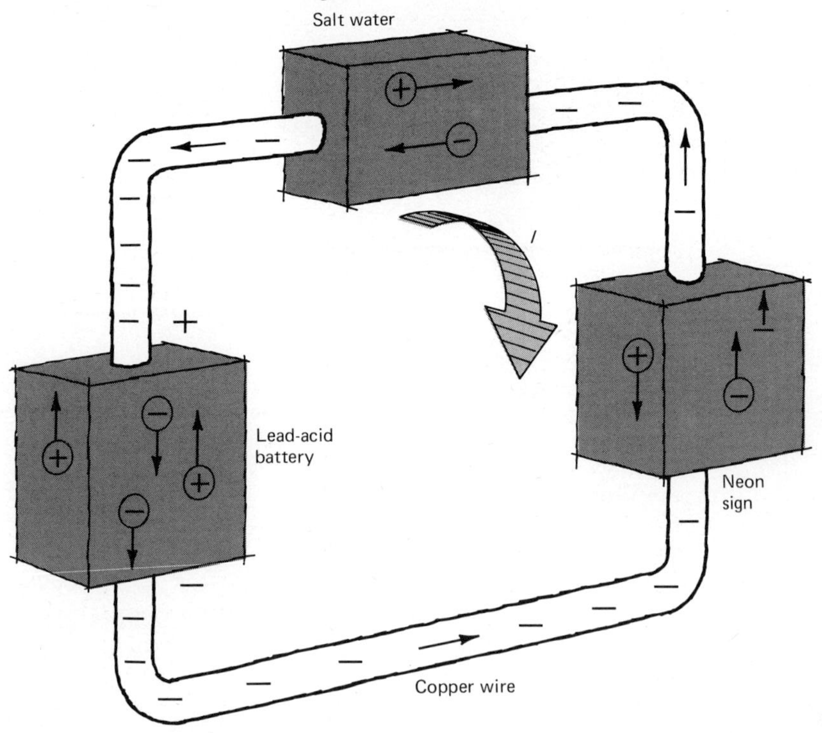

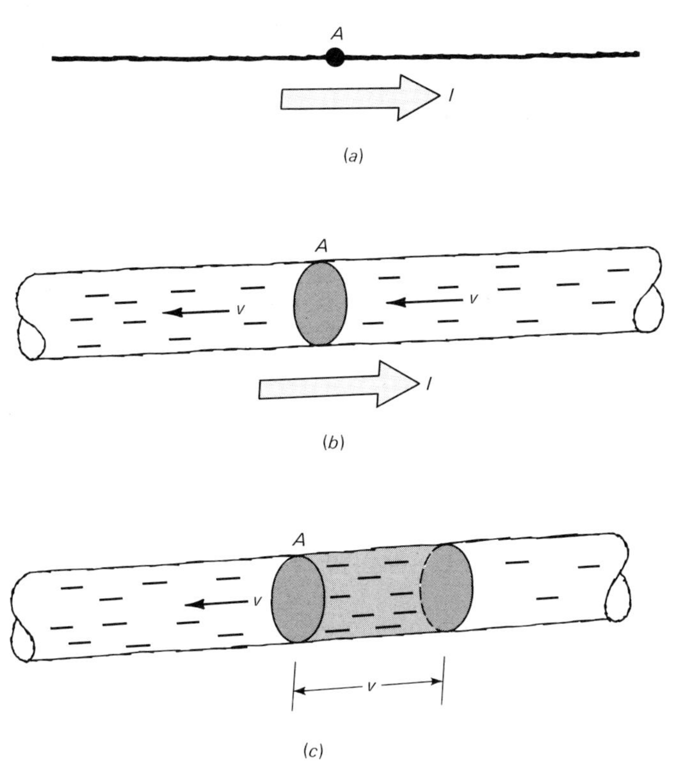

Figure 11-3-3 Current past a point A in a wire. In (b) and (c) the wire is shown magnified, and in (c) the shaded region shows electrons which will pass the pass the point A in 1 sec.

plate to the other in Fig. 11-3-1. For a charge 10^{-6} C, calculate the mass of electrons or positive copper ions which would be transferred.

Figure 11-3-3 emphasizes the meaning of the current at a point in a circuit. Figure 11-3-3a shows part of a circuit, a wire carrying a current I, with one point labeled A. Figure 11-3-3b shows an enlarged view of the wire at the point A, with a shaded area to indicate an imaginary surface cutting across the wire. Electrons are indicated, with velocity vectors to indicate an average speed and direction of motion. At any given instant, the electrons actually have various velocities, in all directions, due to random thermal motion. The average v is the *drift velocity* due to an electric field which causes the current to flow.

The current at A is defined as the charge per second which crosses the shaded area in the figure, given by the number of electrons which pass A per second, times the charge of one electron:

$$I = \text{(electrons per second)(charge of each electron)} \qquad 11\text{-}3\text{-}1$$

The unit of current is the coulomb per second, which is given the name *ampere,* abbreviated "amp." When a current of one ampere flows at a point in a circuit, one coulomb of charge passes that point each second.

How fast do electrons drift in wires which carry current? To begin to answer this question, suppose all the electrons move at the same speed v (the average drift velocity, in a real wire). Then, during a one-second interval after the time for which Fig. 11-3-3*c* is drawn, all the electrons in the shaded volume with length equal to v will have passed point A. The number of electrons per second can be written as the product of the shaded volume times the number of electrons per unit volume in the wire:

$$\text{Electrons per second} = \text{(shaded volume)(electrons per unit volume)} \qquad 11\text{-}3\text{-}2$$

The value of the shaded volume in Fig. 11-3-3*c* is equal to the product of its length, v, and its cross sectional area, so that

$$\text{Electrons per second} = (\text{area})(v)(\text{electrons per unit volume}) \qquad 11\text{-}3\text{-}3$$

The current I can be found by multiplying electrons per second by the charge of the electron e:

$$I = (\text{area})(v)(\text{electrons per unit volume})(e) \qquad 11\text{-}3\text{-}4$$

And, finally, Eq. 11-3-4 can be solved for the velocity v:

$$v = I/(\text{area})(\text{electrons per unit volume})(e) \qquad 11\text{-}3\text{-}5$$

In order to evaluate the number of electrons per unit volume in Eq. 11-3-5, we will assume a copper wire, with one "free" conduction electron per atom. This leads to a result of about 10^{23} electrons per cubic centimeter. For a typical case, assume a current of 1 amp and a wire 0.1 cm in diameter. For an approximate calculation, the area is then about 0.01 cm^2. Substituting these values in Eq. 11-3-5:

$$v \approx \frac{1}{(0.01)(10^{23})(1.6 \times 10^{-19})} \text{ cm/sec}$$

$$v \approx 0.6 \times 10^{-2} \text{ cm/sec}$$

This estimate for v looks ridiculous! Electrons are found to flow like molasses, but we know that when we turn on an electric gadget there is not a long delay before current flows. Is there a contradiction? The answer is "No." Although the drift velocity of electrons in good conductors is very low, when a switch is closed this low-speed motion starts everywhere in the circuit almost simultaneously. The first motion anywhere in the circuit generates electric fields, which cause the motion to start all around the circuit before the individual electrons have drifted any appreciable distance. In analyzing simple electric circuits we often neglect the very brief *transient conditions* when the current begins to flow and consider only *steady-state conditions* after transients are over.

PROBLEM 4

Electroplating results from current flow by motion of positive metal ions. A silver nitrate solution contains silver ions which have a charge $+2e$. One ampere flows in the circuit in Fig. P-11-4, as indicated. On which plate does silver become deposited? How long must the current flow to deposit 10^{-3} gram of silver?

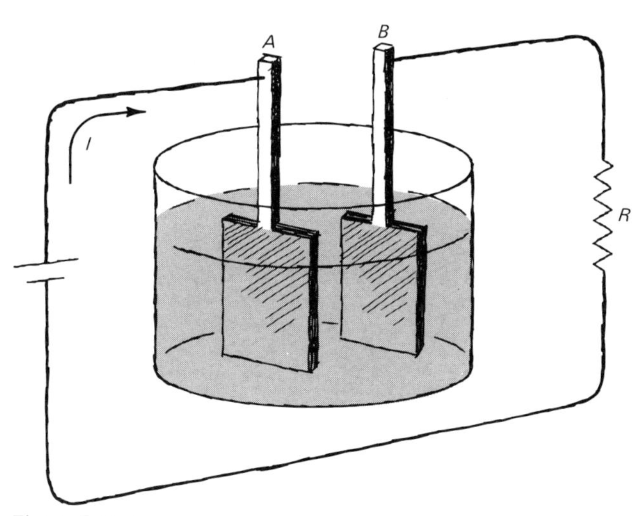

Figure P-11-4

11-4 VOLTAGE

The motion of charges through circuit elements—the current—requires that electric forces act upon the charges (except in perfect conductors, called superconductors). For example, in order to keep the current flowing through the circuit element schematically shown in Fig. 11-4-1, an electric field in the direction indicated must exert a force on the moving charges.

This force does mechanical work on the charges, and this work represents energy given to the charges. The energy is transferred to the circuit element as the charges move through it. For a light bulb this energy appears in the form of heat and light; for an electric motor much of the energy appears in the form of mechanical work done by the motor.

We can account for the overall effect of the electric fields in a circuit element by giving the *voltage* across the element, defined as the work done by the electric fields on one coulomb of charge which moves through the element. In Fig. 11-4-1, for example, the work done in moving one coulomb from A to B is called the voltage across the circuit element, or the *potential difference* between points A and B, because it is also a potential-energy difference per coulomb of charge. This potential-energy difference is converted in the circuit element to heat or mechanical work.

The unit of voltage is energy per unit charge, or joule per coulomb, which is given the special name, *volt*. The most important relation involving the voltages in an electric circuit is that *the sum of the voltage differences around any closed loop is zero.* This is a consequence of the fact that the electric field gives rise to a conservative force, so that the work done by electric fields around a closed path is zero. As a concrete example, the series circuit ori-

Figure 11-4-1 Electric field inside a circuit element, and resultant voltage difference V_{AB} between the two ends of the element.

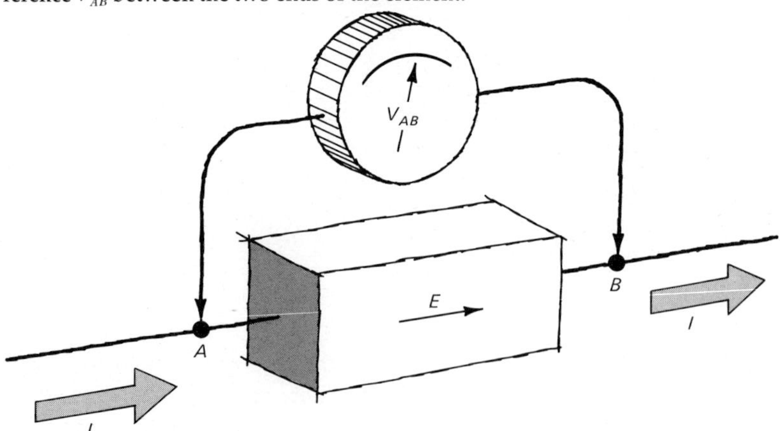

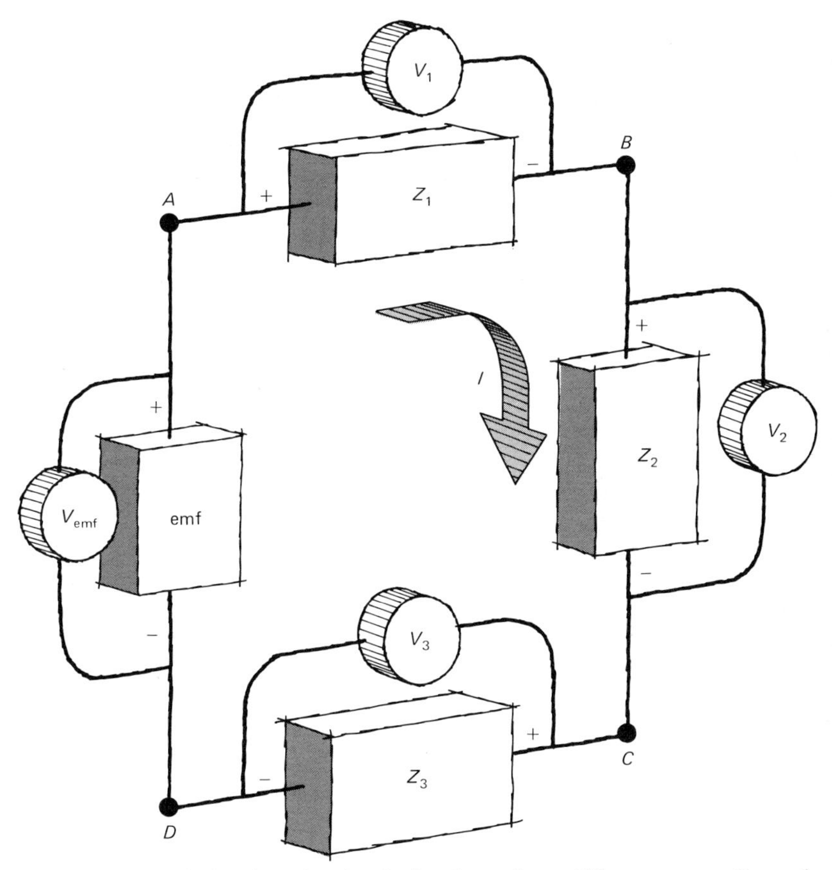

Figure 11-4-2 A simple series circuit showing voltage difference across the emf and circuit elements.

ginally shown in Fig. 11-2-1 is drawn again in Fig. 11-4-2. The plus and minus signs for each circuit element indicate the higher and lower potential-energy terminals of each element.

Consider the work done by electric fields on one coulomb of charge which moves around the circuit in the direction of I, along the path $ABCDA$.

$$W_{ABCDA} = V_1 + V_2 + V_3 + V_{emf} = 0 \qquad 11\text{-}4\text{-}1$$

where the V's in this equation stand for the voltage drops, or potential differences, across each circuit element, and across the emf. The work, W_{ABCDA}, done around a closed loop, must be equal to zero because the electric force is conservative. Notice that the voltage V_{emf} is negative, because the path from D to A goes from the $-$ to the $+$ sign of the source of emf. (Charges gain potential energy in traversing the emf.) The quantity $-V_{emf}$ is positive, often represented by the symbol $\mathcal{E}$, the voltage of the source of emf, and Eq. 11-4-1 can be rearranged in the form:

$$-V_{emf} = \mathcal{E} = V_1 + V_2 + V_3 \qquad \text{11-4-2}$$

If the source of emf is a battery, then $\mathcal{E}$ is the battery voltage. The equation above sets the sum of the voltage drops across the circuit elements equal to the battery voltage. The values of the voltages V_1, V_2, and V_3 generally depend upon the amount of current flowing throughout the circuit elements, while the battery voltage is approximately independent of the current (provided the current is not too great). In the next section we will discuss the current-voltage relationships for various kinds of circuit elements, in order to be able to study some circuits in detail. For a given value of $\mathcal{E}$, Eq. 11-4-2 can be used to determine I if the current-voltage relationships for the circuit elements are known.

11-5 CURRENT-VOLTAGE RELATIONSHIPS

The simplest circuit element is the *resistor*, and the relationship between voltage and current for an ideal resistor is called Ohm's law:

$$V = IR \qquad \text{or} \qquad I = V/R \qquad \text{11-5-1}$$

where R is a constant for a given resistor independent of V or I, called the *resistance.* The unit of resistance is the *ohm*, which stands for "volts per ampere." Equation 11-5-1 is shown graphically in Fig. 11-5-1, along with the symbol for a resistor. For a real resistor the current-voltage relation may in fact show a little curvature, but Eq. 11-5-1 is an accurate approximation in many cases.

The most common example of a circuit element which obeys Ohm's law is a metallic conductor. Although the conduction electrons in a metal can move freely under the influence of an electric field for distances much greater than the size of an atom, after some *mean collision time* the electron is likely to collide with one of the metal atoms so as to give up the energy it has gained while being accelerated by the field. The mean drift velocity of the electron is proportional to the product of this mean time and the value

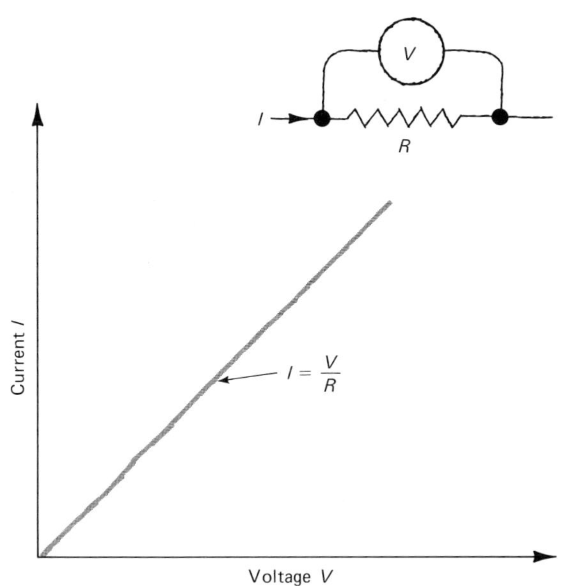

Figure 11-5-1 Current-voltage relation for an ideal resistor. The inset at top right shows the resistor symbol.

of the electric field:

$$v_{\text{mean}} = CE \qquad \text{(mean time)} \tag{11-5-2}$$

where C is some constant.

Recalling that the current is proportional to the mean drift velocity, we find therefore, from Eq. 11-5-2, that I is proportional to E. For a given conductor with voltage V across it, E is proportional to V, so that, finally, I is proportional to V. Calling the constant of proportionality between V and I the resistance R, Ohm's law results, $V = IR$.

When a metal is heated, the atoms vibrate with more energy and, as a result, the mean time that a conduction electron may drift between collisions is found to be reduced. As a consequence, the resistance rises. The current-voltage relationship for an incandescent light bulb is an extreme example of this phenomenon, shown in Fig. 11-5-2. Clearly this bulb is not an ohmic circuit element, mainly because its temperature changes so much as the current through it increases.

While the resistor and light bulb conduct identically when cur-

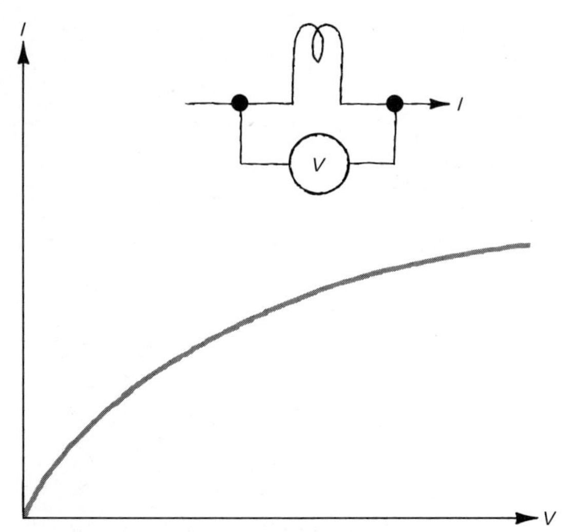

Figure 11-5-2 Current-voltage curve for an incandescent light bulb. Ohm's law is not obeyed, mainly due to the change of resistance with temperature.

rent flows through them in either direction, some circuit elements do not behave this way at all. A *diode*, originally an electron tube with two electrodes, now is more likely a semiconductor device with a current-voltage curve like that shown in Fig. 11-5-3*a*. Chapter 10 contains a detailed discussion of the semiconductor diode. In thinking of the main feature of a diode, we can approximate the curve as in Fig. 11-5-3*b*. This is the curve of an idealized diode, which conducts perfectly in one direction and not at all in the other. The diode symbol, also shown in Fig. 11-5-3, embodies an arrow pointing in the direction of easy current flow.

PROBLEM 5

A crystal radio is diagrammed below (Fig. P-11-5*a*). The semiconductor diode, in a very primitive form involving a lead sulfide crystal, gave the device its name back in the 1920s.

When an AM (amplitude modulated) radio station is received, the voltage between points A and B might vary with time as shown in Fig. P-11-5*b*.

Sketch a graph which shows how the current through the earphone varies with time, for the radio signal above (assume the diode is ideal, as in Fig. 11-5-3*b*). Suppose the earphone can only respond to the average current, averaged over a time of about 10^{-3} sec. Sketch the average in two cases: with the diode in the circuit, and with the diode replaced by a piece of plain wire.

Figure 11-5-3 Current-voltage curves for a real semiconductor diode, *(a)*, **and an ideal diode,** *(b)*.

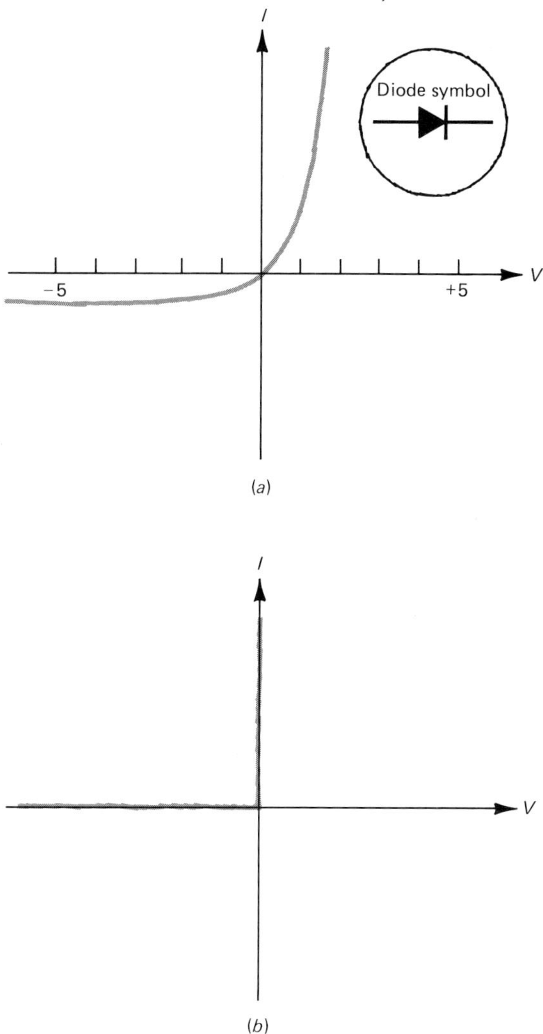

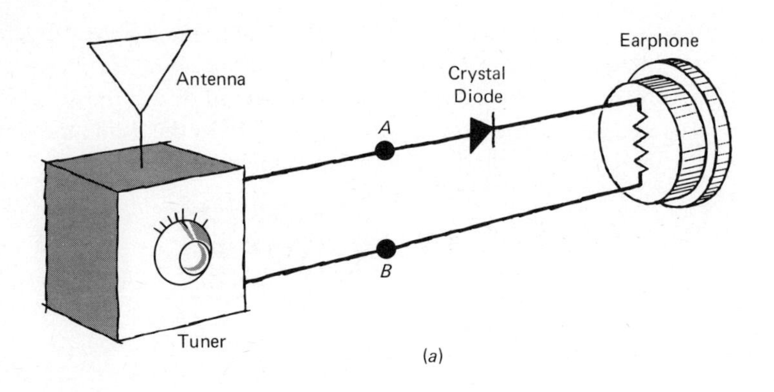

(a)

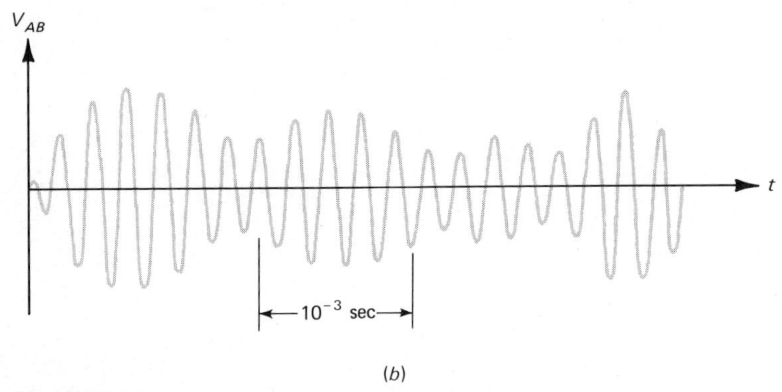

(b)

Figure P-11-5

There are also circuit elements such as the *variable resistor* whose current-voltage curves can be made to vary. The symbol of the variable resistor is shown in Fig. 11-5-4 as well as its current voltage characteristics. A standard way to build one is to provide a sliding contact which can move along a long length of wire. The long wire can be in the form of a coil wrapped around a form. The resistance of a wire of a given diameter is proportional to the length of the wire. Therefore, as the slider is moved, varying the length of wire between it and one end of the coil, the resistance between the two points changes.

The current-voltage curves in Fig. 11-5-4 are labeled according to the slider position in percent of the total length along the coil. The curve labeled 50 means that the slider is 50 percent of the way along the coil, leading to a resistance 50 percent of the total coil resistance. The family of curves shown in Fig. 11-5-4 indicates how the properties of the device depend on slider position, not by giving the results for all slider positions, but by giving a sampling from which one can reliably deduce how the device will perform with slider settings not on the graph. In the case of the ordinary variable resistor the family of curves is quite simple, but other devices, like the transistor, are characterized by more complex families of current-voltage characteristics.

The variable resistor is a two terminal electrical device (it has two places where one connects to it), whose properties depend upon a mechanical variable, the slider position. There are multi-terminal electrical devices in which the characteristics of one pair of terminals depend upon an electrical variable, e.g., the current through some other terminals. A very simple and important example of such a device is called a *relay*.

A basic relay has four terminals and is symbolically represented as in Fig. 11-5-5. The relay is a magnetically operated swtich, in which the current through a coil gives rise to a magnetic field, which exerts a force on an iron piece which is free to move. The

Figure 11-5-4 Current-voltage curves for a variable resistor. The numbers on the curves indicate the percentage of the total resistor which is in the circuit for each curve. The inset figure shows the variable resistor symbol, and the effective resistance for the slider in the position shown.

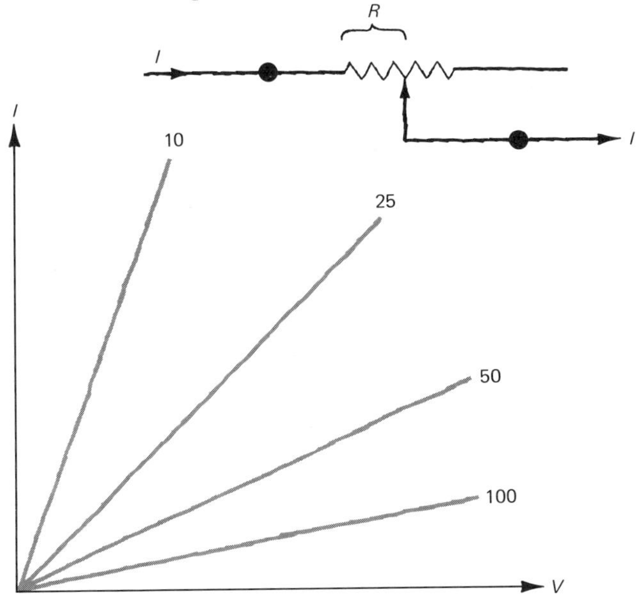

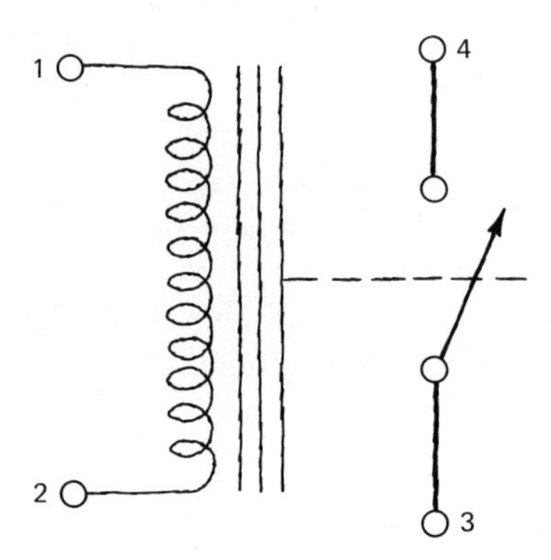

Figure 11-5-5 Symbolic diagram for a relay. Terminals 1 and 2 are for the coil. Current through the coil causes a magnetic field to act on the switch, closing the contacts between terminals labeled 3 and 4.

movement of the iron can then open or close the switch contacts. In a complicated relay there may be many switch contacts all operated by one coil.

The pair of terminals associated with the relay coil have a current-voltage curve like a resistor. At a certain value of coil current the magnetic field is sufficient to operate the relay contacts. The current-voltage curves for a given pair of contacts are like those of a switch which is either opened or closed, depending upon the value of coil current. The relay, and more sophisticated devices such as the transistor and vacuum tube, are examples of electrical control of the electrical properties of a device. Devices of this type provide the foundations of modern electronic technology.

11-6 DC CIRCUITS

In this section examples of circuits will be analyzed to show how the material discussed so far in the chapter is applied to real situations. The circuits will all be battery-powered, employing emf's with steady voltage (not varying with time). Such *steady emf* circuits are called dc for *direct current* circuits, in contrast with ac, or *alternating current* circuits, in which the emf oscillates like a sine curve.

EXAMPLE 1: AN AUTO STARTER RELAY

The starter motor of an automobile has to crank the engine, and therefore requires a large amount of electric power. At a battery voltage of 12 V, it requires a large current, perhaps more than

100 amperes. A relay makes it possible for a small current, controlled by a key-operated switch in most cars, to turn on the starter motor. In Fig. 11-6-1*a* two circuits are drawn with separate batteries, but they would actually both be powered by the same battery as in Fig. 11-6-1*b*. Such an arrangement is sometimes called

Figure 11-6-1 **A starter motor operated by a relay. In *(a)* the relay coil and starter motor are in separate circuits, while in *(b)* they are in parallel circuits connected to a single battery.**

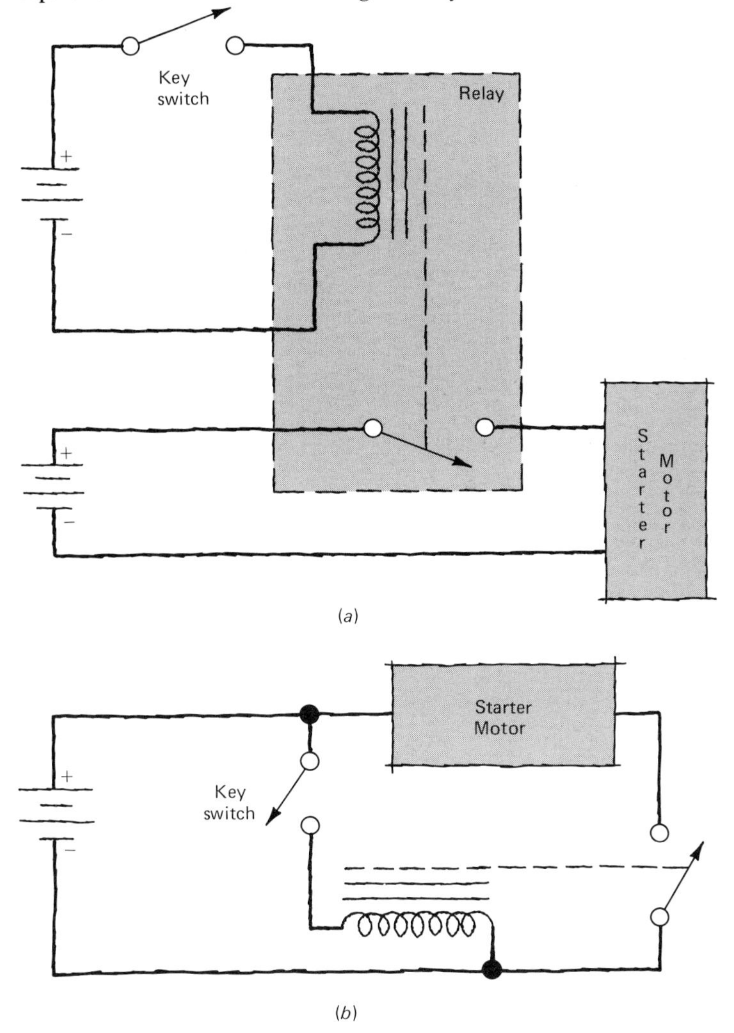

a *parallel circuit,* to emphasize the fact that there are two (or more) *parallel* paths that current can take rather than one *series* path.

Each branch of a parallel circuit like the one in Fig. 11-6-1*b* is independent of the others, provided the battery voltage is independent of the current through it. We always assume this as a first approximation, and in fact parallel circuits are often shown as in Fig. 11-6-2, where the battery is omitted and replaced by two wires labeled 0 V and +12 V. The significance of these labels is the voltage difference of 12 V between the two wires, which implies that the voltages across the circuit elements of each individual parallel branch must add up to 12 V.

We could have labeled the battery leads 0 and −12 instead of +12 and 0, and the difference would be the same. Often the choice of which lead is called zero volts is a matter of taste, but sometimes one wire or the other is connected to some massive conductor, like the chassis of the car in this example. In that case, the lead connected to the massive conductor is usually called *ground* and is conventionally taken to be at 0 V. Either terminal of the battery could be connected to the car chassis, with no change in the circuit analysis, but the label on the *ungrounded* terminal of the battery would be +12 or −12 V, depending upon which one was grounded.

Figure 11-6-2 The circuit of Fig. 11-6-1*b*, shown with the battery replaced by two leads at +12 and 0 V.

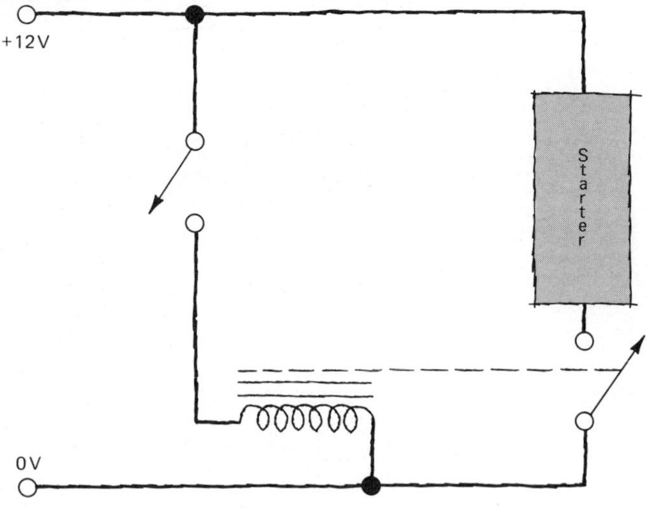

EXAMPLE 2: AN ELECTRIC STOVE UNIT

For this example we will be interested in the heating of a resistor by electric power. The voltage across a circuit element is the electrical work done on a unit charge passing through it, while the current is the charge flow per second. Thus, the product IV is the total electrical work done per second on all the charges passing through the element, which, by conservation of energy, equals the work per second delivered by the charges to the circuit element. *Work per second is defined as power,* so the formula for electrical power is

$$P = I \times V \qquad \text{watts} \tag{11-6-1}$$

where the unit joule/sec is more commonly called the *watt.* Note that Eq. 11-6-1 applies to *any* circuit element, not just to a resistor.

PROBLEM 6

Given two batteries with voltages 4.0 and 6.0 V and a single resistor with resistance 10 ohms, what powers can be dissipated in the resistor by the batteries individually. How can you combine the batteries so as to effectively obtain an emf of either 2 or 10 V? What power dissipations are then made possible?

For a resistor, electric power is converted to heat when the charges collide with atoms of the resistor. In this problem we will assume, as a simple first approximation, that the resistance of the heating elements of an electric stove doesn't change very much over their operating temperature range. Ohm's law will then be used to study how two resistors and some switches can provide various heat outputs.

Suppose an electric stove has two separate heating elements for one "burner," with resistances R_1 and R_2. Figure 11-6-3 shows the four possible circuits which are different electrically. If the circuits were actually for a stove, then the source of power would be an alternating emf, supplied by the power lines. From a heat generation point of view, this emf is equal to a steady battery voltage of about 115 V, which we will simply call V_0 in this example.

Consider the heat generated by the stove for the four different circuits shown in Fig. 11-6-3, assuming the voltage V_0 remains fixed. For circuit (a), the power is easily found in terms of V_0 and R_1:

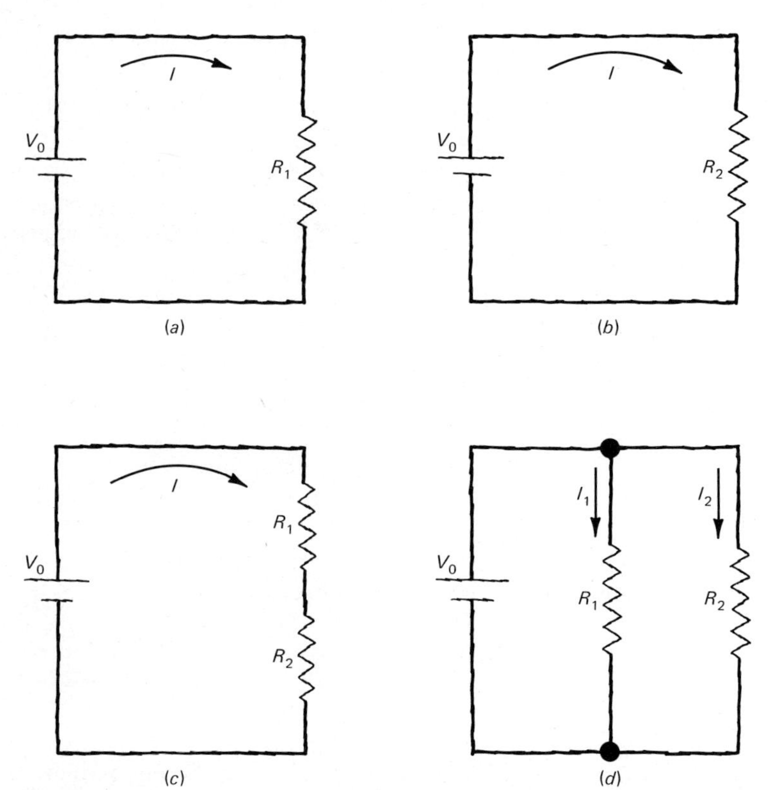

Figure 11-6-3 Four electrically different arrangements involving two different resistors R_1 and R_2 and an emf V_0.

$$P_a = V_0 I$$

and from Ohm's law, $I = V_0/R_1$. Therefore,

$$P_a = \frac{V_0^2}{R_1} \qquad 11\text{-}6\text{-}2$$

Similarly,

$$P_b = \frac{V_0^2}{R_2} \qquad 11\text{-}6\text{-}3$$

The power P_d can be recognized as simply the sum of the preceding two powers:

$$P_d = P_a + P_b \qquad \text{or} \qquad P_d = V_0^2\left(\frac{1}{R_1} + \frac{1}{R_2}\right) \qquad 11\text{-}6\text{-}4$$

Finally, for the series combination, circuit (*c*), the total power to the two resistors will be

$$P_c = V_1 I + V_2 I = (V_1 + V_2)I \qquad 11\text{-}6\text{-}5$$

where V_1 and V_2 are the individual voltages across the resistors, and I is the same current for both (since it is a series circuit).

To solve for P_c in terms of V_0, R_1, and R_2, we have three relations:

$$V_0 = V_1 + V_2 \qquad 11\text{-}6\text{-}6$$

which follows from equating the sum of the voltage drops to the emf, and two Ohm's law equations:

$$V_1 = IR_1 \qquad V_2 = IR_2 \qquad 11\text{-}6\text{-}7$$

Substituting in Eq. 11-6-6 from Eq. 11-6-7:

$$V_0 = IR_1 + IR_2 = I(R_1 + R_2)$$

and rearranging:

$$I = \frac{V_0}{R_1 + R_2} \qquad 11\text{-}6\text{-}8$$

Equation 11-6-8 is in the form of Ohm's law applied to a resistor with $R = R_1 + R_2$. We have thus found that resistors in series behave like a single resistor with resistance equal to the sum of the individual resistors. From Eqs. 11-6-5 and 11-6-6,

$$P_c = V_0 I \qquad 11\text{-}6\text{-}9$$

Substituting for I the value found in Eq. 11-6-9:

$$P_c = \frac{V_0^2}{R_1 + R_2}$$

Now we can summarize the power converted to heat in the four possible circuits:

$$P_a = V_0^2/R_1$$
$$P_b = V_0^2/R_2$$
$$P_c = V_0^2/(R_1 + R_2)$$
$$P_d = \frac{V_0^2}{R_1} + \frac{V_0^2}{R_2} = V_0^2\left(\frac{1}{R_1} + \frac{1}{R_2}\right)$$

Note that P_c and P_d are the lowest and highest values of P, even though the formulas look similar. A simple case is $R_1 = R_2$, which

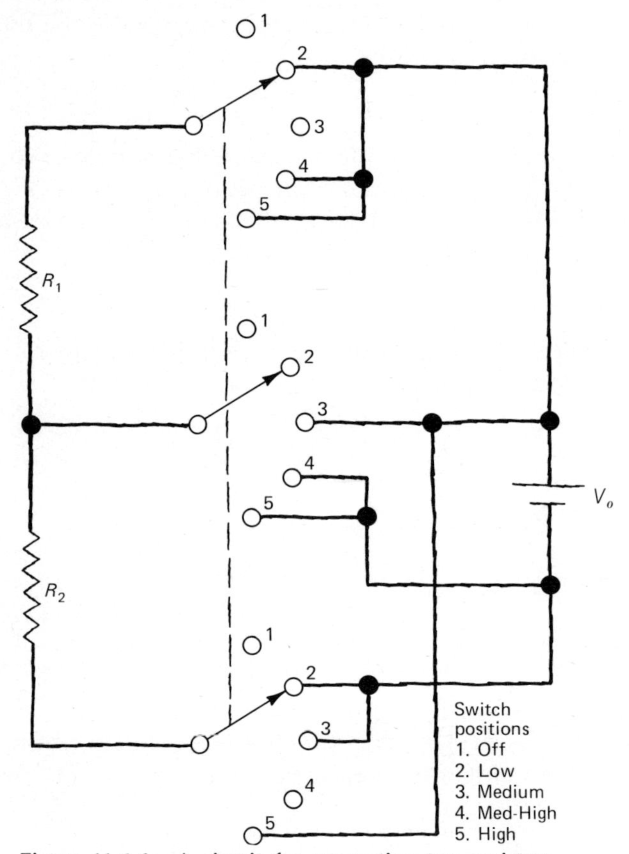

Figure 11-6-4 A circuit for connecting two resistors so as to dissipate successively greater power. The switches are ganged so that all three are connected so that their contactors move together when a single knob is rotated.

leads to three values of power in the ratios 1:2:4. For $R_1 = 2R_2$ there are four powers in the ratios 1:1.5:3.0:4.5.

It is interesting to consider how one switch could be used to change the circuit from configuration *c* to *b* to *a* to *d,* which would give steadily larger powers. The type of switch which would be easiest to use is a rotary switch with several contactors *ganged* so they move together. Figure 11-6-4 shows one way, which may be easy to understand, but probably is not the one which uses the fewest parts.

PROBLEM 7

For two resistors in parallel, show that an effective resistance R is given by

$$\frac{1}{R} = \frac{1}{R_1} + \frac{1}{R_2}$$

where R determines the current in the circuit shown in Fig. P-11-7 from Ohm's law, $I = V/R$.

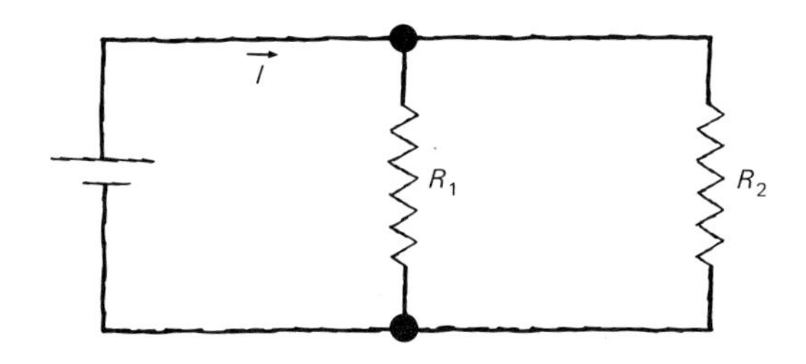

Figure P-11-7

PROBLEM 8

A variable resistor usually has the terminals at either end of the coil, plus the terminal of the slider, accessible, as shown in Fig. P-11-8. If a 6-V battery is connected to terminals 1 and 2, how does the voltage measured between terminals 3 and 2 change as the slider is moved.

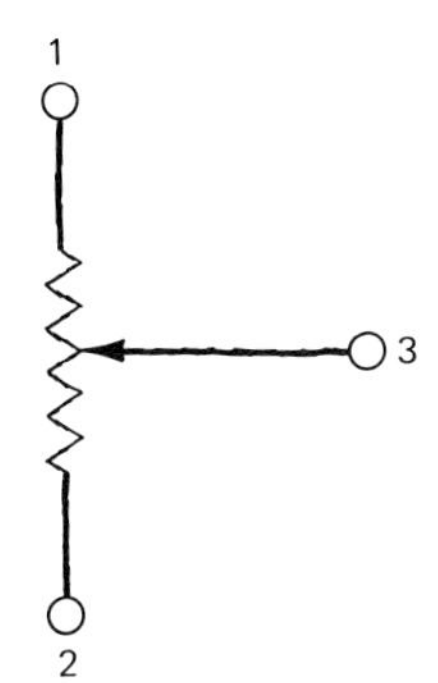

Figure P-11-8

PROBLEM 9

A small relay has a coil with a resistance of 100 ohms. When the coil voltage exceeds 10 V the relay operates, closing a pair of contacts. Calculate the power needed to operate the relay. If the contacts can carry currents up to 2 amperes, flowing in a circuit powered by a 100 V emf, how much power can be controlled by the relay?

PROBLEM 10

A quiz show is a contest between two teams with four members on each team. Each one has a switch he can close when he knows the answer. Draw a circuit utilizing switches and relays that performs the following functions.

1 A single red light is lit when any member of team A closes his switch.
2 A single green light is lit whenever any member of team B closes his switch.
3 Whichever light is lit first, the other is prevented from being lit, thereby eliminating the need for a sharp eyed referee, and making a tie impossible.

EXAMPLE 3: A BUZZER

A very simple relay circuit can be used to make a buzzer. It is shown in Fig. 11-6-5. When the push-button switch is closed, current can flow momentarily through the relay coil, passing through the relay's own "normally closed" contact. However, after the current flows for a very short time, this contact will open as the magnetic field of the relay increases. When the contact opens, the magnetic field decreases, and the contact will then close, which again turns on the current in the relay coil, which again opens the contact, and so on. This kind of thing is not too good for relays which are intended for more calm operation, but buzzers and doorbells are examples of special relays designed to function in this way.

The buzzing circuit is one of the simplest examples of a class of circuits called oscillators. The frequency of oscillation of the buzzer can be controlled by the mechanical properties of the relay, while electronic oscillators can be built whose frequencies are controlled by electric circuit elements. Electric organs and other

Figure 11-6-5 A relay connected as a buzzer. When the push button is closed, the relay buzzes.

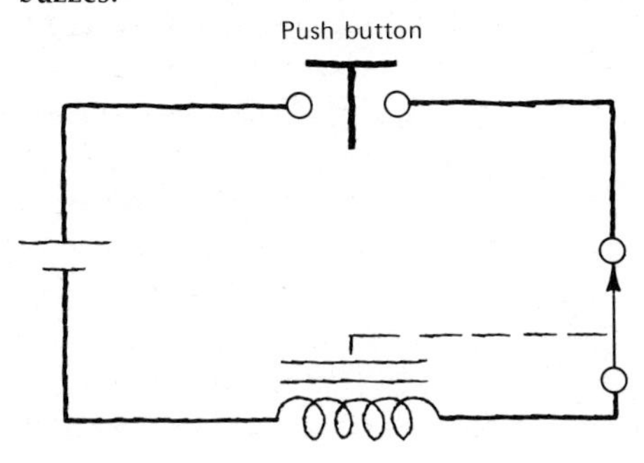

more esoteric electronic instruments, such as the Moog synthesizer, are examples of devices in which electrical oscillations are converted to sound by means of a loudspeaker. When electric currents change rapidly enough, they are effective for broadcasting electromagnetic waves, so that high-frequency oscillators are used to produce radio signals. The electric currents in a buzzer often change suddenly enough to generate radio waves, which are picked up as static on a receiver.

11-7 LOGIC CIRCUITS

The branch of modern electronics which is simplest to understand, and yet probably the most fascinating, is the use of semiconductor devices in the kind of circuits used in digital computers. These circuits utilize transistors as electronic switches, which act very much like relays. However, the small size of the transistor, and its low power needs, makes it possible to put literally hundreds of these electronic switches into a space smaller than an aspirin tablet. Furthermore, the transistor is not only small, but fast. Its moving parts are fleetfooted electrons, and as a result it can act as a switch whose contacts can open or close in a time as short as 10^{-9} sec.

In Chap. 10, the transistor was discussed as an example of solid state physics. Here we will consider its properties purely as a circuit element, without discussing why it behaves the way it does. The transistor has three terminals, the *collector*, *base*, and *emitter*, labeled c, b, and e on the transistor symbol in Fig. 11-7-1. This figure shows a transistor connected so that a battery with voltage V_{cc} (*cc* for *collector circuit*) tends to establish the current labeled I_{ce} which goes through a resistor R and through the transistor from collector to emitter. Also, a battery labeled V_{bc} (*bc* for *base circuit*) is connected so as to cause a current I_{be} to flow through a resistor R_b and through the transistor from base to emitter.

The properties of a transistor connected as shown in Fig. 11-7-1 are conveniently displayed as a set of curves like those in Fig. 11-7-2. These curves, for a typical junction transistor, indicate how the collector-to-emitter current, I_{ce}, depends on collector-to-emitter voltage, V_{ce}, for various values of base-emitter current, I_{be}. (A similar sort of graph, Fig. 11-5-4, described the variable resistor in Sec. 11-5.) In the region for V_{ce} greater than about 1 V, I_{ce} depends mainly on the value of I_{be}, such that, for the example

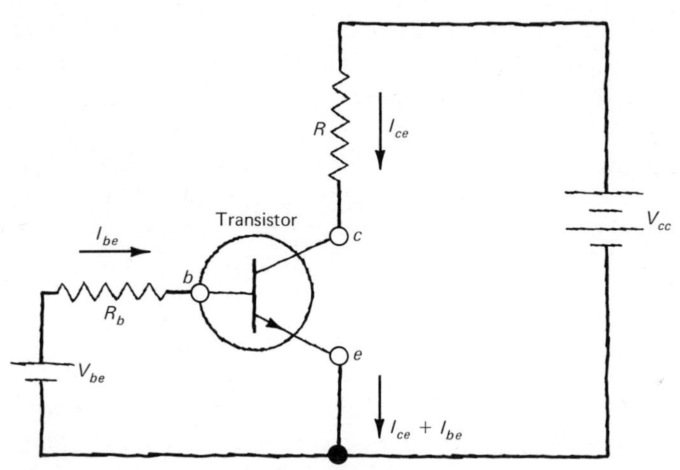

Figure 11-7-1 A transistor connected so that a small base current, I_{be}, controls a large collector current, I_{ce}.

shown, $I_{ce} \approx 50 I_{be}$. The usefulness of the transistor comes mainly from the fact that a small base current (I_{be}) can control a much larger collector current (I_{ce}).

For a large range of I_{ce} and V_{ce} the collector and base currents

Figure 11-7-2 Characteristic curves for a typical junction transistor. Note that for V_{ce} greater than about 1 V, the collector current is about 50 times the base current, which is given in milliamperes (10^{-3} amp) next to each curve.

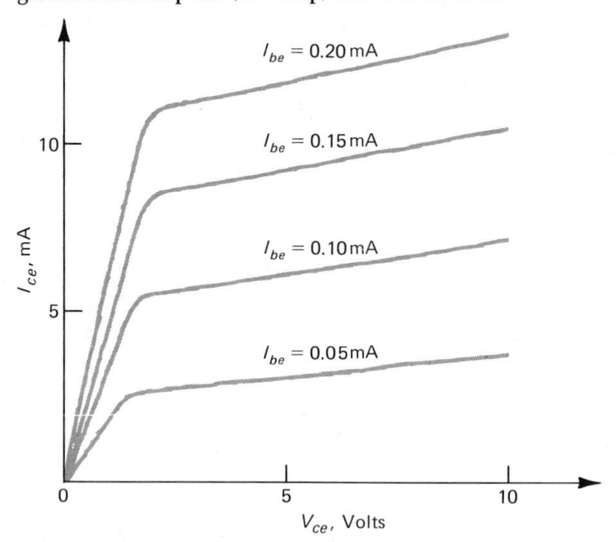

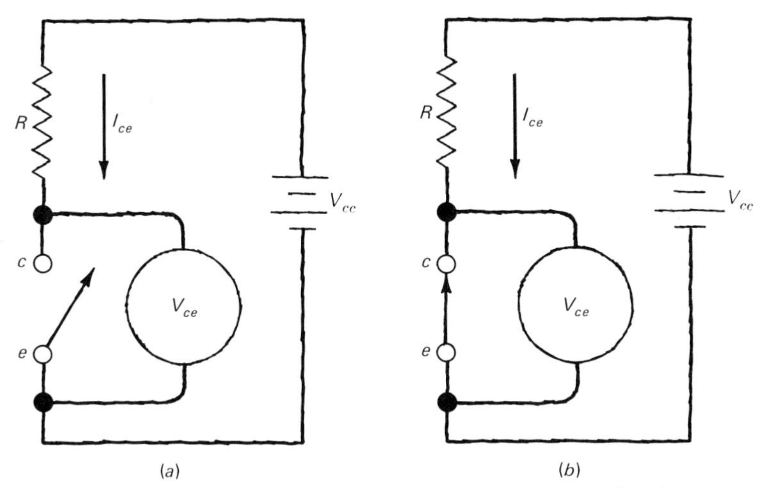

Figure 11-7-3 The collector circuit part of Fig. 11-7-1, assuming the transistor were an ideal switch.

are quite accurately proportional, which is very useful in producing what are called *linear amplifier circuits*, for radio receivers, phonograph amplifiers, and other types of electronics. However, for logic circuits, the transistor is used as a switch and operates outside this range. When the transistor in Fig. 11-7-1 acts as a switch, the collector-emitter current and voltage should duplicate as nearly as possible the two ideal cases shown in Fig. 11-7-3.

Case *a* Switch *opened* $I_{ce} = 0$
$V_{ce} = V_{cc}$

Case *b* Switch *closed* $I_{ce} = V_{cc}/R$
$V_{ce} = 0$

Figure 11-7-4 again shows the same transistor curves as Fig. 11-7-2, with the current and voltage combinations for case *a* and case *b* above also indicated. Sample values, $V_{cc} = 5$ V and $R =$ 1,000 ohms, have been assumed. The transistor is an essentially perfect open switch if we reduce I_{be} to zero. However, for closed-switch operation we approximate the requirement $V_{ce} = 0$ by $V_{ce} \approx$ 0.5 V. For $V_{cc} = 5$ V, this small voltage drop across the transistor is a good approximation to an ideal closed switch. The region of closed-switch operation, with $V_{ce} \lesssim 0.5$ V, is called the *on condition* of the switching transistor.

In order to turn the transistor on it is necessary to furnish a sufficiently large base-emitter current. However, at $V_{ce} = 0.5$ V,

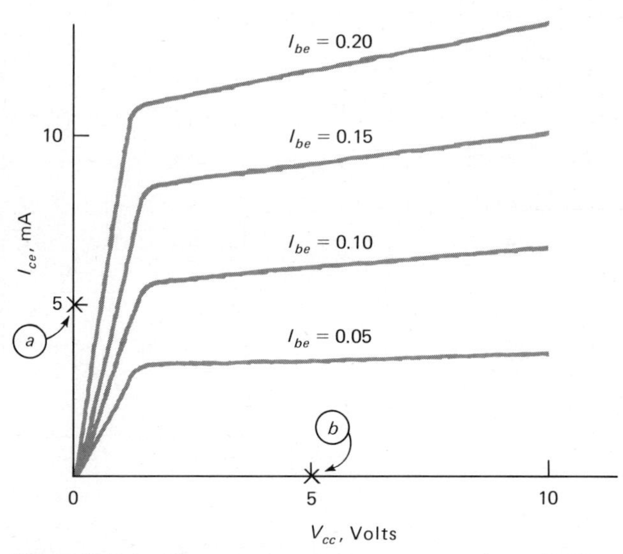

Figure 11-7-4 The same transistor curves as shown in Fig. 11-7-2, with the operating points for an ideal switch indicated by points marked *x* and labeled *(a)* and *(b)*.

the ratio of collector current to base current is still much greater than 1, perhaps about 10 or 20. Thus, the transistor can be turned on comparatively easily, by supplying a relatively small base current. In particular, the two switching states *on* and *off* can be reached by using one transistor connected as in Fig. 11-7-1 to drive a similar one, as shown in Fig. 11-7-5*a*. This works because of the voltage-current characteristics for the base-emitter circuit of a transistor, which so far have not been discussed.

The base-emitter current of a transistor flows through a silicon semiconductor diode within it, in the forward direction. As a result, the current is very small until the applied voltage V_{be} is about 0.8 V (see Fig. 11-5-3), and above this voltage the current increases very rapidly. As a result, when the first transistor switch is on, with $V_{ce} \lesssim 0.5$ V, the second has $V_{be} \lesssim 0.5$ V and is off.

When the first transistor is off, then the base-emitter current to the second is determined by a voltage drop of about 4 V ($V_{cc} - V_{be}$) across the resistor R of the first transistor in series with R_b for the second. Typically, like R, R_b might be about 1000 ohms. As a result, there will be a base current of about 2 mA into the second transistor. Only a fraction of a milliampere would suffice to turn on the second transistor, so it is thus turned on very effectively when the first transistor is off. In fact, a single transistor can control the switching of a large number of transistors like itself, instead of just one as shown in Fig. 11-7-5.

In Fig. 11-7-5*b* the symbol —◁▷— is used to represent each switching transistor and its resistors R_b and R. This is a

Figure 11-7-5 A transistor connected as in Fig. 11-7-1 driving another one. A switch is used to control the first transistor. Each transistor functions as a logical *inverter,* **symbolized by a triangle with a little circle at the input, as in** *(b).*

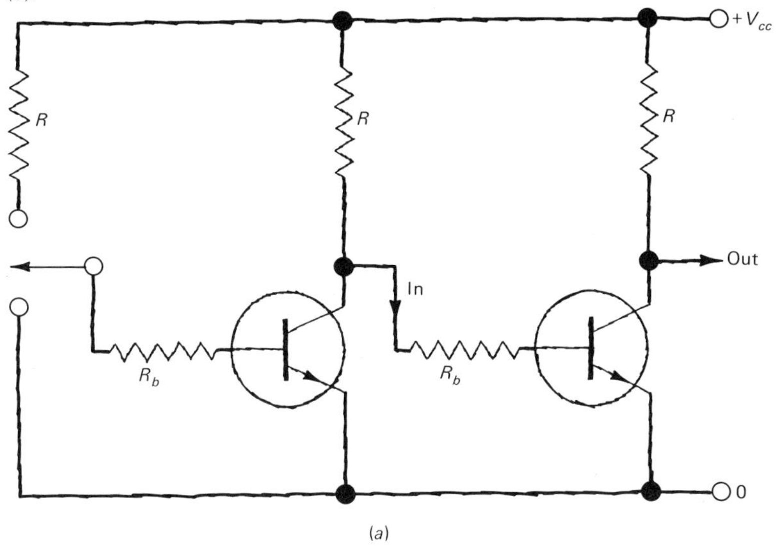

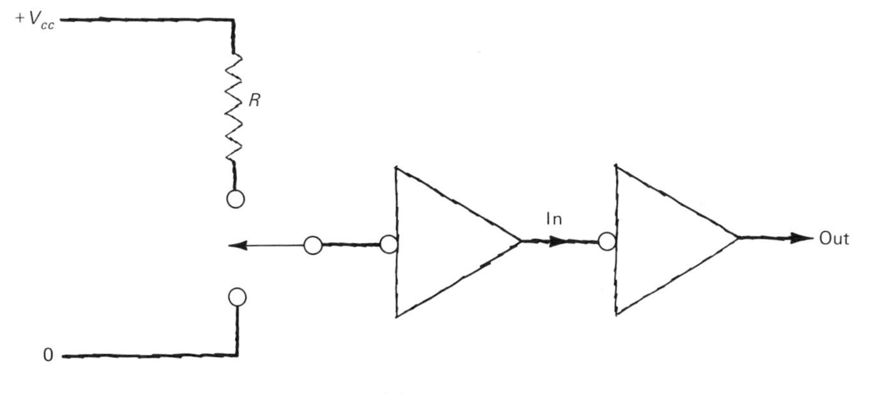

standard electronic symbol for an *inverter*. It describes the circuit we have been discussing because a high-input voltage, at the base, produces a low-output voltage, at the collector, and vice versa. Although our inverter consists of a particular junction transistor circuit, its important attributes for logic electronics are summarized below and can be realized in a variety of ways.

Properties of the inverter:

1. It has two output states, high (H) and low (L) voltages, corresponding respectively to input states L and H.
2. It can be driven by a similar inverter.
3. It can drive the inputs of several similar inverters.
4. It can form an OR *circuit* with one or more similar inverters (to be discussed below).

Figure 11-7-6 Two inverters with their outputs connected. The *truth table* shows the output state for the four possible input combinations.

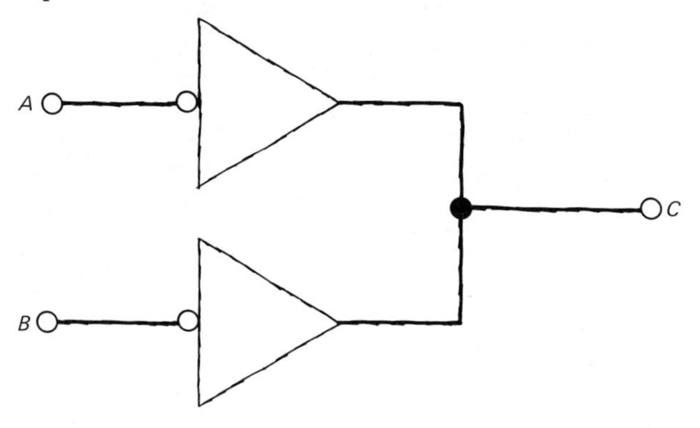

Inputs A	Inputs B	Output C
L	L	H
L	H	L
H	L	L
H	H	L

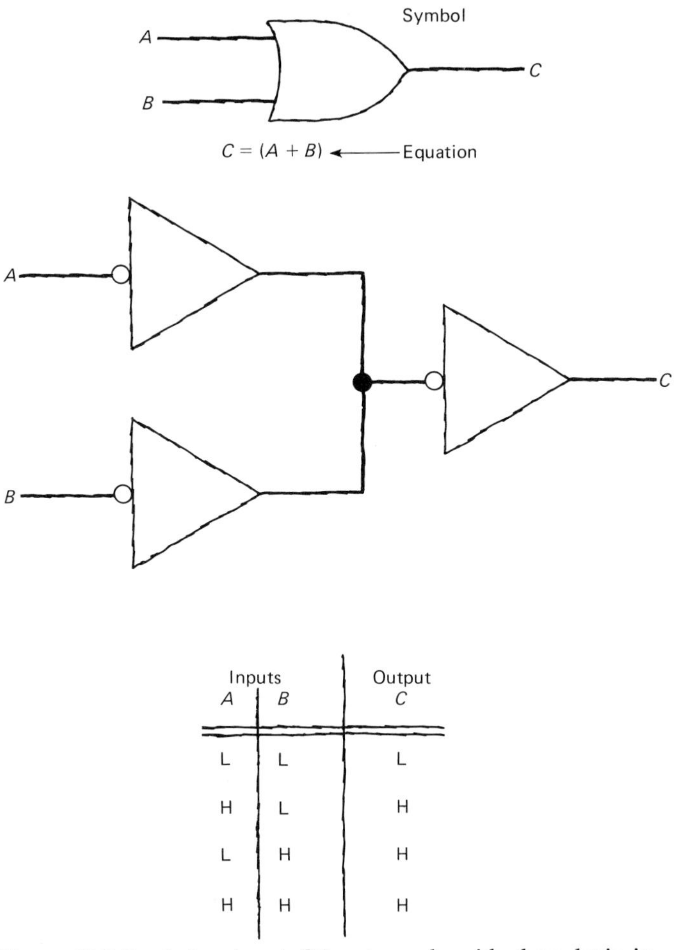

Inputs A	Inputs B	Output C
L	L	L
H	L	H
L	H	H
H	H	H

Figure 11-7-7 A two-input OR gate made with three logic inverters. The logic equation for the gate and its truth table summarize its properties.

Perhaps surprisingly, with this simple inverter as the only basic building block, all the logic circuits needed for a *complete* digital computer, or for *any* electronic control device, can be constructed. To illustrate, we will first combine inverters to produce two other larger building blocks, an OR *circuit* and an AND *circuit.* Figure 11-7-6 shows two inverters with their outputs connected together (i.e., a wire connects the two collector terminals), along with a little table called a *truth table* which summarizes the logical properties of the combination. In the table, H stands for "high" voltage level and L stands for "low." The truth table shows output L for either input H. For our inverter, if two outputs are connected, *either* transistor can pull the output voltage from H to L, without interference from the other.

The definition of an OR *circuit* is one which produces an output signal for any one, or more, of a number of input signals. We will define an input signal as an H voltage level at an input terminal. The truth table for Fig. 11-7-6 would correspond to an OR circuit if the outputs were inverted, so that a simple modification, shown in Fig. 11-7-7, is necessary. This figure is a *two-input* OR *gate* in logic circuit terminology. The truth table indicates the OR function, as does the little equation $C = A + B$, which translates as: "Output C is in its H state if either input A *or* input B is in its H state, or both are." The $+$ sign signifies an OR.

The OR gate has a symbol, shown in Fig. 11-7-7, which tells the number of inputs by the number of input wires and the direction of logic signal flow by its arrowlike shape. The logical OR function can easily be performed for more than two inputs by combining more than two inverters into a single OR gate, or by using combinations of more than one OR gate.

PROBLEM 11

Suppose OR gates with a maximum of four inputs are available. Draw a circuit for producing an output for any of 16 inputs.

PROBLEM 12

The circuit shown in Fig. P-11-12 has two inputs, S and R, and two outputs Q and T. Beginning with both inputs L, Q is H and T is L.

Figure P-11-12

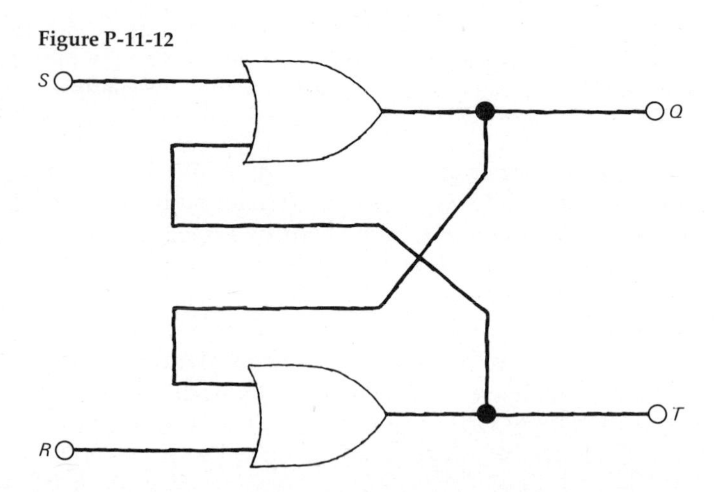

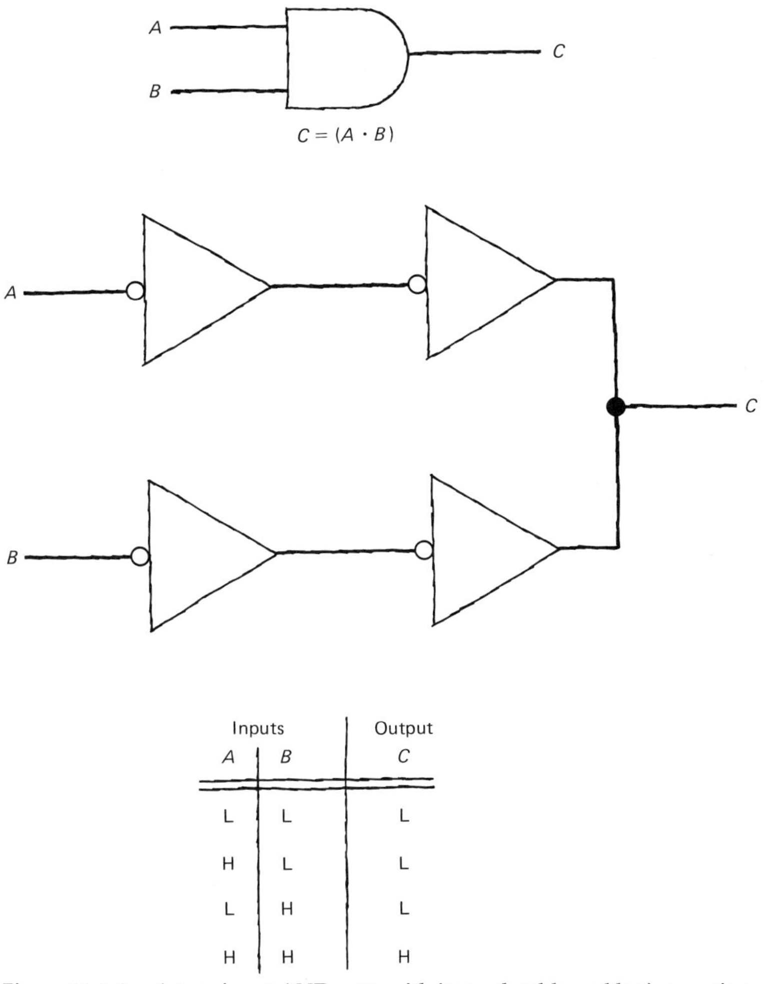

Inputs		Output
A	B	C
L	L	L
H	L	L
L	H	L
H	H	H

Figure 11-7-8 A two-input AND gate, with its truth table and logic equation, made with four logic inverters.

> Let S be made momentarily H with R still L. Describe the outputs during the time S is high and afterward. Make R momentarily H, while S is L, and describe what happens at the outputs. (Hint: This circuit is called a *flip-flop*, or a *latch*.)

The second logic function for which it is convenient to have a building block is the AND *function*, which requires that an output signal exists if, and only if, *all* inputs have signals. Figure 11-7-8 shows a *two-input* AND *gate* made up of inverters, with truth table and the gate symbol. The equation $C = A \cdot B$ translates as "Output C is high if inputs A *and* B are high."

The AND and OR logical functions combined with the inverting function can be used to accomplish a great variety of tasks, which basically involve producing one or more output signals for specified combinations of input signals. Electronic control circuits, electronic counters, and digital computers all perform functions which can be reduced to this sort of process. As an example of the use of logic circuits we will discuss addition of binary numbers.

PROBLEM 13

Design a burglar alarm which detects the opening of a garage door and closes a relay contact if, and only if, the front and rear house doors are locked. Assume suitable electrical signals are available to signify the states of the various doors.

Any three digit decimal number, 378 for example, can be broken up as follows:

$$\begin{array}{rr} 8 \times 10^0 = & 8 \\ +7 \times 10^1 = & +70 \\ +3 \times 10^2 = & \underline{+300} \\ & 378 \end{array}$$

Each digit of the three digit number stands for a contribution of 0 to 9 times the appropriate power of ten. This describes a *base 10* number system. A *base 5* number system would, for example, only allow each digit to range from 0 to 4, standing for 0 to 4, times the appropriate power of 5. The number 378 doesn't exist in a base 5 system, since only numbers 0 to 4 are used in each digit. Let's look at the number $(134)_5$, where the subscript specifies the base 5 system:

$$\begin{array}{rrrcr} (134)_5 = & 4 \times 5^0 = & (4)_5 & & (4)_{10} \\ & +3 \times 5^1 = & (30)_5 & = & (15)_{10} \\ & +1 \times 5^2 = & \underline{(100)_5} & = & \underline{(25)_{10}} \\ & & (134)_5 & & (44)_{10} \end{array}$$

Clearly, we can count as high as we want with a base 5 system, but it does tend to require more digits for a given number than our usual base 10 does. In electronic digital computers, we are willing to drop down to a *base 2*, or *binary*, system, with many digits, in order to gain the enormous simplification that results because each digit now stands simply for either *0 or 1* times the appropriate power of two.

An example of a three-digit base 2 number is $(101)_2$. Again decomposing the contributions of the three digits:

$$\begin{aligned}(101)_2 = 1 \times 2^0 &= \quad (1)_2 = (1)_{10} \\ 0 \times 2^1 &= \quad (00)_2 = (0)_{10} \\ 1 \times 2^2 &= \underline{(100)_2} = \underline{(4)_{10}} \\ & \quad (101)_2 \quad (5)_{10}\end{aligned}$$

It may seem that 101 is a hard way to write 5, but it's the easiest way when dealing with electronic circuits that have just two states, *H* and *L*. A switching circuit can easily represent the value of a binary digit, with 1 for *H* and 0 for *L*.

Binary addition of two numbers takes place like decimal addition, except that a sum of 2 or more results in a carry, rather than 10 or more. For example, adding 5 + 3 in binary numbers:

$$\begin{array}{rr} \underline{111\ } & \leftarrow \text{carries} \\ 101 & 5 \\ +\ 011 & \underline{3} \\ \hline 1000 & 8 \end{array}$$

The answer is $1 \times 2^3 = 8$, which checks. Each time the sum $(1 + 1)$ occurs, it generates a carry, since 2 is the base of the counting system. Another example is more varied:

$$\begin{array}{rr} \underline{111\ } & \leftarrow \text{carries} \\ 111 & 7 \\ +\ 011 & \underline{3} \\ \hline 1010 & 10 \end{array}$$

Note that the answer is $1 \times 2^3 + 1 \times 2^1 = 10$, which checks. For the second digit the sum $1 + 1 + 1$ generates a carry plus a 1, just as in base 10 arithmetic $8 + 7$ generates a carry and a 5 in its own digit.

PROBLEM 14

Use binary arithmetic to *multiply* the two binary numbers 101 and 011. Proceed by analogy with decimal multiplication of three-digit numbers.

In order to electronically perform binary addition of two numbers, *A* and *B*, all that is needed is a basic circuit for adding the three inputs of one digit: the value of that digit, either 0 or 1, from either *A*, *B*, or the carry from the adjacent lower order digit. A truth table for this circuit, called a *full adder*, is as follows:

INPUTS			OUTPUTS	
A	*B*	CARRY	SUM	CARRY
0	0	0	0	0
1	0	0	1	0
0	1	0	1	0
0	0	1	1	0
1	1	0	0	1
0	1	1	0	1
1	0	1	0	1
1	1	1	1	1

Since there are three possible inputs, each with two possible values, there are eight rows in the truth table. However, the answer has only four possible values. The column "Sum" means the binary number in the answer for the digit, or *bit*, being added, while the column "Carry" is passed on to the *carry input* of the next higher bit.

Figure 11-7-9 shows the complete circuitry for adding two 3-bit binary numbers, starting with switches to specify the numbers *A* and *B* and ending with the answer in the form of a 4-bit binary number indicated by lights. The circuit with the truth table discussed above is shown in Fig. 11-7-9 simply as a block labeled "adder," with three inputs and two outputs. In Fig. 11-7-10 one way of implementing a full adder, with two combinations of AND, OR, and inverter gates, called *half adders*, is shown.

PROBLEM 15

Use OR gates, AND gates, and inverters as needed, and create a single circuit which functions as a full adder, obeying the truth table at the top of this page.

The simple inverter, and the circuits which perform the logical AND and OR functions, are extremely powerful electronic building blocks. Furthermore, modern semiconductor technology can now produce whole computers, containing thousands of AND and OR gates, memory elements, etc., all in a space of cubic inches. While we have begun this discussion with a single transistor connected as an inverter, logic circuitry is now usually made with many transistors on one tiny semiconductor chip, plus all the necessary resistors and wiring to make a complex cir-

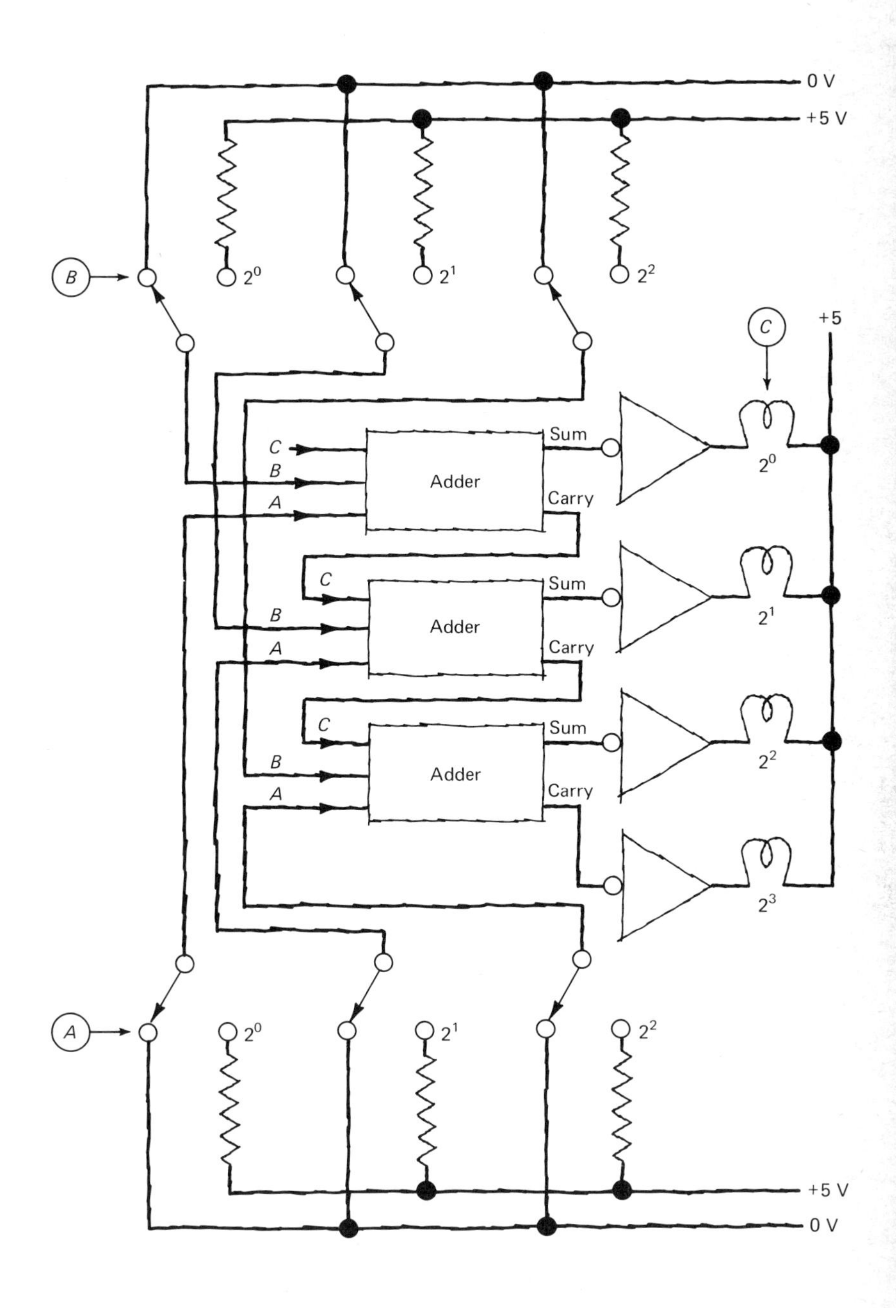

Figure 11-7-9 A three bit (or three digit) binary adder, $C = A + B$. The inputs A and B are set with switches. The output C appears in lights.

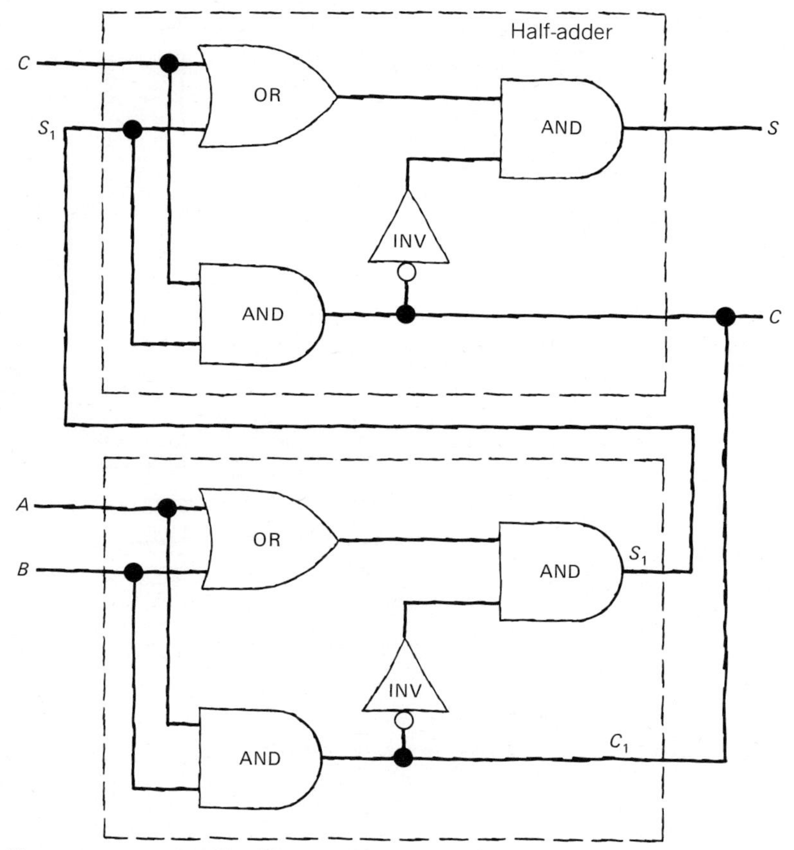

Figure 11-7-10 A full adder, producing sum, S, and carry, C, outputs from inputs A, B, and C. Each circuit enclosed in a dashed box is called a half adder, and sums two inputs. The two half adders are easy to connect as a full adder, and each half adder can be seen to consist of one OR gate, one inverter, and two AND gates.

cuit. Also, while the junction transistor has been our starting point, newer semiconductor technologies are in use, and may dominate the electronic logic circuits of the future.

SUMMARY

Electric current flows in a circuit as a result of the action of an *emf.* The current flows through *circuit elements,* carried by electrons or ions of either sign of charge, but is always assumed to be a *conventional current* of *positive charges.*

The two general laws used in analyzing the behavior of circuits are:

1 The net current into any point of a circuit must be zero.
2 The sum of the voltage differences around any closed path must be zero.

With the aid of these laws plus a knowledge of the *current-voltage relationships* of the circuit elements any circuit can be analyzed.

For a *resistor,* the current-voltage relationship is simply given by *Ohm's law*:

$$V = IR$$

Other circuit elements have more complex relations. The *power* delivered to any circuit element is given by

$$P = IV$$

Logic circuits can be built using transistors as switches. From a simple *inverter* circuit, combinations such as AND and OR circuits, with characteristic *truth tables* appropriate to their names, can be built. From these, plus inverters, a great variety of logic functions can be performed electronically, including *digital computation.*

OPTICS

12-1 INTRODUCTION

The fundamental principles of optics have almost all been introduced in Chap. 6, on light. However, there we were primarily interested in exploring the basic nature of light, while here we will study some of the interesting aspects of applied optics. For example, we will see how the nature of light makes possible optical instruments as varied as the telescope, the diffraction grating, and the laser.

In Chap. 6 we discovered the wave-particle duality of light: the phenomena of interference (as exhibited in Young's two-slit experiment) and of light quanta (as in the photoelectric effect). Most of the material of this chapter can be characterized as *classical optics*, meaning optics which can be understood in terms of light as an electromagnetic wave. Only in the final section, on the laser, will quantum effects play a significant role. However, even within the realm of classical optics we will find there is a duality which has been touched on in Chap. 6. Sometimes, as in the behavior of lenses and prisms, we can best think of light propagating simply as straight-line rays. Other times, as in the phenomena like Young's interference experiment, we must recognize that light is a wave motion.

To help in clarifying the distinction between *ray optics* and *wave optics*, consider the two views of water waves seen from above, shown in Fig. 12-1-1, which might, for example, be an opening in a sea-wall at the entrance to a harbor. In Fig. 12-1-1*a*, the gap is only about twice as wide as the distance between wave crests, which we call the wavelength. For such a narrow gap, the waves which pass through spread out afterward in circles. In contrast, Fig. 12-1-1*b* represents the case of a gap much wider than a wavelength. Here, the waves which pass through still have essentially straight wavecrests, with perhaps a bit of curving at the edges.

For the two cases illustrated in Fig. 12-1-1, rays are also drawn in the direction of motion of the waves. In Fig. 12-1-1*a* the passage through the gap changes the rays from parallel arrows, pointing to the right, to arrows radiating out in a broad range of directions. However, in Fig. 12-1-1*b*, the rays after the gap are to a large extent the same as before. In this case straight-line rays passing through the gap are a good approximation to the behavior of the water waves.

When light waves pass through apertures in an optical system, effects similar to those in Fig. 12-1-1 occur. If the apertures are all

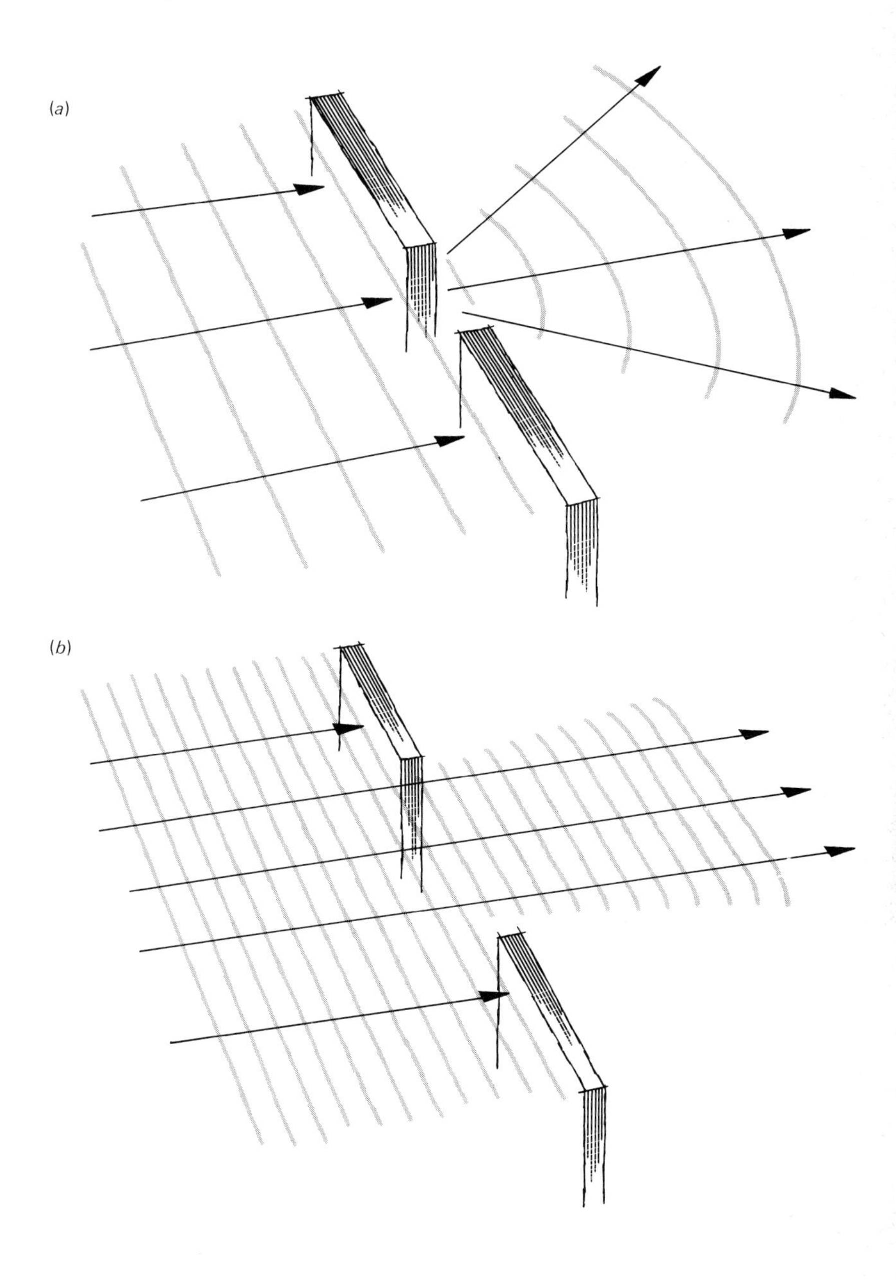

Figure 12-1-1 Two views of water waves passing through an aperture. In *(a)*, the aperture width is of order of the wavelength; in *(b)* the aperture is much wider than a wavelength.

extremely large compared to the wavelength of light, then the wave motion can be very well characterized by drawing straight-line rays. We describe the optical techniques which apply in this case as *ray optics,* or *geometrical optics.* While the rays are imaginary lines, telling us the direction of motion of wavefronts, we treat them as if they are real and express the basic optical laws of *reflection* and *refraction* in terms of the bending of rays. Since the wavelength of visible light is less than 10^{-4} cm, ray optics is usually a very good approximation for the behavior of common lens systems, which have aperture sizes of the order of centimeters, or greater.

In optical instruments with narrow apertures, the effects of waves spreading, as in Fig. 12-1-1, do become important. This spreading, called *diffraction,* makes it necessary to modify the results of ray optics. Furthermore, the wave nature of light can sometimes give rise to *interference* effects, regardless of aperture sizes. In such cases we must at least extend the results of the simple ray optics to include a description of the maxima and minima of light intensity which arise from the interference.

Ray optics is such a useful approximation that it is the most basic of all optical techniques. We will therefore begin with a discussion of the basic laws which describe the behavior of rays. The next two sections will then be devoted to applications of ray optics, such as the magnifying lens and the telescope. Afterward, we will turn to applications of diffraction and interference, in which we will again recall that light is, after all, wavelike.

Rays follow straight lines in any transparent medium, such as air, a vacuum, clear water, etc. However, at a boundary between two transparent media, the phenomena of *reflection* and *refraction* are manifested, as indicated in Fig. 12-1-2. Two laws describe, respectively, the relation of the angles of the reflected and refracted rays to the incident ray. For reflection,

$$\theta_1' = \theta_1 \qquad \text{12-1-1}$$

the angle of incidence equals the angle of reflection. For refraction, the law, called *Snell's law,* uses the *sine function* of trigonometry:

$$\frac{\sin \theta_1}{v_1} = \frac{\sin \theta_2}{v_2} \qquad \text{12-1-2}$$

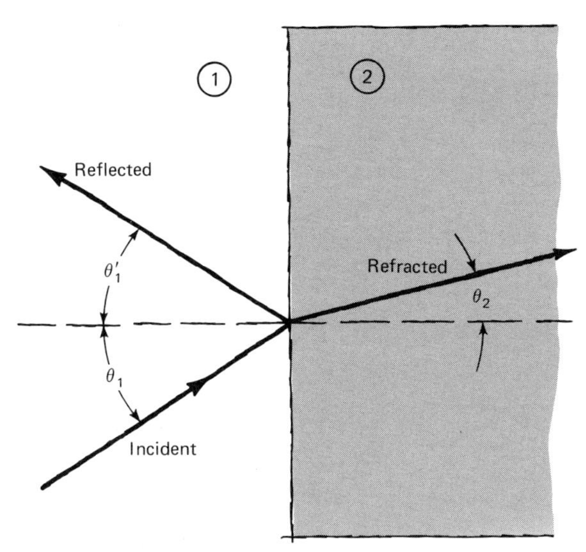

Figure 12-1-2 Reflection and refraction of light at a boundary between two different media.

where v_1 and v_2 are the velocities of light in the media traversed by the incident ray and the refracted ray, respectively. For a given value of θ, the value of "sin θ" (the sine of θ) can be found from a graph of a sine curve, or from a table. Figure 12-1-3 shows a sine graph and also gives a table for angles between 0 and 90°, which are the only ones relevant to Eq. 12-1-2.

It is customary to specify the velocity of light in various materials by giving the ratio c/v, where c is the velocity in vacuum and v is the velocity in the material. This ratio is called *index of refraction* and is conventionally represented by the letter n. For a vacuum, $n = 1.00$, while for water $n = 1.33$, implying that $v = c/1.33 = (0.67)c$. For optical glass the available range of n is from about 1.4 to 1.8. In terms of n_1 and n_2, the refractive indices of the two media, Snell's law takes the form:

$$n_1 \sin \theta_1 = n_2 \sin \theta_2 \qquad 12\text{-}1\text{-}3$$

Both Eqs. 12-1-1 and 12-1-2 can be derived by a technique known as *Huygens construction* which is used to correctly predict many aspects of wave propagation. The idea is that each point in space which is exposed to an electromagnetic wave radiates a spherical *Huygens wavelet*, and that all these Huygens wavelets

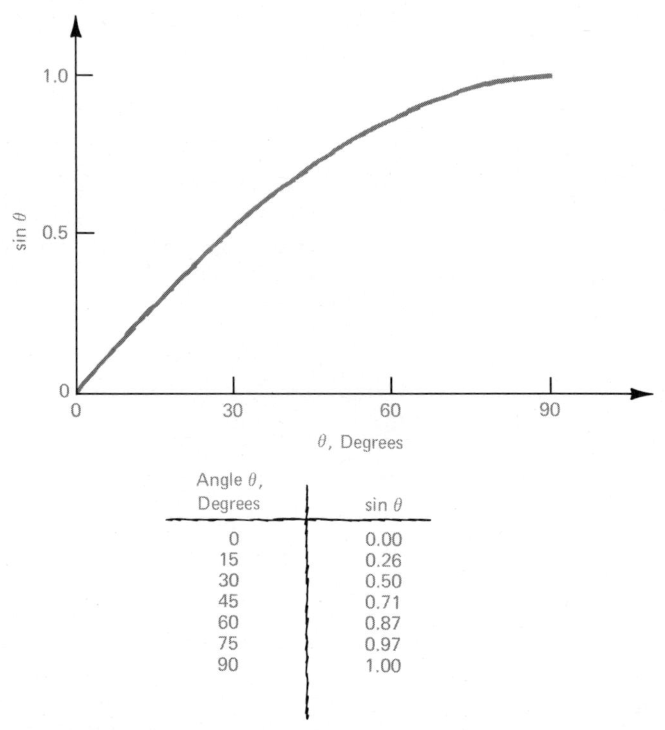

Angle θ, Degrees	$\sin\theta$
0	0.00
15	0.26
30	0.50
45	0.71
60	0.87
75	0.97
90	1.00

Figure 12-1-3 Graph and table of the quantity sin θ, for θ between 0 and 90°.

add up to produce the ongoing wave. The idea might be thought to relate to real physics, inasmuch as the electrons in matter do each reradiate a scattered electromagnetic wave when exposed to an incident wave. However, Huygens principle also is used in a vacuum and should more correctly be interpreted as a graphical method for getting solutions to the mathematical equation of wave propagation.

PROBLEM 1

A typical wavelength of visible light, in the green region of the spectrum, has λ equal to 0.5 micron. What is its frequency? In glass with refractive index 1.5 what is the frequency and wavelength of this light?

For the case of reflection, Fig. 12-1-4 shows that the Huygens wavelets lead to Eq. 12-1-1. For refraction, Fig. 12-1-4 is drawn as if

v_2 were less than v_1 and shows qualitatively that the rays are bent toward the direction perpendicular to the refracting surface. Application of trigonometry to the figure leads exactly to Snell's law, Eq. 12-1-2.

If the law of refraction is applied twice to a ray as it passes through a prism, there is a net bending of the ray, as shown in Fig. 12-1-5. For a lens, the shape is such that the bending, analogous to that by a prism, varies along the lens in exactly the right way to produce focusing of rays, as in Fig. 12-1-6*a*. Parallel rays

Figure 12-1-4 Reflected and refracted waves, found with Huygens' construction. Each wavelet was generated by the incident wave shown with a solid line. Waves and wavelets are shown at a given time. Wavelets in medium 2 have smaller radii because it is assumed that the velocity of light in medium 2 is less than in medium 1.

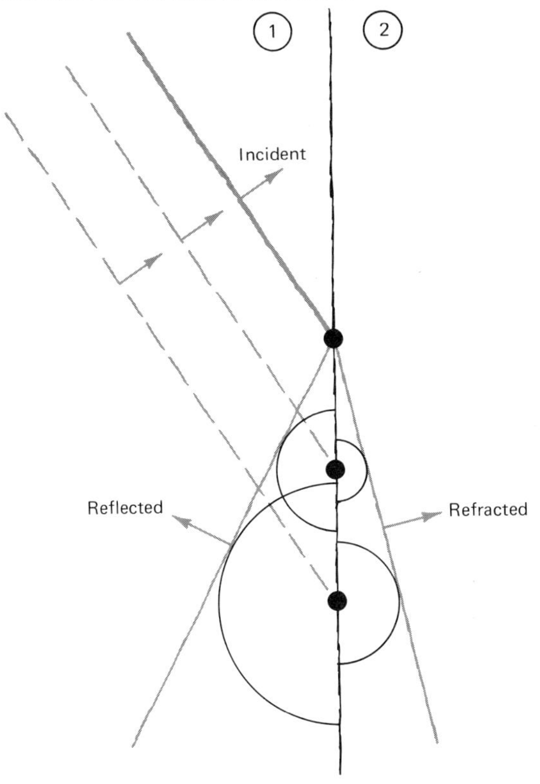

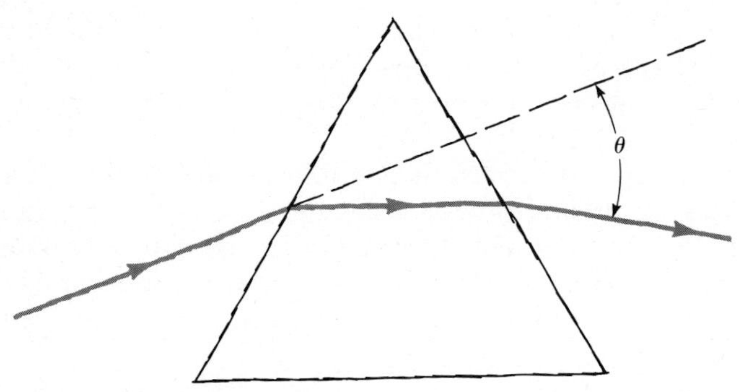

Figure 12-1-5 Net bending of a light ray through an angle θ by passage through a prism.

(from a plane wave) are shown focused by a lens to a point at a distance behind the lens labeled f, called the *focal length* of the lens. If the diameter of the lens is much less than the distance f, then the thickness of the lens can be neglected in comparison with f. In this case we call the lens a *thin lens*, and in our succeeding applications of geometrical optics we will always restrict ourselves to thin lenses. For a thick lens nothing is different in principle, but accurate calculations are more complicated.

PROBLEM 2

Find the net bending of light by a 60° prism when the ray inside is parallel to one face, as shown in Fig. P-12-2. (Use symmetry to save trouble in finding the answer.)

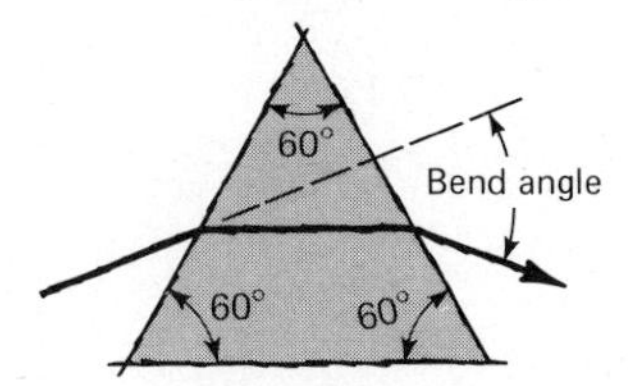

Figure P-12-2

PROBLEM 3

Consider light striking a glass-air interface *from the glass side.* Make a graph of the refracted angle against the incident angle, using Snell's law with $n = 1.5$ for glass and 1.0 for air. When the refracted angle reaches 90°, *total internal reflection* sets in and there is *no refracted ray.* This phenomenon is useful for guiding light down long "pipes" provided the rays don't make too large an angle with the sides of the pipe. Sketch a few rays being transported down a glass rod in this fashion.

For parallel rays at an angle to the axis of symmetry of the lens,

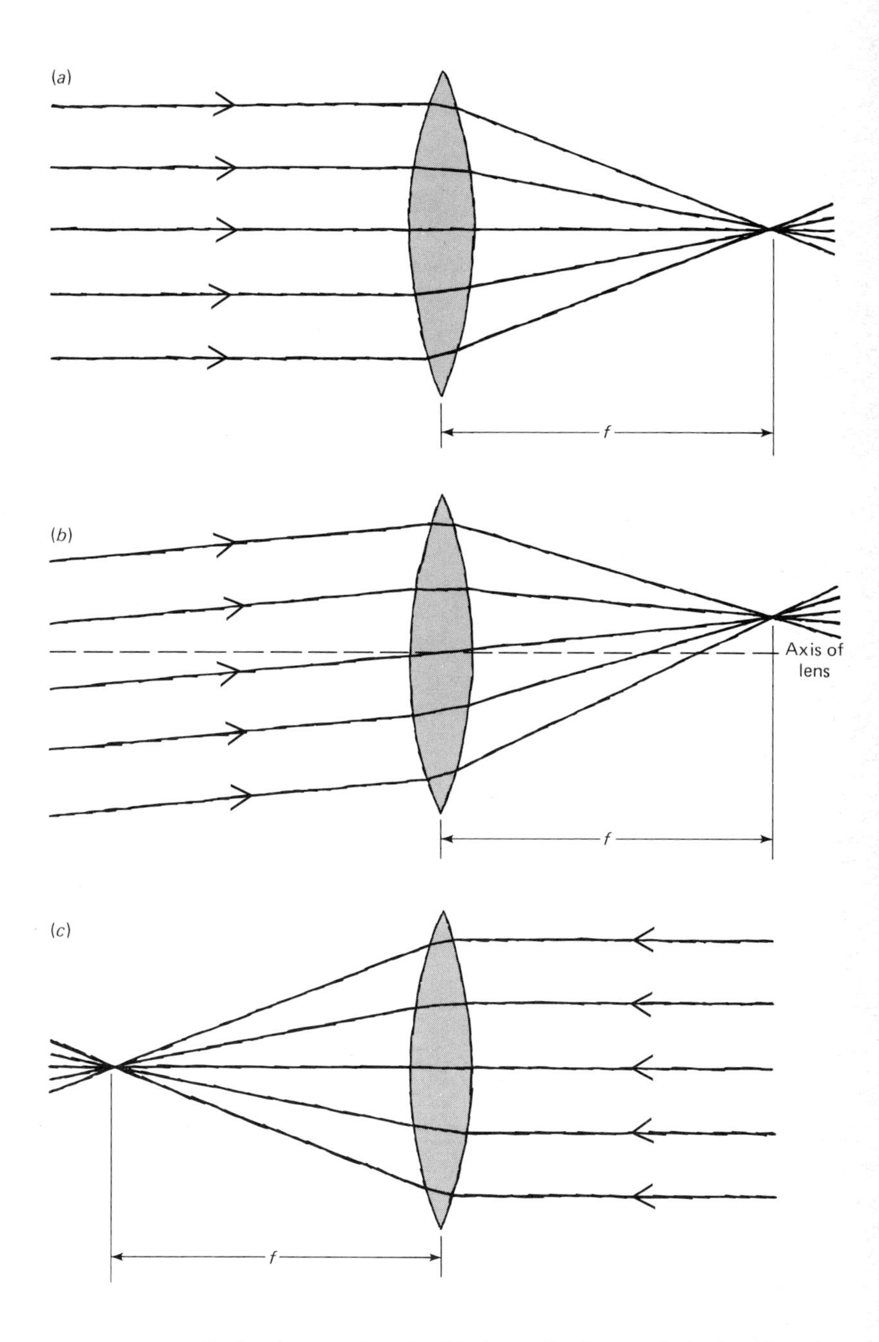

Figure 12-1-6 The focal properties of a thin lens. The distance f, the focal length, is the same for the three cases illustrated above for a given lens.

there is still a good focus, as indicated in Fig. 12-1-6*b*, provided the rays do not make a large angle with the lens axis. The distance from the lens to the focus, as shown in Fig. 12-1-6*b*, is still equal to f, to a very good approximation. Finally, as indicated by the comparison of Fig. 12-1-6*a* and *c*, parallel light incident from either side of a thin lens is focused at the same distance f. The focal length of the thin lens is the same looking through it either way.

For real lenses, the ideal focusing properties, which we will assume, are not completely achieved for a variety of reasons. The various causes of imperfect focus are called lens *aberrations*, and the way in which the lens designer minimizes aberrations is an interesting subject which we will not be able to go into here. It is, however, possible to design thick lenses, composed of more than one thin lens, and often utilizing more than one kind of glass, such that the performance of the lens is mainly limited by the small irreducible diffraction effects.

12-2 REAL AND VIRTUAL IMAGES

The image-forming properties of a thin lens are determined by its focal length. Figure 12-2-1 illustrates the formation of an image of an arrow by a lens of focal length f. The arrow might be a luminous neon sign, or a simple arrow-shaped object made visible by the fact that it scatters some of the light with which it is illuminated. In either case, we will abstract from the arrow what is, in effect, a point source of light at its tip. The light from the entire arrow may be thought of as arising from a very great number of such point sources arrayed along it.

Figure 12-2-1 Formation of a real image by a lens. The three rays drawn are all simple to use for a graphical construction.

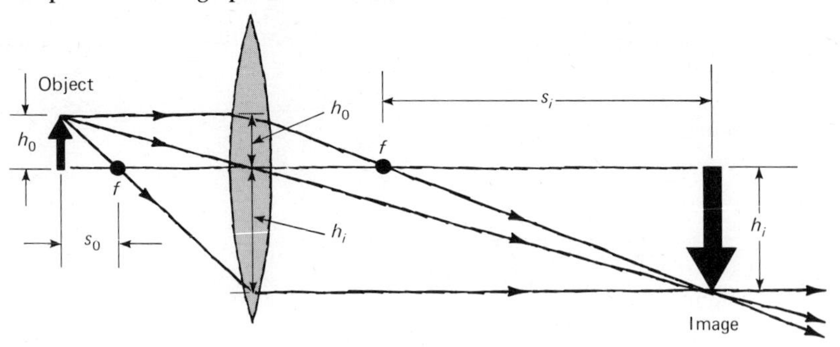

The figure shows three rays from the tip, the three which are easy to draw utilizing the basic ideas which have already been presented. One ray leaves the arrow tip parallel to the lens axis, and it therefore must be refracted by the lens so that it passes through the focal point on the other side. (The definition of the point f is that all such incoming parallel rays pass through it.) Another ray goes from the arrow tip through the near focal point and is therefore refracted so as to be parallel to the lens axis on the other side. If this ray were going from right to left in the figure, then the result just stated would follow from the definition of f. Ray trajectories are reversible, in that the path of a ray from right to left is also the path from left to right. This principle of *optical reversibility* follows from the fact that the laws of both refraction and reflection are independent of the ray direction.

Finally, a third ray drawn in Fig. 12-2-1 passes through the center of the lens. By symmetry, this ray is undeflected, and for a thin lens it is undisplaced. It proceeds straight through the lens. In graphically constructing images formed by lenses and lens systems, these three rays are the easiest to draw. Notice that *any two* are sufficient to define the point where all three rays, plus other rays from the same point, cross at the image. The image acts like another point source, from which a cone of light rays appears to diverge. To an observer to the right of the image of the arrow point, the rays appear to radiate from the image just as if it were a real arrow point.

The fundamental formula of lenses and systems of lenses relates the object and image distances for a lens of a given focal length f. By making use of similar triangles, one form of that formula is easy to derive from Fig. 12-2-1. The object and image heights are labeled h_o and h_i, respectively, and two distances equal to h_o and h_i are also indicated in the plane of the lens. These distances are known because they represent places where parallel rays from the object and image arrowpoints strike the lens.

The ray diagram is redrawn in Fig. 12-2-2 for clarity, without the undeviated ray through the center of the lens and with two pairs of similar right triangles indicated by shading. For the similar triangles on the object side of the lens, the ratios of the corresponding sides must be equal, so that we can write

$$\frac{h_o}{h_i} = \frac{s_o}{f} \qquad \text{12-2-1}$$

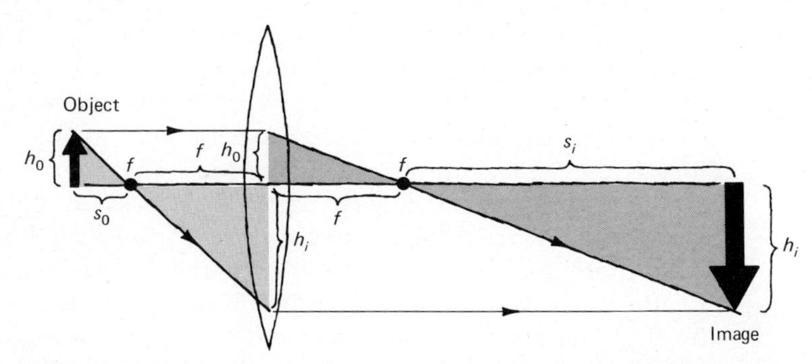

Figure 12-2-2 Similar triangles, redrawn from Fig. 12-2-1. There are two pairs of triangles, one pair on either side of the lens. Corresponding sides, whose ratios must be equal, are those along the lens axis and those perpendicular to it.

In the same way, the similar triangles on the image side of the lens lead to the equation

$$\frac{h_o}{h_i} = \frac{f}{s_i} \qquad 12\text{-}2\text{-}2$$

If these equations are combined by equating the two ratios which are equal to h_o/h_i, the result is as follows:

$$\frac{s_o}{f} = \frac{f}{s_i} \qquad 12\text{-}2\text{-}3$$

Equation 12-2-3 can be rearranged to give the equation known as Newton's form of the lens equation:

$$s_o s_i = f^2 \qquad 12\text{-}2\text{-}4$$

From Eq. 12-2-4, given f, we can find s_i from s_o, or vice versa, where s_i and s_o locate the object and image with respect to the focal points. Note that Eq. 12-2-4 does not involve h_o or h_i, so that, for example, it applies to any point of the object arrow. Therefore, the points of the image arrow all lie at a given distance s_i, which can be found from f and s_o. Returning to a picture which makes use of undeviated rays through the center of the lens, Fig. 12-2-3, we can construct at the known s_i and s_o distances a set of corresponding pairs of points on the object and image arrows, labeled (1,1′), (2,2′), etc. Perhaps the result was obvious, but in this figure we can see rigorously that the image of the arrow formed by the lens is an inverted and magnified, but properly proportioned, replica of the arrow.

If we prefer to work with object and image locations defined by their distances from the lens, then we can call the object distance p and the image distance q, as defined below and shown on Fig. 12-2-3,

$$\begin{aligned} p &= s_o + f \\ q &= s_i + f \end{aligned} \qquad 12\text{-}2\text{-}5$$

From Eq. 12-2-5, we can solve for s_o and s_i in terms of p and q, and substitute into Eq. 12-2-4. The algebra and the final result are shown below:

$$s_o = p - f$$
$$s_i = q - f$$
$$s_o s_i = pq - (p + q)f + f^2 = f^2$$
$$f = \frac{pq}{p + q}$$
$$\frac{1}{f} = \frac{1}{p} + \frac{1}{q} \qquad 12\text{-}2\text{-}6$$

Equation 12-2-6 is called Gauss' form of the lens equation and is more commonly used than Newton's form in elementary texts. In working problems, either equation will yield the same result, and there is no general reason to prefer either. We will, following tradition, standardize on Gauss' form.

The image shown in Fig. 12-2-3 is described as *real*, because it can be viewed by light scattered off a screen placed at the location of the image. Suppose we imagine a focusing lens, like the one we've been considering so far, with a given value of f, and let p and q vary so as to find out when a real image is produced. The

Figure 12-2-3 The same as Figs. 12-2-1 and 12-2-2, with the images of several points along the arrow indicated. In addition, the object and image distances used in Gauss' lens formula are shown.

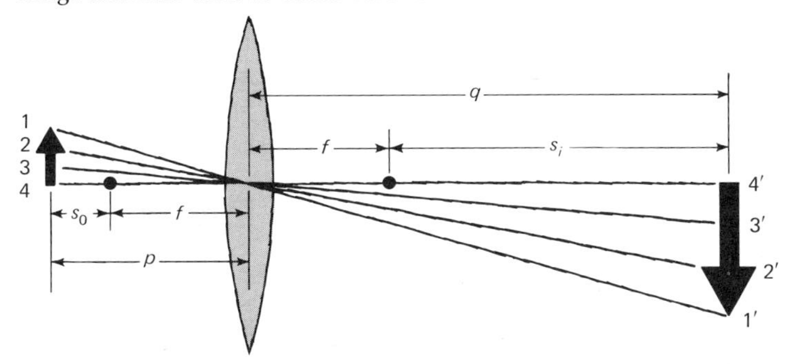

graph shown in Fig. 12-2-4 shows how the image distance q varies from f to infinity as the object distance p varies from infinity to f. When the object distance p is less than f, the lens equation requires that q be negative. This means that the rays from an object point too close to the lens cannot be focused by the lens to form a real image point behind the lens. The rays still appear to diverge from a point on the same side of the lens as the object.

To be more specific, a ray diagram for an arrow with object distance less than f is shown in Fig. 12-2-5. Three special rays, like those originally considered in Fig. 12-2-1, are shown. They are bent by the lens so as to diverge less than when they left the tip of the arrow, but they aren't bent enough to converge and form a real image. Instead, they appear to come from what is called a *virtual image*, shown dashed in the figure. The lens formula can be rederived for the somewhat different geometry of Fig. 12-2-5. The result is that Eq. 12-2-6 remains valid, provided the distance to the virtual image is taken to be negative. The main attribute that differentiates a virtual from a real image is that it *cannot* be viewed on a screen. The rays *appear* to come from an image point at which they do *not* actually cross.

Up to now, we have considered the real and virtual images formed by a *positive lens*, one which *converges* the light rays incident on it, and which brings parallel light to a *real* point image.

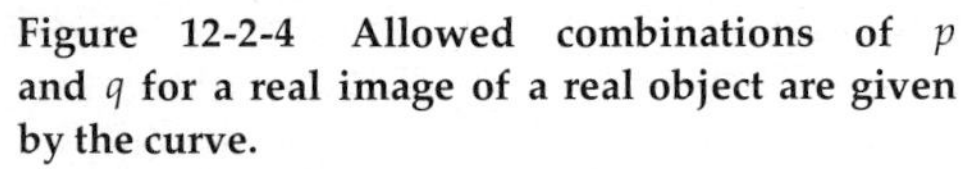

Figure 12-2-4 Allowed combinations of p and q for a real image of a real object are given by the curve.

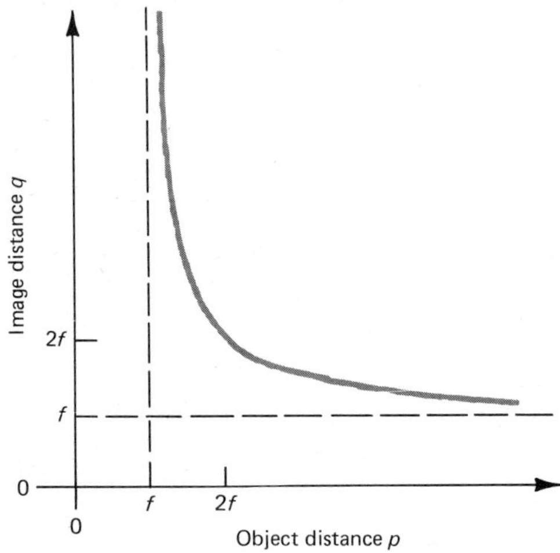

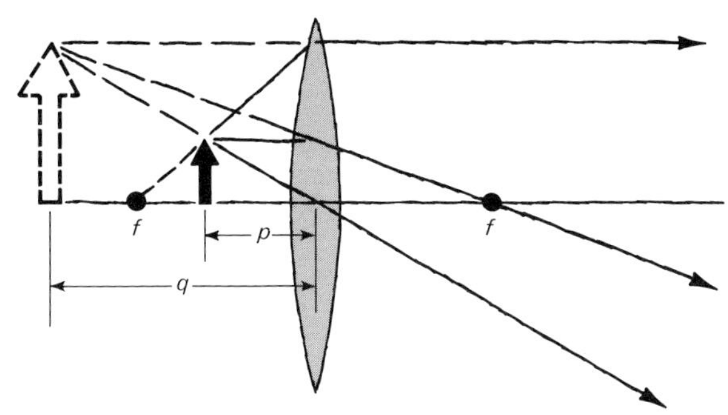

Figure 12-2-5 Formation of a virtual image of an object closer to the lens than the focal length f.

A lens that is thinner at the center than at the edges is called a *negative lens* and behaves differently. Figure 12-2-6 shows that parallel rays incident on a negative lens are made to *diverge.* A *virtual* image is produced, and the focal length of the negative lens is defined as the distance to this virtual image, with a minus sign. Thus if f in the figure were 10 cm from the lens, the focal length would be -10 cm.

If we substitute in the lens equation the values $p = \infty$ (for a plane wave) and $f = -10$ cm, then $q = -10$ cm. In the case shown in Fig. 12-2-6, the usual lens equation therefore gives an image distance of correct magnitude, which is negative, correctly signifying a virtual image, provided we use the negative focal length for the negative lens. Without rederiving the lens equation for negative lenses, one can state that the equation works the same as for positive lenses, *provided f is taken negative*. Notice that for a negative lens and a real object, the image is always virtual, which is simpler than for a positive lens. The sign conventions and the results discussed in this section are summarized below:

LENS TYPE	OBJECT DISTANCE	IMAGE
Positive	$p \geq f$	Real, $q \geq f$
Positive	$f \geq p \geq 0$	Virtual, $q \leq -f$
Negative	$p \geq 0$	Virtual, $q \leq 0$

All the combinations of f, p, and q are correctly related by the same lens equation, Eq. 12-2-6. The *sign conventions* are that f becomes negative for negative lenses, and q becomes negative when the image is virtual, i.e., on the same side of the lens

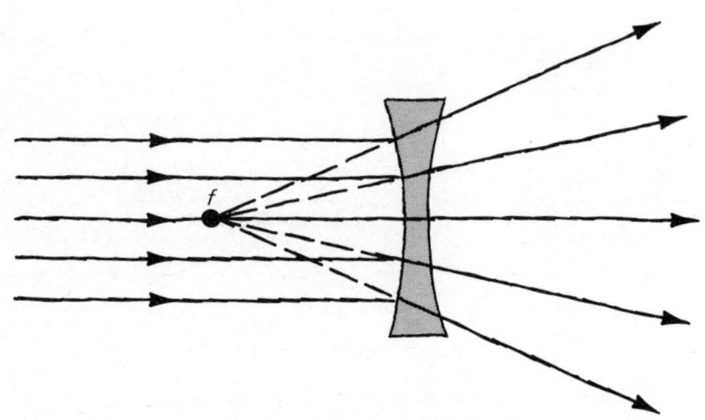

Figure 12-2-6 Parallel light incident on a negative, or diverging, lens produces a virtual focus, as shown. The distance from the lens to the point labeled f is still called the focal length.

as the real object. Can p become negative, signifying a *virtual object*? The answer is yes, but we will leave virtual objects as the subject of a problem.

PROBLEM 4
A lens with 10-cm focal length is used to produce an image the same size as the object. What are the object and image distances.

PROBLEM 5
A pinhole camera consists of a light-tight box with film on one side and a pinhole opposite it. The pinhole replaces a lens. Illustrate with ray diagrams. Does the pinhole have a focal length?

PROBLEM 6
Graphically construct a virtual image of an arrow, formed by a negative lens. Check your result with the lens formula.

PROBLEM 7
A camera lens has a focal length of 5 cm. What range of distances from lens to film must be allowed in order to focus on objects from 50 cm to infinity?

PROBLEM 8
Suppose in Fig. 12-2-1 a negative lens with focal length $s_i/2$ is placed between the lens and the image, a distance $s_i/4$ away from the image. Can you discover the location of the new image? The old image is said to act as a *virtual object* for the negative lens in

this case. Perhaps you can also discover the correct sign convention so that the usual lens equation works in his case.

12-3 MICROSCOPES AND TELESCOPES

One lens, used to obtain a magnified image, is usually called a simple *magnifier.* Both telescopes and microscopes utilize such magnifiers, in conjunction with other lenses, and the study of the magnifier is an appropriate preliminary to understanding more complex optical systems. Figure 12-3-1 shows a positive lens form-

Figure 12-3-1 Production of magnified real and virtual images. From the shaded similar triangles, the magnification is found to be q/p in both cases.

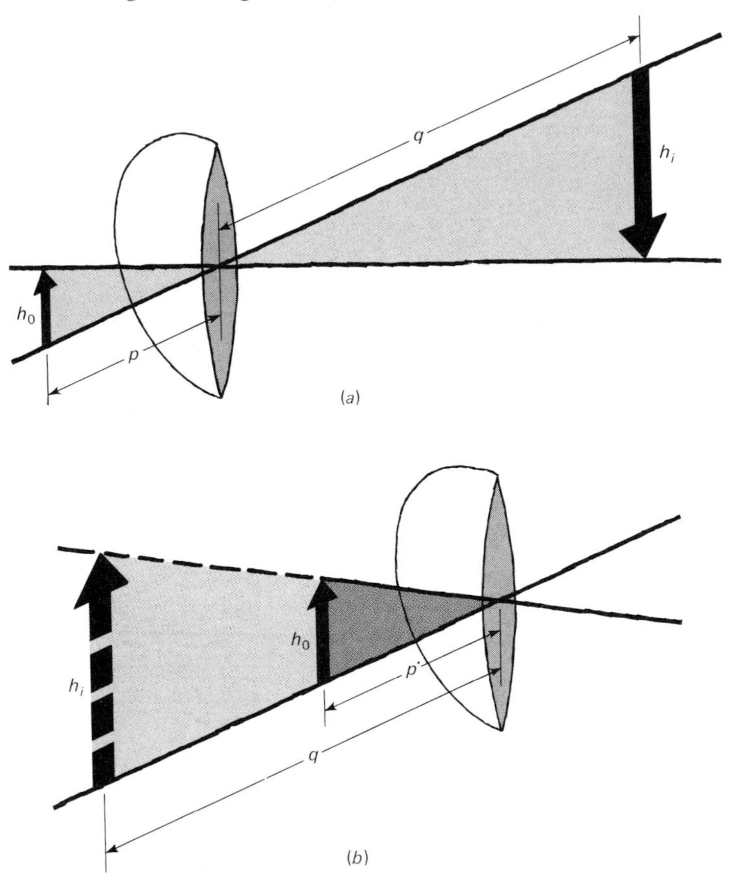

ing both a real and a virtual image in two different instances. In both sketches the undeviated ray is shown, passing through the center of the lens, and in both there are a pair of similar triangles indicated with shading. The ratio of the heights of image to object, h_i/h_o, called the *linear magnification* M, can be related to the distances p and q by means of the similar triangles. For both cases,

$$M = \frac{h_i}{h_o} = \frac{q}{p} \qquad 12\text{-}3\text{-}1$$

We know that in both cases indicated in Fig. 12-3-1, as p approaches f, q approaches infinity, and therefore the linear magnification can be made arbitrarily large. However, the magnifier is not solely evaluated on the basis of its linear magnification. Instead, the *apparent* size of the image to a viewer, the *angular size,* which is related to the distance of the image from the eye as well as the size of the image, is of paramount importance. The closer the image to the eye, the larger the effective magnification, provided we can still see the image in sharp focus. Although the closest distance at which the eye can focus varies from one individual to the next, it is often conventionally taken to be 25 cm. Nearsighted people, who can focus closer than this, have the equivalent of built-in magnification.

In addition to its magnification, the magnifier is also evaluated in terms of its field of view. The eye's entrance aperture, or pupil, is a few millimeters in diameter, and in order for us to see an image the rays must enter this aperture. The magnified image in Fig. 12-3-1*a* can be very large, and the eye can be placed 25 cm to the right of it; but the rays from the various points of the magnified image will all be nearly parallel to the lens axis, if q/p is large and if the lens is thin so that its diameter is much less than f. As a result, only a small part of the image, of the order of the diameter of the pupil of the eye, can be viewed at a given time. Furthermore, the part of the image that can be viewed only occupies a small fraction of the field of view of the eye. The system of Fig. 12-3-1*a* is therefore not useful by itself as a simple magnifier. It can, however, be used to advantage in combination with another lens, as it is in the micrscope. It can also be useful with photographic film in the image plane, for making photomicrographs.

The ordinary simple magnifier is a lens used as in Fig. 12-3-1*b*. Here, the best magnification occurs, and good use is made of the field of view of the eye, when the eye is close to the lens and the virtual image is 25 cm from the eye. If the focal length of the lens is not exceedingly short, we can approximate these conditions by assuming that the eye is essentially at the lens and the object is at the distance p which leads to $q = -25$ cm. The magnifier used in this way is simply a convenient means for becoming temporarily nearsighted!

To calculate the magnification for the simple magnifier, we set it equal to $25/p$, the factor by which we have effectively moved the object closer to our eye. This is sometimes called the *angular magnification.* For $q = -25$, we apply the lens equation and find

$$\frac{1}{p} = \frac{1}{f} - \frac{1}{q} = \frac{1}{f} + \frac{1}{25} \qquad \text{12-3-1}$$

noting that q must be taken negative since we have a virtual image. The magnification is then found,

$$M = \frac{25}{p} = \frac{25}{f_{(\text{cm})}} + 1 \qquad \text{12-3-2}$$

Since Eq. 12-3-2 involves the nearest focus for the eye, expressed as 25 cm, the focal length must also be in centimeters. The highest power simple magnifiers in common use utilize focal lengths of order 1 cm, yielding a magnification of about 25. Shorter focal length lenses are difficult to use.

PROBLEM 9

An aging professor finds that the closest distance he can focus on with his unaided eyes is 50 cm. Prescribe the correct focal length eyeglass lenses to restore him to a youthful 25 cm near point.

PROBLEM 10

A reading glass is in use which has a 12.5-cm focal length, is 10 cm from the book being read, and is 30 cm from the eye of the reader. What is the magnification? What would be the maximum magnification with this lens, and how would it be attained?

In order to achieve magnifications of several hundred, the microscope utilizes two stages of magnification. First, a very short focal length *objective* lens is used as in Fig. 12-3-1*a*. The enlarged real image formed by this lens then serves as the object for a magnifying *eyepiece,* which produces another stage of magnification and also matches the rays to the pupil of the eye. In its most basic form, the eyepiece is a simple magnifier, and the microscope is as shown in Fig. 12-3-2. The objective can give a magnification of up to about 100, and the eyepiece up to about 30, for an overall calculated magnification of up to 3,000. At the highest calculated magnifications, the limitations of diffraction lead to smaller useful magnifications, such that the finest visible details are of order the wavelength of visible light.

The common astronomical telescope is related to the microscope. It utilizes an objective and an eyepiece in much the same way as the microscope. However, while the microscope is used for observing small objects close to the lens, the telescope is used for observing very large objects at enormous distances. The magnification of a telescope is therefore defined by the increase in the angle subtended by the object when viewed through the telescope. A telescope is sketched in Fig. 12-3-3, with two sets of parallel rays incident, differing in angle by an amount α. We can think of the rays as coming from opposite extremities of the moon, for example. The objective focuses each bundle of parallel rays to a point in its focal plane, and the eyepiece produces a virtual image of the two focal points which are seen by the eye to be separated by the angle α'.

The magnification of the telescope, M, is given by the ratio of

Figure 12-3-2 The compound microscope, at low magnification for ease of sketching. The object and two images, along with just a single ray, are shown.

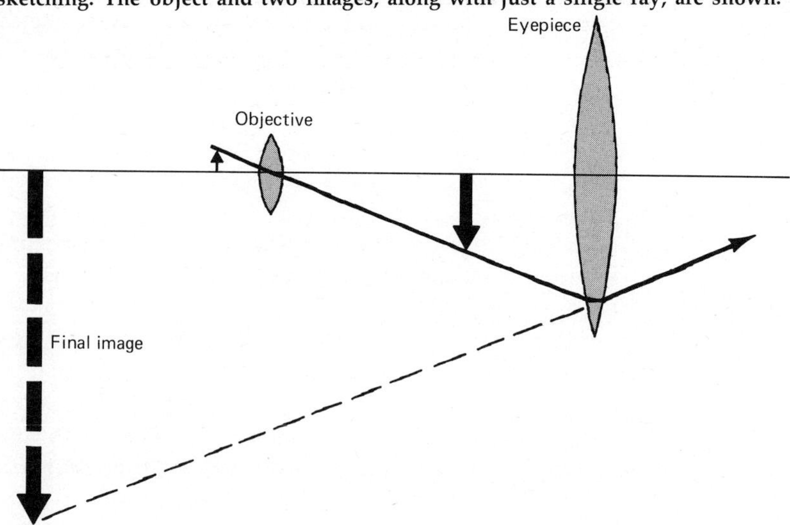

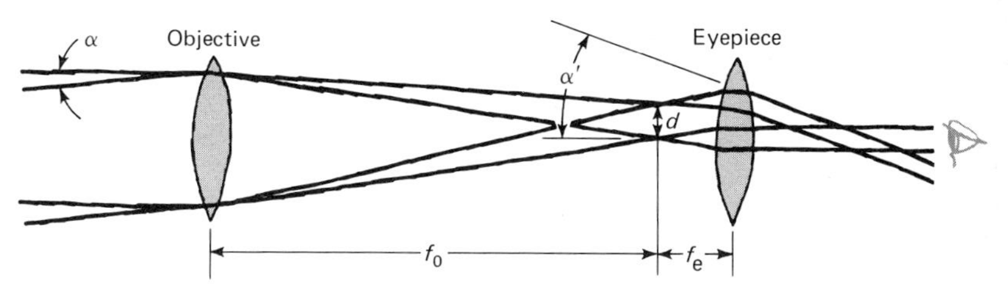

Figure 12-3-3 An astronomical telescope. The magnification is the ratio of the angles α'/α equal to f_o/f_e for small angles.

the apparent angle to the real angle, α'/α. To maximize M the eyepiece magnifier should be as close as possible to the image plane of the objective, at a distance essentially equal to the eyepiece focal length. For α small, the ratio α'/α is simply equal to the ratios of $1/f$ for the two lenses. This is because, for small angles, the relation illustrated in Fig. 12-3-4 leads to the results

$$\alpha = 57.3\,\frac{d}{f_o} \qquad \alpha' = 57.3\,\frac{d}{f_e} \qquad \text{12-3-3}$$

so that $M = \dfrac{\alpha'}{\alpha} = \dfrac{f_o}{f_e}$ where f_o and f_e are the focal lengths of objective and eyepiece, respectively.

According to Eq. 12-3-3 it would appear as if the magnification of a telescope might be made arbitrarily large if the objective were made to essentially disappear so that f_o approached infinity! In practice, the telescope performance is ultimately determined by diffraction and the wavelike nature of light. As a result, the optimum magnification is found to be just large enough so that the minimum angular separation which can be achieved (dictated by diffractive effects) can be made comfortably visible to the eye of the observer, or can be resolved on the photographic plate which commonly takes the place of the human eye.

PROBLEM 11

Suppose you have a microscope objective with a focal length of 0.2 cm and a diameter of 0.15 cm. Design a compound microscope with an overall magnification of 500. Specify the distance from objective to eyepiece, the eyepiece focal length, and the diameter of the eyepiece lens. What is the field of view which can be observed without moving the sample relative to the objective?

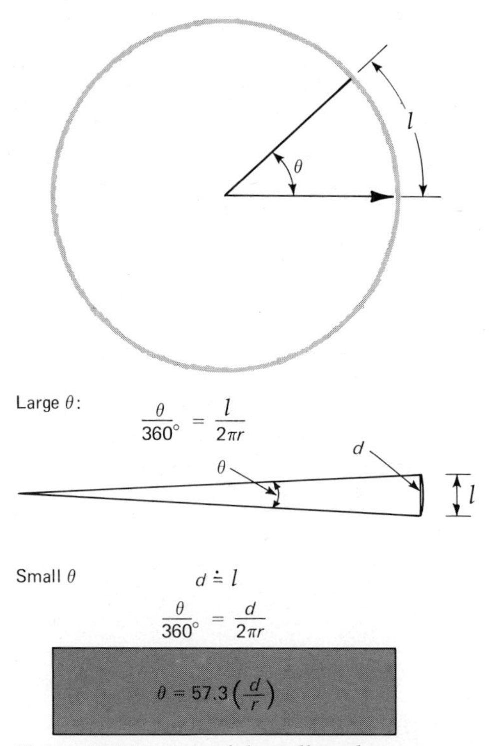

Figure 12-3-4 A useful small-angle approximation, valid when the straight line d equals the arc length l to good accuracy, which is true for θ less than about 10°.

PROBLEM 12

The telescope discussed in the text produces an inverted image. Can you describe how to add an *erecting lens* that will result in an upright final image, sometimes desirable for terrestrial use, as on shipboard.

PROBLEM 13

Galileo's telescope had a *negative* lens for an eyepiece, located *between* the objective and the image formed by it in the absence of an eyepiece. Describe, with ray diagrams, how this telescope works, and show that its magnification is the ratio of the focal lengths of objective and eyepiece, just as for the astronomical telescope we have discussed. Opera glasses are still made this way, but not telescopes. Can you figure out why?

In telescopes it becomes essential to construct optics that are as perfect as possible for producing sharp images of point sources. A major problem, which degrades the performance of telescopes whose objectives are made from lenses, is *chromatic aberration.* Because the refractive index of glass varies with wavelength, the focal length of a glass lens also varies with the wavelength of the light being focused. It is possible to correct for chromatic aberration to some extent by using a doublet lens, made of two elements which utilize different types of glass. More complex combinations of lenses can do a really excellent job of correcting for chromatic aberration but are not practical for the objective of a large telescope. As a result, the *reflecting* telescope, which utilizes a mirror for its objective, has become almost completely dominant for astronomical instruments. (Light-gathering ability is so important in astronomy that it is not acceptable to run a refracting telescope with a selective color filter, which would indeed reduce the chromatic aberration.)

The focal properties of a concave mirror are illustrated in Fig. 12-3-5, as if the mirror were spherical, with radius of curvature R. According to the law of reflection the incident ray is reflected at an angle θ to the radius line if it is incident at an angle θ. For small angles, the focal distance f may be found to be approximately

Figure 12-3-5 The focal length of a concave spherical mirror, illustrated with the undeviated ray (on axis) plus one other ray which satisfies the law of reflection. The value $R/2$ is approximate, for small values of the angle θ.

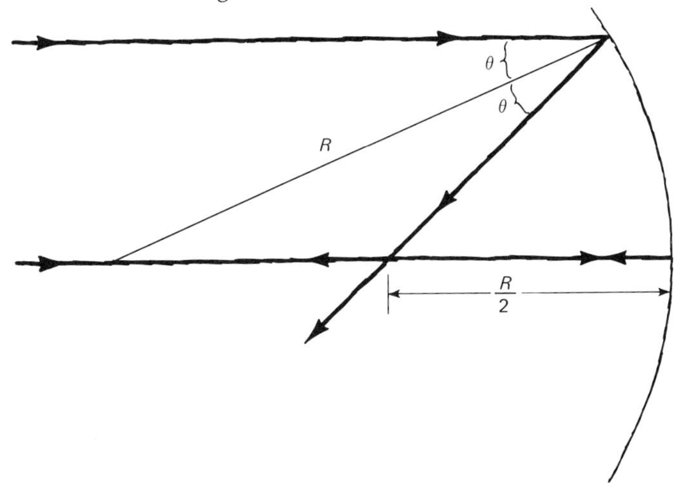

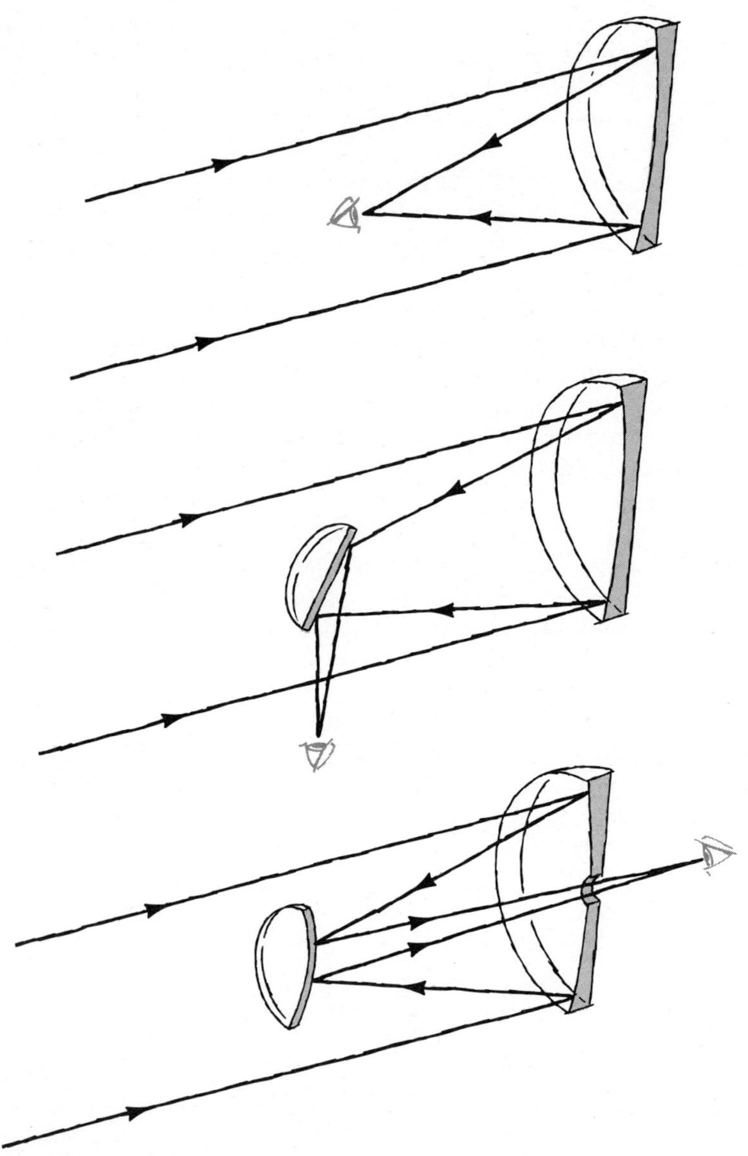

Figure 12-3-6 Three locations for the primary image of a reflecting telescope, the Hale *(a)*, the Newtonian *(b)*, and the Cassagrainian *(c)*. The drawings are not in scale, and no eyepieces are indicated.

equal to $R/2$. For larger values of θ the focal distance moves toward the mirror, so the sphere does not have the exactly correct shape to accurately focus all parallel rays at one point. This *spherical aberration* of a spherical mirror also has a counterpart for spherical lenses.

In the case of an astronomical telescope objective, the optical requirements are so stringent that the spherical aberration must be removed by making a mirror with the theoretically correct profile, that of a parabola. This involves first making a spherical mirror and then correcting it by removing a little extra glass, mainly near the center. Ultimately, the curve of the mirror must be correct to within a fraction of a wavelength of light.

One problem with reflecting telescopes is how to view the image formed by the objective while not seriously obstructing the incoming light. Figure 12-3-6 illustrates some of the arrangements that are used. The main mirror, and all the necessary auxiliary mirrors, viewing platform, etc., are mounted so that the telescope can be aimed at any desired astronomical object. The Hale geometry (named for the Hale 200 in. telescope, the world's largest) works in the rare instance when the telescope mirror is large compared with the size of a man, and produces the same optics as the ordinary refracting telescope. The Newtonian and Cassegrainian arrangements are indispensable for small telescopes and important on the largest ones when the star images are to be analyzed with a bulky piece of apparatus, such as a high-precision spectrograph.

To conclude this section, it is worth emphasizing that when it was first invented the telescope provided one of those technological breakthroughs which have exerted enormous influence on the development of pure science. Beginning with Galileo's observations of the moons of Jupiter, the telescope revolutionized astronomy. At present, telescopes are essential for studies of the general theory of relativity, stellar evolution, cosmology, and other subjects of intense interest to both physicists and astrophysicists. Radio-wave telescopes share the limelight with optical telescopes. The giant telescopes and the giant particle accelerators today occupy complementary positions on the frontiers of physics, probing for knowledge of the most enormous and the most diminutive aspects of the universe.

12-4 INTERFERENCE AND DIFFRACTION

In Chap. 6 the two-wave interference phenomenon observed in Young's experiment has been discussed. In this section, we will examine other interference effects, and, in particular, we will develop a technique for finding the result of the interference of many waves, from a great number of slits. With the aid of this technique two interesting examples will then be examined: the diffraction grating, and the diffracted intensity from a single slit.

To begin, recall that the basic idea which leads to wave interference is the principle of *superposition* of wave amplitudes. Suppose, for example, that two sources of monochromatic light with the same frequency both illuminate a particular region of space.

Figure 12-4-1 Superposition of pairs of waves with equal amplitudes and various phase differences $\Delta\phi$: *(a)* $\Delta\phi = 0$; *(b)* $\Delta\phi = T/4$; *(c)* $\Delta\phi = T/2$.

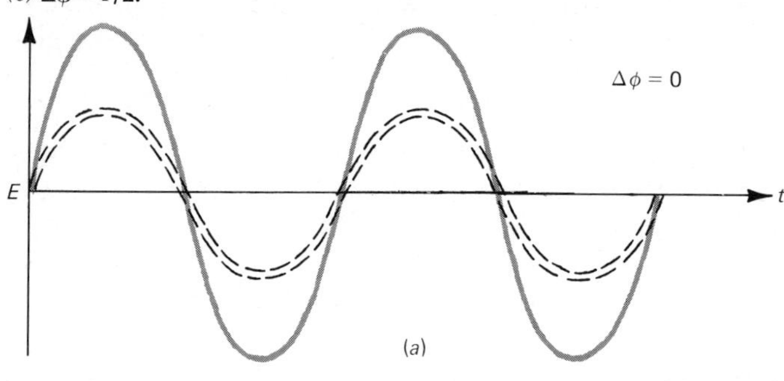

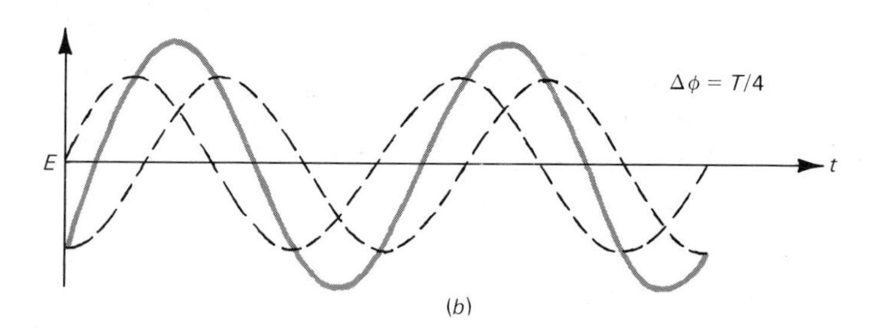

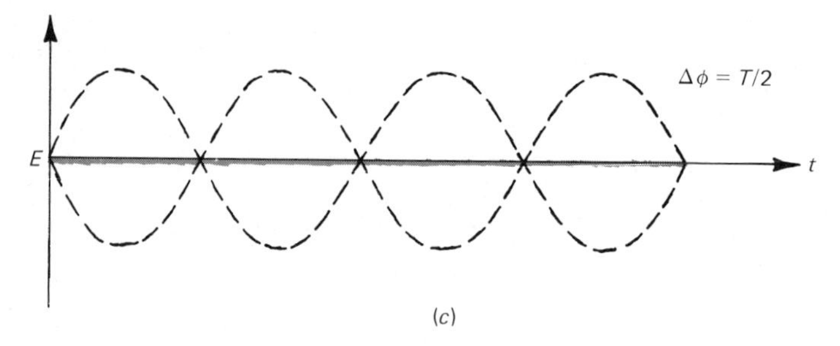

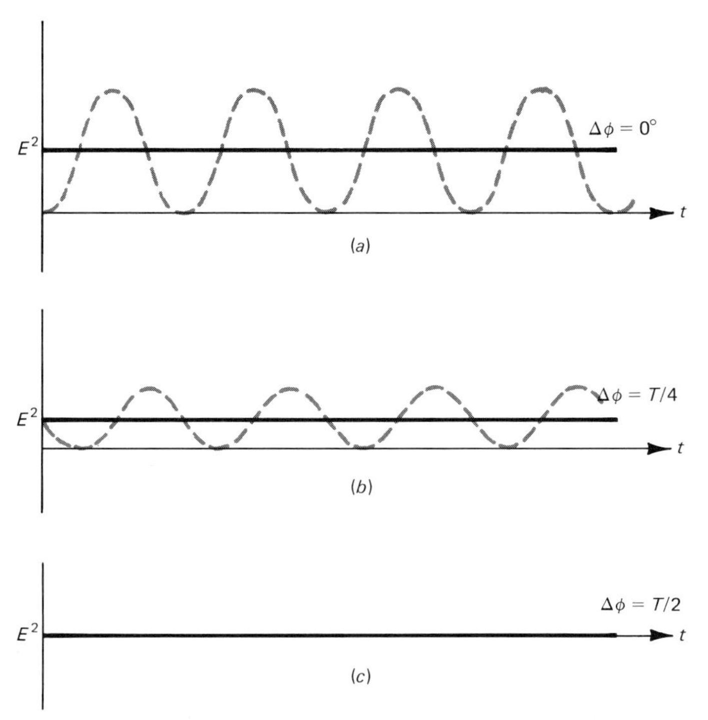

Figure 12-4-2 Instantaneous intensities, proportional to E^2, for the cases shown in Fig. 12-4-1, shown by dashed curves. The time average intensities are shown by solid curves.

Then the combined wave from the two sources is found by adding, or superposing, two component waves, taking account of their *phase difference.* In Fig. 12-4-1 examples are shown of the addition of pairs of waves with equal amplitudes and various phase differences. The resultant amplitudes vary from twice the amplitude of each wave to zero.

The light intensity, the energy delivered per unit area by the light wave, is proportional to the square of the resultant amplitude. The instantaneous and time-averaged intensities are shown in Fig. 12-4-2 for waves of equal amplitude and various phase differences. When the waves are in phase (zero phase difference), the energy delivered is four times as great as for a single wave, or twice that for the two waves individually. (However, energy conservation remains valid because in other regions of space, destructive interference reduces the energy flow below that for the two waves individually.)

In order to study interference with a minimum of algebra and trigonometry it is useful to utilize a kind of vector diagram called a *phasor* diagram. A phasor is a graphical representation of a monochromatic sine wave. To introduce the idea of a phasor, Fig. 12-4-3 shows four "snapshots" of a *rotating vector* with length A, which rotates at constant angular velocity such that it makes one revolution in a time T. The projection of this vector on the axis labeled E is then a *sine wave* of the form

$$E = A \sin \theta \qquad 12\text{-}4\text{-}1$$

where the angle θ is that shown on the figure which changes with time.

The angle θ is called the *argument* of the *sine function,* which simply means that the expression in Eq. 12-4-1 is evaluated for a given value of θ by reading the value of $\sin \theta$ off a graph such as that of Fig. 12-4-4. Notice that the sine function is periodic, repeating itself after each 360° change in the argument θ. If the phasor rotates with a period T, then the time T corresponds to a change in θ of 360°. We can write a relation between θ and T which expresses this fact mathematically:

Figure 12-4-3 Four "snapshots" of a rotating *phasor*. The angle θ increases uniformly with time. The projection of a phasor of length A onto the E axis is $A \sin \theta$.

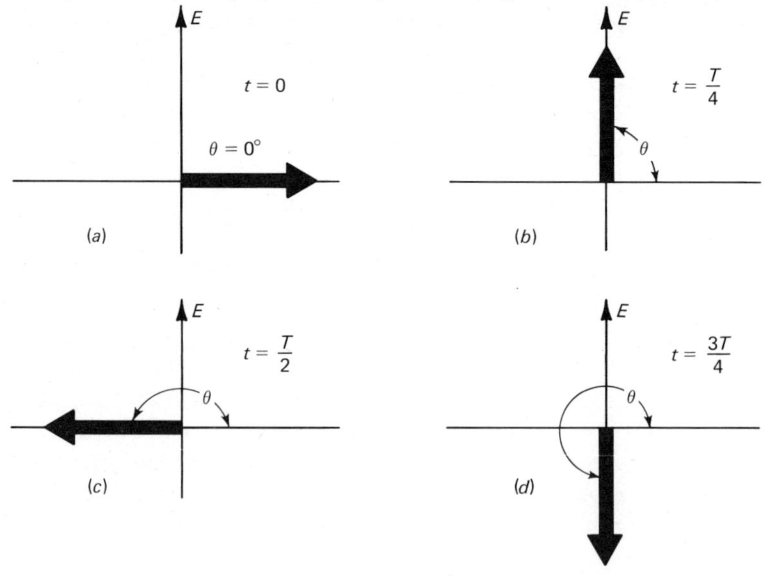

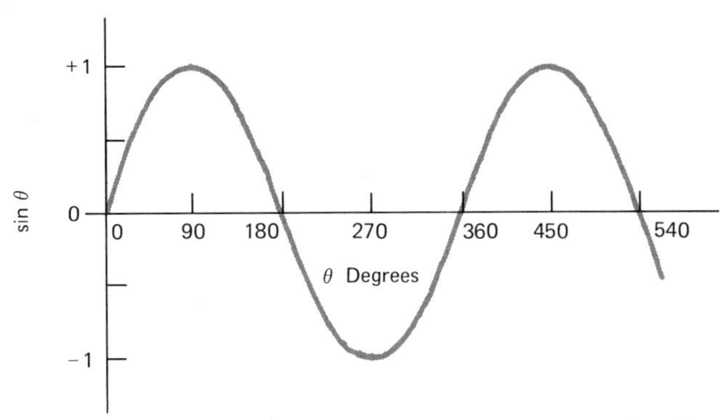

Figure 12-4-4. Graph of mathematical function sin θ plotted against argument θ. The function repeats after 360°.

$$\theta = 360\left(\frac{t}{T}\right) \quad 12\text{-}4\text{-}2$$

Substituting this expression for the argument in Eq. 12-4-1,

$$E = A \sin\left(360\,\frac{t}{T}\right) \quad 12\text{-}4\text{-}3$$

which is the way to write down a sine wave variation with period T.

Figure 12-4-5 Phasor diagram for two waves with equal amplitudes, having a phase difference of 90°.

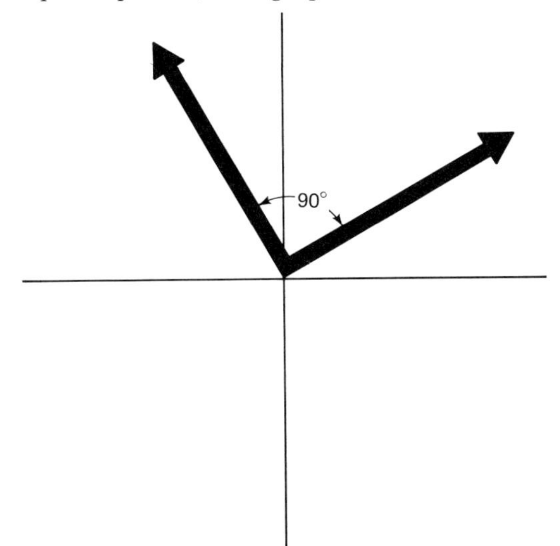

The rotating vector in Fig. 12-4-3 is a graphical representation, called a *phasor*, which is especially convenient for combining two or more sine waves with different phases. Figure 12-4-5 shows a phasor diagram at an arbitrary time for two sine waves of equal amplitude and period which have a phase difference $T/4$. The angle between them is 90°, which represents the difference in their arguments at any time. As the phasors rotate they will always stay 90° apart.

A diagram such as Fig. 12-4-5 represents two coherent waves which are "90° out of phase." The term *coherent* means that they have the same frequency, and that they will therefore maintain their phase difference indefinitely at the same value. For interference effects to be easily observed we need coherent light,

Figure 12-4-6 Phasor diagrams corresponding to the waves in Fig. 12-4-1 at $t = 0$. The lengths of the solid vectors give the resultant amplitudes, in agreement with those found by direct addition in Fig. 12-4-1.

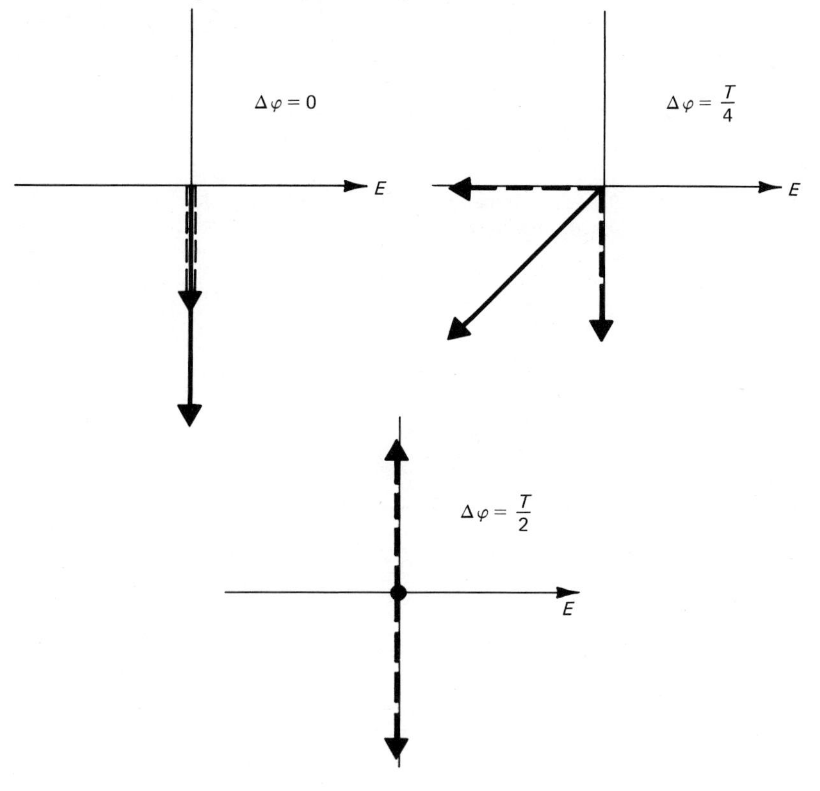

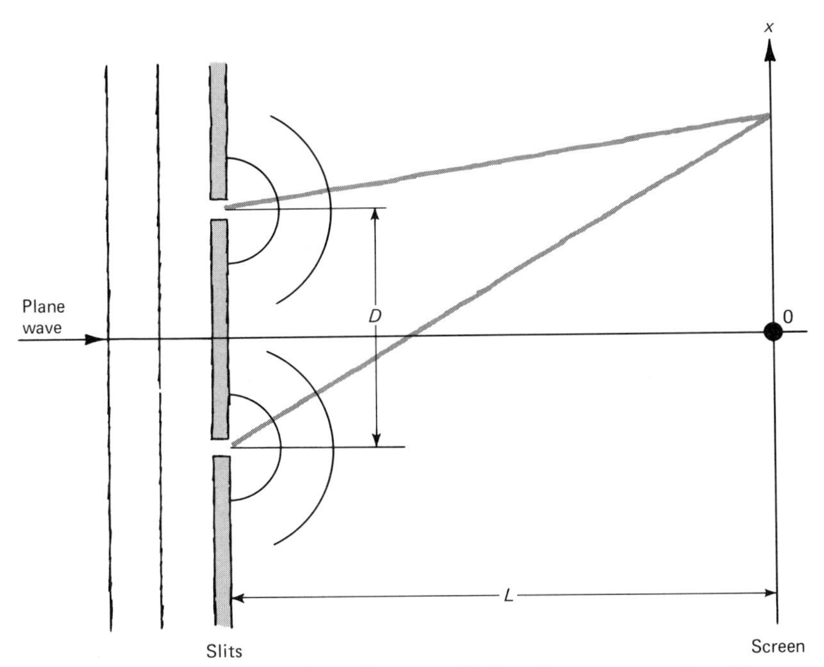

Figure 12-4-7 Schematic view of a two-slit interference apparatus. The resultant wave at any point on the screen is found by superposition of the two waves from the slits.

and in this section we will limit the discussion entirely to the ideal case of perfectly coherent waves.

When the phasor diagram for two waves is shown at a given time, the resultant wave can be found by *vector addition of the two phasors* according to the usual rules for combining vectors. The examples in Fig. 12-4-1 are represented by phasor diagrams in Fig. 12-4-6. For each pair of waves with a given phase difference, one phasor diagram, at some arbitrary time, is shown. The vector addition gives the amplitude of the resultant wave which arises from superposition of the two individual waves. The phasor diagrams are not drawn as "movies," the way the introductory diagram in Fig. 12-4-3 was shown, but it is understood that all the phasors are rotating with period T or frequency f, without explicitly drawing more than one phasor diagram.

As a first example of the analysis of an interference pattern, consider the familiar two-slit interference experiment, shown in Fig. 12-4-7. Let each slit diffract the light into spreading wavefronts, as indicated on the figure, and consider the way the waves from the two slits add at various points on the screen. At $x = 0$, the waves emitted in phase at the slits arrive in phase because the distance to each slit is the same. The phasor diagram looks like that of Fig. 12-4-8*a*. At a value of x such that the path length to the nearer slit is 1/4 wavelength less than that to the farther slit, the two waves are 1/4 period out of phase when they reach the screen. This is shown on the phasor diagram in Fig. 12-4-8b.

The relation between path-length difference and distance x from the center of the screen can be found, to a good approximation, as shown in Fig. 12-4-9. From the similar triangles,

$$d = D\frac{x}{L} \qquad 12\text{-}4\text{-}4$$

The path-length difference d is seen to be proportional to x and the value of x for $d = 1/4$ wavelength is found to be

$$x = \frac{\lambda}{4}\frac{L}{D} \qquad 12\text{-}4\text{-}5$$

where λ stands for the wavelength of the incident light.

Figure 12-4-8 Phasor diagrams for the resultant amplitudes of pairs of waves with phase difference 0, 90, 180, and 360°.

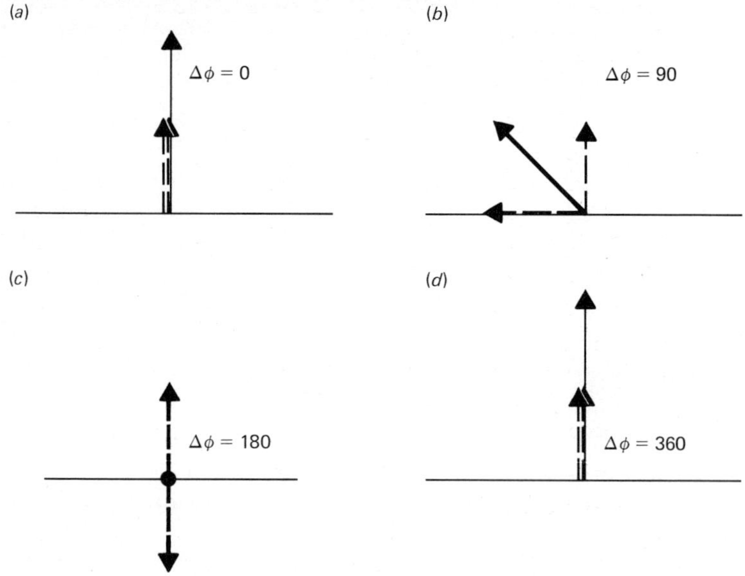

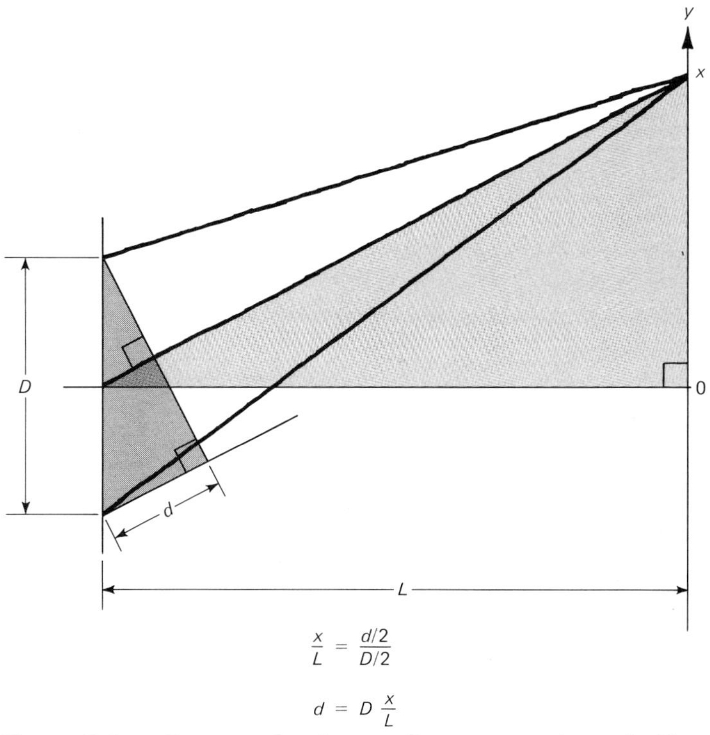

Figure 12-4-9 **Geometry for the two-slit apparatus shown in Fig. 12-4-7. The shaded triangles are similar, leading to the formula given for the distance** d. **In the small-angle approximation, when** D **and** x **are both** $\ll L$, **the distance** d **is a good approximation for the path-length difference to point** x **from the two slits.**

PROBLEM 14

Suppose there are two monochromatic light sources emitting in phase. Prove that at a point for which there is a path-length difference Δl to the sources there is a phase difference given by:

$$\Delta\gamma = \frac{\Delta l}{\lambda} 360^\circ$$

PROBLEM 15

A slightly curving lens is placed on top of an optically flat glass plate and light is directed down onto the combination, as shown in Fig. P-12-15*a*. This is called a *Newton's rings apparatus* because of the concentric light and dark rings seen when it is viewed from above, schematically shown in Fig. P-12-15*b*.

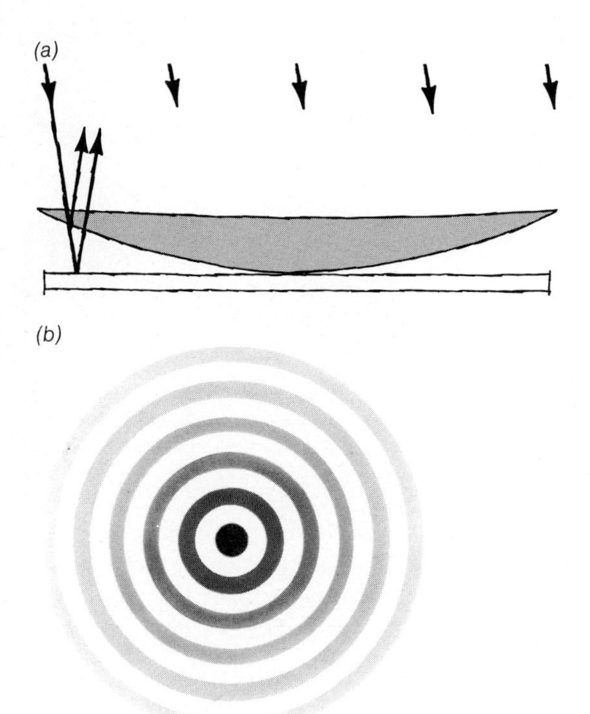

Figure P-12-15

The rings are ascribed to interference between the rays reflected from the top and bottom surfaces of the air space between the lens and the bottom plate. Near the center, when the thickness of the air film is less than one wavelength, there is *destructive* interference, because there is a phase change of 180° for reflection at an air-glass interface and not at a glass-air interface. How thick is the air gap at the location of the third bright ring? Assume illumination with monochromatic light of wavelength 0.50 microns.

PROBLEM 16

For two slits of different widths, such that the light from them has amplitudes in the ratio 2:1, show qualitatively how the two-slit interference pattern differs from the equal amplitude case which has been discussed.

PROBLEM 17

A nonreflecting coating for a lens looks as follows, greatly magnified.

AIR	COATING	GLASS
$n = 1.50$	$n = 1.25$	$n = 1.5$
Light →		

Can you explain qualitatively how such a coating might work by means of destructive interference? Will it work at all wavelengths simultaneously?

As x increases and the path-length difference approaches $\lambda/2$, the phasors add up to produce a decreasing resultant, until completely destructive interference takes place, as in Fig. 12-4-8*c*. For larger values of x the angle between the phasors continues to increase, and they are again completely in phase when the phase difference is one period, or 360°, as in Fig. 12-4-7*d*.

The phasor diagrams in Fig. 12-4-8 provide results for the variation in amplitude of the resultant sine wave for two slits. The intensity distribution on the screen varies proportional to the square of this amplitude, and Fig. 12-4-10 shows how I varies with x, with the results for the four values of x corresponding to Fig. 12-4-8 shown by points marked x. To fully trace out the curve shown in Fig. 12-3-10, it would only be necessary to consider how the phasors add for a variety of phase differences, finding the

Figure 12-4-10 Intensity distribution at the screen of the two-slit interference apparatus. Points marked x are proportional to the squares of the resultant amplitudes found with the phasor diagrams in Fig. 12-4-8.

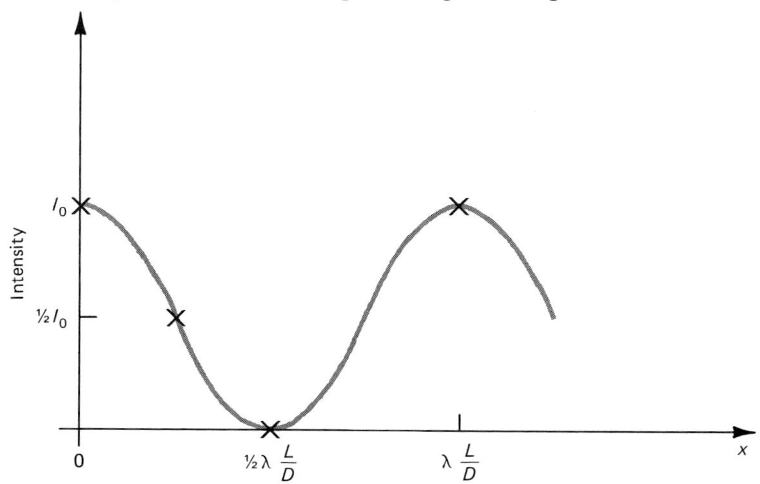

results either graphically or with trigonometry. The unit of intensity in Fig. 12-4-10 is arbitrary, chosen so that the maximum intensity has some value I_0.

The value of phasor diagrams is particularly evident when it is necessary to add many interfering waves instead of two. A very important application of interference and diffraction, called the diffraction grating, is composed of a large number of parallel slits, as indicated schematically in Fig. 12-4-11. For a real grating the number of slits may be 100,000 instead of the 6 shown in the figure, and making such gratings, with 10,000 or more slits per inch, is a technological tour de force. Diffraction gratings are almost indispensible for measuring the wavelengths of light in atomic spectra with high precision. The many-slit interference gives the grating its important property, and diffraction is only incidentally important, for making each slit a source of a spreading wave.

Figure 12-4-11 Schematic view of a six-slit diffraction grating. For a plane wave incident from the left, six diffracted rays interfere to give the resultant intensity at a point x.

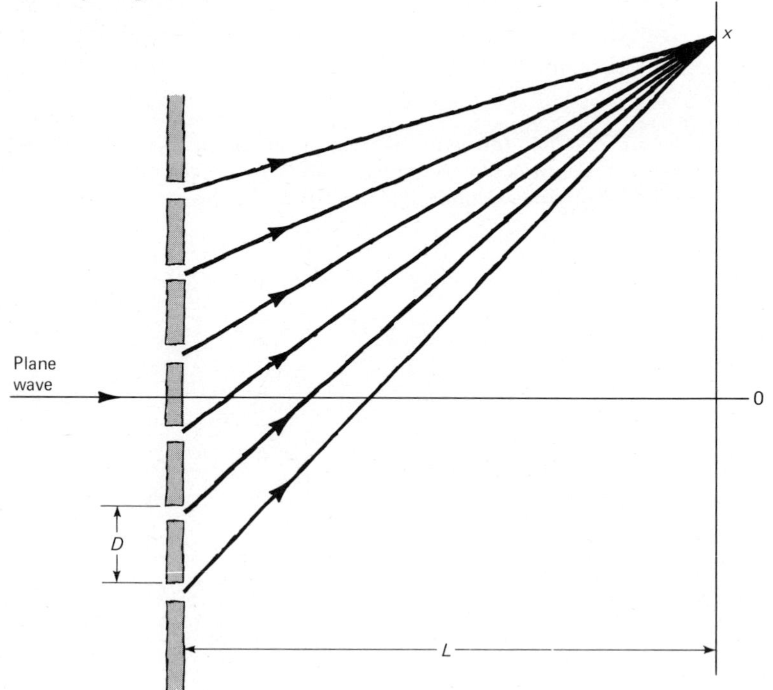

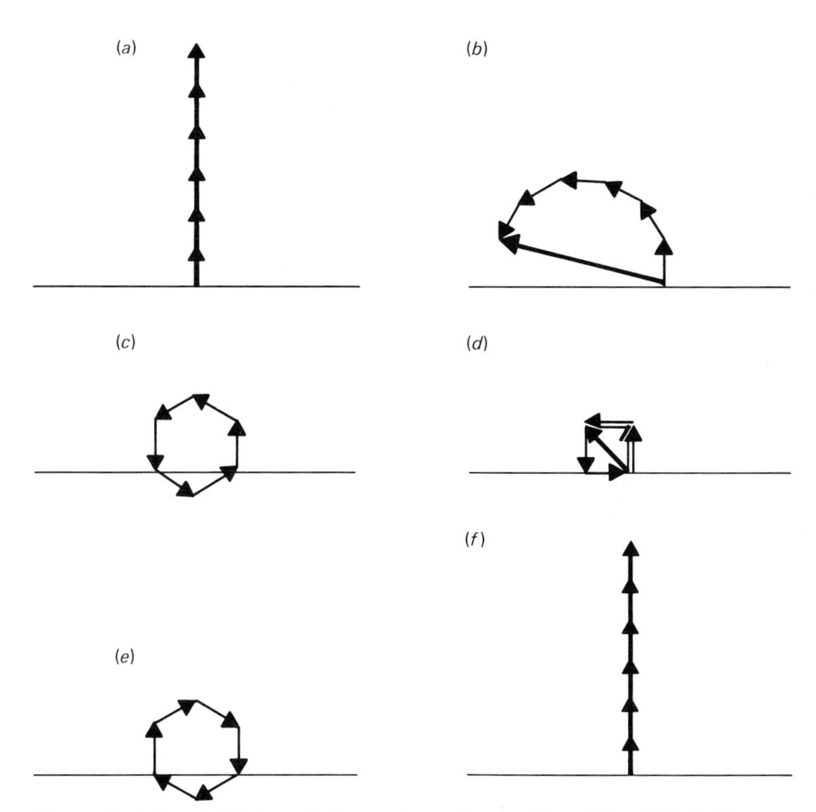

Figure 12-4-12 Addition of phasors from the six slits of Fig. 12-4-11. Figures *(a)* to *(f)* correspond to phase differences for the waves from neighboring slits of 0, 30, 60, 90, 300, and 360°.

Some six-slit phasor diagrams, appropriate to Fig. 12-4-11, and assuming equal light intensity from each slit, are shown in Fig. 12-4-12. At a given value of x, each of the six phasors is spaced from its neighbors by a phase difference which depends on the path-length difference d. As for the two-slit case, d depends on x according to the formula

$$d = \frac{D}{L} x \qquad \text{12-4-6}$$

For six slits, when the path-length difference is 1/6 wavelength and the phase difference is 1/6 period, the resultant then goes to zero, as shown in Fig. 12-4-12c. As x increases past this point the phasors get scrambled up as in Fig. 12-4-12d in such a way

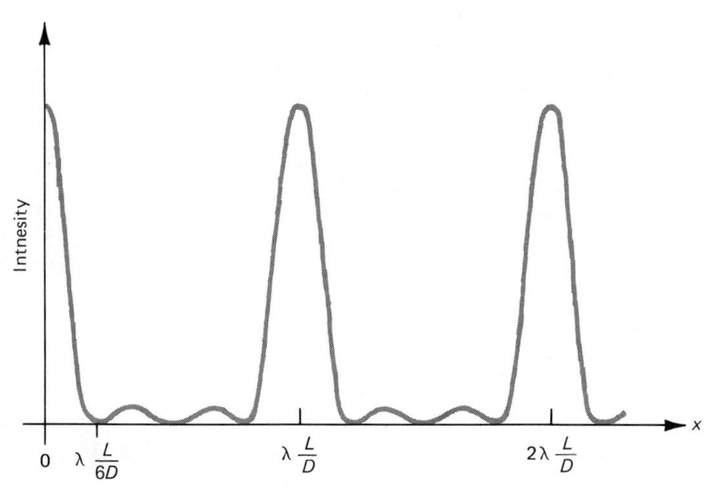

Figure 12-4-13 Intensity distribution for a six-slit interference apparatus, such as that in Fig. 12-4-11.

that secondary small maxima and minima in resultant amplitude occur. Finally, when the path difference d reaches 5/6 wavelength there is again a zero of amplitude, as shown in Fig. 12-4-12*e*, and then a steady increase to a big maximum when $d = \lambda$, shown in Fig. 12-4-12*f*.

The variation of intensity with x for the six-slit interference is thus found to be a series of sharp peaks separated by broad, slightly bumpy valleys, as shown in Fig. 12-4-13. Each peak has a "width" corresponding to a path difference of roughly $\lambda/6$.

PROBLEM 18

Sketch qualitatively the variation of amplitude with x and intensity with x for the interference pattern of three identical slits, spaced at equal distances.

The six-slit result can be generalized to a grating with a very large number of slits, N, for which, by analogy, the width of the big peak corresponds to a path-length difference of about λ/N. If the grating were illuminated with light containing one wavelength, the interference pattern would be as schematically indicated in Fig. 12-4-14*a*. More significantly, the sharpness of the maxima makes it possible to see or photograph two distinct interference patterns for two nearly equal wavelengths as in Fig. 12-4-14*b*. The *resolution* of the grating expresses its ability to distinguish closely equal wavelengths and also indicates the precision with which it can be used to determine the value of a given wavelength. The resolution, determined by the narrowness of the interference peaks, is approximately equal to $1/N$. This means that two wavelengths differing by 1 part in 10^5 can be distinguished by a 100,000-line grating, with $N = 10^5$.

Up to now, the light emerging from each slit has been assumed to be diffracted so that the intensity in all directions of interest was the same. The actual distribution of light diffracted from a slit is called the *single-slit diffraction pattern,* which we are now ready to examine. To begin, the plane wave in the aperture of a single slit is assumed to propagate by means of the superposition of Huygens wavelets, imagined to be emitted across the aperture of the slit. This is illustrated schematically in Fig. 12-4-15. To find the diffracted intensity at a position x on the screen, we really

Figure 12-4-14 Interference pattern from a diffraction grating, with a very large number of slits at a spacing D. In *(a)* the variation of intensity with x is shown for one wavelength of incident light. In *(b)* the pattern for two wavelengths is shown.

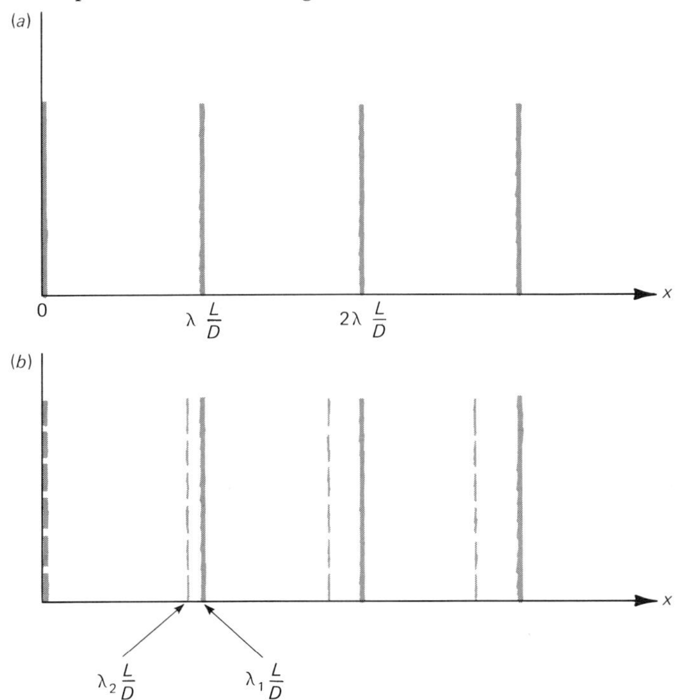

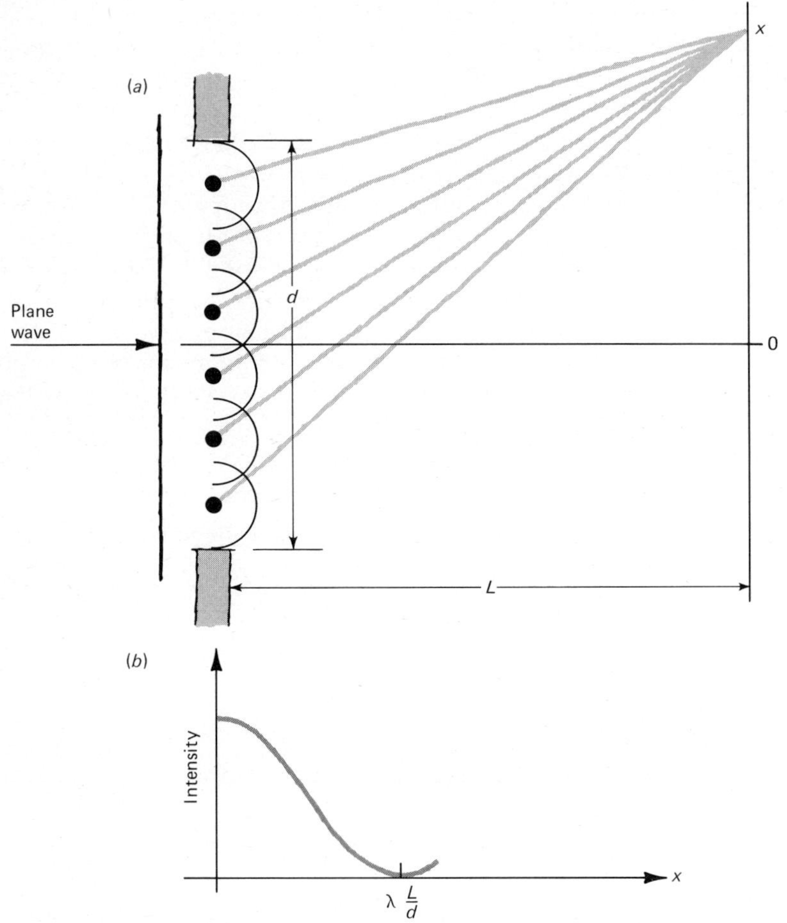

Figure 12-4-15 Single-slit diffraction, approximated with six Huygens wavelets. The observed intensity variation is shown.

need to add up the contributions of an infinite number of infinitesimal Huygens' wavelets. However, as a beginning approximation we will consider only six, as in Fig. 12-4-15.

The six wavelets add just as we have seen for the six-slit grating, as x increases from zero to the value for the first zero of intensity. In order to find the true result for the diffraction pattern, we must now, however, carefully consider the result of increasing the number of interfering waves while keeping the overall slit width the same. This is different than the case for the grating, where we left the original six slits and let the number increase by adding more with the same spacing. In Fig. 12-4-16, the addition of Huygens' wavelets at the first zero is again shown for six wavelets and also for larger numbers, 12, 24, and an infinite number. You can see that increasing the number produces a perfect circle in the limit. The sum of all the wavelets is such that the length around the circle is the same as the length along a straight line at $x = 0$.

Figure 12-4-17 shows the variation of the sum of all the wavelets with x when there is an infinite number of them. After the

Figure 12-4-16 Phasor diagrams showing the sums of the Huygens wavelets for single-slit diffraction at $x = \lambda\,(L/d)$, the first minimum. Beginning with *(a)* the number of wavelets increases from 6 to 12 to 24 to an infinite number. As the number increases, the amplitude of each wavelet is reduced proportionately, and its phase difference with the neighboring wavelet is also reduced proportionately. The pattern thus continues to represent a single slit of fixed width d.

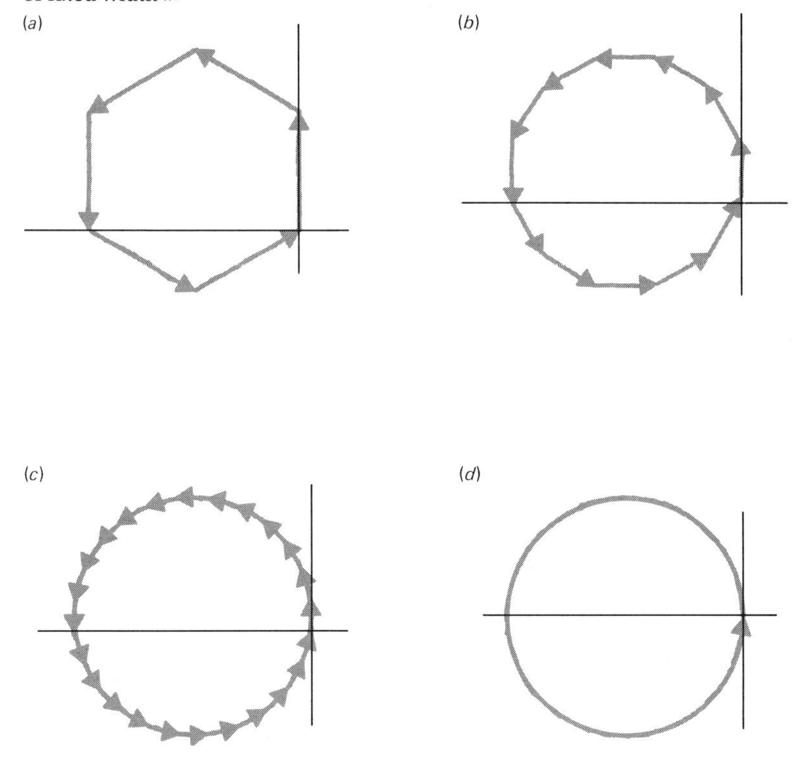

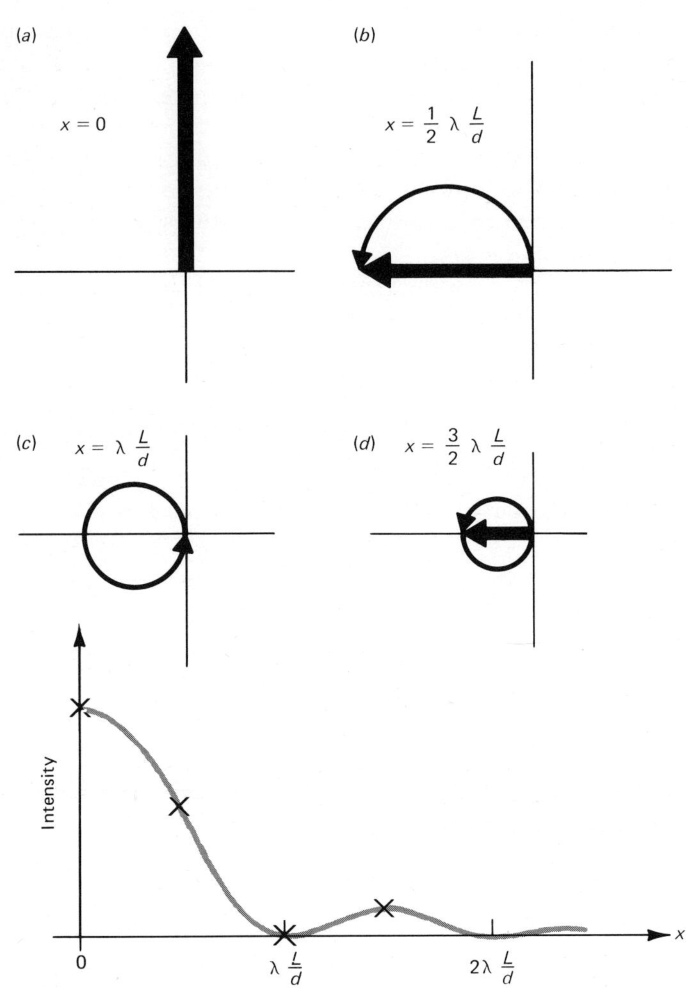

Figure 12-4-17 Phasor addition of an infinite number of Huygens' wavelets to find single-slit diffraction pattern. Points labeled × on the graph of intensity correspond to *(a)* through *(d)* above.

first zero, the largest resultant occurs at 3/2 that value of x, shown in Fig. 12-4-17*d*. At twice the value of x for the first zero there is another one, followed by a still smaller maximum, another zero, etc. The graph of Fig. 12-4-17 shows how the single-slit intensity varies with x.

From Fig. 12-4-17 we can conclude that the discussions of two-slit interference and the diffraction grating have tacitly assumed that the slits were much narrower than the distance between them. In that case, the central maximum in the diffraction pattern covers a wide angle compared to the angular range for the interference maxima and minima. As an example, the true interference pattern for two slits of width d and separation $4d$ is shown in Fig. 12-4-18. The characteristic two-slit maxima and minima are seen to occur inside an *envelope curve*, which is the single-slit diffracted intensity for either slit.

Quantitatively, the diffracted intensity from a single slit is mainly contained in an angular interval of $57.3(\lambda/d)$ degrees, where d is the slit width. This angular interval corresponds to a distance on the screen extending from $x = -\frac{1}{2}L(\lambda/d)$ to $x = +\frac{1}{2}L(\lambda/d)$, which is the region of the central maximum of the diffraction pattern on the screen. Were we to consider a circular aperture instead of a long slit, the pattern would again be found to spread out over a similar region, so that the diffraction pattern of a circular aperture is as shown in Fig. 12-4-19.

Returning now to geometrical optics, we can see that the ideal point images which we have calculated are really, at best, diffraction patterns of the aperture through which light enters the optical system. The diffraction limit for a telescope objective is the angle given by the expression:

Figure 12-4-18 Two-slit interference and diffraction pattern, for slits with dimensions as shown. The dashed single-slit diffraction curve defines the way the intensity of the two-slit interference pattern varies with x.

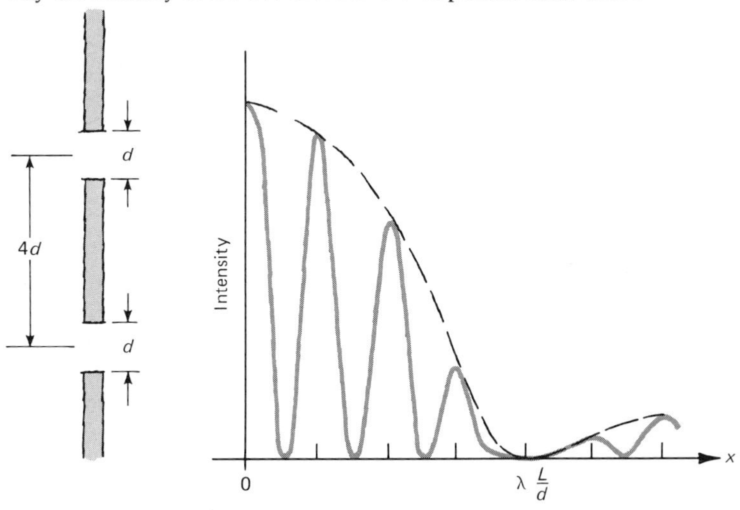

$$\theta = (1.22)57.3\frac{\lambda}{d} \qquad \text{12-4-7}$$

when d is the diameter. For two stars whose light arrives in directions differing by this angle, the images at the focal plane of the objective look like the two overlapping diffraction patterns shown in Fig. 12-4-20. An increase in magnification merely gives a larger view of the overlapping diffraction patterns.

PROBLEM 19

A laser 1 cm in diameter can at best produce a wave which is like a plane wave diffracted by a circular aperture 1 cm in diameter. For such a laser, what would be the beam size at the moon if no extra optics were added? (Distance to the moon is roughly 250,000 m.)

Figure 12-4-19 The image on a screen of the diffraction pattern of a circular aperture of diameter d a distance L from the screen, assuming a plane wave incident.

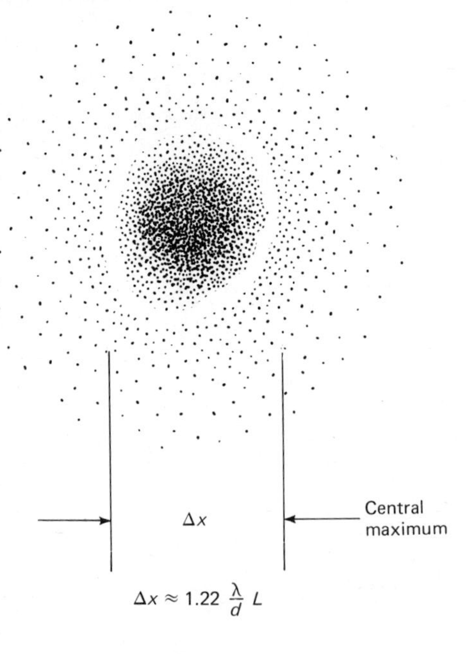

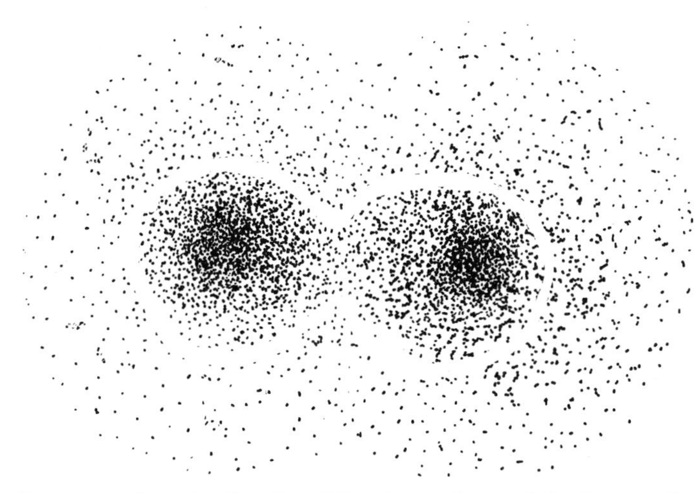

Figure 12-4-20 Overlapping diffraction patterns of the images of two barely resolved stars or point light sources.

PROBLEM 20

Describe how the intensity distribution shown in Fig. 12-4-18 changes if the slit spacing is kept the same but the slits are made narrower. Describe the effect of keeping the slit width the same and increasing the spacing.

12-5 THE LASER

The name laser stands for *l*ight *a*mplification by *s*timulated *e*mission of *r*adiation. Since its invention in 1961, the laser has opened up a very large number of possibilities for basic experiments and for new applications of optics. The uses of the laser tend to capitalize on two of its properties: It is a source of extremely monochromatic light, and it provides, in pulsed operation, enormous instantaneous light intensities. While the name laser contains an "a" that stands for amplification, in its usual use it is more an *o*scillator than an *a*mplifier. It creates optical power at a particular frequency from input power in a different form. (Perhaps the unappealing result of changing the "a" in laser to an "o" has discouraged a name change.)

To understand the laser, it is first necessary to explore some

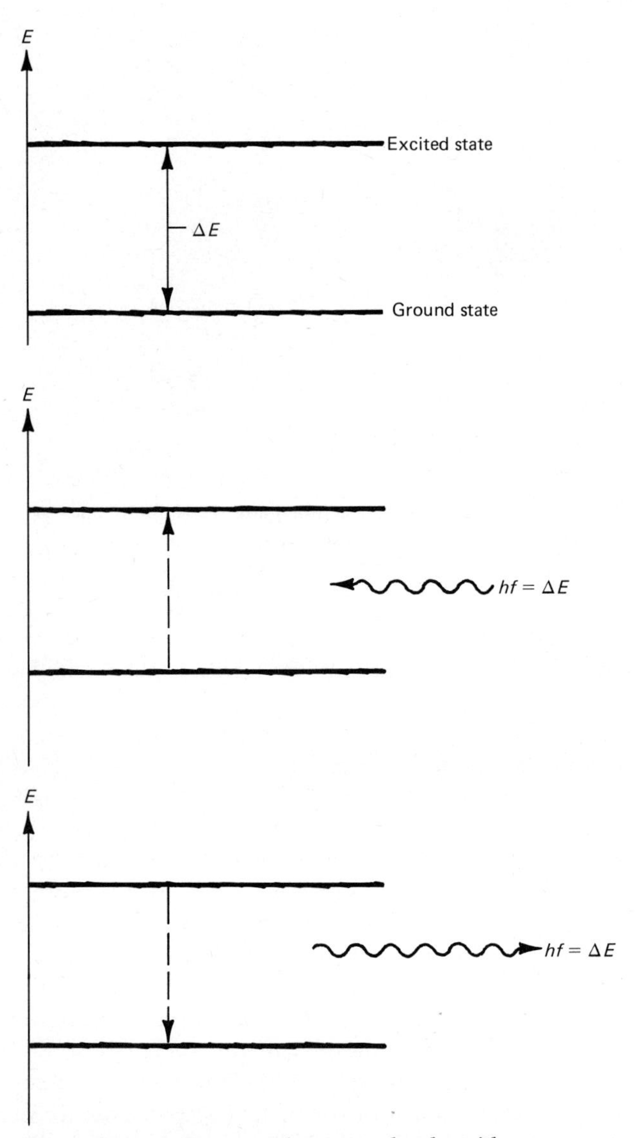

Figure 12-5-1 Two atomic-energy levels with separation ΔE. Transitions from the ground state to the excited state can occur by photon absorption and from the excited state to the ground state by photon emission.

basic atomic physics. Consider two states of an atom: its lowest energy state, or ground state, and an excited state, whose energy is higher by an amount ΔE, as shown in Fig. 12-5-1. Provided ΔE is large compared with the thermal energy kT, which we will assume throughout this discussion, a large number of undisturbed identical atoms with energy levels as shown will almost all be found in the ground state. If for any reason a particular atom gains energy ΔE and makes a transition up to the excited state, it will drop back to the ground state by *spontaneous emission* of a photon with $hf = \Delta E$ in a time which is usually of order 10^{-8} sec. Atomic spectra are normally observed as a result of such spontaneous emission.

If an excited atom is illuminated by photons with energy very nearly equal to the energy ΔE, then it will drop to the ground state sooner than if it were left to emit a photon spontaneously. The presence of an environment of *right energy photons* enhances the probability of the transition to the ground state and leads to an additional probability for *stimulated emission* over and above the probability for spontaneous emission. Furthermore, a photon emitted by stimulated emission prefers to travel in the same direction as the stimulating photon, and in phase with it.

The phenomenon of stimulated emission means that if there is one photon in a given state, a second one is more likely to enter the state. The more photons there are in a state, the more likely another will join them. While electrons obey an exclusion principle, which forbids many in one state, photons do not. In fact, just the opposite is true for photons. They appear to be gregarious, and they like to congregate together.

Particles which exhibit this gregarious photonlike behavior are called *bosons,* while those which exhibit the exclusiveness of electrons are called *fermions* (after two physicists, Bose and Fermi, who studied this behavior). In nature, all the basic particles are found to belong to either one or the other of these classes.

For a population of atoms with energy levels as in Fig. 12-5-1, in thermal equilibrium, with almost all of them in the ground state, the likelihood of a photon being absorbed is much greater than the probability that it will induce stimulated emission of another photon. This is purely because there are many more ground-state atoms to absorb the photon compared with the number of excited atoms which it could stimulate to emit. However, if there were *more* excited atoms than ground-state atoms, spontaneous emission would become more likely. In such a case one photon would be likely to produce a second, these two to produce four, and so on in an avalanche, all traveling in the *same direction* with the *same phase.*

When there is an excess of atoms in the higher of two energy states, which can lead to an avalanche of stimulated emissions, we say there is a *population inversion.* This is a necessary condition for laser action. The term *inversion* is used to signify relative populations unlike the natural state of affairs. Left to come to equilibrium, any group of physical systems will tend to their lowest energy states, so that, normally, higher energy states are less populated than those of lower energy. For the first successful laser, a population inversion was achieved between two states of chromium ions in a material known as synthetic pink ruby. Pink ruby consists of a small percentage of such ions embedded in a crystal of aluminum oxide. The ions behave almost like free atoms.

The energy-level diagram for the chromium ion in ruby is very

Figure 12-5-2 Three chromium ion energy levels used for the ruby laser. Absorption of light pumps the atoms to level 2, from where they rapidly drop to level 3. Laser action accompanies the transitions from 3 to 1, with emission of photons with energy E_{13}.

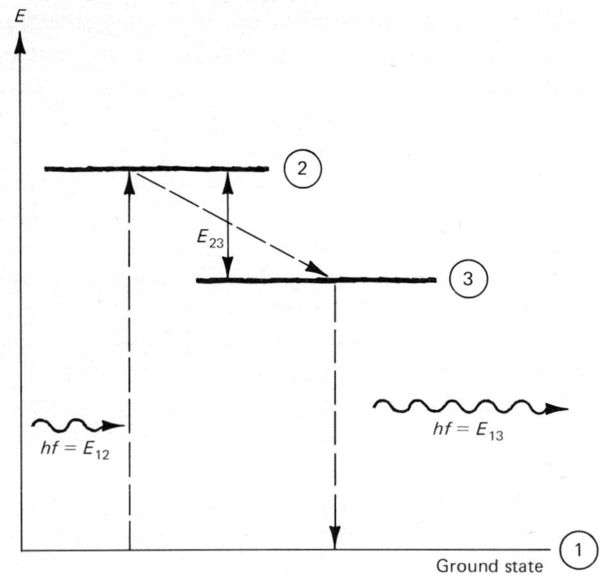

complicated, but in Fig. 12-5-2 the three levels which are essential to laser action are shown. Beginning with a sample of chromium ions in the ground state, many of them can be excited to the state numbered 2 by energy absorption from an intense light source which provides photons of the appropriate energy, E_{12}. Such excited ions tend to make a rapid transition to level 3, by giving the energy E_{23} to the crystal in the form of vibrational energy. The transition from level 2 to level 3 is much more likely than that from 2 back to 1.

When the ions have reached level 3, they cannot make easy spontaneous transitions to the ground state, or to any other state. The state numbered 3 is called a *metastable* state, "almost stable" because the probability for radiation to lower states is very small. The reason for this small transition probability is subtle and depends upon details of the wave functions for state 3 and the lower states, which we will not discuss here.

T. H. Maiman studied the energy levels in ruby, described above, in 1960, and realized that there was an opportunity to create a population inversion between levels 1 and 3 and to create the first working laser. The idea of the laser had been previously explored in some detail following the successful creation in 1954 of a similar device, the maser, which produced radio waves, rather than light. Maiman succeeded in creating a ruby laser in the manner shown schematically in Fig. 12-5-3. The flashtube emitted a very large amount of light for a time of a few milliseconds. During this time Maiman succeeded in exciting more than half the chromium ions out of the ground state, exciting them to a whole group of states like level 2 of Fig. 12-5-2. From these states they quickly dropped to the metastable state, level 3.

When the number of ions excited to level 3 exceeded the number in level 1, an intense flash of monochromatic light, consisting of photons with energy E_{13}, was observed. Even though the spontaneous transition rate from level 3 was very low, stimulated emission resulted in a very rapid transition rate after the population inversion was established. In order to enhance the likelihood that a photon would produce others by stimulated emission, Maiman utilized a ruby rod with *highly reflecting* ends. Those photons emitted longitudinally along the rod therefore made many round trips back and forth through the rod, and they were almost certain to induce stimulated emission of still more photons moving in the same direction.

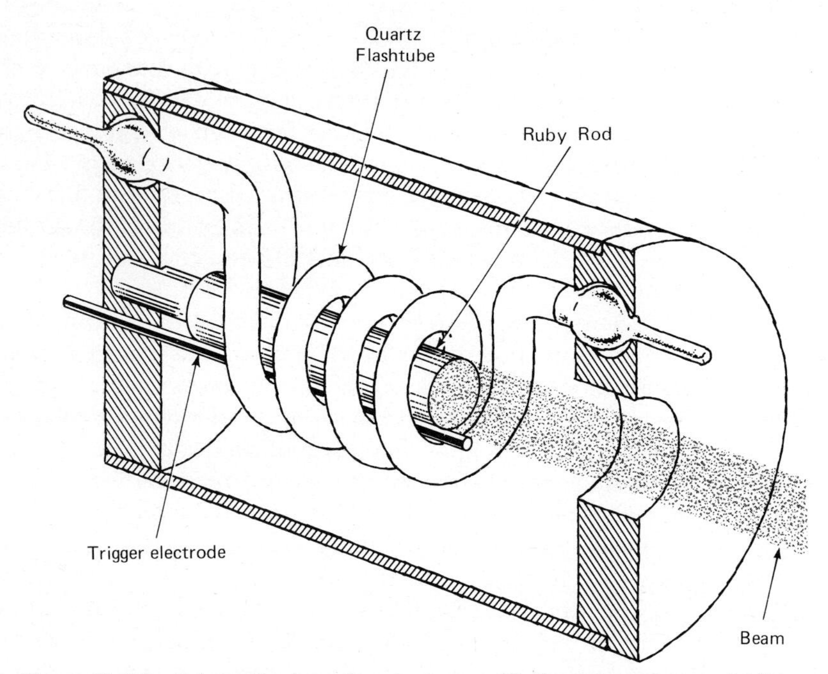

Figure 12-5-3 Schematic view of a ruby laser, with the power source and trigger electronics for the flashtube not shown. The ruby rod is typically about 1 cm in diameter and 10 cm long. The ends are made highly reflective by vacuum evaporated coatings.

PROBLEM 21

Explain why, if a laser is excited with light, the frequency of emitted light must be less than the frequency of the exciting light.

The rod with reflecting ends made the stimulated emission from the laser consist almost entirely of photons with very nearly the same direction and frequency. A very high intensity pulse of extremely monochromatic light, in the form of a nearly perfect plane wave, was produced inside the rod. By making the mirror at one end slightly transmitting, some of this light could leak out, creating an intense beam of extremely coherent light. The coherence of this beam has two aspects. First, because it is so monochromatic, it varies with time almost as a perfect sine wave. As a result, different parts of the laser beam can produce interference with each other for path-length differences of many inches, corresponding to more than 10^5 wavelengths. Second, because the light emerges almost as a nearly perfect plane wave, it can be focused to a point whose size is a very small fraction of a millimeter in diameter.

The good coherence in both space and time has made the laser a very important research tool. For the ruby laser, the time coherence is not impressive compared with other types of lasers, but the spatial coherence is useful for producing enormous optical intensities. In modern ruby lasers, the laser action can be inhibited until a large population inversion has been achieved, and then switched on (a process called Q switching). In this case, the stimulated emission occurs in a time of 10^{-9} sec, or even less. The emission of more than a joule of light energy can be achieved in 10^{-9} sec, which corresponds to an instantaneous optical power of more than 1 billion watts! Furthermore, this energy can be focused into a volume of much less than a cubic millimeter, leading to light waves with electric fields intense enough to literally rip matter apart. The ability to propagate this extremely high-intensity light through matter, albeit for very short times, has produced a great deal of knowledge of hitherto unexplored physical phenomena.

For some purposes, the pulsed nature of the ruby laser is not advantageous, but soon after Maiman's success other workers succeeded in making a laser which is a source of steady light. This is the neon gas laser, and the basic reason for its success is that a transition is stimulated from a metastable state to a state *other than* the ground state (i.e., to another excited state with lower energy than the metastable state.) This state is normally not appreciably populated. As a result, there is a population inversion almost as soon as any metastable atoms exist. For ruby, where the transition is to the ground state, enormous excitation energy is needed to establish the population inversion. For neon, since most of the atoms can be left in the ground state, a very modest excitation power can create the necessary inversion.

PROBLEM 22

Sketch an energy-level diagram for a helium-neon gas laser like that previously given for the ruby, except that you add the important fourth level which makes population inversion so easy to attain. Indicate the transition responsible for laser action.

Neon gas can be excited to satisfactorily produce the necessary population inversion by passing a moderate electric current through it, as in a neon sign. Actually a mixture of neon and helium is most efficient. A neon-gas laser is illustrated schematically in Fig. 12-5-4, and, as in the ruby laser, light emitted along a particular direction is reflected by mirrors back and forth through the neon. The useful beam exits through one mirror, usually a few millimeters in diameter, which is made slightly transmitting. The spatial coherence is excellent, and beautiful interference effects can be easily observed by placing slits in the beam of the helium-neon laser.

Figure 12-5-4 Schematic view of a helium-neon gas laser. The RF generator powers the gas discharge which excites the neon atoms. The bellows allow motion of the mirrors for precise alignment. Some gas lasers have the end mirrors external to the gas discharge tube.

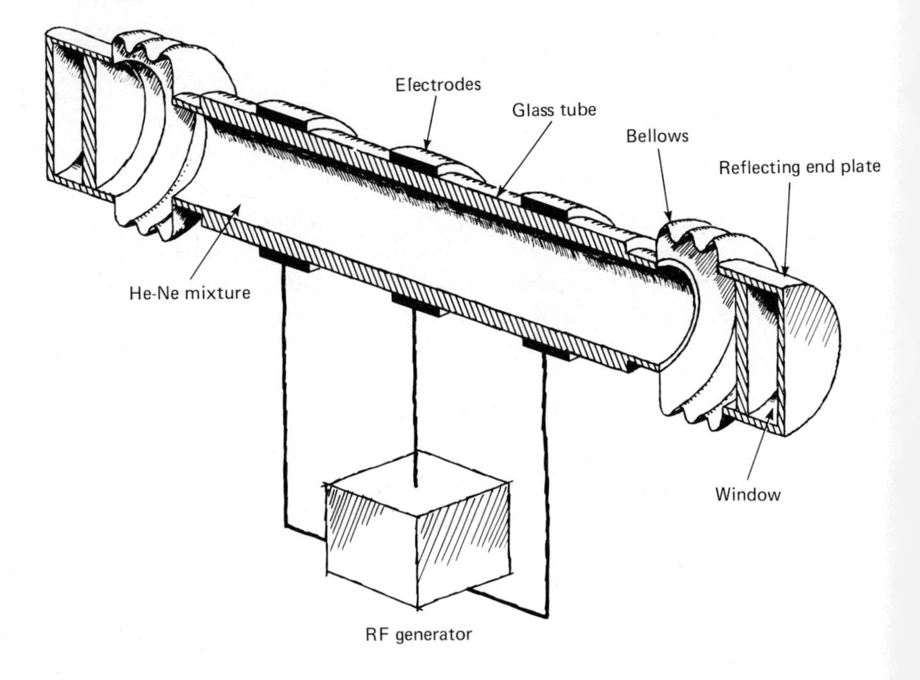

The time coherence, or monochromaticity, of a neon laser is such that for times of order 1 sec the frequency may be constant to within one part in 10^{13}. The light from even a simple and inexpensive gas laser has sufficiently constant frequency that steady interference effects can be observed for path-length differences of a meter or more, greater than 10^6 wavelengths. Furthermore, the gas laser can be built so easily and cheaply that it enjoys a variety of relatively mundane technical uses, as a kind of super flashlight.

One fascinating, and sophisticated, area of optics, known as *holography*, has been made really practical by the development of the gas laser. Holography provides a means for recording on film a record of both the amplitude *and phase* of wavefronts. A wavefront can be the resultant of all the light scattered off a three-dimensional object illuminated by a laser beam, as in Fig. 12-5-5.

Figure 12-5-5 Schematic view of a setup for making a hologram. The incident plane wave is split, with one half illuminating the object and the other half being deflected so that it overlaps the scattered light from the object at the photographic plate. Interference between the scattered wave and the *reference* wave, deflected by the prism in the figure, produces the hologram image. The coherence required for the incident light beam is easily provided by a gas laser, but would otherwise be hard to achieve.

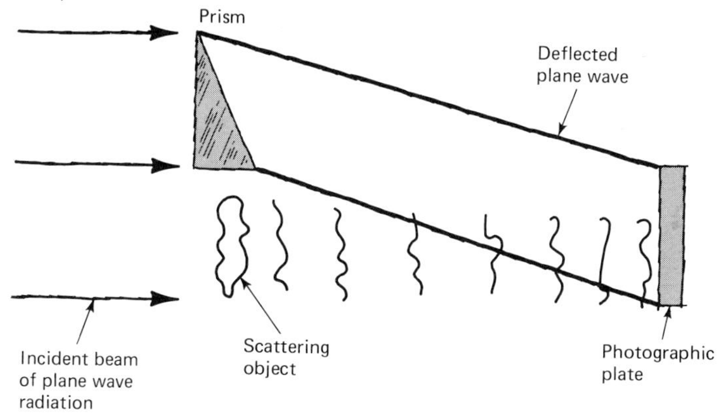

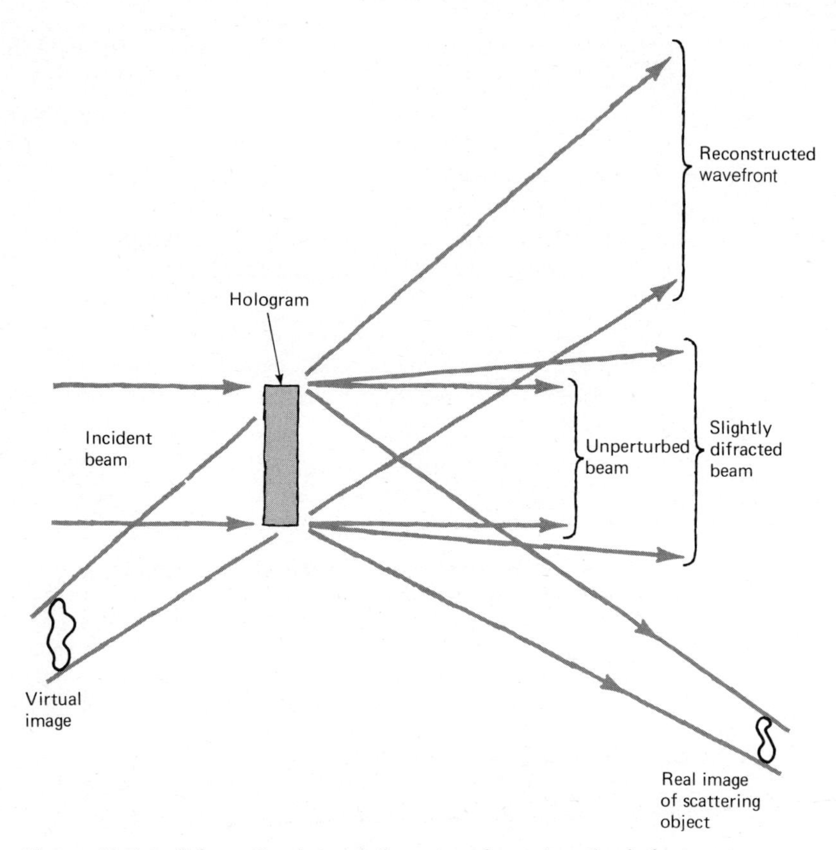

Figure 12-5-6 Schematic view of the setup for using the hologram to produce a reconstructed wavefront and an image. The incident beam must be coherent over the whole area of the hologram and is typically generated by a laser. Diffraction by the hologram produces the image and the reconstructed wavefront. The reconstructed wavefront comes from a virtual image behind the hologram.

A reference beam and the scattered light from the object produce an innocuous looking image on film, which is actually a very complicated interference pattern. This interference pattern is generated by combining the scattered light with the reference beam and contains information about both the amplitude and phase of the scattered light. When a laser beam is passed through the film image, called a *hologram,* diffraction by the interference pattern which constitutes the hologram can recreate a nearly perfect copy of the original scattered wave, called a *reconstructed wavefront.*

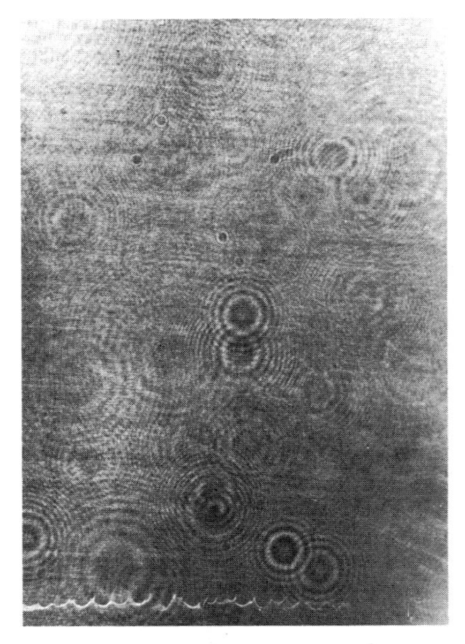

Figure 12-5-7 Images of a three-dimensional scene reconstructed from a hologram (at left) according to Leith and Upatnieks, and Stroke. Images were photographed by a camera looking through the hologram, focused and inclined to reveal the nature of the reconstruction. Photographs at top were taken with a small camera aperture, at two angles, to show parallax; those below, with a large camera aperture, at the same angle with respect to the hologram, to show the three-dimensional field depth. *(Courtesy of Prof. G. W. Stroke.)*

The "playback" of a hologram is schematically shown in Fig. 12-5-6. The holographic replica of the scattered wavefront can re-create for an observer an almost perfect sensation of viewing the original object in three dimensions. As the observer moves, the object appears in different perspectives, just as if it were real! In Fig. 12-5-7, a hologram and photographs taken of its reconstructed wavefront are shown. However, viewing the photographs cannot begin to reproduce the sensation of viewing the actual holographic image.

SUMMARY

Ray optics, based on *Snell's law* of refraction, can be used to understand lens systems, except for small effects due to the wave nature of light. The *simple lens* is characterized by its *focal length* f, the distance at which parallel rays are brought to a focus. For object and image distances p and q, respectively,

$$\frac{1}{p}+\frac{1}{q}=\frac{1}{f}$$

Depending on the values of f and the object distance, the image can be *real* or *virtual.*

Both the *compound microscope* and the *telescope* consist of combinations of an *objective* lens which produces a real image and a magnifying *eyepiece.* The eyepiece, which is in essence a simple magnifier, produces an enlarged virtual image of the image formed by the objective. The *reflecting* telescope utilizes a *parabolic mirror* as objective, to avoid chromatic aberration.

Narrow apertures in optical systems exhibit behavior dominated by *diffraction.* For a multislit system called a *diffraction grating,* the interference of a large number of diffracted rays can be analyzed with *phasors.* For N slits, the interference pattern is a series of *sharp maxima,* such that patterns for wavelengths differing by a fraction $1/N$ can be differentiated. Since N can be 10^5 or greater, this leads to an instrument which is extremely useful for high resolution spectroscopy.

By using multislit interference theory, the diffraction pattern of a *single slit* can be inferred. The diffracted light mainly emerges at angles up to about *57.3(*λ/d*) degrees,* where d is the aperture width.

The *laser* is an optical instrument based on the phenomenon of *stimulated emission*. By exciting a large number of atoms to a *metastable state*, a *population inversion* can be achieved, such that stimulated emission leads to *highly coherent light* consisting of a very large numbers of photons emitted with almost the same direction and phase. The coherent light leads to the possibility of interference effects with very *large path differences* for the beams and to the production of *extremely high-intensity* focused beams of light.

THE SPECIAL THEORY OF RELATIVITY

13-1 GALILEAN RELATIVITY

A theory of relativity describes how the laws of physics depend upon an observer's *frame of reference*. For example, in the case of classical mechanics, Newton's laws are used to describe how forces affect the motion of a body. The frame of reference is what we think of as fixed, what we measure motion with respect to. In discussing relativity we often speak of *observers*, each thinking of *himself* as fixed, as ways of expressing different frames of reference, fixed with respect to the observers. Suppose, for example, we consider a stone freely falling from the roof of a tall building. Its motion might be viewed by three observers: one stationary on the ground, a second driving by in a car, and a third—for the sake of argument only—freely falling next to the stone. Each of these observers represents a different frame of reference to which the motion of the stone might be referred. For example, to the freely falling observer, the stone is at rest.

In the example above, two observers, those on the ground, are in uniform motion with respect to each other (assuming the car is not accelerating). The third is accelerating (downward) with respect to either of the others. The *special* theory of relativity, which we will discuss in this chapter, describes how the laws of physics are related for observers in uniform motion with respect to each other, while the *general* theory, which we will not discuss, includes accelerated relative motion. The word *special* really stands for *restricted*, in the context of relativity theory.

While Einstein's special theory of relativity may perhaps have "put relativity on the map," it was not the first relativity theory. After the development of classical mechanics, as described by Newton's laws, a relativity theory called classical, or galilean, relativity was thought to be correct. Later, the development of electromagnetic theory, particularly as applied to electromagnetic waves, caused a crisis in relativity which was resolved by Einstein. Einstein's theory, and some of its quite startling consequences, will be the main subject of this chapter, but to help grasp some of the concepts of relativity while on familiar ground we will begin with a discussion of galilean relativity.

Galilean relativity relates to mechanics, as described by the second law:

$$\vec{F} = m\vec{a} \qquad 13\text{-}1\text{-}1$$

The simplest everyday application of Eq. 13-1-1 is the motion of a freely falling body. A force F, arising from the gravitational attraction of the earth, causes a body released from rest to accelerate toward the center of the earth. In a small region of space, like a laboratory, the gravitational force is constant, and therefore produces a constant downward acceleration.

It is part of our experience, and it was also part of Galileo's, that if an object is dropped in the cabin of a ship moving with constant speed on a smooth sea (or perhaps sometimes in a jet airliner for us), Eq. 13-1-1 still seems true. Precise quantitative measurements verify this impression. However, on a ship that is bouncing in a rough sea, or on an airplane that is changing its speed or direction, "free fall" does not seem to occur in a normal way. The evaluation of motion takes place by a comparison of the moving object with its environment. In a rough sea the ship may be rapidly accelerated in various directions, which leads to an impression of "peculiar" paths for a body in free fall, curving toward one wall or another, for example.

On a ship we might simply describe the cabin as our frame of reference, but a convenient, more specific frame would be a rectangular (x, y, z) coordinate system fixed with respect to the cabin. The coordinate system could, for example, have its origin in one corner of the cabin, as shown in Fig. 13-1-1. Motion of a falling body is then described with respect to this frame of reference by specifying the time variation of its coordinates.

Figure 13-1-2 shows graphically the time variation of the coordinates of a freely falling body in the cabin frame of reference

Figure 13-1-1 An idealized sketch of a ship's cabin. A convenient frame of reference, fixed with respect to the cabin, would be the rectangular coordinate axes shown.

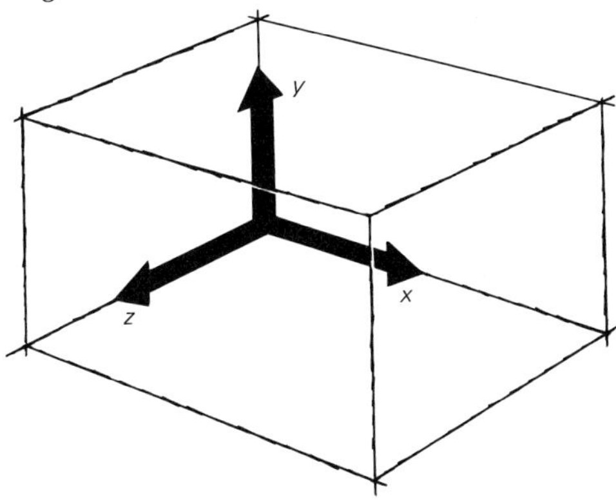

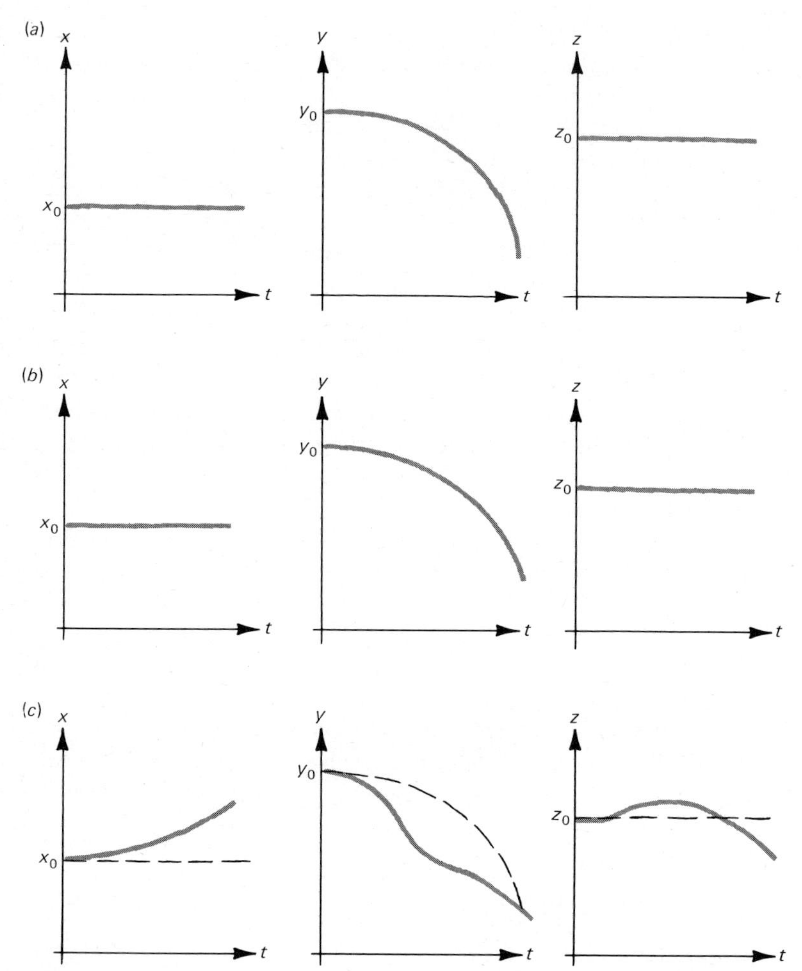

Figure 13-1-2 **Motion of a body in the cabin coordinate system, released from rest at x_o, y_o, z_o. Case (a), ship at rest with respect to earth; case (b), ship in uniform motion; and case (c), ship accelerating with respect to earth. The ship is observed to be an inertial frame in case (a) and case (b). In these cases the motion is described correctly by the vector equation $\vec{F} = m\vec{a}$ with F being the downward force of gravity. In case (c) spurious accelerations apparently occur.**

under three conditions: when the ship is anchored in a smooth harbor, in smooth motion on a calm sea, and on a rough sea. The body in free fall is seen to obey Newton's second law with respect to the cabin reference frame so long as the frame is either at rest with respect to the earth or in uniform motion.

PROBLEM 1

Suppose the ship to which Fig. 13-1-2 refers is in a rough sea and momentarily has an acceleration $a_y = -g$. What does Fig. 13-1-2*a* look like if the ball is released at the time of such an acceleration?

In classical mechanics, a reference frame in which newtonian mechanics is valid is called an *inertial frame*. The galilean principle of relativity can be summed up as follows: "The laws of mechanics are true in every inertial frame. All frames in uniform motion with respect to an inertial frame of reference are themselves inertial frames." The reason for calling this a *relativity* theory is that if it is true, the physical laws do not permit the selection of any special reference frame, with respect to which *absolute motion* could be defined. The laws are equally true, or *invariant,* in all inertial frames, correctly describing motion relative to the frame in use, and giving no basis for choosing between one inertial frame or another.

We will now discuss the motion of a freely falling body as simultaneously observed in two reference frames, shown in Fig. 13-1-3, a laboratory frame (with x, y, z axes) and a rocket frame, rockets being more modern than ships (with x', y', z' axes). The

Figure 13-1-3 A ball dropped from rest in the laboratory frame, viewed in that frame and in a rocket frame moving with relative velocity v_o as indicated. At $t = 0$, the ball was released from the origin of the laboratory frame, and at that time the origin of the rocket frame coincided with the origin of the lab frame.

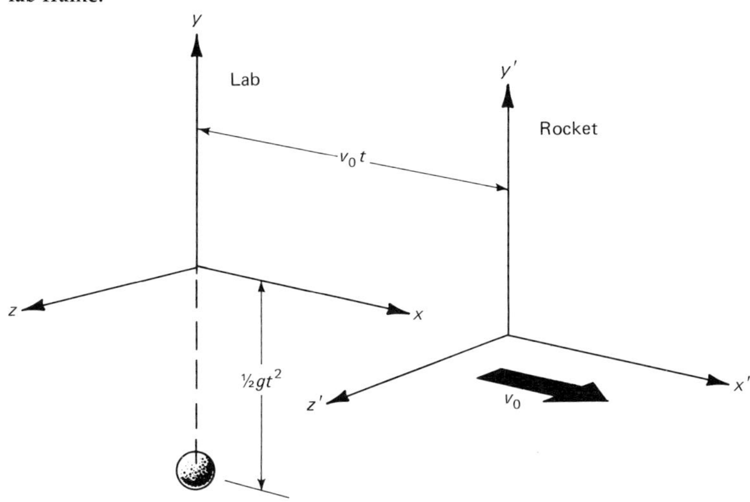

laboratory observer and the rocket observer are hypothetical experimenters who record motion with respect to coordinate systems fixed in their respective frames of reference. As shown in Fig. 13-1-3, a ball is dropped from rest in the laboratory frame. The rocket is in uniform motion with respect to the laboratory with velocity v_o in the x direction, so that the ball is apparently moving with a velocity $-v_o$ in the x' direction of this reference frame. Considering this example, we will explore the ideas of galilean relativity by means of a *thought experiment*, which is a powerful technique for studying the consistency of physical ideas (but not necessarily the best way to find out how the world really behaves).

Since the ball is dropped from rest in the laboratory, the laboratory observer can describe the motion as follows:

$$\begin{aligned} x &= 0 \\ z &= 0 \\ y &= -\frac{1}{2}gt^2 \end{aligned} \qquad 13\text{-}1\text{-}2$$

These are the solutions of the newtonian vector equation $\vec{F} = m\vec{a}$ for motion starting from rest with a force $F = mg$ in the $-y$ direction.

The rocket observer sees the ball initially moving to the left in the $-x'$ direction with velocity v_o. Then he sees an acceleration in the $-y'$ direction caused by the gravitational force after the ball is released. He finds the following equations of motion for the ball, with t being the time measured from when the ball is released in the laboratory:

$$\begin{aligned} x' &= -v_o t \\ z' &= 0 \\ y' &= -\frac{1}{2}gt^2 \end{aligned} \qquad 13\text{-}1\text{-}4$$

These are solutions of the vector equation $\vec{F}' = m'\vec{a}'$, where quantities measured in the rocket reference frame are denoted by primes ($'$), *provided* the following assumptions are made:

$$\begin{aligned} F' &= F \\ t' &= t \\ m' &= m \\ (v'_x)_{t'=0} &= -v_o \end{aligned} \qquad 13\text{-}1\text{-}5$$

To the inventors of Galilean relativity, all these assumptions seemed intuitively reasonable. Besides, their relativity theory worked. As we shall see later, Einstein's theory boldly rejected some of this intuition.

The overall consistency of galilean relativity can now be checked, by comparing the answer, Eq. 13-1-4, with the result found directly from Eq. 13-1-2 by a *coordinate transformation.* There is a simple intuitive relationship between the x, y, z, t coordinates and the x', y', z', and t' coordinates of any point in space at a given time:

$$\begin{aligned} x' &= x - v_0 t \\ y' &= y \\ z' &= z \\ t' &= t \end{aligned} \qquad 13\text{-}1\text{-}6$$

The first equation of 13-1-6 tells us the two coordinate systems are moving with relative velocity v_0 along the x direction, and the next two say that there is no relative motion along the y and z directions. The final equation says clocks don't change due to relative motion. At $t = 0$, $x' = x$, $y' = y$, $z' = z$, and $t' = 0$, signifying that the origins of the two coordinate systems were at the same point at $t = 0$ and the clocks both read zero. The coordinate transformation, Eq. 13-1-6, is called the *galilean transformation,* and can be used to find the equations of motion as seen in the rocket frame directly from Eq. 13-1-2 for the laboratory frame. The results are indeed the same as those of Eq. 13-1-4.

We can now sum up the conclusion of our thought experiment as follows: The galilean *principle of relativity* was found to be consistent with the galilean *coordinate transformation,* provided force and mass, or their ratio, are *invariant* in different inertial frames. The italicized terms in the preceding statement are the important aspects of the relativity theory. The *principle of relativity* is that newtonian mechanics holds true in all inertial frames, and the *coordinate transformation* between such frames is the intuitively obvious one for uniform relative motions. Galilean relativity can be described briefly by the statement "Physics is invariant under galilean transformations," but the physics it clearly applies to is simply mechanics. The fact that "physics is invariant" means the laws (the equations) stay the same. In addition, the invariance of some quantity, like mass, constitutes a detailed discovery about nature, implied by the success of relativity theory.

PROBLEM 2

Verify that applying the galilean transformation to Eq. 13-1-2 leads to Eq. 13-1-4.

PROBLEM 3

A jet airliner is in smooth flight at an airspeed of 600 miles per hour. However, it has a headwind of 100 miles per hour. Find the ground speed by a galilean transformation.

There is still another way to establish the connection between Newton's law, the galilean transformation, and galilean relativity, independent of any specific example or thought experiment. The second law, $\vec{F} = m\vec{a}$, can be shown to be invariant *in general*, for any galilean transformation with relative velocity $\vec{v}_0$. If $\vec{a}$ is the rate of change of $\vec{v}$, then $\vec{a}'$, the rate of change of $\vec{v}'$, is the rate of change of $\vec{v} + \vec{v}_0$. Since $\vec{v}_0$ is constant, its rate of change is 0, and $\vec{a}' = \vec{a}$. Thus, if $\vec{F}' = \vec{F}$ and $m' = m$, $\vec{F}' = m'\vec{a}'$, and Newton's law thus holds under any galilean transformation.

The fact that a particular, plausible, coordinate transformation can be used to establish the invariance of physical laws in different frames is a powerful proof of the *internal* consistency of a relativity theory. However, the theory, the physical laws, and the coordinate transformation may all still be wrong. For velocities much less than the speed of light, galilean relativity is experimentally found to be correct. However, for velocities near the velocity of light, the theory is found to be wrong in almost every way! It thus turns out to be only an approximation of the truth, usually valid in the realm of everyday experience.

13-2 THE INVARIANCE OF c

Electromagnetic theory put galilean relativity to a test which it ultimately failed. The trouble began in 1861 when James Clerk Maxwell predicted the existence of traveling electromagnetic waves, with velocity equal to the velocity of light. In the years following this prediction, it was well verified in many experiments. With the confirmation of Maxwell's equations describing electromagnetism, the scope of established physical laws was extended beyond newtonian mechanics. And Maxwell's equations were found *not* to be invariant under galilean transformations. A general proof of invariance, such as the one presented for Newton's second law in the previous section, could not be constructed.

PROBLEM 4

According to Maxwell's equations, two parallel currents exert an attractive force on each other, and antiparallel ones repel, as shown in Fig. P-13-4*a*. Coulomb's law is also part of Maxwell's equations. Consider the arrangement of charges shown in Fig. P-13-4*b*, at rest in the laboratory. Then consider them as viewed in a moving frame, moving along the direction of the lines of charge. According to galilean relativity the electromagnetic force should be the same in both frames. Is it, according to Maxwell?

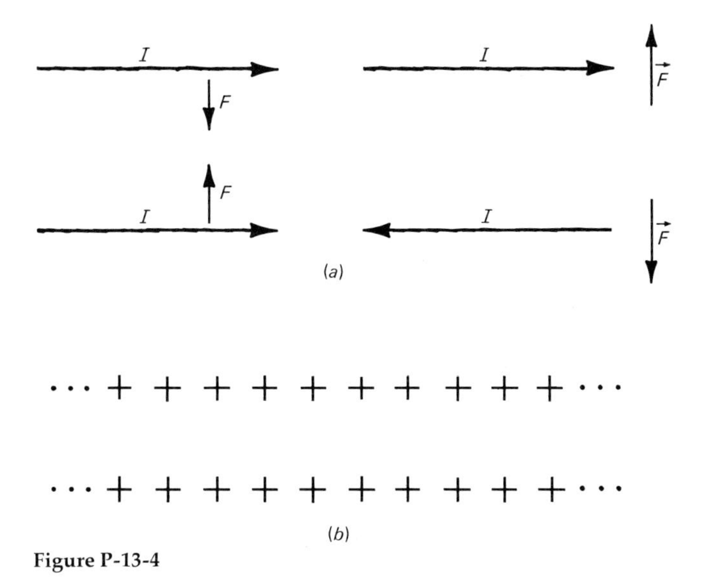

Figure P-13-4

The principle of relativity embodied in the notion that there was no preferred inertial frame was lost, and it appeared that the preferred frame was distinguishable as the one in which Maxwell's equations were true. The velocity of light could be used as an indicator for this special *Maxwell frame of reference*, since it would really be the velocity with respect to this preferred frame.

Figure 13-2-1 shows a light flash in the *Maxwell frame*, with a wave propagating as predicted, in all directions with $v = c$. Figure 13-2-2 shows a flash emitted in the same frame by a source moving with velocity v_0. According to Maxwell's equations the wave again travels out from the origin with $v = c$, independent of the source velocity. There is a reasonable, intuitive model for this kind of behavior. We can imagine that electromagnetic waves

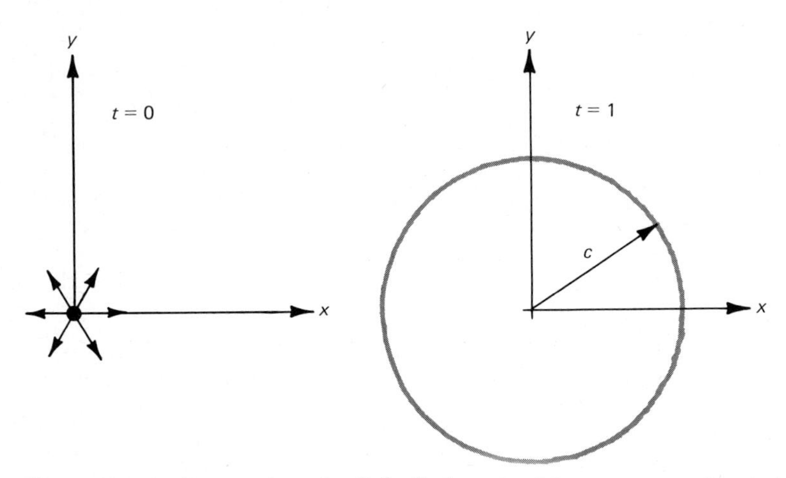

Figure 13-2-1 Propagation of a light flash emitted from a source at rest at the origin, according to Maxwell's equations.

propagate in some substance, which physicists called the *ether*, and that the frame of reference with respect to which the ether is at rest is the frame in which Maxwell's equations are correct. Water waves on a pond spread out independently of the velocity of the projectile which made them, just as the light waves do in Fig. 13-2-2.

The ether was at best a peculiar substance, whose existence

Figure 13-2-2 Propagation of a light flash emitted from a moving source at the origin, according to Maxwell's equations. The motion of the source has no influence on the motion of the light wave.

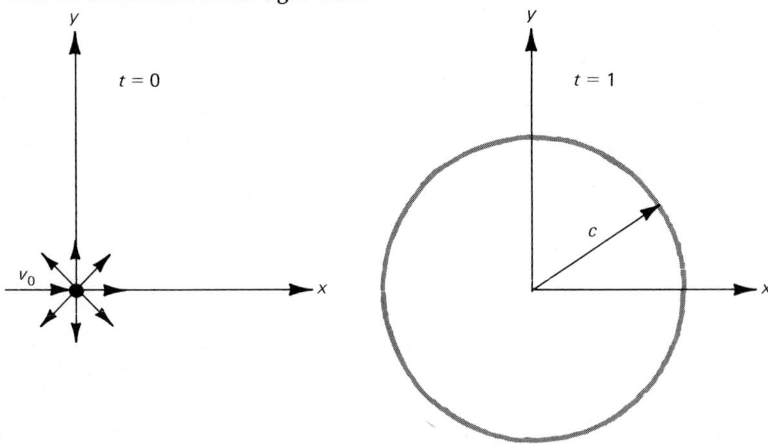

seemed only to be sensed by electromagnetic waves. One could rebel at this ether and call it, as some did, a fiction merely invented to make the predictions of Maxwell's equations more plausible. However, Fig. 13-2-2 does represent a prediction of Maxwell's theory, ether or not. Suppose now that we are in a laboratory moving with velocity $\vec{v}_0$ with respect to the ether, or the preferred Maxwell frame. Figure 13-2-3 shows a light flash emitted at the origin by a source at rest in our laboratory, and the resulting wave. The observed velocity in the lab in the direction of $\vec{v}_0$ would be expected to be less than c by an amount v_0, and it would be expected to be greater by an amount v_0 in the direction opposite. In directions perpendicular to $\vec{v}_0$ the velocity would still be very close to c.

A young American naval ensign, A. A. Michelson, devised a very sensitive way for comparing the velocities of light in two directions at right angles to each other and set out to detect the motion of the earth with respect to the ether. He performed a series of experiments with greater and greater precision, and by 1887 he and his collaborator, E. W. Morley, had determined that if the earth were moving through the ether, its speed with respect to it was, *at most*, much less than its orbital speed around the sun. In fact, there was no significant evidence for *any* motion relative to an ether.

Figure 13-2-3 A light flash propagating in the laboratory frame, assuming a relative velocity v_0 of about 0.2c between the laboratory frame and the ether or Maxwell frame.

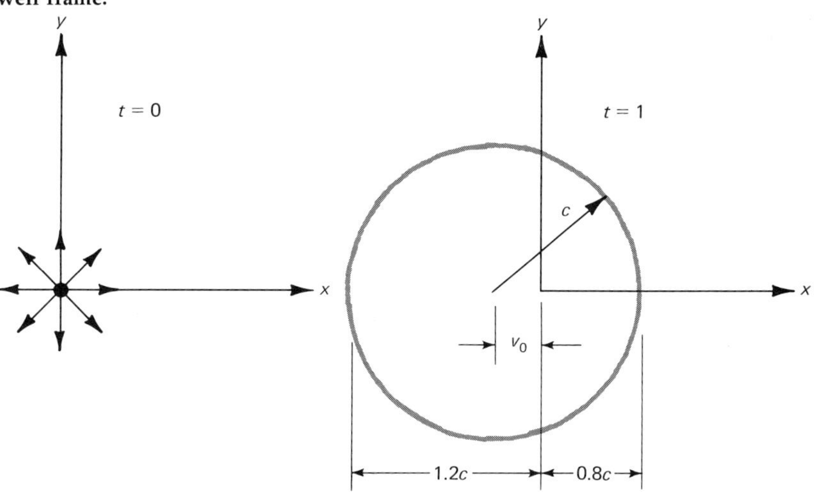

PROBLEM 5

Estimate the sensitivity required in the Michelson-Morley experiment. Assume the velocity of light along the direction of the earth's orbital motion is compared with the velocity at right angles. Compare the orbital speed with the speed of light to see how well this comparison had to be made. The radius of the earth's orbit is about 10^8 miles.

The all-pervading ether, or *Maxwell frame,* seemed to move exactly in the special way the earth does in orbiting the sun, an implausible and unattractive idea. An alternative, equally hard to take, was that nature conspired to prohibit men from detecting their velocity with respect to the ether, for example, by distorting Michelson's apparatus appropriately. Fitzgerald and Lorenz each independently invented the same rule for transforming the coordinates of a laboratory apparatus, depending on its relative motion with respect to the ether, in just such a way that measurements would not show any apparent relative motion. No justification was given for this trick, however.

In 1905, Einstein proposed the special theory of relativity, eliminated the ether by ignoring it, and made two hypotheses which to date appear to be correct:

1 The laws of physics are the same in all inertial frames.
2 The velocity of light is a universal constant, the same in all inertial frames.

The first hypothesis asserts a principle of relativity, which applies to all physical laws, and the second asserts that at least part of Maxwell's electromagnetic theory, the part which leads to electromagnetic waves, is a correct physical law. A stronger second hypothesis, that Maxwell's equations are correct laws of physics, is also consistent with Einstein's theory. In asserting the truth of his hypotheses, Einstein had to reject, as only approximations, those two cornerstones of galilean relativity: newtonian mechanics, and the galilean transformation.

Although the results of Michelson and Morley predated his theory, Einstein's main motivation was not their unexplained lack of success in detecting the earth's motion with respect to the ether. He was impelled mainly by a philosophical desire to maintain the idea of a relativity principle consistent with electromagnetism. Since newtonian mechanics had to be rejected, the old definition of an inertial frame was changed by Einstein to one in which the velocity of light was c and in which a new *relativistic mechanics* was true. As before, however, all frames moving with constant velocity with respect to an inertial frame were also asserted to be inertial frames. Therefore, Einstein's theory could then be summed up by the statement, "Physics is invariant in all frames of reference with constant relative velocities."

To begin our study of implications of Einstein's relativity, we will first find that, if Einstein is correct, time intervals are *not* the same for two observers in relative motion. Consider the physical situation in which a flash of light is emitted, reflected from a mirror, and then detected, as shown in Fig. 13-2-4. Let this be observed by both a laboratory observer and a rocket observer moving with constant velocity v_0 in the x direction. We are thus considering a light-propagation experiment viewed in two inertial frames,

Figure 13-2-4 A light ray bounced off a mirror, in laboratory and rocket frames. In *(a)* a light flash takes place at the origin when both coordinate systems coincide. In *(b)* the ray reflected by the mirror is detected at the origin at time t_B. In *(c)* the ray reflected off the mirror is detected in the rocket frame at the x axis but displaced from the origin, at a time t'_B.

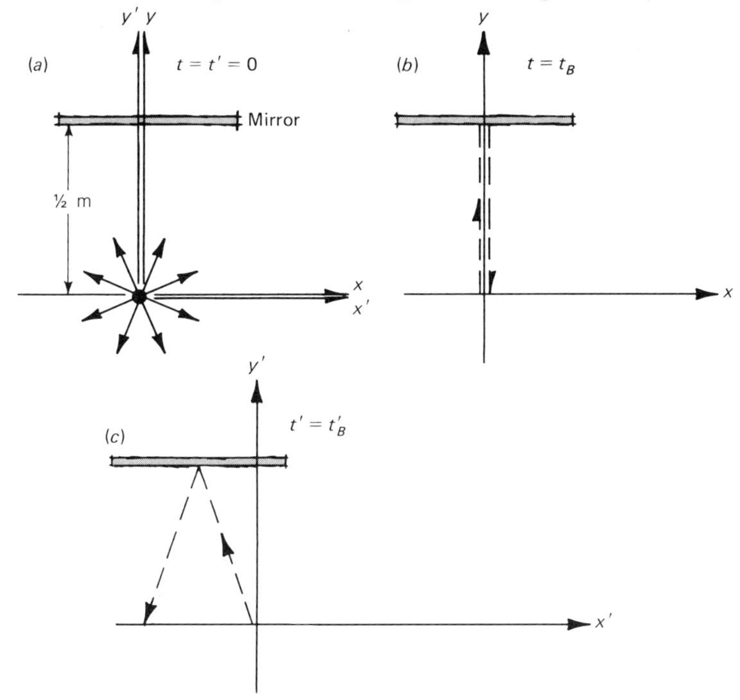

similar to the observation of free fall in two inertial frames discussed in the section on galilean relativity.

Suppose the two observers describe the emission and detection of the light flash by giving space coordinates, and times, for events A (emission) and B (detection). In the laboratory reference frame:

$$\begin{aligned} x_A &= 0 \\ y_A &= 0 \\ z_A &= 0 \\ t_A &= 0 \end{aligned} \qquad \text{13-2-1}$$

signifying that the emission of the flash is used to define the zero of time and that the flash is emitted at the origin of the laboratory coordinate system. For event B in the laboratory:

$$\begin{aligned} x_B &= 0 \\ y_B &= 0 \\ z_B &= 0 \\ t_B &= \frac{1}{c} \end{aligned} \qquad \text{13-2-2}$$

signifying that the flash is detected at the origin of the coordinate system and that the speed of light is c, leading to a time $1/c$ to cover a total flight path of 1 m. (The mirror is at $y = 1/2$ m in the laboratory.)

For the observer in the rocket, the space coordinates and the times of events A and B are as follows, where primed quantities describe measurements with respect to a coordinate system fixed with respect to the rocket:

$$\begin{aligned} x'_A &= 0 \qquad & x'_B &= -x_o t'_B \\ y'_A &= 0 \qquad & y'_B &= 0 \\ z'_A &= 0 \qquad & z'_B &= 0 \\ t'_A &= 0 \qquad & t'_B &= \sqrt{1 + \frac{(x'_B)^2}{c}} \end{aligned} \qquad \text{13-2-3}$$

The values for event A signify that at the time of the light flash the origin of the rocket coordinate system coincided with the origin of the laboratory system, and that the flash was used by the rocket observer to define the zero of his time measurement. Note, however, that the time measurement of the rocket observer is "primed." We will not make the tacit assumption of galilean relativity that time is invariant from one inertial frame to another.

For event B in the rocket frame, the space coordinates are those which define uniform motion of the rocket frame relative to the laboratory frame with velocity v_0 in the x direction. The time coordinate is that time interval which must be measured in order for the apparent velocity of light in the rocket frame to be equal to c. This time must therefore be equal to the path length of the light in the rocket frame divided by c. In calculating the length of the light path in the rocket frame, we will assume that the y' coordinate of the mirror is the same as the y coordinate.

In general, distances perpendicular to the direction of relative motion must remain the same from one inertial frame to another. Were they to differ, that fact could be used to distinguish a preferred inertial frame. For example, let the observer in each frame mark off a distance of one unit on his y or y' axis, and then let them compare these distances at the time $t = 0$ when the two axes coincide. They will then see if either one distance or the other is shorter, thereby distinguishing inertial frames by the amount of shrinkage of their perpendicular dimensions.

By using the phythagorean formula, the path length of the light in the rocket frame, shown by the dashed line in Fig. 13-2-4c, can be found, and divided by c to determine the time t'_B given in Eq. 13-2-3. Note that the observed times of event B in the two reference frames are *not* identical. *Requiring that c be invariant has eliminated the invariance of time.* Now we are forced to what is called a *four-dimensional space-time coordinate system.* As in Eq. 13-2-3, an event must be specified by three space coordinates and one time coordinate. Just as relative motion introduces a transformation of the x coordinate, in Einstein's relativity it also must introduce a transformation of the time coordinate.

13-3 THE LORENTZ TRANSFORMATION

In the previous section we saw that requiring the velocity of light to remain invariant in all inertial frames ruled out the galilean coordinate transformation, which assumed that time was invariant. The transformation which works is called the Einstein-Lorentz transformation, or more usually, simply the Lorentz transformation. It embodies the same mathematical formulas "cooked up" by Lorentz and Fitzgerald to explain the null result of Michelson's interferometer experiments. Einstein independently discovered these formulas and showed that they were the required coordinate transformation for this theory of relativity.

Suppose we consider two frames for which the origins of the

x, y, z and x', y', z' coordinate systems coincide at a time we call $t = t' = 0$. This is the case in the example discussed in the previous section. Let the primed frame be moving in the $+x$ direction, also as in the preceding example. Then, the Lorentz transformation equations, which relate primed to unprimed spacetime coordinates, are as follows:

$$\begin{aligned} x' &= \gamma x - \beta\gamma ct \\ y' &= y \\ z' &= z \\ t' &= \gamma t - \beta\gamma\frac{x}{c} \end{aligned} \qquad 13\text{-}3\text{-}1$$

where c is the velocity of light, and the relative velocity v_o of the two frames of reference is contained in the quantities β (beta) and γ (gamma), defined as follows:

$$\beta \equiv \frac{v_o}{c}$$

$$\gamma \equiv 1/\sqrt{1 - \frac{v_o^{\,2}}{c^2}} = \frac{1}{\sqrt{1-\beta^2}}$$

For relative motion in the opposite sense, with the primed frame moving in the $-x$ direction, the sign of v_o is taken negative and β is then negative. Equations 13-3-1 remain correct. We won't attempt to prove the Lorentz transformation, but will check its consistency and examine its consequences. First, consider the limit of low velocities, where galilean relativity ought to be a good approximation. In this limit, specified by $\beta << 1$ and $\gamma = 1$, we find from Eq. 13-3-1,

$$\begin{aligned} x' &= x - v_o t \\ y' &= y \\ z' &= z \\ t' &= t \end{aligned} \qquad 13\text{-}3\text{-}2$$

which is indeed the galilean transformation.

PROBLEM 6

Verify the correctness of Eq. 13-3-2 by taking the appropriate limit of the Lorenz transformation equations. Do you have to add some extra qualifications to get $t' = t$?

Suppose we next consider the light flash example of the previous section, and use the Lorentz transformation equations to transform the coordinates of event B (detection of the light flash) from the laboratory frame to the rocket frame. It is left as an exercise to verify that the space and time coordinates found in the rocket frame are the same as Eq. 13-2-3. The Lorentz transformation is seen to satisfy at least two requirements which we demand of it. Assuming it to be correct in general, we may now examine some of its further consequences. If these are treated as predictions, and verified experimentally, we will become more confident of the correctness of the transformation and of Einstein's relativity theory.

PROBLEM 7
Verify that applying a Lorenz transformation to Eq. 13-2-2 leads to Eq. 13-2-3.

The rather complicated dependence of the parameter γ on the velocity v_0 is a nuisance in doing numerical calculation. For convenience, Fig. 13-3-1 is a graph of how γ varies with v_0/c, or β, for β between 0 and 0.99. The transformation equations predict that the *limiting relative velocity* possible in the world, for which γ becomes infinite, is the *velocity of light.* For a velocity greater than

Figure 13-3-1 Variation of γ with β.

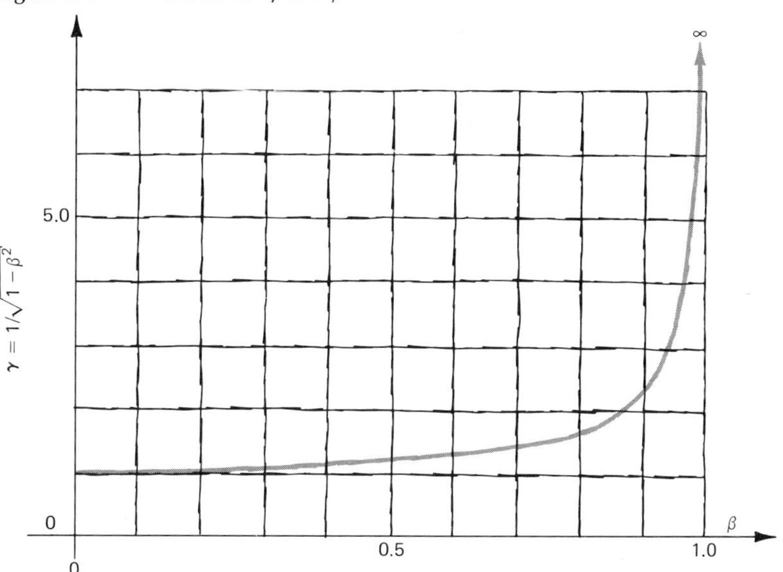

c the quantity within the square root in the expression for γ becomes negative, and square roots of negative numbers lead to uninterpretable values of the coordinates when used in the transformation equations.

The acceleration of atomic and nuclear particles to very high energies permits a direct check of the prediction that c is the ultimate speed. For example, at the Stanford linear accelerator, electrons are accelerated down a straight tube two miles long, in an average electric field of more than 6 million V/m. Their final energy is about 20×10^9 electron volts (20 GeV). The operating conditions of the machine verify that after the first 10 m of acceleration, the electron's speed is greater than $0.99c$, but that the entire remaining 3,000 m of accelerator does not increase the speed beyond the value c.

PROBLEM 8

Suppose an object built like a scissors, schematically indicated in Fig. P-13-8, is snapped shut quickly. Can the velocity of the point where the blades touch, indicated by P on the sketch, be greater than c? Does this violate the results of relativity theory?

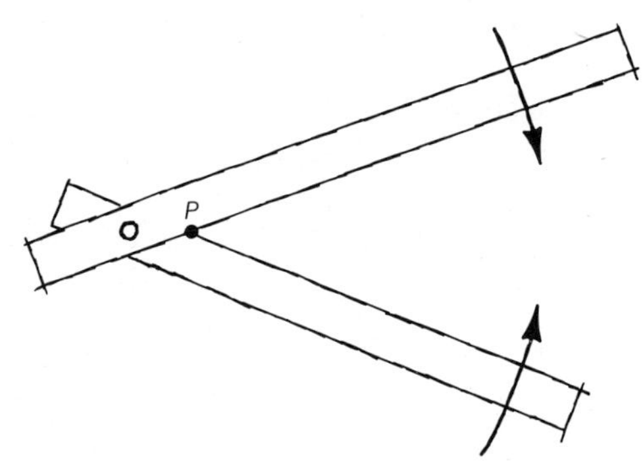

Figure P-13-8

13-4 TIME DILATION AND SPACE CONTRACTION

In this section we will consider two of the most dramatic consequences of the Lorentz transformation. However, it will first be worthwhile to obtain the transformation equations for the coordinate and time *differences* between any two arbitrary events, which we'll call A and B. Applying Eq. 13-3-1 to event A, and then to event B:

$$
\begin{aligned}
x'_A &= \gamma x_A - \beta\gamma c t_A & x'_B &= \gamma x_B - \beta\gamma c t_B \\
y'_A &= y_A & y'_B &= y_B \\
z'_A &= z_A & z'_B &= z_B \\
t'_A &= \gamma t_A - \beta\gamma \frac{x_A}{c} & t'_B &= \gamma t_B - \beta\gamma \frac{x_B}{c}
\end{aligned}
\qquad \text{13-4-1}
$$

In the primed reference frame, the coordinate differences between events A and B are $x'_B - x'_A$, written as $\Delta x'$; $y'_B - y'_A$, written as $\Delta y'$, etc. By subtracting the corresponding relations for events A and B in Eq. 13-4-1 above, we can find that the coordinate and time differences transform as follows:

$$\Delta x' = \gamma(\Delta x) - \beta\gamma c(\Delta t) \qquad \text{13-4-2}a$$

$$\Delta y' = \Delta y \qquad \text{13-4-2}b$$

$$\Delta z' = \Delta z \qquad \text{13-4-2}c$$

$$\Delta t' = \gamma(\Delta t) - \beta\gamma\left(\frac{\Delta x}{c}\right) \qquad \text{13-4-2}d$$

It is important to be clear about how space and time differences in a given reference frame are defined. The three components of the spatial distance between events are simply measured with a good ruler and other necessary surveying instruments. The time difference between two events is measured, in principle, by noting the times of the events recorded by *synchronized clocks* at their locations. Any clock in a reference frame can be synchronized with one master clock, which might be located at the origin, by using a light flash. The flash is sent out from the master clock, and specifies an agreed calibration time (like high noon), and the clock to be synchronized is set when the light flash is received. However, it is set to a time *later* than the agreed calibration time by an amount equal to the *transit time* of light from the master clock to the clock being set. Thus, all clocks are synchronized in such a way that *the delay time due to propagation of the synchronizing signal is taken account of and eliminated.*

PROBLEM 9

Describe in detail how you would synchronize a clock at $x = 3 \times 10^3$ m, $y = 4 \times 10^3$ m, $z = 0$, with one at the origin.

Simultaneous events are defined as those which occur at the same time on clocks synchronized as described above. Notice, from Eq. 13-4-2d, that simultaneous events in one frame do not remain simultaneous in another one. Let the events be simultaneous in the unprimed frame, so that $\Delta t = 0$ in Eq. 13-4-2d. Then, depending upon β and Δx, the time interval in the primed system can have various values, positive, negative, or zero. Sometimes this is referred to as the *relativity of simultaneity.* (An interesting aspect of the time interval transformation, left as a problem, is whether it leads to ambiguity in assigning causal relationships between events.)

PROBLEM 10

Consider two events in the laboratory, separated in space and time by Δx, Δy, Δz, and Δt. Transform these differences to a rocket frame, and verify that the following equation is true:

$$(\Delta x)^2 + (\Delta y)^2 + (\Delta z)^2 - c^2(\Delta t)^2 = (\Delta x')^2 + (\Delta y')^2 + (\Delta z')^2 - c^2(\Delta t')^2$$

Thus, a quantity, which is called the *interval* is found to be invariant under Lorenz transformations. If the second event in the laboratory could have been caused by the first, then we must have:

$$\text{Interval} \quad (\Delta x)^2 + (\Delta y)^2 + (\Delta z)^2 - c^2(\Delta t) \leqslant 0$$

Explain why this must be true, if c is the ultimate speed.

What is the sign of the invariant interval for the case when the first event causes the second to occur? What is the sign when the first could not have caused the second? What does a value of zero for the interval mean?

Probably the most surprising prediction of Eq. 13-4-2 is described as *time dilation.* Suppose a clock at rest in the unprimed frame is viewed from the primed frame, and consider the two events which are successive ticks of this clock. In the unprimed frame the two ticks are separated by a time interval Δt and a space interval of zero. In the primed system, therefore (from Eq. 13-4-2d), $\Delta t' = \gamma(\Delta t)$. For a relative velocity close to c, the factor γ can be a very large number. As a result, the ticks in the primed frame can occur much more slowly than in the unprimed frame. For example, a 1-sec interval between ticks can *dilate,* or stretch, to 10 sec, 100 sec, or more, as observed in a moving frame with sufficiently high relative velocity.

The prediction of time dilation may perhaps raise the question of what is meant by a clock. Some examples of clocks are: a light flash bouncing between mirrors a known distance apart, a swinging pendulum, or the oscillating electric field of the light wave emitted in a particular atomic transition. These clocks are all examples of devices which specify a time interval by utilizing some law of physics. *Any* such clock must appear to run slow as viewed from a reference frame in relative motion, if Einstein's theory is correct.

A most convincing verification of the time dilation predicted by the Lorentz transformation is the observed mean lives of unstable elementary particles which move at high speed. For example, elementary particles called pi-mesons decay with a mean lifetime of 2.6×10^{-8} sec when they are at rest in the laboratory (with the decay scheme $\pi^{\pm} \rightarrow \mu^{\pm} + \nu$). At a velocity equal to c, the meson would travel about 8 m in a time equal to its mean life at rest. Yet high-energy pi-mesons are routinely studied in experiments where they travel 10 times this far, with few decaying. The energies of the pions are so high that β approaches 1, and the time dilation factor γ can be as large as 100. The time dilation predicted by the Lorentz transformation has been quantitatively verified to high accuracy by observing the apparent lengthening of elementary particle lifetimes. Another confirmation of the phenomenon is the apparent "reddening" of light emitted by fast moving atoms, which corresponds to a reduction in the observed frequency, or an apparent slowing down of the atomic clocks.

PROBLEM 11

Show that the time dilation of a clock on a rocket, as viewed from the laboratory, is the same as that of a clock in the laboratory as viewed from the rocket. If this were not true, then the inertial frames could be distinguished as "moving" or "fixed," violating the principal of relativity.

PROBLEM 12

The twin paradox has sometimes been thought to violate the principle of relativity, but it does not. (Hence, the name paradox.) Let there be two twins, and let one take off in a rocket and travel at high speed away from the other for ten years by the stay-at-home's clock. Then, let him reverse his trip and return home to his brother, arriving twenty years after he left, by his brother's clock. Assume that the periods of acceleration and deceleration required in this story are short compared with the total time, and the aging during them can be neglected.

It is asserted that the brother who went off on the rocket is younger, because of time dilation, which violates the symmetry between frames required by relativity. One could clearly distinguish the brother who had been moving from the one who remained at rest. It is believed that the rocket traveler would indeed be younger but that this does not violate the principle that all *inertial* frames are equivalent. Explain.

A companion to time dilation is *length contraction.* The length of a body is found by *simultaneously* determining the location of its ends. If a meter stick is at rest in the unprimed frame, lined up along the x axis, two events which determine its length have a certain separation Δx, equal to 1 m, and a time separation $\Delta t = 0$. Using Eq. 13-4-2a, we find the x' separation of these events in a moving frame to be

$$\Delta x' = \gamma(\Delta x) - \beta\gamma c(\Delta t) \qquad 13\text{-}4\text{-}3$$

However, this is *not* the length as measured in the primed system, because the two simultaneous events which measured the length in the unprimed system are *not* simultaneous in the primed system. They have a time interval

$$\Delta t' = -\beta\gamma\frac{\Delta x}{c} \qquad 13\text{-}4\text{-}4$$

so events which determine the length in one frame do not determine the length in another.

We can use the two Eqs. 13-4-3 and 13-4-4 to find the length in the primed system by reasoning out the following formula:

$$\text{Length}' = \Delta x' - (\beta)c(\Delta t') \qquad 13\text{-}4\text{-}5$$

where $\Delta x'$ and $\Delta t'$ refer to the values given by Eqs. 13-4-3 and 13-4-4. The second term "corrects" the spatial separation by the distance moved by the rod in the time difference $\Delta t'$. By substituting for $\Delta x'$ and $\Delta t'$ the values given in Eqs. 13-4-4 and 13-4-5, and using the relation between β and γ, Eq. 13-4-5 can be simplified to the form:

$$\text{Length}' = \frac{\Delta x}{\gamma} = \frac{\text{length}}{\gamma} \qquad 13\text{-}4\text{-}6$$

In case you feel uneasy about the "correction term" in Eq. 13-4-5, it is possible to establish the length in the primed system directly. We first write the *inverse transformation* corresponding to Eq. 13-4-2a:

$$\Delta x = \gamma(\Delta x') + \beta\gamma c(\Delta t') \qquad 13\text{-}4\text{-}7$$

This equation is the same as Eq. 13-4-2*a*, with primed and unprimed quantities exchanged and the sign of β reversed. This is because the relative velocity of the unprimed frame seen from the primed frame is in the opposite direction to that of the primed frame seen from the unprimed frame. The events which determine the length in the primed frame must have $\Delta t' = 0$, so that Eq. 13-4-7 directly gives the result

$$\Delta x = \gamma(\Delta x')$$

where Δx is, as before, the length in a frame in which the meter stick is at rest. Rearranging,

$$\Delta x' = \frac{\Delta x}{\gamma} \qquad 13\text{-}4\text{-}8$$

which is the same as that obtained before, Eq. 13-4-6.

The relation in Eq. 13-4-8 describes an apparent length contraction for objects in relative motion, where the contraction only applies to the *x coordinate*, the coordinate *along* the direction of motion. The dimensions in the coordinate directions *perpendicular* to the direction of relative motion, Δy and Δz, are invariant. Hence, a moving sphere is expected to flatten into an ellipsoid. Direct observation of this phenomenon, analogous to the detection of time dilation for unstable particles, has not been achieved. However, powerful indirect evidence for space contraction exists. For example, the relativistic transformations of electric and magnetic fields have been well verified, and these depend upon both space contraction and the invariance of Maxwell's equations.

PROBLEM 13

Two events occur with the following space-time coordinates in the laboratory frame:

$x = 0$ $y = 0$ $z = 0$ $t = 0$	Event A	$x = L$ $y = 0$ $z = 0$ $t = 0$	Event B

Are events A and B simultaneous as observed in a rocket frame moving at $v/c = 0.5$ in the $+x$ direction? If not, what is their time difference?

Explain the following paradox: A rocket moves along the x axis in the laboratory frame, with $v/c = 0.5$. At $t = 0$, two bullets traveling in the y direction, a distance L apart, just miss the head and tail of the rocket, as in Fig. P-13-13. In the frame moving along at the speed of the rocket, the distance L would be Lorentz-*contracted*. Yet the bullets miss the rocket as observed in this frame too.

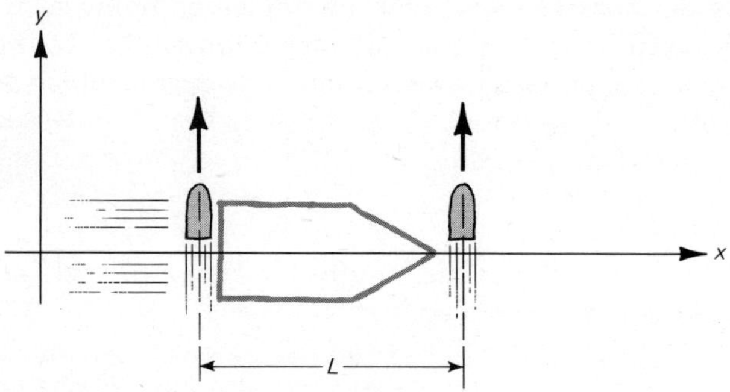

Figure P-13-13

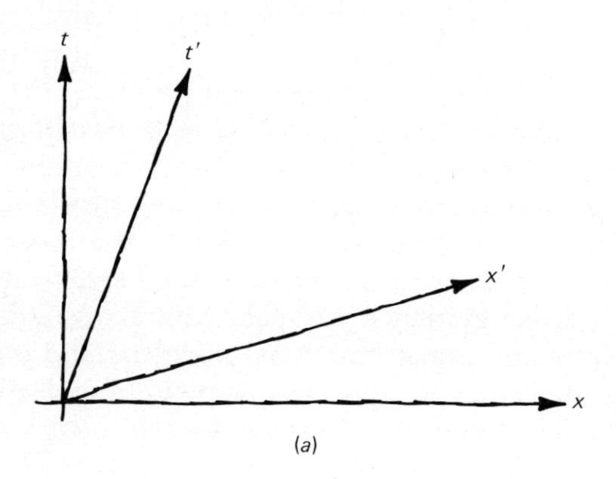

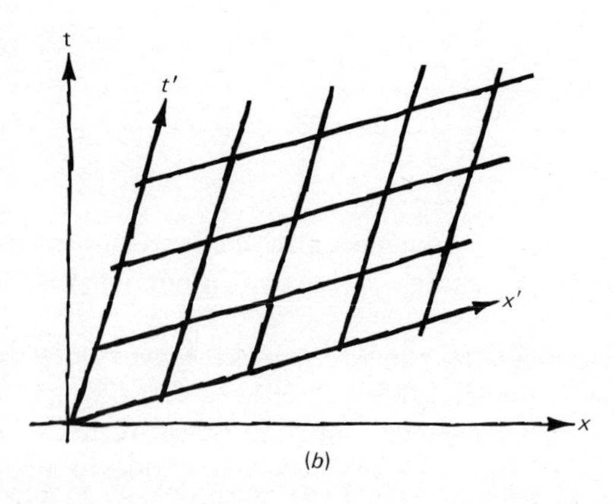

Figure P-13-14

PROBLEM 14
In order to make a geometrical interpretation of a Lorenz transformation, consider motion only along the x axis, so that the four dimensions of space-time reduce to two, x and t. Draw a pair of perpendicular axes for x and t, the way you would normally draw xy axes. Consider a rocket frame x', t'. Draw on your xt axes the line for all events which occur at $x' = 0$, which can be called the t' axis. Similarly, draw on your xt axes the line for all events with $t' = 0$, the x' axis. You now have a graph which shows how xt *axes* are changed to $x't'$ axes by a Lorenz transformation, and it should look approximately like the one in Fig. P-13-14*a*. Try to see the effects of time dilation and space contraction on your graph. You will have to figure out how to calibrate your $x't'$ axes with a grid of lines, as in Fig. P-13-14*b*, which are in proper scale with the xt grid.

PROBLEM 15
Electrons at the halfway point of the Stanford linear accelerator have $\gamma = 20{,}000$. Suppose they are allowed to coast the remaining mile through the accelerator pipe. How long does this mile of straight pipe look to an observer moving with the velocity of the electron (who is said to be in the electron's "rest frame.")

13-5 RELATIVISTIC MECHANICS

When the galilean transformation of coordinates is abandoned, the equations of classical mechanics are no longer invariant with respect to transformations between frames in uniform relative motion with respect to each other. Einstein, who stuck to his guns that "Physics was invariant" with respect to such transformations, was thus faced with the problem of creating a new dynamics to replace Newton's. This new dynamics is largely based on preserving the conservation laws for momentum and energy.

For Einstein's theory, both momentum and energy must be redefined, and the dynamical equation $\vec{F} = m\vec{a}$ must be replaced with the equation

$$\vec{F} = \frac{\Delta\vec{p}}{\Delta t} \qquad \text{13-5-1}$$

force equals the time rate of change of momentum. With the new relativistic definition of momentum, this equation still reduces to Newton's $\vec{F} = m\vec{a}$ for velocities much less than c, as it should. The dynamical equation 13-5-1 works in all inertial frames, but the force F is not invariant, and must be found from a transformation relation.

As a preliminary to the discussion of relativistic momentum, we will first consider the transformation of velocity. Let the Lorentz transformation, Eq. 13-4-2, be applied to two events in the unprimed system characterized by the space and time intervals,

$$\begin{aligned} \Delta x &= 0 \\ \Delta y &= v_y(\Delta t) \\ \Delta z &= 0 \\ \Delta t &= \Delta t \end{aligned} \qquad 13\text{-}5\text{-}2$$

These two events are points along the path of a particle moving with velocity v_y in the y direction. If we transform these events to the primed system, with relative velocity $v_x = v_o$, we find the results:

$$\begin{aligned} \Delta x' &= -v_o(\gamma)(\Delta t) \\ \Delta y' &= v_y(\Delta t) \\ \Delta z' &= 0 \\ \Delta t' &= (\gamma)(\Delta t) \end{aligned} \qquad 13\text{-}5\text{-}3$$

The velocities in the primed system can now be found from the above equations:

$$\begin{aligned} v_x' &\equiv \frac{\Delta x'}{\Delta t'} = -v_o \\ v_y' &\equiv \frac{\Delta y'}{\Delta t'} = \frac{v_y}{\gamma} \\ v_z' &\equiv \frac{\Delta z'}{\Delta t'} = 0 \end{aligned} \qquad 13\text{-}5\text{-}4$$

We have found that v_x is given by $-v_o$, the relative motion of the two frames, as expected, but that v_y changes by a factor $1/\gamma$ when viewed in the primed frame. This is the time-dilation factor. A bouncing ball, moving back and forth in the y direction between two walls, could be used for a clock, so this factor must result for consistency with time dilation.

A more interesting velocity transformation is for the case when $v_y = v_z = 0$, but v_x is nonzero. The space and time intervals between two events in the unprimed frame, analogous to Eq. 13-5-2, are then

$$\begin{aligned} \Delta x &= v_x(\Delta t) \\ \Delta y &= 0 \\ \Delta z &= 0 \\ \Delta t &= \Delta t \end{aligned} \qquad 13\text{-}5\text{-}5$$

and the transformed differences, for a frame in relative motion with velocity v_o (from Eq. 13-4-2), are:

$$\begin{aligned} \Delta x' &= v_x(\gamma)(\Delta t) - v_o(\gamma)(\Delta t) \\ \Delta y' &= 0 \\ \Delta z' &= 0 \\ \Delta t' &= (\gamma)(\Delta t) - v_o v_x(\gamma)\frac{\Delta t}{c^2} \end{aligned} \qquad 13\text{-}5\text{-}6$$

The velocities in the primed frame are then:

$$\begin{aligned} v_x' &= \frac{\Delta x'}{\Delta t'} = \frac{v_x - v_o}{1 - v_x v_o/c^2} \\ v_y' &= \frac{\Delta y'}{\Delta t'} = 0 \\ v_z' &= \frac{\Delta z'}{\Delta t'} = 0 \end{aligned} \qquad 13\text{-}5\text{-}7$$

The first of Eqs. 13-5-7 is often called the rule for "adding" velocities, in the sense that the velocity v_x is combined with the reference frame velocity v_o along the same axis. For our example both v_o and v_x are along the $+x$ direction. The numerator of the expression for v_x' is just what would be expected from galilean relativity, the intuitive way of combining velocities that works for $v_o << c$. Also, for $v_o << c$, the denominator becomes equal to 1, so the correct nonrelativistic equation results.

Another special limiting case is $v_x = c$. Then, Eq. 13-5-7 says $v_x' = c$. The velocity of light is indeed invariant. To see another important application of Eq. 13-5-7, consider relative motion of the two reference frames so that $\vec{v}_o$ is in the $-x$ direction, thereby making the observed velocity in the x' direction greater than v_x. For this case, where velocities v_o and v_x "add" instead of cancel, the signs change in Eq. 13-5-7, leading to the result

$$v_x' = \frac{v_x + v_o}{1 + v_x v_o/c^2} \qquad 13\text{-}5\text{-}8$$

Suppose v_x and v_o are both very nearly equal to c. Then the denominator is close to 2 and the "sum" of these two velocities is still *less than* c. Thus, the velocity addition gives the result that c is the ultimate speed.

Relativistic expressions for momentum and energy cannot really be "derived" from the Lorentz transformation, but are more or less invented, in order to lead to momentum and energy conservation in all inertial frames. To first find the expression for momentum, we will consider a highly symmetric elastic collision of two particles of equal mass, moving toward each other with equal velocities, as in Fig. 13-5-1*a*. Then let the collision be viewed in two other moving frames so that it appears as in Fig. 13-5-1*b* and *c*.

We will require that momentum be conserved in the three reference frames shown in Fig. 13-5-1*a* to *c*. Initially, momentum is defined only as some vector quantity which points along the direction of a particle's velocity. We can write $\vec{p} = "m"\vec{v}$, where now "m" can be any complicated algebraic expression, involving the magnitude of the particle velocity, for example, instead of its classical value which is simply the mass of the particle. By symmetry, momentum is conserved in Fig. 13-5-1*a*, along both the x and y directions. In either Fig. 13-5-1*b* or *c*, the x component of momentum is conserved by symmetry, but the y component will only be conserved for a particular choice of an expression for "m".

In order to conserve the y component of momentum in Fig. 13-5-1*c*, we require:

$$"m(1)"v_y = "m(2)"v_{oy} \qquad 13\text{-}5\text{-}9$$

where the quantities "m" for the two particles are labeled with the numbers of the particles. Each component of momentum must be proportional to the corresponding component of velocity, to satisfy our requirement that momentum be a vector in the direction of the velocity vector. Hence v_y and v_{oy} appear in Eq. 13-5-9. The transformation between the frames shown in Fig. 13-5-1*b* and *c* involves a relative velocity which is simply v_x. For particle 1, the y component v_{oy} in Fig. 13-5-1*b* then transforms into v_y in Fig. 13-5-1*c* according to the velocity transformation equation derived earlier in this section:

$$v_y = \frac{v_{oy}}{\gamma} = v_{oy}\sqrt{1 - \frac{v_x^{\,2}}{c^2}} \qquad 13\text{-}5\text{-}10$$

Now we can find the relativistic expression for "m" from Eqs. 13-5-9 and 13-5-10. Let the velocity v_{oy} be so low as to be nonrelativistic, which corresponds physically to a *grazing collision*. Then, the component of momentum in the y direction for particle 2 in Fig. 13-5-1*c* is known to be given by the classical momentum,

$$"m(2)"\ v_{oy} = m_o v_{oy} \qquad 13\text{-}5\text{-}11$$

where m_o is the *classical inertial mass* of the particle, called in relativity the *rest mass*, because it is the mass at rest, or in the limit of small velocity.

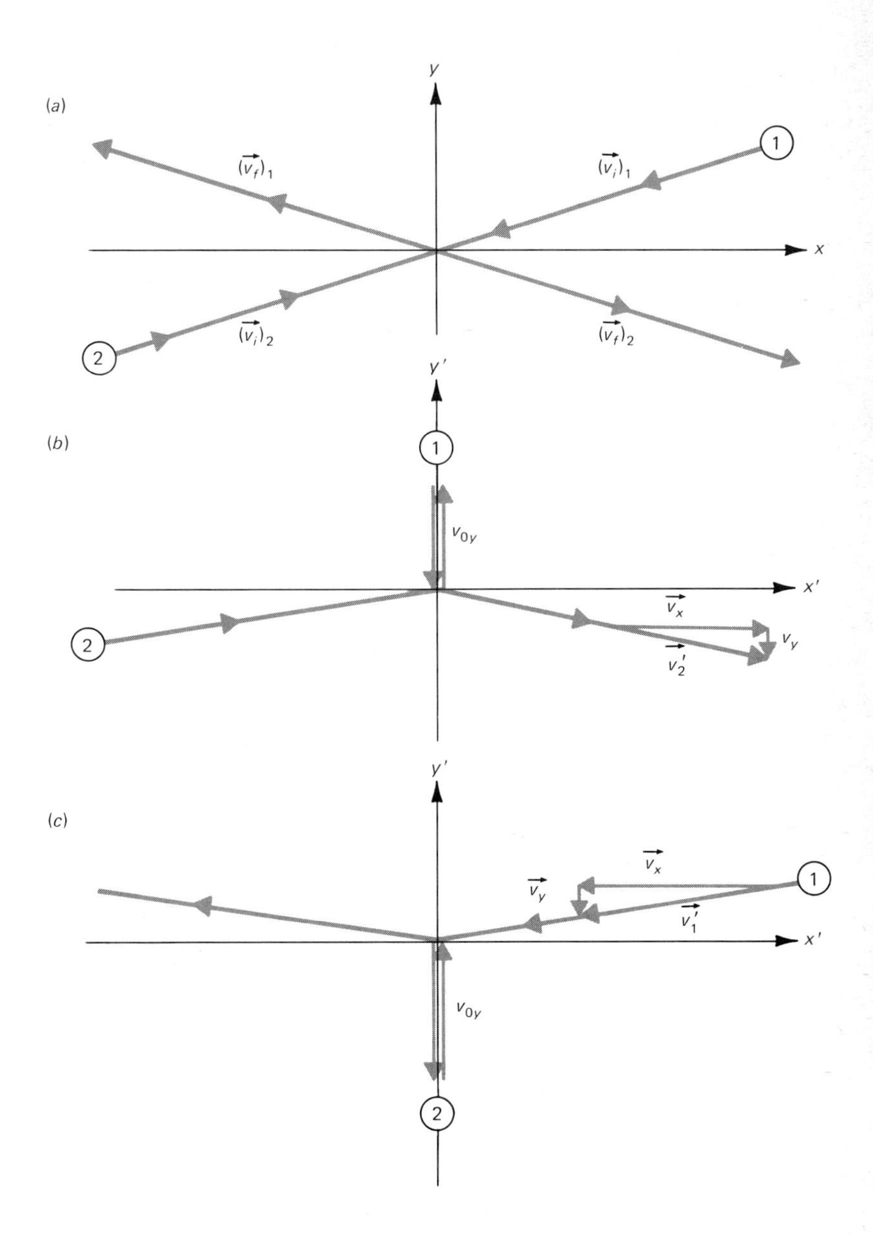

Figure 13-5-1 An elastic collision between two particles with equal masses, as viewed in three frames. In *(a)* the collision is highly symmetric, seen in the frame where the total momentum is zero. In *(b)* and *(c)* either one or the other particle has its x component of velocity transformed to zero. Requiring momentum to be conserved in *(b)* and *(c)* leads to a relativistic expression for momentum.

While particle 2 is moving nonrelativistically, let particle 1 be moving very rapidly in Fig. 13-5-1c, but almost along the x direction (again appropriate for a grazing collision). Its "m" will be the desired relativistic expression, and its velocity $(v_x^2 + v_y^2)^{1/2}$ can be set equal simply to v_x to a good approximation.

On the left side of Eq. 13-5-9, v_y can be replaced by the value found from Eq. 13-5-10. On the right-hand side of Eq. 13-5-9 "$m(2)$" can be replaced by m_o. With these two substitutions, Eq. 13-5-9 takes on the form:

$$\text{"}m(1)\text{"}\ v_{oy}\sqrt{1 - \frac{v_x^2}{c^2}} = m_o v_{oy} \qquad \text{13-5-12}$$

Solving for "$m(1)$", and setting $v_x = v$, we find the relativistic analog of m_o, the *relativistic mass* m,

$$m = m_o/\sqrt{\left(1 - \frac{v^2}{c^2}\right)} \qquad \text{13-5-13}$$

The velocity v, the speed of the particle, has replaced v_x, correct for the approximation of an extremely grazing collision.

The relativistic expression for the momentum of a particle with rest mass m_o moving with velocity v has now been found:

$$p = mv = \frac{m_o v}{\sqrt{1 - v^2/c^2}} \qquad \text{13-5-14}$$

Even though v is limited to values less than c, as v approaches c, the momentum can increase without limit, since the denominator in Eq. 13-5-14 approaches zero. If we keep pushing on the electron in the Stanford linac, it does keep gaining momentum, even though its speed changes only infinitesimally.

In order to find the relativistic expression for energy, we will look for a *scalar* quantity which is conserved under Lorentz transformations. Figure 13-5-2a and b show a particularly simple *inelastic* collision viewed in two frames. Two particles of rest mass m_o collide and stick, forming a particle of rest mass M_o, at rest in Fig. 13-5-2a. Suppose, perhaps by analogy with classical mass conservation, we investigate whether the relativistic mass m $(m = m_o/\sqrt{1 - v^2/c^2})$ is conserved. To test for conservation of total m, we have two equations. From Fig. 13-5-2a,

$$M_o = \frac{2m_o}{\sqrt{1 - v^2/c^2}} \qquad \text{13-5-15}$$

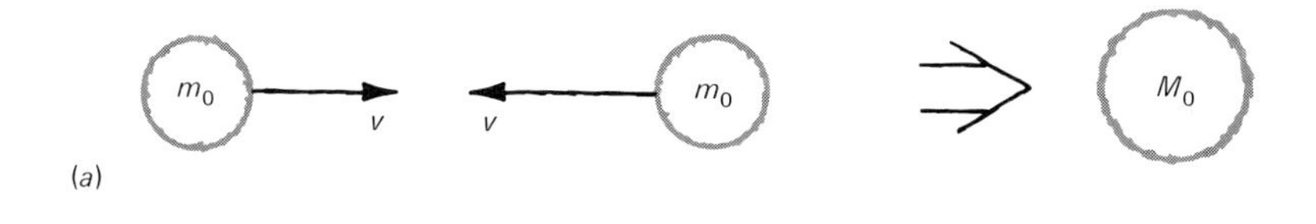

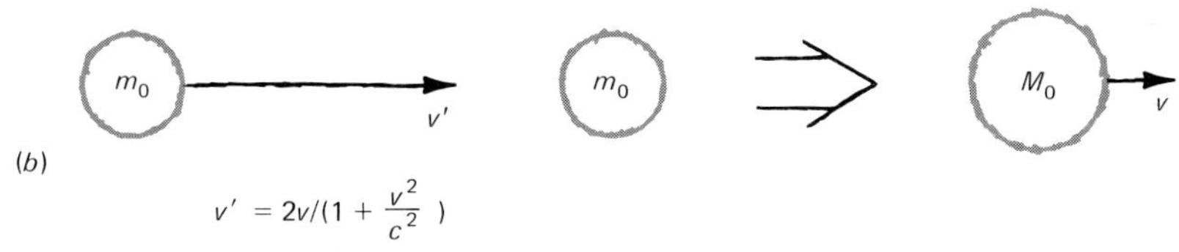

Figure 13-5-2 An inelastic collision in which two particles of mass m_o and velocity v collide and stick together. In *(a)* the total momentum is zero, and the final mass M_0 is at rest. In *(b)* the collision is viewed in a frame moving with velocity v with respect to the frame of *(a)*.

while from Fig. 13-5-2*b*,

$$\frac{M_o}{\sqrt{1 - v^2/c^2}} = m_o + \frac{m_o}{\sqrt{1 - \frac{1}{c^2}\left(\frac{2v}{1 + v^2/c^2}\right)^2}} \qquad 13\text{-}5\text{-}16$$

We can substitute the value of M_o from Eq. 13-5-15 into Eq. 13-5-16, to obtain the following equality, *if* total relativistic mass is conserved,

$$\frac{2m_o}{\left(1 - \frac{v^2}{c^2}\right)} \stackrel{?}{=} m_o + \frac{m_o}{\sqrt{1 - \frac{1}{c^2}\left(\frac{2v}{1 + v^2/c^2}\right)^2}} \qquad 13\text{-}5\text{-}17$$

The question mark over the equal sign in Eq. 13-5-17 means we want to check whether the equality is true. By a reasonable amount of algebraic manipulation the equality can be shown to hold, so that we have found a *conserved scalar quantity*, at least for the test example. To make this mass into an energy, we need to multiply it by a constant with the right dimensions. To find the constant, we can investigate the behavior of m for v/c very small and require that the relativistic energy be consistent with classical mechanics in this limit. Approximating m in the classical limit,

$$m = \frac{m_o}{\sqrt{1 - \frac{v^2}{c^2}}} \doteq \frac{m_o}{1 - \frac{1}{2}\frac{v^2}{c^2}} \doteq m_o\left(1 + \frac{1}{2}\frac{v^2}{c^2}\right) \qquad 13\text{-}5\text{-}18$$

Equation 13-5-18 has a term that can be made into the classical kinetic energy, by defining the relativistic energy E as follows:

$$E = mc^2 = \frac{m_o c^2}{\sqrt{1 - v^2/c^2}} \qquad 13\text{-}5\text{-}19$$

Then, for $v << c$,

$$E \doteq m_o c^2 + \frac{1}{2} m_o v^2 \qquad 13\text{-}5\text{-}20$$

Equation 13-5-20 shows the new energy E has *two parts* in the classical limit, an energy proportional to *rest mass* plus the usual kinetic energy.

While classical physics originally was based on separate conservation of mass and energy, the relativistic energy includes both and requires only conservation of their *sum*. Thus, relativity implies that kinetic energy and mass may be exchanged. For example, the mass M_o in Fig. 13-5-2*a* is *greater* than the sum of the two masses m_o because they hit with kinetic energy.

The *relativistic kinetic energy* at any velocity is defined by the total energy E given by Eq. 13-5-19, minus the *rest energy* $m_o c^2$,

$$\text{KE} = m_o c^2 \left(\frac{1}{\sqrt{1 - v^2/c^2}} - 1 \right) \qquad 13\text{-}5\text{-}21$$

The kinetic energy in Eq. 13-5-21 has the property that it reduces to the classical value for $v << c$, and that when the kinetic energy becomes appreciable compared to $m_o c^2$, the classical formula $\frac{1}{2}mv^2$ becomes inaccurate. As v approaches c, the kinetic energy can increase without limit. Thus, as we accelerate a particle toward the ultimate speed, we can add unlimited kinetic energy as well as unlimited momentum.

PROBLEM 16

Consider a particle in motion with momentum in the x direction and a certain value of v/c. Use the Lorenz transformation to transform p and E to a frame in which the particle is at rest, assuming that the components of p transform like x, y, and z, while E transforms like t. Does your answer make sense? The coordinates and time are said to constitute a *four-vector* because they transform according to the Lorenz transformation. The momentum and energy also make up a four-vector.

PROBLEM 17

Consider the high-energy particle reaction:

$$e^+ + e^- \rightarrow \pi^+ + \pi^-$$

where e^- stands for electron, e^+ for positron, and π^+ and π^- stand for pi mesons. The mass energies at rest, m_0c^2 are as follows:

e^- and e^+	0.51 MeV	each
π^- and π^+	140. MeV	each

Find the electron momentum needed for the reaction to take place, as observed in two frames:

1 The frame in which e^- and e^+ approach each other with equal velocities.
2 The frame in which the e^+ is at rest before being struck by the e^-.

Hint: Simplify your calculations by assuming the rest energy of the electron is negligible compared to its kinetic energy. Utilize the fact that the momentum-energy four-vector transforms according to the Lorenz transformation.

PROBLEM 18

The kinetic energy of an electron in a TV picture tube is about 20,000 eV. It rest energy m_0c^2 is about 500,000 eV. Compare its actual velocity with that which would be calculated classically.

When the kinetic energy becomes much greater than m_0c^2, it is easy to verify that the kinetic energy and momentum are approximately related in a simple but highly nonclassical way:

$$pc \doteq \text{KE} \qquad 13\text{-}5\text{-}22$$

According to Maxwell's theory, this is the relation which is exact for the energy and momentum carried by electromagnetic waves, just as if the waves corresponded to a stream of particles (photons) moving with $v = c$!

The definition of relativistic energy has been presented basically as a guess, but it appears to work in a variety of ways. Historically, like most radical ideas, it did begin as a guess. The direct conversion of mass to energy was not quantitatively checked until many years after the formula was suggested. In fact, the widest possible appeal to experiment was called for in order to verify the general truth of Einstein's initially shocking theory. However, since 1905, the predictions of special relativity have been verified so well that Lorentz invariance and special relativity now constitute one of the most firmly established areas of physics.

PROBLEM 19

Figure out how much mass the sun must convert to radiant energy per year, given the following:

1. At the earth's orbit, the intensity of *radiant energy* is about 1.5×10^3 joules per second on an area of 1 m^2 facing the sun.
2. The earth-sun distance is about 10^8 mi.

Compare your answer with an estimate for the mass of the earth, using an earth radius of 8,000 mi and an average density of 5 g/cm^3.

PROBLEM 20

Mu mesons are unstable cosmic ray particles, typically originating at about 3×10^4 m above the earth. About 30 percent of the mu mesons traveling vertically decay before they strike the earth's surface. The mean life of a mu meson measured at rest in the laboratory is 2.2×10^{-6} sec. Find the mean value of the parameter γ for the cosmic-ray mu mesons. The energy of a mu meson at rest, m_0c^2, is about 106 MeV. What is the mean energy of the cosmic-ray mesons?

SUMMARY

Galilean relativity asserts that:

1. The laws of mechanics are true in all frames in uniform motion with respect to an *inertial frame*, which is one in which $\vec{F} = m\vec{a}$.
2. The coordinate transformation between frames in uniform relative motion is the *galilean transformation*:

$$x' = x - v_0 t \qquad y' = y \qquad z' = z$$

As a direct consequence of the transformation, $\vec{a}' = \vec{a}$. It is also necessary that $\vec{F}' = m'\vec{a}'$ if *the laws of newtonian mechanics are invariant*, so it is tacitly assumed that $F' = F$ and $m' = m$, which does give $F' = m'a'$.

Maxwell predicted the existence of electromagnetic waves with velocity c, where c is the measured velocity of light. *The Maxwell equations are not invariant under galilean transformation*, and therefore it was expected that only in one particular inertial frame would the velocity of light equal c. The *Michelson-Morley experiment*, and others, have however *not detected* any relative velocity between the earth's motion and the hypothetical preferred Maxwell frame.

Einstein's theory of special relativity is derived from two postulates:

1 *Physics is invariant in all inertial frames* (frames moving with respect to any one inertial frame).
2 *Maxwell's theory is correct*, and thus the velocity of light is invariant in all inertial frames.

These two postulates imply a rejection of the galilean transformation, in order that Maxwell's equations be invariant under the correct transformation. Furthermore, the correct transformation does not preserve Newton's laws in all inertial frames, so classical mechanics is found to be incorrect, except as an approximation for velocities much less than c. *The Lorentz transformation* is found to fulfill the two postulates above. It relates the *space-time coordinates* of an event x, y, z, t in one frame to those in a frame at relative velocity v_o, (x', y', z', and t'), as follows:

$$x' = \gamma x - \beta\gamma ct \qquad y' = y \qquad z' = z$$

$$t' = \gamma t - \beta\gamma\frac{x}{c}, \text{ where } \beta = \frac{v_o}{c}, \text{ and } \gamma = \frac{1}{\sqrt{1 - v_o^2/c^2}}$$

Some important aspects of the Lorentz transformation are:

1 *The relativity of simultaneity*, emphasizing the fact that t does not generally equal t', and furthermore that *two events at the same time t do not in general occur at the same time t'.*
2 *Time dilation.* A time interval Δt in the rest frame of a clock is found to be lengthened to $\Delta t' = \gamma(\Delta t)$ as observed in a frame in relative motion with respect to the clock.
3 *Space contraction.* An object of length l, as measured in its rest frame, is found to contract, such that $l' = l/\gamma$, in a frame in relative motion with respect to it, in a direction along the length l.
4 *Velocity addition.* $v' = \dfrac{v \pm v_o}{1 \pm v(v_o/c^2)}$

   The velocity v' can never exceed c, if v and v_o are less than c. There is always a v_o possible such that v' is zero. The primed frame is then called the *rest frame* of the object.
5 *Relativistic Mechanics:* Newton's second law is replaced by $\vec{F} = \Delta\vec{p}/\Delta t$. The relativistic momentum $\vec{p}$ is given by $p = \gamma\beta m_o c$, and the total energy E is given by $E = \gamma m_o c^2$. For a particle at rest, $E = m_o c^2$, where $m_o c^2$, *the rest energy*, comes from the rest mass. The energy of motion, or kinetic energy, T is then given by:

   $$T = E - m_o c^2 = (\gamma - 1)m_o c^2$$

ELEMENTARY PARTICLES

14-1 PARTICLES AND FIELDS

From the days of the ancient Greek philosophers to the present, the highest goal of physics has been to discover underlying order and simplicity in nature. One form of order which has had a powerful appeal to physicists is the idea that all of the varied manifestations of matter are made of a few basic building blocks. The first major step in discovering these building blocks was the realization that all material substances are made of molecules and that molecules are in turn composed of atoms of the different chemical elements. The first *elementary particles* were 92 different kinds of atoms, corresponding to the chemical elements.

We now understand that the different kinds of atoms are merely different arrangements of electrons bound to positively charged nuclei. Furthermore, we know that the different nuclei are simply combinations of varying numbers of protons and neutrons. Thus, we might be tempted to point with some pride to our understanding of the seemingly infinite variety of matter in terms of combinations of just three elementary particles. Table 14-1-1 lists them, with their masses and charges.

Table 14-1-1 Elementary Particles

PARTICLE NAME	SYMBOL	MASS, MeV	CHARGE, e
Electron	e^-	0.51	−1
Proton	p	938.2	+1
Neutron	n	939.5	0

The unit of mass in Table 14-1-1 is an energy unit, the MeV, which is an abbreviation for a million electron volts. The Einstein equation, $E = mc^2$, is used to relate a given rest mass m to an energy E. In this equation the energy E is in joules if the mass m is in kilograms, and the velocity of light, c, is in meters per second. However, in elementary particle physics, the convenient energy unit is MeV rather than joules, and the convenient mass unit is the mass of the electron, m_e, rather than the kilogram. The Einstein equation gives, for an electron of mass m_e, a mass energy m_ec^2 equal to 0.511 MeV. For a particle with another mass (for example the proton, whose mass is $1{,}838m_e$), the rest-mass energy is then $(1{,}838)(0.511) = 938$ MeV. The unit of charge in Table 14-1-1 is the magnitude of the charge of the electron, 1.60×10^{-19} C, so that the electron charge is -1 and the proton charge is $+1$.

Having described an atom as an assemblage of protons, neutrons, and electrons, it is natural to ask what holds it together. Quantum mechanics in conjunction with the *electric force* has been very successful in describing the way in which the electrons in a given atom are bound to the positively charged nucleus. However, the binding of neutrons and protons within the nucleus is not well understood. We do know that there is a *strong nuclear force* which holds the protons and neutrons together, a force so strong that at short distances, of the order of the diameter of a nucleus, it completely overpowers the electrostatic repulsion of the protons. In trying to study this force, much of the complication—and the challenge—of elementary particle physics has arisen.

In addition to the proton, neutron, and electron, there are two other stable particles: the photon and the neutrino. The neutrino is emitted in nuclear beta decay and interacts with matter only via the *weak nuclear force.* We will postpone further discussion of the neutrino and of this weak interaction. The photon is of special interest because it is a *field quantum.* In a correct quantum theory, the electromagnetic interaction is not represented by classical electric and magnetic fields. Instead, the fields are replaced by electromagnetic quanta, photons. This field quantization is yet another manifestation of the by now familiar wave-particle duality of nature. Pictorially, we can represent a single electron moving in a straight line as in Fig. 14-1-1*a*, where the wavy loops are photons emitted and reabsorbed by the electron. These take the place of its classical electric field and lead to a force on other charges.

Figure 14-1-1 *(a)* **Electric field of an electron represented by emitted and reabsorbed photons (wavy lines).** *(b)* **Electron-electron scattering as a result of a photon exchange.**

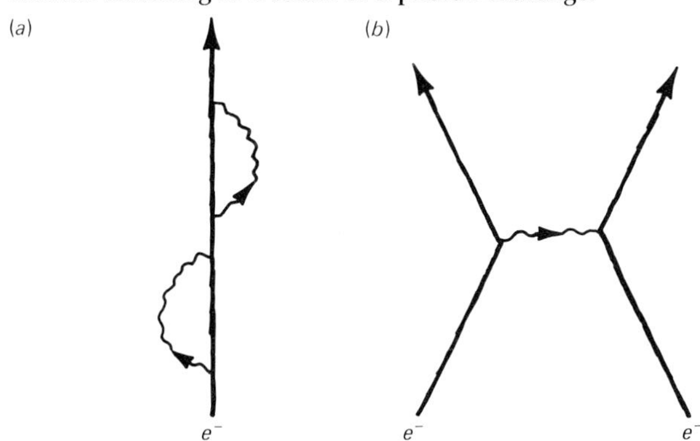

When an electron interacts electrically with another charge, for example, when it collides with another electron, the force between them results from an exchange of one or more photons, as in Fig. 14-1-1*b*. The electric field of one electron, represented by the emission of a photon, is shown here to scatter another electron because it absorbs the photon. Classically you might imagine this *exchange force* to be like the result of one ice skater throwing a heavy snow ball at another. Momentum conservation leads to forces on the thrower and receiver of the ball, such that they seem to have interacted by means of a repulsive force between them. Classically, we cannot imagine how an exchange force can lead to an *attraction*, but quantum mechanically such attractive exchange forces are indeed possible. For like charges, photon exchange leads to repulsion, and for unlike ones to attraction.

If a great number of photons are exchanged, the average effect is the same as that of a classical field. However, in the realm of elementary particle physics, where particles often interact for very short times, exchange of a single photon is the dominant electromagnetic interaction. Diagram *b* of Fig. 14-1-1 is therefore actually a symbolic representation of all that must be calculated to quite accurately predict the scattering of one high-energy electron by another. Such calculations, and many other very precise predictions of the theory we have been describing, called *quantum electrodynamics*, have been completely verified by experiments. In spite of some of its nonintuitive aspects, the theory is accepted as our most correct description of the electromagnetic interaction between charged particles.

With the success of the theory of quantum electrodynamics, a natural question was whether all types of forces actually arose from exchange of field quanta. In 1935, a Japanese physicist, H. Yukawa, proposed that this was at least true for the strong nuclear force. He predicted the existence of elementary particles, called *mesons*, which would be the field quanta of this force. Yukawa knew that the strong nuclear force is *short-range*, dominating the electric force at short distances but not at large distances. Evidence for this is that protons are held together in an individual nucleus, yet nuclei of adjacent atoms in a molecule exert negligible nonelectric forces on each other. The short-range nuclear force is characterized by field quanta with nonzero mass (in contrast to the photon). Yukawa estimated the mass of a nuclear force quantum to be about 300 times that of an electron.

Over ten years after Yukawa's hypothesis, the particles he

predicted were first definitely identified and were named pi mesons, or pions. With field quanta known for the electromagnetic and strong interactions, we may ask about the existence of others. At present there are only two other known fundamental interactions: the gravitational interaction and the weak nuclear interaction. Although field quanta called gravitons and W particles have been hypothesized for these interactions, they have not been observed. The extreme weakness of both interactions makes experiments to detect the field quanta very difficult. The present lack of evidence does not rule out the possibility that such quanta exists.

We have so far included as elementary particles the basic constituents of matter and the field quanta of the basic interactions. An additional class of elementary particles consists of the *antiparticles.* In relativistic quantum mechanics, the theory predicts an antiparticle for every particle—the antiparticle being identical to the particle in mass but with opposite charge. In 1928, the first successful relativistic quantum theory was developed, by Dirac, for electrons. However, the predicted *antielectron* was not taken seriously until it was discovered experimentally several years later by C. D. Anderson. This first antiparticle has its own name, the *positron*. Others, such as the antiparticles of protons and neutrons, which have also been observed, are described simply as antiprotons, and antineutrons, and so on.

A general technique for creating an antiparticle is to produce a particle-antiparticle pair with a high-energy photon. Only energy need be supplied, equal to $2mc^2$, where m is the particle mass, since the net charge of a particle-antiparticle pair is zero. However, the particle-antiparticle pair can only be created in the vicinity of a third particle (such as a nucleus) which can receive a relatively small amount of momentum during the pair-production process. Otherwise it is impossible to conserve momentum. Except in special instances, we will not consider the antiparticles as distinct and separate elementary particles.

Finally, particles like neutrinos are also considered to be elementary. They aren't field quanta, and they also aren't building blocks of atoms, because they do not interact via either the strong or electromagnetic interactions. However, along with other weakly interacting particles, including gravitons (the gravitational field quanta) if they exist, we consider neutrinos to be a basic constituent of the universe as a whole, and just as elementary as any other particles.

14-2 MESONS, BARYONS, AND LEPTONS

The particles predicted by Yukawa are called mesons, from a Greek root "meso" meaning middle, because their mass is intermediate between the mass of the electron and of the proton and neutron. Mesons were first discovered by physicists studying the cosmic radiation, which consists of fast nuclear particles striking the earth from outside its atmosphere. We can write down three typical meson production reactions, for collisions between a cosmic ray proton and an individual neutron of an air nucleus:

$$\begin{aligned} p + n &\rightarrow p + p + \pi^- \\ p + n &\rightarrow n + n + \pi^+ \\ p + n &\rightarrow p + n + \pi^0 \end{aligned} \qquad 14\text{-}2\text{-}1$$

The reactions in Eq. 14-2-1 show production of pi mesons with three different charges, -1, $+1$, and 0. The field quantum of the strong interaction was thus found to differ from that of the electromagnetic interaction in that it was both massive and charged.

For each of the reactions 14-2-1, the mass energy of the final particles is greater than that of the initial particles by one pion mass, about 140 MeV. Therefore, in order to conserve energy, 140 MeV of kinetic energy must be converted to mass energy in such interactions. Individual protons must therefore be accelerated to energies in excess of 140 MeV, to be able to conveniently study mesons in the laboratory. In fact, because momentum as well as energy must be conserved, the reactions shown in Eq. 14-2-1 require a minimum proton kinetic energy of about 280 MeV, assuming that the target neutron is at rest. In order to attain such high proton energies, elementary particle physicists were led to develop and use high energy particle accelerators. The term *high energy physics* has become a synonym for elementary particle physics.

Two types of pion production reactions which are *never* observed are shown below, with a bar across the arrow to indicate that the reaction does *not* take place:

$$\begin{aligned} p + n &\not\rightarrow p + p + \pi^0 \\ p + n &\not\rightarrow n + \pi^+ \end{aligned} \qquad 14\text{-}2\text{-}2$$

Were the first reaction of Eq. 14-2-2 observed, net charge would be created, since the charge of the initial particles is $+1$ while that of the final ones is $+2$. This would violate the law of *charge conservation*, which, so far as we know, is never violated.

The second reaction in Eq. 14-2-2 shows a proton disappearing.

Of the particles we have so far discussed, the two heaviest, the proton and neutron, are called *baryons*, where the root "bary" means massive. Just as the total charge of the universe is believed to be invariant, we also believe the total number of baryons in the universe is constant. If we assign baryon number +1 to every proton and neutron, and −1 to their antiparticles, nature appears to require that *baryon number is always conserved* (like charge). The second reaction in Eq. 14-2-2 is never observed because it would violate *baryon conservation.*

PROBLEM 1
What are the allowed reactions for production of a single pion by protons incident on protons?

PROBLEM 2
Explain why conservation of momentum and energy forbids the reaction $\pi^- + p \rightarrow n$ but allows the reactions:

$$\pi^- + p \rightarrow n + \nu$$
$$\pi^- + p + p \rightarrow n + p$$

Obviously, from allowed meson production reactions like $p + n \rightarrow p + p + \pi^-$, there is no law of meson conservation. Mesons are created and absorbed freely by baryons, much as photons are created and absorbed by electric charges. Meson absorption reactions like $\pi^- + p + n \rightarrow n + n$ and $\pi^0 + p + n \rightarrow p + n$ are allowed, so that mesons tend to interact in nuclear matter and disappear soon after they are produced. Even if the mesons are kept in vacuum, they disappear for another reason. They are unstable and decay radioactively.

Charged pions decay according to the reactions:

$$\begin{aligned} \pi^+ &\rightarrow \mu^+ + \nu \\ \pi^- &\rightarrow \mu^- + \nu \end{aligned} \qquad 14\text{-}2\text{-}3$$

where ν (nu, pronounced "noo") stands for neutrino, and μ (mu, pronounced "mew") represents a particle called a muon. This *mu decay* of a pion takes place with a mean life (average lifetime) of 2.6×10^{-8} sec. The exact lifetime of any individual pion is not precisely predictable because the pion decays by random chance, but lifetimes more than twice the average are unlikely.

For a large sample, of N_0 pions, in existence at time $t = 0$, the number remaining varies with time according to the exponential curve of Fig. 14-2-1. The property of such a curve is that in a time

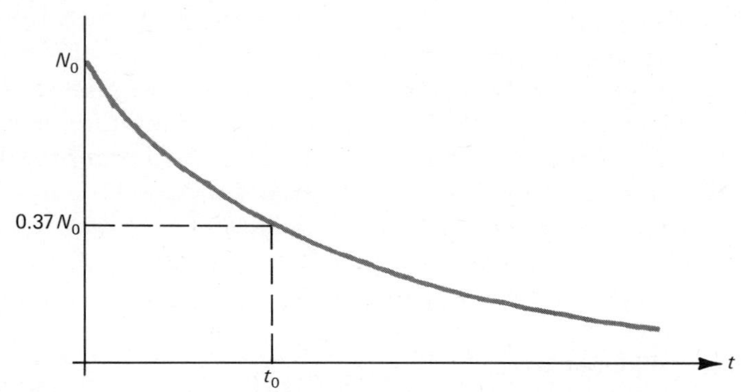

Figure 14-2-1 Exponential decay curve for N_o particles with a mean life t_o.

interval of one mean life, indicated by the interval t_0 on the figure, the number of pions is always reduced by a factor 0.37, regardless of when the time interval is taken to start. A majority of the known elementary particles are found to randomly decay with very short mean lives, making it necessary to study them just after they are produced, at the high-energy accelerator laboratories.

The pi-mu decay is called a *weak decay*, which is one manifestation of the weak interaction. The particles involved in weak interactions are called leptons, and of all elementary particles, leptons are those of which the fewest examples are known. There are only four known leptons, the electron, muon, electron-neutrino, and muon-neutrino. There is also a law of *lepton conservation*, which requires that in all processes in nature the numbers of *electron leptons* (electrons plus electron-neutrinos) and of *muon leptons* (muons plus muon-neutrinos) are separately conserved. As a result, the pi-mu decay reactions are properly written as follows:

$$\begin{aligned} \pi^+ &\rightarrow \mu^+ + \nu_\mu \\ \pi^- &\rightarrow \mu^- + \overline{\nu}_\mu \end{aligned} \qquad 14\text{-}2\text{-}4$$

The μ^+ is the antiparticle of the μ^-, and the $\overline{\nu}_\mu$ is the antiparticle of the ν_μ, which is the muon-neutrino. On the right side of each equation we thus have one lepton and one antilepton. We assign lepton number +1 to the leptons and −1 to the antileptons. The sum of the lepton numbers on the left and right of both reactions in Eq. 14-2-4 is then zero, and the conservation of leptons means conservation of this total lepton number. As far as we know, from many experiments, this is an absolutely conserved quantity, like charge and baryon number.

A more complex decay takes place for the muon. With a mean life of about 2.2×10^{-6} sec, muons decay as follows:

$$\begin{aligned} \mu^+ &\rightarrow e^+ + \nu_e + \overline{\nu}_\mu \\ \mu^- &\rightarrow e^- + \overline{\nu}_e + \nu_\mu \end{aligned} \qquad 14\text{-}2\text{-}5$$

Here the muon-lepton number is -1 on both sides of the first reaction equation and $+1$ on both sides of the second, satisfying lepton conservation for muon-leptons. The electron lepton number, which must separately be conserved, is seen to be 0 for both sides of both reactions.

PROBLEM 3

If a star produces energy by fusing hydrogen into helium, does it produce neutrinos or antineutrinos?

PROBLEM 4

High-energy neutrino beams are produced by the decay of high-energy pions according to the reactions of Eq. 14-2-4. The interactions of ν_μ and $\overline{\nu}_\mu$ can then be studied, with some difficulty because they are so weak. Which of the following processes is allowed and which forbidden? Explain why. The nonoccurrence of one of these reactions led to the conclusion that there were two kinds of neutrinos. Which reaction?

$$\begin{aligned} &\nu_\mu + p \rightarrow \mu^+ + n \\ &\overline{\nu}_\mu + p \rightarrow \mu^+ + n \\ &\nu_\mu + n \rightarrow \mu^- + p \\ &\nu_\mu + n \rightarrow e^- + p \\ &\nu_\mu + p \rightarrow e^+ + n \end{aligned}$$

We have so far ignored the decay of the neutral pi meson. There are only + and − charged muons, so mu decay of the π^0 cannot take place because of charge conservation. However, the neutral pion can decay into photons. One-photon decay is not allowed because it cannot take place while conserving momentum. (Consider, as a simple example, decay of a π^0 at rest.) Therefore, with a very short mean life of about 10^{-16} sec, the π^0 decays into two photons:

$$\pi^0 \rightarrow \gamma + \gamma \qquad 14\text{-}2\text{-}6$$

Summing up, we have introduced three categories of elementary particles: *mesons*, *baryons*, and *leptons*. These are all there seem to be, except for the photon, which describes one unique particle. In addition, we have seen that the pi mesons and the mu leptons both decay so that only stable particles like photons, electrons, and neutrinos remain at times much greater than about 10^{-6} sec after a pion is produced. Table 14-2-1 lists all the elementary particles so far discussed and gives some of their important properties.

Table 14-2-1

PARTICLE	MASS, MeV	DECAY	MEAN LIFE, sec
Photon, γ	0		
LEPTONS			
Neutrino, ν	0		
Electron, e^-	0.51		
Muon, μ^-	106	$\mu^- \rightarrow e^- + \bar{\nu}_e + \nu_\mu$	2.2×10^{-6}
MESONS			
Pion, $\pi^\pm$	140	$\begin{Bmatrix} \pi^- \rightarrow \mu^- + \bar{\nu}_\mu \\ \pi^+ \rightarrow \mu^+ + \nu_\mu \end{Bmatrix}$	2.6×10^{-8}
Pion, π^0	135	$\pi^0 \rightarrow 2\gamma$	$\sim 10^{-16}$
BARYONS			
Proton, p	938		
Neutron, n	940		

PROBLEM 5

Which of the four interactions does each of the following particles participate in: electron, proton, neutrino, pion, muon?

Table 14-2-1 shows zero mass for the photon and neutrino, indicating particles which move at the velocity of light and carry energy and momentum in spite of their lack of mass. (We have used mass here consistently to stand for what would be called *rest mass* in relativity, the inertial mass of the particle when it is at rest.) The table also gives the mean lifetimes of the unstable particles and their dominant mode of decay. A particle can usually decay in more than one way. For example, the π^+ meson can undergo electron decay as well as mu decay:

$$\pi^+ \rightarrow e^+ + \nu_e$$

However, the mu decay is much more likely, taking place more than 99.9 percent of the time.

Finally, we have introduced a number of new conservation laws and have recalled some that have been discussed previously. We can now give a complete list of all the quantities which are believed to be absolutely conserved in nature:

Energy
Momentum
Angular momentum
Charge
Baryon number
Lepton number
TCP

The last quantity of the preceding list stands for the product of three quantum numbers, called T, C, and P, which can each only have the allowed values +1 or −1. These numbers describe the properties of a quantum-mechanical state under three hypothetical transformations: reversing the sense of time, changing particles into antiparticles, and reflecting the system in a mirror. If the special theory of relativity is true, then it can be shown that *TCP* is an absolutely conserved quantity.

In elementary particle physics we have so little detailed theoretical knowledge that principles such as the conservation laws represent most of our important physical insights. The presence of a conserved quantity can be shown to be equivalent to what we call a *symmetry* in nature. For example, the laws of physics which describe collisions of two billiard balls are the same for an experimenter in an unaccelerated laboratory at any location. Moving from one place to another is called a *translation*, and we believe that the laws of billiard ball collisions—and in fact all physical laws—remain the same with respect to translations. There is "symmetry (or invariance) of physical laws with respect to translations." Perhaps surprisingly, this symmetry can be shown to require that momentum is conserved. The search for new symmetries in nature and the testing of old ones constitute very important parts of our attempts to understand elementary particle physics.

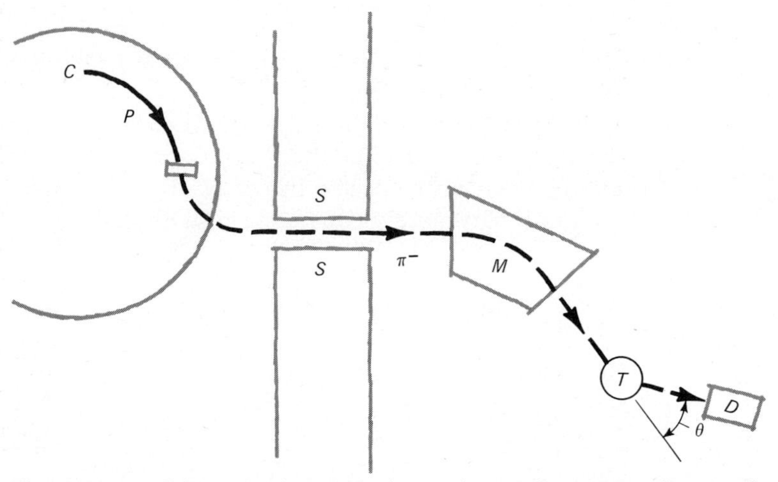

Figure 14-3-1 Schematic plan view of an early experiment to study π^--p elastic scattering. Protons *(P)* accelerated in a cyclotron *(C)* strike an internal target and produce π^- mesons, which are deflected out of the cyclotron by its magnet, and pass through a hole in a shielding wall *(S)*. A beam magnet *(M)* deflects beam particles of the desired momentum toward a liquid hydrogen target *(T)*. Pions scattered through an angle θ are counted in a detector *(D)*.

14-3 EXPERIMENTAL TECHNIQUES

In order to create and detect elementary particles, physicists have had to develop many aspects of a complex and sophisticated technology. In a typical meson production reaction, for example, $p + p \rightarrow p + n + \pi^+$, the kinetic energy required to produce the pion is supplied by a high energy proton from an accelerator which strikes a stationary target proton. The production and study of elementary particles requires accelerator beams of both high energy and high intensity, where intensity refers to the number of beam particles per second incident on the target. This is because most experiments are concerned with rare processes involving *secondary beams.* A secondary beam consists of particles such as pi mesons, produced when the primary accelerator beam strikes a target. Pion-proton elastic scattering will provide a good example of a typical high-energy physics experiment. In general, scattering experiments are one of the most important ways in which elementary particle properties are studied.

Figure 14-3-1 is a schematic view of a very important early π-p elastic scattering experiment, performed at the University of Chicago around 1950 by Enrico Fermi and his collaborators. Modern high-energy experiments differ from the setup shown in Fig. 14-3-1 mainly in increased size and sophistication of each component of the apparatus, but the essential features remain the same. An accelerator is used, usually to accelerate protons, in order to generate a secondary beam of pions or of other particles. The secondary beam is then aimed at a target, usually a proton target in the form of liquid hydrogen. Some type of interaction of the secondary beam with the target protons is then studied, usually by detecting outgoing particles which differ from the beam particles in direction, energy, or identity.

In Fig. 14-3-1 the accelerator is a cyclotron with a maximum energy of 450 MeV, which was considered high energy in 1950. By 1969 the maximum available accelerator energy was 70,000 MeV, at Serphukov in the U.S.S.R. In 1972 it is expected that an energy about 1,000 times that of the old Chicago cyclotron will be reached at the U.S. National Accelerator Laboratory (NAL) in Batavia, Illinois. Each increase in accelerator energy has so far produced important new knowledge.

As the push to higher energies has taken place, the cyclotron has been supplanted by the synchrotron as the dominant type of accelerator. The major constituent of a synchrotron is a ring of electromagnets which causes particles to travel in a circular orbit while they are accelerated. Figure 14-3-2 shows the synchrotron now being built at the National Accelerator Laboratory. The main ring of the synchrotron appears successively enlarged in Fig. 14-3-2*a* to *c*. In the first figure, the entire ring is shown, $1\frac{1}{4}$ mi in diameter; in the second, the pattern of individual magnets along part of the vacuum pipe in which the beam circulates is shown; and in the third figure a typical cross section of magnet and beam pipe is shown, with the magnetic field direction indicated.

The synchrotron magnet exerts a magnetic force on a moving charge, as discussed in Chap. 4. If the magnetic field direction is perpendicular to the direction of motion of a charge, the force is given by

$$F = qvB \qquad \text{14-3-1}$$

where $\vec{F}$ is perpendicular to both $\vec{v}$ and $\vec{B}$. If particles are in motion around the ring and the magnetic field is perpendicular to the plane of the ring, then the force, at right angles to both v and B, points along a radius of the ring. In Fig. 14-3-2*a*, if positively charged protons are moving in the direction of the arrow around the main ring, and if the magnetic field is upward out of the paper,

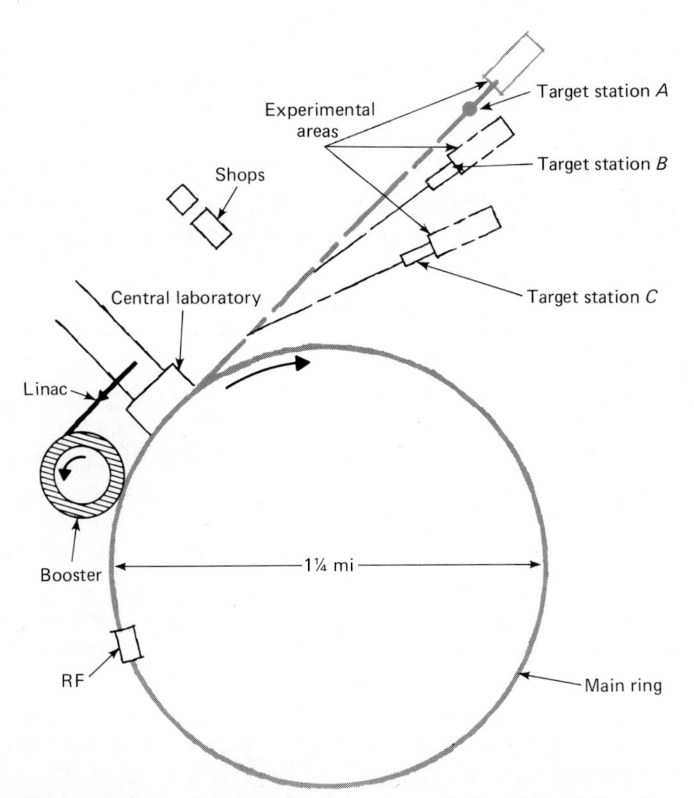

Figure 14-3-2*a* An overall plan view of the U.S. National Accelerator Laboratory. Arrows indicate the path of protons, from linac to booster to main ring, as they are accelerated to energies up to over 400 GeV (4×10^5 MeV). The main ring is 1¼ mi in diameter, and the accelerator cost is about 250 million dollars.

then the magnetic force will be inward, toward the center of the ring. Such an inward, or *centripetal*, force is discussed in Chap. 3, in connection with circular motion.

We have seen in Chap. 3 that circular motion requires an inward acceleration, given by an inward force. For a particle of mass m moving with constant speed v around a circle of radius R, we found a force F given by

$$F = \frac{mv^2}{R} \qquad 14\text{-}3\text{-}2$$

Combining the two equations for force, Eqs. 14-3-1 and 14-3-2, we

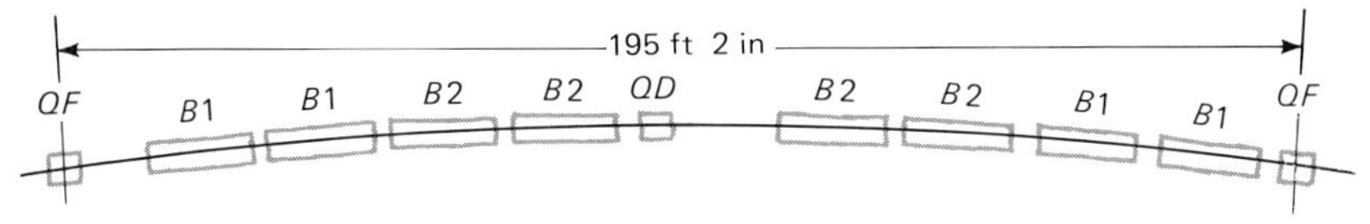

Figure 14-3-2*b* Schematic plan view of the magnets along a typical section of the main ring. In a length of 195 ft there are eight bending magnets (*B*1 and *B*2) and three focusing magnets (*QF* and *QD*).

can find the relation between magnetic field B and radius R for a moving charged particle (always assuming B is perpendicular to v as in the synchrotron):

$$\frac{mv^2}{R} = qvB \qquad \text{or} \qquad R = \frac{mv}{qB} \tag{14-3-3}$$

Figure 14-3-2*c* Cross section of a main ring bending magnet. The magnet is mainly iron, with water-cooled coils through which large currents flow. The coils on the left side of the figure carry current in the direction pointing out of the paper; those on the right in the opposite direction; and a magnetic field in the direction labeled $\vec{B}$ results. The beam circulates inside a vacuum chamber centered in the magnet.

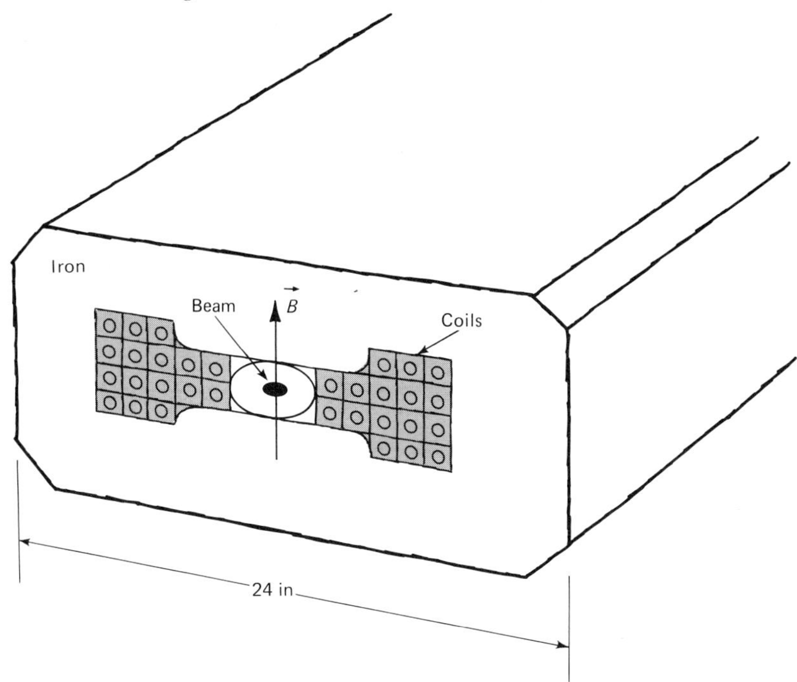

The derivation of Eq. 14-3-3 uses Eq. 14-3-2, which is derived from classical (nonrelativistic) mechanics. However, if we replace the quantity mv by the momentum, symbolized by the letter p, the equation remains correct for relativistic motion. We have

$$R = \frac{p}{qB} \qquad 14\text{-}3\text{-}4$$

If the particle is very relativistic, meaning that $v/c \approx 1$, then $pc \approx E$, where E is the particle's total energy. In high-energy physics this is often true, so it is convenient to write Eq. 14-3-4 in terms of pc:

$$R = \frac{pc}{qcB} \approx \frac{E}{qcB} \qquad 14\text{-}3\text{-}5$$

PROBLEM 6
Show that, for the charge $q = +e$, and for the energy pc in MeV, Eq. 14-3-5 can be written:

$$R = \frac{(pc)_{\text{MeV}}}{300\ B}$$

According to Eq. 14-3-5, the larger B, the smaller R. However, if we use iron electromagnets, the maximum value of B is about 2.0 W/m^2. At present, iron synchrotron magnets are dictated by economic reasons, so the minimum radius of a synchrotron magnet is connected to the accelerator energy through Eq. 14-3-5, with $B = 2.0$. For the very high energies which are presently interesting, radii approaching a mile result. In addition to its role with regard to accelerators, the relation given by Eq. 14-3-5 is important in elementary particle experiments for determining the value of pc for particles with initially unknown momentum. In such a case, the path of the particle in a region of known B is determined, thereby determining R. From the known values of B and R, pc is then found.

PROBLEM 7
What is the radius of curvature of a 10-GeV pion in a magnetic field of 2.0 W/m^2? For a path of 1 m, through what angle does the pion bend in such a magnetic field? (Calculate the angle, or find it with a protractor from a scale drawing.)

To return to the proton synchroton, the acceleration of the protons remains to be discussed. Ordinarily this takes place in several stages. First, slow hydrogen ions are formed in an electric discharge and accelerated in an *ion gun* somewhat similar to the electron gun of a cathode-ray tube. This gun is often run at about 1 million volts.

In the NAL machine, shown in Fig. 14-3-2, protons from the ion source and gun are injected into a linear accelerator, or *linac*, in which they are accelerated along a straight line by radio frequency electric power. The particles emerge from the linac at 200 MeV and are then ready for injection into a small synchrotron called the booster. At injection into the booster, the electromagnets which form the booster ring are run at a relatively low magnetic field so that R for 200 MeV protons is equal to the ring radius. Were the protons instead injected into the large main ring, the appropriate magnetic field would be too low to conveniently establish with the required precision. The booster is used to raise the energy to a convenient value for injection into the main ring.

Injection into the booster takes place when a pulsed magnet bends the linac beam onto the synchrotron orbit for a short time and is then turned off. The booster then contains about 10^{13} circulating protons. Spaced around the circumference of the booster are about 20 small accelerating *cavities*, formed of copper and powered by radio frequency power sources. The circulating protons pass through these cavities and are accelerated by an electric field each time they traverse a cavity.

In going around the booster orbit once, the protons gain, on the average, about 1 MeV. Beginning at 200 MeV, they make about 10^4 turns in about 1/30 sec, reaching a final energy of about 10,000 MeV, or 10 GeV. (The unit GeV is equal to 1,000 MeV, and stands for "giga" electron volt.) At 10 GeV, the booster synchrotron has finished its job, and the particles are ready for transfer to the main ring, where they are accelerated from 10 to 500 GeV.

The acceleration cycle of the main ring takes several seconds, and before beginning a cycle, 10 or more cycles of booster operation are used to inject up to 10^{14} protons into the main ring. The main ring then increases the proton energy to 500 GeV, the maximum allowed by the radius and the available B field, by the same technique the booster used. When the protons reach 500 GeV, they are ejected by a pulsed magnetic field along a tangent to the main ring and sent to the experimental areas. Here they can be shared among several possible targets, creating secondary particles which can be used to perform five to ten experiments at any given time. After each acceleration cycle, the main ring magnetic field is reduced to the field required for injection, protons are injected from the booster, and another cycle begins.

PROBLEM 8
What is the power in watts of a beam such as that from the NAL accelerator?

The name synchrotron refers to two kinds of synchronization necessary during the acceleration cycle. First, the magnetic field of the ring must increase, so that as the particles are accelerated their radius remains unchanged, and, second, the motion of the protons and the radio frequency power of the accelerating system must be in proper synchronization with each other. Two physicists, E. M. McMillan of the U.S.A. and V. I. Veksler of the U.S.S.R., shared a Nobel prize for inventing the synchrotron and for showing that the necessary synchronization was possible.

In order to do high-energy experiments, particle detectors are needed as well as an accelerator. All particle detection ultimately depends upon the electromagnetic interactions of charged particles. When a fast charged particle goes through matter, its electric field disrupts the atoms along its path. Some atoms are *ionized* by having one or more of their electrons completely ejected, while others are *excited* by energy given to one or more electrons which remain bound.

Ionization or excitation created along the path of a fast particle can be used to detect the passage of the particle through a sensitive part of a detector, or particle counter. The most widely used form of counter is the *scintillation counter*, which utilizes one of a variety of transparent materials that are especially efficient in emitting

Figure 14-3-3 A typical scintillation counter. A sheet of plastic scintillator (seen on edge) emits light when a fast charged particle traverses it. Some of this light reaches the photomultiplier tube, after successive reflections at the walls of the scintillator and the transparent plastic light pipe.

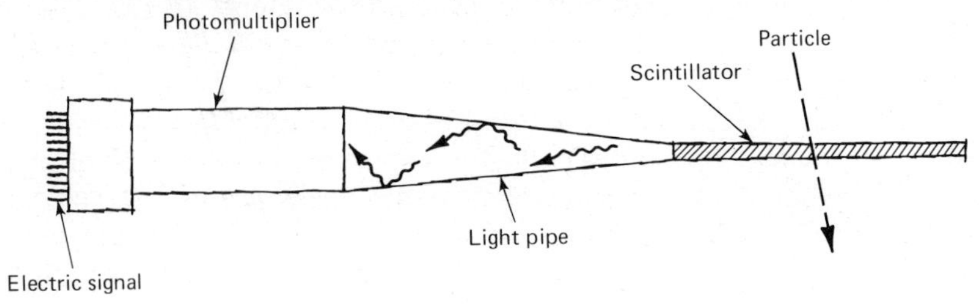

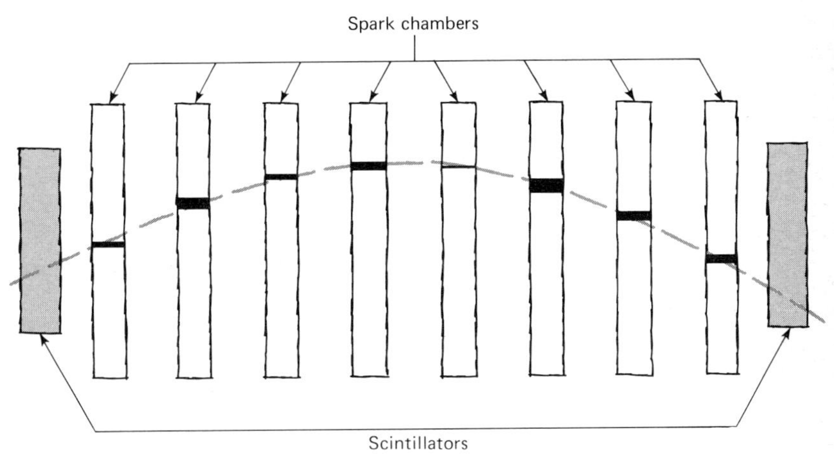

Figure 14-3-4 Sparks in a series of eight spark chambers record the curved trajectory of a charged particle in a magnetic field. Scintillation counters are shown which provide information that a particle has passed through the spark chambers.

light as a result of excitation. The light-emitting material, called the scintillator, is viewed by a very sensitive photon detector, called a photomultiplier tube. When one or more photons are detected, the photomultiplier tube and its associated electric circuitry produce an electrical signal which can be in the form of a voltage pulse lasting for a time as short as 10^{-9} sec. A sketch of a scintillation counter is shown in Fig. 14-3-3.

Other types of particle counters detect the ions created by fast particles in traversing a gas. These ions can be collected, and sometimes multiplied in number, by establishing an electric field in the gas, and the collected charge generates electrical signals. The *Geiger counter* and *proportional counter* are both examples of gas counters. However, at present a more important use of ionization in gases for particle detection is the *spark chamber.* In the spark chamber, a very strong electric field is established between two parallel electrodes separated by a gas, usually neon. The electric field is so strong that when free electrons and ions are created in the gas by the passage of a fast particle a spark results, which can be photographed to record precisely where the ionization was produced. Figure 14-3-4 shows a schematic view of a particle trajectory, curving in a magnetic field made visible by an array of spark chambers.

It is sometimes valuable to have a photographic record of a particle trajectory, but in experiments where tens of thousands, or even millions, of trajectories must be evaluated to find a small percentage of interesting events, photographic records are expensive and hard to analyze. Instead, the plates of spark chambers can be formed from grids of wires, and the particular wires struck by the spark can be recorded and read directly into an on-line computer. Typically, such a *wire spark chamber* might be 1 m square, with 1,000 horizontal wires forming one plate and 1,000 vertical wires the other. If, for example, a spark occurs between horizontal wire number 856 and vertical wire number 352, these numbers are determined by electronic circuitry connected to the spark chamber wires. Trajectory information obtained from wire spark chambers is recorded in a form most suitable for computer analysis of a large amount of data.

The combination of counters and spark chambers can be used to make high energy particles somewhat "visible." However, the ultimate visual detector is the *hydrogen bubble chamber.* Hydrogen liquefies at a temperature of about 15° absolute (about −430°F) at atmospheric pressure, and liquid hydrogen is the usual proton target for high-energy experiments. If a volume of liquid hydrogen is at a temperature above its boiling point when a fast particle produces ions in the liquid, then boiling takes place preferentially along the track of ions. A visible trail of bubbles is formed, which can be photographed to record the particle trajectory. Liquid hydrogen can thus be used not only as a source of target protons but also to detect the tracks of particles which arise from interactions in the liquid.

The operation of a bubble chamber involves rapidly lowering the pressure on the hydrogen, which causes it to be above its boiling point, then photographing the bubbles within milliseconds, and finally raising the pressure to stop bulk boiling from occurring. A typical bubble chamber photograph is shown in Fig. 14-3-5. This is a view looking into a chamber about 30 in in diameter and 15 in deep, showing antiproton beam particles with pc about 1.5 GeV, and one antiproton-proton interaction which produces four charged secondary particles. The particle tracks in Fig. 14-3-5 are curved because the bubble chamber is operated in a strong magnetic field, so as to be able to determine pc from the track curvatures. The chamber is photographed by three cameras, in stereo (with only one view shown in Fig. 14-3-5), so that the actual particle trajectories in three dimensions can be found.

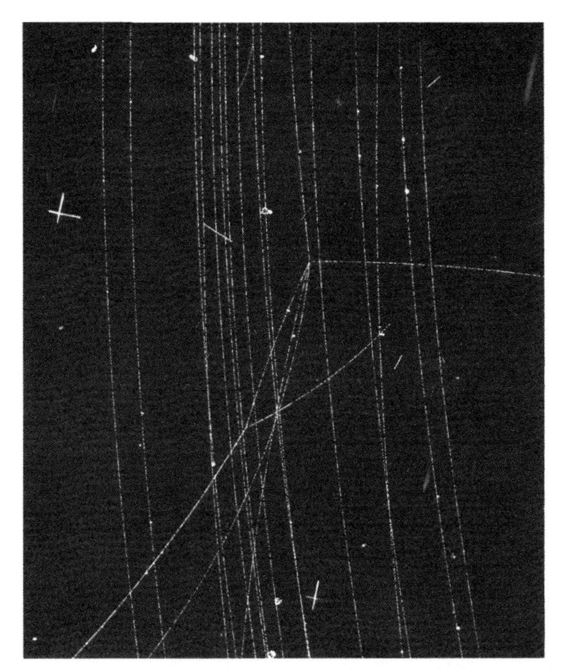

Figure 14-3-5 Bubble chamber photograph showing the reaction $\bar{p} + p \rightarrow \pi^+ + \pi^- + \pi^+ + \pi^-$. An antiproton beam enters from the left, and a magnetic field along the line of sight of the camera curves the particle paths. One of the pi mesons scatters nearly backward off a proton of the liquid hydrogen. *(Courtesy of Argonne National Laboratory.)*

PROBLEM 9

In Fig. 14-3-5, how can we determine that no π^0 meson was emitted? (A high-energy gamma ray does not have a high probability of producing a visible secondary charged particle in a bubble chamber the size of that shown.)

PROBLEM 10

What is the magnetic field direction in Fig. 14-3-5?

Up to now, nothing has been said about particle identification, although we know how to determine the sign of a particle and its momentum from its path in a magnetic field. To identify the particle we need to measure its mass, which is usually done by measuring its velocity in addition to its momentum. From a knowledge of both velocity and momentum, the mass can be found. Velocity can be measured in a variety of ways, the most direct being *time of flight*, where two counters are used to measure the transit time of the particle over a measured distance. For very high energies, when all particle velocities, regardless of mass, are very close to c, mass determinations become very difficult.

14-4 *I* SPIN AND STRANGENESS

The second section of this chapter described the three kinds of particles: *baryons*, *mesons*, and *leptons*. The *strong nuclear interaction* only involves the first two types of particles and represents the major area of research in elementary particle physics. The remainder of this chapter will therefore emphasize baryon and meson physics. In this section, two important quantities will be described, which are conserved in strong interactions but not in electromagnetic and weak processes.

Pion-proton scattering experiments, such as the one mentioned at the start of the last section, were undertaken soon after pion production was first observed with cyclotrons. Figure 14-4-1 shows in more detail the actual apparatus for measuring the scattering in the experiment previously diagrammed in Fig. 14-3-1. The momentum of the pion beam was defined by requiring a particular bending angle in traversing the *beam magnet* shown in

Figure 14-4-1. Target and counters for an elastic-scattering experiment. Counters *S*1 and *S*2 detect the arrival of a beam particle, while *S*3, *S*4, and *S*5 help determine that an elastic scattering has occurred.

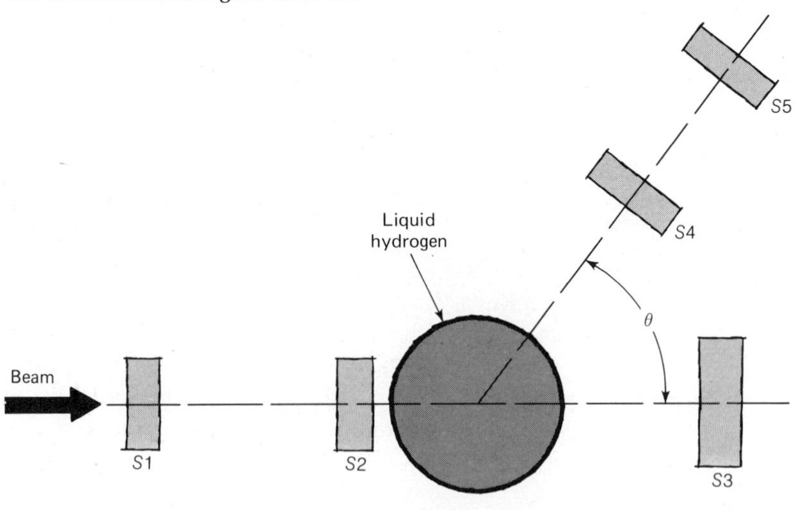

Fig. 14-3-1. An incident-beam particle was then defined by simultaneous counts in the two counters labeled $S1$ and $S2$, which eliminated single counts due to stray radiation in the laboratory.

More than 99 percent of the incident-beam particles passed through the target without any strong interaction. The rare scattered mesons were detected by simultaneous counts in $S4$ and $S5$, along with the absence of a count in $S3$. This indicated that the beam particle had not passed straight through the target but had scattered through the angle θ. The experiment consisted of determining how the scattering probability varied with the scattering angle and energy of the incident pions.

At the relatively low pion-beam energy of this experiment, pions observed in counters $S4$ and $S5$ could only have scattered elastically. At higher energies it is necessary to distinguish the elastic scattering reactions such as $\pi^- + p \rightarrow \pi^- + p$ from inelastic reactions such as $\pi^- + p \rightarrow \pi^- + \pi^+ + n$. This can be done by using an analyzing magnetic to measure the charge and momentum of the scattered pion. A further simplification at low pion energy is that the recoil proton from an elastic scattering reaction has so little energy it is not confused with the pion.

From earlier experiments in which p-p and p-n scattering were measured it was inferred that the nuclear force, which we now call the strong interaction, was independent of the electric charge. The proton and neutron appear to be two states of a single kind of strongly interacting particle, called the *nucleon.* As a hypothesis, this *charge independence* of nuclear force was then extended to include the pions. While the angular dependence of the pion-nucleon scattering could not be successfully explained by any simple force law, the comparison between scattering for different charges of meson and nucleon did establish the correctness of the hypothesis of charge independence.

The observed symmetry of the strong interaction with respect to changes of charge results in a new conserved quantity called *I spin* (short for "isotopic spin"). The *conservation of I spin* can be briefly summed up as follows:

1. Each type of strongly interacting particle is assigned an I spin quantum number, called I, such that the number of charge states of the particle is equal to $2I + 1$.
2. In strong interactions the total I spin of the final particles must equal the total I spin of the initial particles. Special rules, like those for combining angular momentum, are used to find the combined I spin of two or more particles from their individual I spins.

The number *I* is called *I* spin because of its mathematical analogy to angular momentum, which is associated with spinning objects.

Applying these notions to pions and nucleons, we find that the nucleon has *I* spin 1/2 because of its two charge states, proton and neutron. The pion has *I* spin 1 because of its three charge states π^+, π^0, and π^-. The combining rule for *I* spin, which will be discussed later in Sec. 14-6, results in just two possibilities for a pion-nucleon system: $I = 1/2$, and $I = 3/2$.

Charge independence actually means that the pion-nucleon interaction can depend on the total *I* value of the two particles, but not on their individual charges. Because of charge independence, the six possible interactions of three kinds of pion and two kinds of nucleon are found to be completely described by two interactions: those for $I = 1/2$ and those for $I = 3/2$ pion-nucleon states. In the next two sections we will hear more about *I* spin. But first, we will introduce another new conserved quantity in strong interactions.

Some particles other than pions and nucleons, first discovered in the cosmic rays, were called "strange particles" because they exhibited certain peculiarities and because they were unexpected. Provided enough energy is available, the strange particles are produced in nucleon-nucleon or pion-nucleon collisions nearly as often as pions. This indicates that they are strongly interacting particles. Some of the simplicity of the Yukawa picture disappeared with the discovery of another field quantum, a *strange meson.*

The strange mesons, which were named *K mesons,* were observed to sometimes decay into two pions, but to decay very slowly compared to the time expected. The typical time for a strongly interacting particle to change into a pair of strongly interacting particles is expected to be of the order of 10^{-23} sec, while the *K*-meson decay time is about 10^{-10} sec for the neutral *K* and longer for the charged *K*. The observed decay times would be expected for a weak nuclear interaction, such as the decay $\pi \rightarrow \mu + \nu$, instead of a strong one. This apparently strong production, yet weak decay, was the major peculiarity of the strange particles.

In addition to strange mesons, strange baryons were also found, which typically decayed into nucleon plus pion, also in a time too long by an enormous factor. One hypothesis, which was beautifully verified by experiment, was that the strong production of strange particles required that *two* be produced at once. Figure 14-4-2 shows a bubble chamber picture of a common kind of

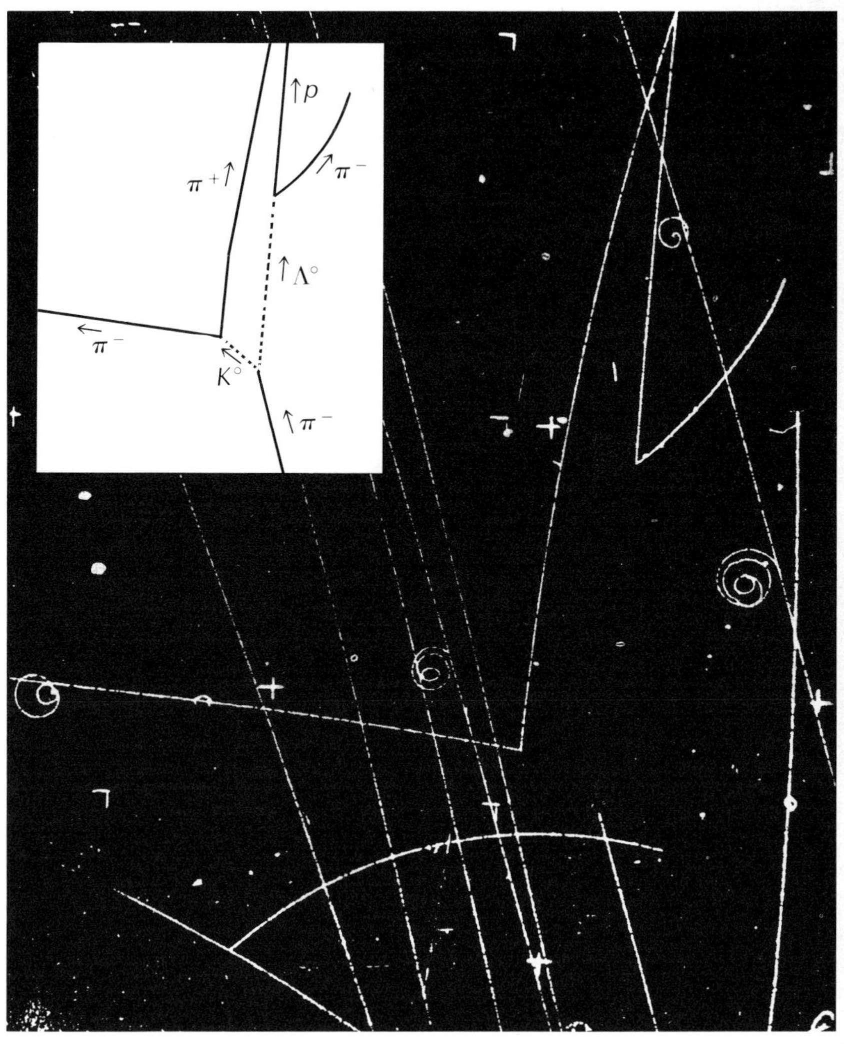

Figure 14-4-2 Bubble-chamber photograph showing associated production of two strange particles: $\pi^- + p \rightarrow K^0 + \Lambda^0$. Both produced particles are neutral and do not make tracks in the chamber, but both are seen to decay into a pair of charged particles. The inset sketch points out the locations of the interaction and decays. (*Courtesy of Prof. Luis W. Alvarez.*)

strange particle event, in which the following reaction takes place:

$$\pi^- + p \rightarrow K^0 + \Lambda^0 \qquad 14\text{-}4\text{-}1$$

The K^0 meson and the Λ^0 (lamda zero) baryon, both strange particles, are seen to be produced together.

Each neutral particle in Fig. 14-4-2 decays after it has moved some distance away from the place where the π^--p production interaction took place. The meson decays into $\pi^+ + \pi^-$ and the baryon into $p + \pi^-$. Thousands of such *associated production* events have been observed, while production of a single strange particle in a π^--p interaction has never been seen. An absolute conservation law operates which requires that at least two strange particles are produced in strong interactions of nonstrange particles. This law, proposed independently in 1953 by an American, M. Gell-Mann, and a Japanese, K. Nishijima, is called *conservation of strangeness.* Each strange meson is assigned a *strangeness quantun number* S and the total strangeness of a group of particles is simply the sum of their strangeness quantum numbers. This total strangeness is conserved in strong interactions.

PROBLEM 11

Assuming that Fig. 14-4-2 is approximately a life-sized picture of the bubble chamber event, how long would each neutral particle have lived if its speed were about one-half the speed of light?

The strange mesons K^+ and K^0 have $S = +1$, and have antiparticles K^- and $\overline{K^0}$ with $S = -1$. The four lightest strange baryons all have strangeness -1. They are the Λ^0, involved in the reaction of Eq. 14-4-1, and the three Σ (sigma) particles Σ^+, Σ^0, and Σ^-. There are many possible reactions for producing the K mesons and the strange baryons, but the only ones which have ever been observed are those which conserve strangeness.

The mystery of the slow decays of strange particles is solved because none of those mentioned above can decay into strongly interacting particles while still conserving strangeness. The decays which are energetically allowed are into pion plus nucleon for the baryons, and into either two or three pions for the mesons. In every case the final state has zero strangeness, while the strange particle has nonzero strangeness. As a result the strong interaction is forbidden, so rigorously that the decay takes place by a weak interaction, a factor 10^{13} more slowly!

PROBLEM 12

The Σ^0 mass isn't great enough to make the decay $\Sigma^0 \rightarrow \Lambda^0 + \pi^0$ energetically allowed. If it were, would the decay be an allowed strong interaction?

PROBLEM 13
Which of the following reactions is allowed and which forbidden by strangeness conservation?

$$\pi^- + p \to K^0 + n$$
$$\pi^- + p \to K^+ + \Sigma^-$$
$$\pi^- + p \to K^- + \Sigma^+$$
$$\pi^- + p \to K^+ + K^- + n$$
$$\pi^- + p \to \overline{K^0} + \overline{\Lambda^0}$$

14-5 RESONANCES

A mechanical system which can vibrate, such as a guitar string, has a natural frequency of oscillation which is often called its *resonant frequency*. Such a system is particularly responsive at its resonant frequency and, for example, is easily excited to large vibrations by weak forces which vary at the resonant frequency. For quantum-mechanical systems we observe resonances at particular values of the total energy of the system, which is somewhat different from the classical case. (Note, however, that the familiar quantum relation, $E = hf$ results in a connection between E and f in quantum mechanics.) If we illuminate atoms of some element with light photons of just the right energy to excite them to the first quantum state above the ground state, then they respond *resonantly* by absorbing the light very strongly.

Elementary particle resonances are said to occur when two or more particles interact particularly strongly at a particular value of their combined total energy. The first elementary particle resonance was discovered in low-energy pion-nucleon scattering experiments. Figure 14-5-1 shows how the *total scattering cross sections* for π^+-p and π^--p elastic scattering vary with pion energy, where the *total cross section* means the total scattering probability for all angles. For pion kinetic energies near 195 MeV there is apparently a strong *resonant* scattering.

Because the ratio of cross sections for π^+-p and π^--p is 3:1 at the resonance energy, it is possible, from the details of I-spin theory, to infer that the resonant interaction is for a single value of isotopic spin, $I = 3/2$. This resonance can be further visualized as a two-step process, as shown in Fig. 14-5-2. The pion and nucleon collide and combine into a resonant state, which then decays (very quickly) back into a pion and nucleon. The observed resonance is characterized by the *mass* of its resonant state in a picture such as that in Fig. 14-5-2. From conservation of energy and momentum

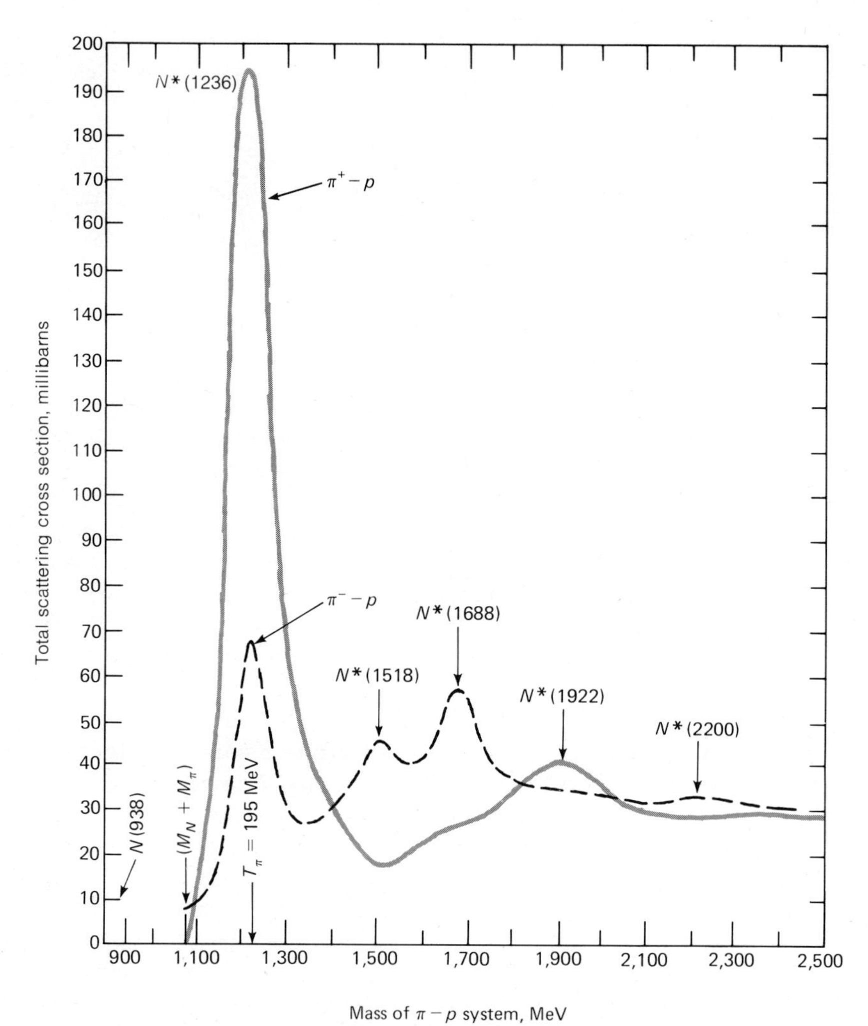

Figure 14-5-1 Total elastic-scattering cross sections for pions on protons, showing their variation with pion energy, or mass of the π-p system. Resonances at various mass values are labeled N^*.

in the pion-nucleon collision, we can determine the resonant-state mass, which is often called simply the mass of the resonance. For the resonance we are discussing the mass is 1,236 MeV.

PROBLEM 14

If the 1,236 MeV pion-nucleon resonance has $I = 3/2$, how many charge states must it be able to occur in? Try to guess what they

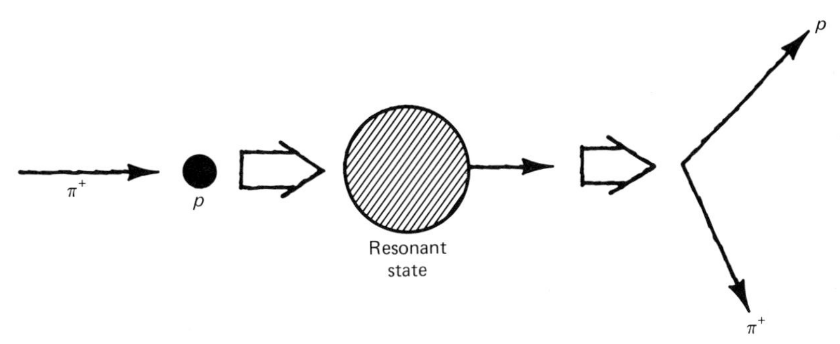

Figure 14-5-2 Pictorial representation of π-p elastic scattering with formation and subsequent decay of a resonance state.

are. (All possible pion-nucleon combinations can form the resonance.)

PROBLEM 15
If a particle has a lifetime Δt, then according to the uncertainty principle its energy must be uncertain according to the relation:

$$(\Delta E)(\Delta t) \geq h$$

Let the wave associated with any particle have the same relation between E and f as the photon,

$$E = hf$$

For a length of time Δt, determine f by the relation:

$$f = \text{number of cycles}/\Delta t$$

If a particle lifetime is Δt and the number of cycles in its wave is uncertain by one, regardless of Δt, show that the uncertainty relation above results.

PROBLEM 16
Estimate the spread in energy of the 1,236 MeV resonance from Fig. 14-5-1. Use the uncertainty relation of the preceding problem to estimate its lifetime.

We now know of many pion-nucleon resonances, mainly inferred from studying elastic-scattering data at higher energies. In addition, there are a number of known strange baryon resonances, which can be formed from K mesons and nucleons. Finally, there are meson resonances and strange-meson resonances which are very clearly observable. One of the first meson resonances discovered was a three-pion resonance, originally observed in bubble chamber pictures of antiproton-proton annihilations in which five pions were produced:

$$\bar{p} + p \rightarrow \pi^+ + \pi^- + \pi^+ + \pi^- + \pi^0 \qquad 14\text{-}5\text{-}1$$

This reaction appeared superficially to produce randomly moving pions, materialized from the nearly 2 GeV of rest energy of the proton plus antiproton. However, following a theoretical

Figure 14-5-3 **Mass distribution for combinations of three pions from the reaction: $\bar{p} + p \rightarrow 2\pi^+ + 2\pi^- + \pi^0$. Three-pion states with net charge = 0 show a striking peak at a mass just below 800 MeV (0.8 BeV). The peak does not appear in combinations with charge +1 or +2, as shown by the curves labeled *(a)* and *(b)*. This three-pion state with charge zero is called the ω^0 meson.**

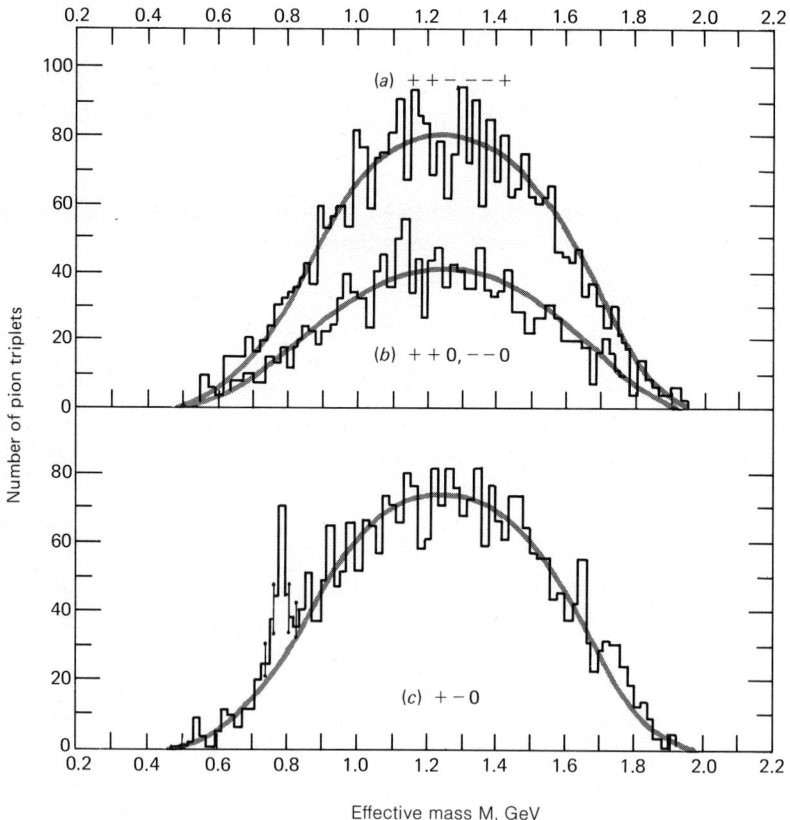

guess that a three-pion resonance should exist, a group of experimenters carefully examined bubble chamber pictures of the reaction given in Eq. 14-5-1. For all possible combinations of three pions, they calculated the total energy and momentum. Just as in the case of the pion-nucleon resonance, the total energy and momentum determine a mass, which can be calculated according to the rules of relativistic motion.

Figure 14-5-3 shows the three-pion experimental data. There is a very definite narrow peak in the mass distribution for combinations with zero net charge, and no peak for charges ± 1 or ± 2. This peak, at a mass of 780 MeV, is now called the ω^0 (omega zero) meson and decays via the strong interaction into three pions. The decay takes place so rapidly that after the ω^0 is produced, it only travels a distance of the order of a proton diameter before it decays. From the fact that the ω^0 is observed in just a single charge state, with $q = 0$, we find that $I = 0$.

We speak of the ω^0 exactly as if it were a particle. Or is it a resonance? This distinction worried elementary particle physicists for a long time, but it is now believed that resonances which decay rapidly via strong interactions are still bona fide elementary particles. Nucleons and pions are not singled out for special treatment, merely because they are the lightest baryons and mesons, and hence cannot decay by strong interactions. If we include all the resonances discovered up to 1970, there are more than 100 known elementary particles. Nuclear structure has apparently turned out to be as complex as the chemistry of the 92 elements.

Perhaps the multitude of elementary particles are built on a simple foundation. In the same spirit with which we call resonances elementary particles, we could also consider, for example, the ground state and first excited state of hydrogen to be different particles. They are, but we now understand the underlying structure of hydrogen so well that we make little of the distinction. Of course, in the days of Bohr, the energy difference between the two states played a very important role in the development and verification of the theory of the hydrogen atom.

14-6 THE EIGHTFOLD WAY

Table 14-6-1 comes from a comprehensive survey of particle properties, published in 1958. A similar table, published 12 years later, has become a 16-page booklet! As elementary particle data has accumulated, the strongest challenge has been to find some

Table 14-6-1 Elementary Particles, November, 1957

	PARTICLE	MASS, MeV	MEAN LIFE, sec
Photon	γ	0	Stable
Leptons	ν	0	Stable
	e^-	0.51	Stable
	μ^-	105.7	2.2×10^{-6}
Mesons	$\pi^\pm$	139.6	2.6×10^{-8}
	π^0	135.0	$\leqslant 10^{-16}$
	K^+	494	1.2×10^{-8}
	K^0	494	0.95×10^{-10}
Baryons	p	938.2	Stable
	n	939.5	1.0×10^3
	Λ	1115	2.8×10^{-10}
	Σ^+	1189	0.8×10^{-10}
	Σ^-	1194	1.7×10^{-10}
	Σ^0	1190	2.1×10^{-10}
	Ξ^-	1320	?
	Ξ^0	?	?

order in a chaos of different masses and decay modes, and of different values of I spin and strangeness. A scheme proposed independently by M. Gell-Mann and Y. Ne'eman in 1961 provides by far the most successful effort to date in grouping elementary particles.

The 1961 Gell-Mann paper was entitled: "The Eightfold Way: A Theory of Strong Interaction Symmetry." The symmetry involved in the eightfold way is similar to the symmetry for different charges that we have previously discussed in connection with isotopic spin but is more inclusive and more complicated. Because the strong interaction is charge-independent, *multiplets* of particles, like π^+, π^0, π^-, containing 1, 2, 3, or more members, all exhibit the same strong interaction properties. Recall, from the previous section, that the number of members of a multiplet, called the *multiplicity*, was related to the value of the quantum number I, called the I spin, and that the multiplicity was given by the quantity $2I + 1$.

A graphical way for representing I-spin multiplets provides a simple beginning for understanding the eightfold way. In Table 14-6-2, successively increasing values of I are listed, with the

Table 14-6-2

ISOTOPIC SPIN, I	MULTIPLICITY	GRAPHICAL REPRESENTATION
0	1, singlet	•
1/2	2, doublet	• •
1	3, triplet	• • •
3/2	4, quartet	• • • •
2	5, quintet	• • • • •

multiplicities in numbers and words, and as dots on a line. Consider, for example, the dot plot for the doublet state $I = 1/2$. In Fig. 14-6-1 it is shown in detail, with the two dots at the values $+1/2$ and $-1/2$ on an axis labeled I_3. The name I_3 is the conventional name for the quantity which varies as the charge changes within an I-spin multiplet. For a given multiplet, the rule for assigning values to I_3 is that the values are each spaced by one unit, and that the multiplet is centered around $I_3 = 0$.

Suppose we want to combine a particle that has $I = 1/2$ with one that has $I = 1$, which would, for example, be necessary to learn about the I spin of a system consisting of a pion plus a nucleon. There is a simple rule for counting up the multiplicity of possible states: Take the dots for either particle and with each as a center lay down the dot pattern of the other. This is easier to explain in pictures than in words, and Fig. 14-6-2 shows three doublets of dots, centered at $I_3 = -1$, 0, and $+1$, the locations of the three dots for $I = 1$. We find a total of $2 \times 3 = 6$ different dots, which are possible states. However, a more subtle result is that from the way the charge states are distributed along the I_3 axis they can be organized into a quartet and a doublet. For the quartet $I_3 = -3/2, -1/2, +1/2, +3/2$; for the doublet $I_3 = -1/2, +1/2$; so that together they give six states like those in Fig. 14-6-2.

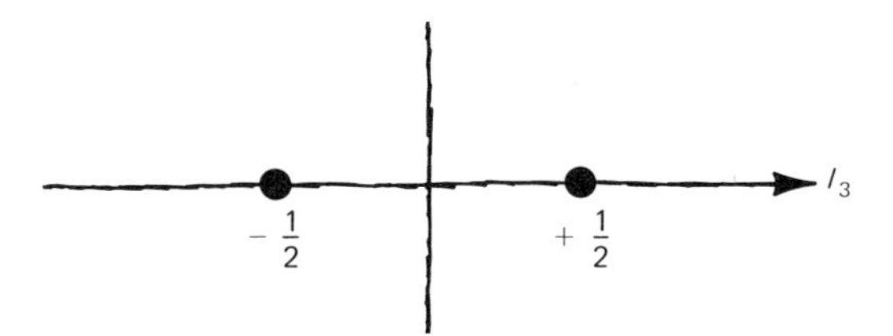

Figure 14-6-1 **Isotopic-spin dot plot for $I = 1/2$.**

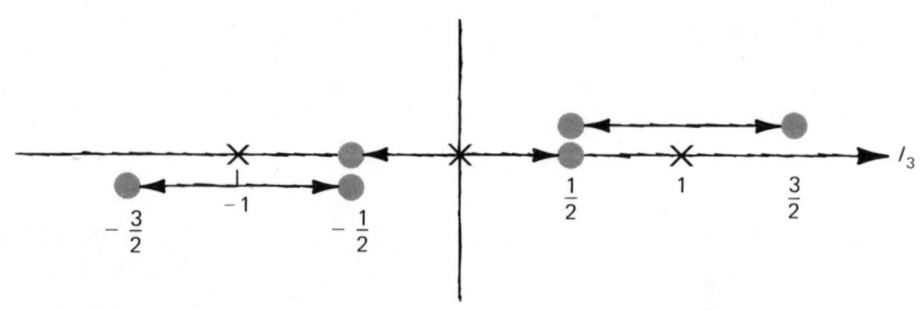

Figure 14-6-2 Dot plot for combining an $I = 1$ multiplet and an $I = ½$ multiplet. From each $I = 1$ state, marked with an x, a set of $I = ½$ states is laid out with the $I = 1$ state as a center. The result is six states, marked with dots, with values of I_3 corresponding to an $I = 3/2$ multiplet plus an $I = ½$ multiplet.

According to a certain set of abstract rules, we have just found that a nucleon doublet and a pion triplet combine into a quartet and a doublet of states. The quartet corresponds to $I = 3/2$, the doublet to $I = 1/2$. The justification for the sort of game we have described is that in nature the strong interaction *actually works* according to these rules. Pions and nucleons *do* form two different kinds of combinations like the ones we have found. The six possible charge states of p, n, and π^+, π^0, π^- interact in *just two* ways, depending upon whether $I = 3/2$ or $I = 1/2$.

At the time when Gell-Mann was working on his paper, "The Eightfold Way," there were eight "stable" baryons known, stable in the sense that they did not decay by the strong interaction. They are all shown in the 1958 particle-properties table, Table 14-6-1. The individual masses of these baryons can be seen to vary by less than about 20 percent from the average of all eight masses, which was one of the most compelling reasons for the conjecture made by Gell-Mann. He proposed that these eight baryons were the states of a *supermultiplet* of strongly interacting particles.

Except for relatively minor effects, Gell-Mann and Ne'eman assumed that the strong interaction was *the same* for all members of such a supermultiplet. This is an extension of what was known about charge multiplets, in the sense that more particles are encompassed: the members of more than one I-spin multiplet. In exchange for a broader sweep, the theory was admittedly somewhat approximate. There are relatively small differences between the strong interactions of the individual I-spin multiplets which make up the supermultiplet.

There is a graphical way to display the baryon supermultiplet, like the I spin dot plots but in two dimensions, as shown in Fig.

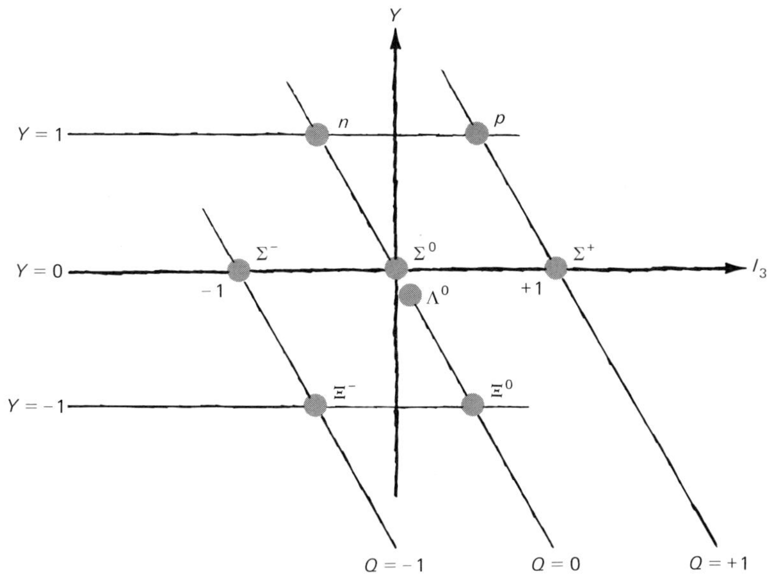

Figure 14-6-3 An octet of baryons, plotted as dots in two dimensions, according to the values of I_3 and Y. Four I-spin multiplets are involved, the sigma (Σ) with $I = 1$, the nucleons (n, p) with $I = ½$, the xi (Ξ) with $I = ½$, and the lamda (Λ) with $I = 0$. Note that the hypercharge Y determines the mean charge of each I-spin multiplet, (Q), such that $(Q) = Y/2$.

14-6-3. Now there are two axes, I_3 and Y, where Y, called the hypercharge, provides the way to integrate strange and non-strange particles into the octet. Y is defined for any particle by the relation

$$Y = B + S$$

where B is the baryon number of the particle and S its strangeness. A wonderful feature of the plot shown in Fig. 14-6-3, which encompasses all eight stable baryons, is not only that it looks symmetrical, but that it is a dot plot which is part of a known mathematical system of two-dimensional dot plots. The one-dimensional I-spin dot plots are graphical representations of the members of a mathematical group. The two-dimensional dot plot of the baryon octet shown in Fig. 14-6-3 is the graphical representation of one of the members of another mathematical group, a group which leads to multiplicities of 1, 3, 6, 8, 10, 15, etc. Figure 14-6-4 shows dot plots, on (I_3, Y) axes, which represent the first five members of that group.

The discovery of an octet of baryons raised the question of whether, at least from the standpoint of strong interactions, all elementary particles fell into the supermultiplets of a particular group. The first conjecture was that, in spite of the relatively wide variation in their masses, the mesons in Table 14-6-1 were also part of an octet. The mesons were grouped as in Fig. 14-6-5, and in order to form an octet it was necessary that there be a neutral

Figure 14-6-4 Dot plots of the first five members of the mathematical symmetry group which includes an octet like that of Fig. 14-6-3.

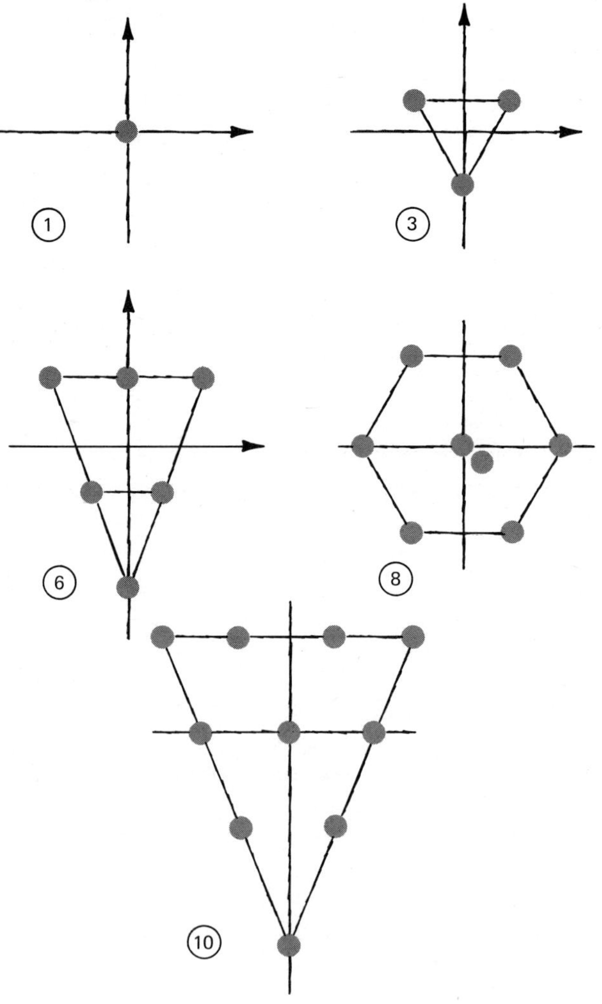

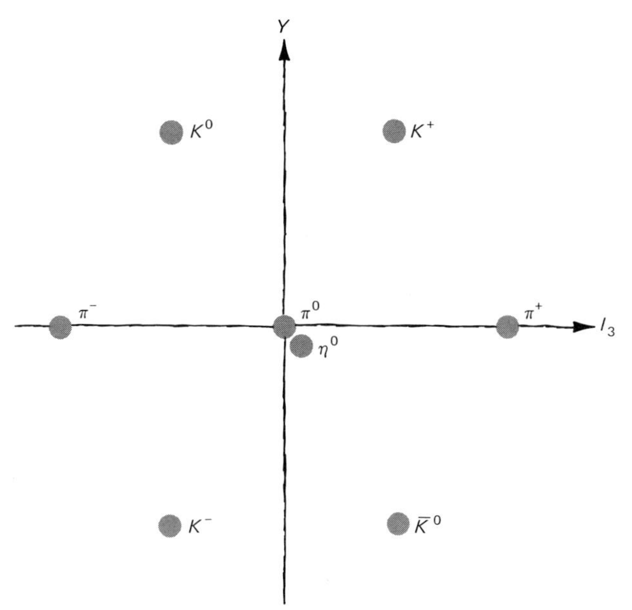

Figure 14-6-5 An octet of spin 0 mesons, showing the same symmetry on the I_3-Y plot as the baryon octet.

meson with $I = 0$, called the η^0 (eta zero), occupying the same place in Fig. 14-6-5 as the Λ^0 does in Fig. 14-6-3. The existence of this meson was suspected, and later experiments confirmed its existence.

In addition to the octet of mesons shown in Fig. 14-6-5, there were other more or less well-established mesons at the time Gell-Mann wrote his paper. The eightfold way was used to predict yet another octet of mesons, which was indeed found to exist.

A more far-reaching experimental test of the eightfold way came when the existence of a supermultiplet containing *10* particles was confirmed. If there were an underlying order in nature, corresponding to the group-theoretical mathematics of Gell-Mann's paper, then one would expect to find examples of other members of the group, perhaps the sextet, decuplet, etc. In fact, a family of baryons was known, as shown in Fig. 14-6-6, which seemed to conspicuously lack just one member of a decuplet. These baryons have large enough masses that they are all unstable resonances. In fact, the Δ, with four different charges, is the $I = 3/2$, mass 1,236, pion-nucleon resonance discussed in the pre-

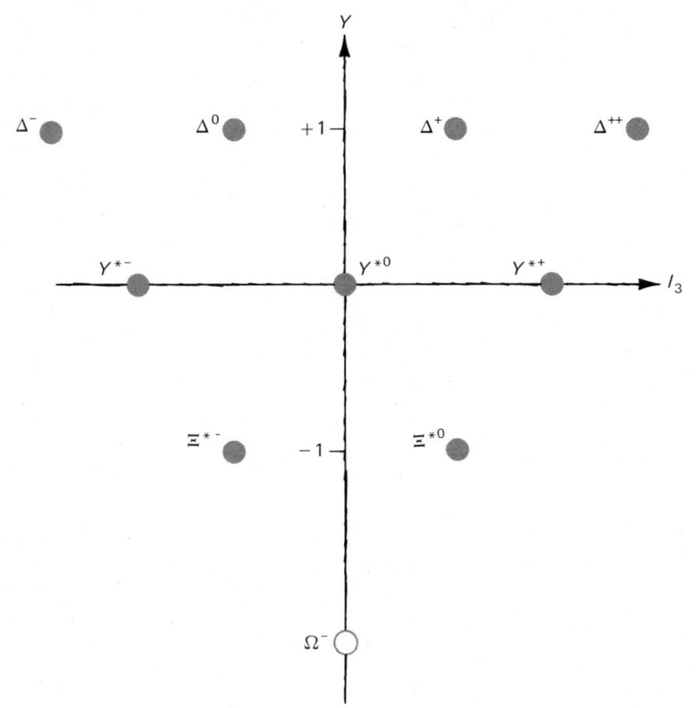

Figure 14-6-6 A decuplet of spin 3/2 baryons, of which the nine with $Y = +1$, 0, and -1 were known when the eightfold way was proposed. The discovery of the $Y = -2$ baryon, the Ω^-, predicted to exist in order to complete this decuplet, was a striking confirmation of the ideas of Gell-Mann and Ne'eman.

ceding section. The properties of the missing member could be specified very completely: $I = 0$, strangeness -3, and mass about 1,680 MeV. After an intensive search in bubble-chamber pictures, the Ω^- (omega minus) baryon, the tenth member of the decuplet, was discovered. This was the most definite evidence that the scheme of Gell-Mann and Ne'eman was followed in nature. Technically, the group they used is called SU(3), which stands for "special unitary group in three dimensions."

PROBLEM 17

Predict the mass of the Ω^- baryon from the masses of the other members of the decuplet:

Δ	1,236 MeV
Y^*	1,385 MeV
Ξ^*	1,530 MeV

PROBLEM 18

Determine a production reaction for making the Ω^- baryon with a strong interaction. You can assume an incident beam of nucleons, pions, or kaons.

PROBLEM 19

From the particles you know of, and their properties, can the Ω^- meson decay by a strong interaction?

PROBLEM 20

From the results of the two preceding problems, make a sketch of the production and decay of the Ω^-, as seen in a bubble chamber.

It is not possible here to discuss all of the detailed SU(3) predictions which have been verified. For example, relations between masses of particles within an SU(3) multiplet have been found, as well as relations between the interactions of various particles. There has been sufficient experimental verification of the eightfold way so that most physicists now believe it represents a true insight into nature. However, there are still vast areas where we lack understanding. We are still far from knowing why only certain SU(3) multiplets seem to exist, and there are instances in which nature seems at odds with SU(3). We still cannot predict the details of strong interaction dynamics.

Among the experimental questions of greatest current interest to elementary particle physicists, one relates directly to SU(3). A basic SU(3) triplet can be used to create all the multiplets which we seem to observe in nature. In Fig. 14-6-7 graphical pictures are shown of a triplet and its corresponding antiparticles. The possible combinations of one member of the particle triplet with one member of the antiparticle triplet lead to $3 \times 3 = 9$ states, consisting of an octet and a singlet, which can be identified with known meson multiplets. Furthermore, the combinations of any three of the triplet particles lead to 27 states, including two octets and a decuplet which correspond to baryon SU(3) multiplets. The members of this single SU(3) triplet, and their antiparticles, can actually be combined to form *all* the known elementary particles.

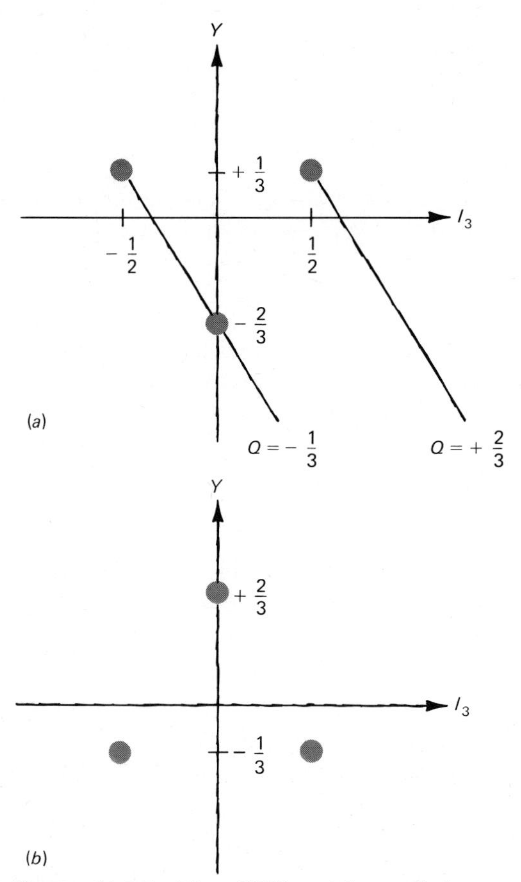

Figure 14-6-7 The SU(3) triplet called quarks by Gell-Mann, shown in *(a)*, and the antiquarks, shown in *(b)*. Note that two quarks have $S = 0$; one has $S = -1$. The baryon numbers, as well as the charges, are *fractional*.

Gell-Mann called the members of the basic triplet *quarks*, taking the word from James Joyce's *Finnegan's Wake.*[1] Hunting for quarks is one of the most exciting activities in modern ele-

[1]Three quarks for Muster Mark
Sure he hasn't got much of a bark
And sure any he has it's all beside the mark . . .
James Joyce, Finnegan's Wake, Viking Press, New York, 1939, p. 383.

mentary particle physics, but to date there has been no success, although the properties of the quarks, in particular their *fractional charges*, would make them extremely conspicuous.

PROBLEM 21

Show that the nine states obtained by combining a quark and an antiquark consist of an octet and a singlet, where the octet is just like the meson octet in Fig. 14-6-5. To combine the quark and antiquark, proceed in a way analogous to the I spin combination shown in Fig. 14-6-2: center the antiquark triplet successively at the location of each of the three quarks.

PROBLEM 22

If baryon number is conserved, can the combination of a quark-antiquark pair be used to explain the baryon octets? Can three-quark combinations give states with baryon number $= 1$, and $S = 0$ or 1?

At present we are forced to conclude that if strongly interacting quarks really exist, they must have very large masses, resulting in such rare production in high-energy interactions that they have eluded detection. There is of course a real possibility that the quark is solely a mathematical building block and not a physical one. This would mean, for example, that the real world only has octets and decuplets which fit the SU(3) mathematics, but not the quark triplet from which the mathematics can be built up.

Since most elementary particles are strongly interacting we have emphasized this aspect of elementary particle physics. However, studying the weak interaction is also of great interest at this time. There is an exciting search in progress for a W particle, which is the quantum of the weak-interaction field. In addition, there remains, for the weak interaction, a very basic question of twenty years standing: Why are there *two* leptons, the muon and the electron, rather than one or many?

If history is a good guide, we may expect that the elementary particles will ultimately be understood in terms of a few basic principles. In the meantime, the theorists have an enormous amount of puzzling data to reckon with, and the experimenters are about to probe a new energy region with the giant NAL accelerator. The 1970s promise to be a crucial decade for high-energy physics.

SUMMARY

Elementary particles include the fundamental building blocks of matter: the *field quanta* of the basic interactions, the *neutrino*, and the *graviton*. For all the particles there are *antiparticles*, characterized by opposite charge and the ability to be created or to annihilate in particle-antiparticle pairs.

All particles can be assigned to one of three classes: *mesons*, *baryons*, or *leptons*, except the photon which is in a class by itself. The *mesons and baryons are strongly interacting*. Mesons can be created and absorbed without restriction, while the net number of baryons is conserved. *Leptons* are the *weakly interacting* particles, and both electron-leptons and muon-leptons are separately conserved. The known *absolutely conserved quantities* can be listed as follows: energy, momentum, angular momentum, charge, baryon number, lepton number, and the quantum number product *TCP*.

Elementary particle physics requires *high-energy accelerators* to produce the many unstable short-lived particles. Today, the dominant type of high-energy accelerator is the *synchrotron*. Particles are detected by the *ionization* and *excitation* their charges produce in matter, either with *counters* or with visual detectors such as the *spark chamber* and *bubble chamber*.

The study of strong interaction physics has led to the discovery of conservation laws for *strangeness* and *I spin* in *strong interactions only*, and to the discovery of a multitude of baryon and meson *resonances*, both strange and nonstrange. By means of the *eightfold way*, all the known elementary particles can apparently be grouped into multiplets consisting of groups of up to ten particles, each of which can include a number of different *I* spin multiplets. All these multiplets can be derived from the mathematical group called SU(3), which can have as its simplest pattern of particles a group of three which are called *quarks*. Apparently *all* the known elementary particles could be built up of quark combinations. However, to date, the search for quarks in nature has been unsuccessful.

INDEX